T0192187

Graduate Texts in Mathematics **198**

Springer
New York
Berlin
Heidelberg
Barcelona
Hong Kong
London
Milan
Paris
Singapore
Tokyo

Graduate Texts in Mathematics

(continued after index)

Alain M. Robert

A Course in p-adic Analysis

With 27 Figures

Alain M. Robert
Institut de Mathématiques
Université de Neuchâtel
Rue Émile-Argand 11
Neuchâtel CH-2007
Switzerland

Mathematics Subject Classification (2000): 11-01, 11E95, 11Sxx

Library of Congress Cataloging-in-Publication Data
Robert, Alain.
A course in p-adic analysis / Alain M. Robert.
p. cm. — (Graduate texts in mathematics ; 198)
Includes bibliographical references and index.
ISBN 0-387-98669-3 (hc. : alk. paper)
1. p-adic analysis. I. Title. II. Series.
QA241, R597 2000
512'.74 — dc21 99-044784

Printed on acid-free paper.

Production managed by Timothy Taylor; manufacturing supervised by Erica Bresler.
Typeset by TechBooks, Fairfax, VA.
Printed and bound by R.R. Donnelley and Sons, Harrisonburg, VA.
Printed in the United States of America.

9 8 7 6 5 4 3 2 1

ISBN 0-387-98669-3 Springer-Verlag New York Berlin Heidelberg SPIN 10698156

Preface

Kurt Hensel (1861–1941) discovered or invented the p-adic numbers[1] around the end of the nineteenth century. In spite of their being already one hundred years old, these numbers are still today enveloped in an aura of mystery within the scientific community. Although they have penetrated several mathematical fields, number theory, algebraic geometry, algebraic topology, analysis, ..., they have yet to reveal their full potential in physics, for example. Several books on p-adic analysis have recently appeared:

F. Q. Gouvêa: *p-adic Numbers* (elementary approach);
A. Escassut: *Analytic Elements in p-adic Analysis*, (research level)

(see the references at the end of the book), and we hope that this course will contribute to clearing away the remaining suspicion surrounding them. This book is a self-contained presentation of basic p-adic analysis with some arithmetical applications.

* * *

Our guide is the analogy with classical analysis. In spite of what one may think, these analogies indeed abound. Even if striking differences immediately appear between the real field and the p-adic fields, a better understanding reveals strong common features. We try to stress these similarities and insist on *calculus* with the p-adics, letting the mean value theorem play an important role. An obvious reason for links between real/complex analysis and p-adic analysis is the existence of

[1] The letter p stands for a fixed prime (chosen in the list 2, 3, 5, 7, 11, ...) except when explicitly stated otherwise.

an *absolute value* in both contexts.[2] But if the absolute value is *Archimedean* in real/complex analysis,

$$\text{if } x \neq 0, \text{ for any } y \text{ there is an integer } n \text{ such that } |nx| > |y|,$$

it is *non-Archimedean* in the second context, namely, it satisfies

$$|nx| = |\underbrace{x + x + \cdots + x}_{n \text{ terms}}| \leq |x|.$$

In particular, $|n| \leq 1$ for all integers n. This implies that for any $r \geq 0$ the subset of elements satisfying $|x| \leq r$ is an additive subgroup, even a subring if $r = 1$. For such an absolute value, there is (except in a trivial case) exactly one prime p such that $|p| < 1$.[3] Intuitively, this absolute value plays the role of an order of magnitude. If x has magnitude greater than 1, one cannot reach it from 0 by taking a finite number of unit steps (one cannot walk or drive to another galaxy!). Furthermore, $|p| < 1$ implies that $|p^n| \to 0$, and the p-adic theory provides a link between characteristic 0 and characteristic p.

The absolute value makes it possible to study the convergence of *formal power series*, thus providing another unifying concept for analysis. This explains the important role played by formal power series. They appear early and thereafter repeatedly in this book, and knowing from experience the feelings that they inspire in our students, I try to approach them cautiously, as if to tame them.

* * *

Here is a short summary of the contents

Chapter I: Construction of the basic p-adic sets $\mathbf{Z}_p$, $\mathbf{Q}_p$ and $\mathbf{S}_p$,

Chapters II and III: Algebra, construction of $\mathbf{C}_p$ and Ω_p,

Chapters IV, V, and VI: Function theory,

Chapter VII: Arithmetic applications.

I have tried to keep these four parts relatively independent and indicate by an asterisk in the table of contents the sections that may be skipped in a first reading. I assume that the readers, (advanced) graduate students, theoretical physicists, and mathematicians, are familiar with calculus, point set topology (especially metric spaces, normed spaces), and algebra (linear algebra, ring and field theory). The first five chapters of the book are based solely on these topics.

The first part can be used for an introductory course: Several definitions of the basic sets of p-adic numbers are given. The reader can choose a favorite approach! Generalities on topological algebra are also grouped there.

---

[2] Both Newton's method for the determination of real roots of $f = 0$ and Hensel's lemma in the p-adic context are applications of the existence of fixed points for contracting maps in a complete metric space.

[3] Since the prime p is uniquely determined, this absolute value is also denoted by $|\,.\,|_p$. However, since we use it systematically, and hardly ever consider the Archimedean absolute value, we simply write $|\,.\,|$.

The second — more algebraic — part starts with a basic discussion of ultrametric spaces (Section II.1) and ends (Section III.4) with a discussion of fundamental inequalities and roots of unity (not needed before the study of the logarithm in Section V.4). In between, the main objective is the construction of a complete and algebraically closed field $\mathbf{C}_p$, which plays a role similar to the complex field $\mathbf{C}$ of classical analysis. The reader who is willing to take for granted that the p-adic absolute value has a *unique* extension $|\,.\,|_K$ to every finite algebraic extension K of $\mathbf{Q}_p$ can skip the rest of Chapter II: If K and K' are two such extensions, the restrictions of $|\,.\,|_K$ and $|\,.\,|_{K'}$ to $K \cap K'$ agree. This proves that there is a unique extension of the p-adic absolute value of $\mathbf{Q}_p$ to the algebraic closure $\mathbf{Q}_p^a$ of $\mathbf{Q}_p$. Moreover, if $\sigma \in \mathrm{Aut}\,(K/\mathbf{Q}_p)$, then $x \mapsto |x^\sigma|_K$ is an absolute value extending the p-adic one, hence this absolute value coincides with $|\,.\,|_K$. This shows that σ is *isometric*. If one is willing to believe that the completion $\widehat{\mathbf{Q}_p^a} = \mathbf{C}_p$ is also *algebraically closed*, most of Chapter III may be skipped as well.

In the third part, functions of a p-adic variable are examined. In Chapter IV, continuous functions (and, in particular, locally constant ones) $f : \mathbf{Z}_p \to \mathbf{C}_p$ are systematically studied, and the theory culminates in van Hamme's generalization of Mahler's theory. Many results concerning functions of a p-adic variable are extended from similar results concerning polynomials. For this reason, the algebra of polynomials plays a central role, and we treat the systems of polynomials — umbral calculus — in a systematic way. Then differentiability is approached (Chapter V): Strict differentiability plays the main role. This chapter owes much to the presentation by W.H. Schikhof: *Ultrametric Calculus, an Introduction to p-adic Analysis*. In Chapter VI, a previous acquaintance with complex analysis is desirable, since the purpose is to give the p-adic analogues of the classical theorems linked to the names of Weierstrass, Liouville, Picard, Hadamard, Mittag-Leffler, among others. In the last part (Chapter VII), some familiarity with the classical gamma function will enable the reader to perceive the similarities between the classical and the p-adic contexts. Here, a means of unifying many arithmetic congruences in a general theory is supplied. For example, the Wilson congruence is both generalized and embedded in analytical properties of the p-adic gamma function and in integrality properties of the Artin-Hasse power series. I explain several applications of p-adic analysis to arithmetic congruences.

* * *

Let me now indicate one point that deserves more *justification*. The study of metric spaces has developed around the classical examples of subsets of $\mathbf{R}^n$ (we make pictures on a sheet of paper or on the blackboard, both models of $\mathbf{R}^2$), and a famous treatise in differential geometry even starts with "*The nicest example of a metric space is Euclidean n-space* $\mathbf{R}^n$." This point of view is so widely shared that one may be led to think that ultrametric spaces are not genuine metric spaces! Thus the commonly used notation for metric spaces has grown on the paradigmatic model of subsets of Euclidean spaces. For example, the "closed ball" of radius r and center a — defined by $d(x, a) \leq r$ — is often denoted by $\overline{B}(a; r)$ or $\overline{B}_r(a)$. This notation comforts the belief that it is the closure of the "open ball" having the same radius and center. If the specialists have no trouble with the usual terminology and notation (and may defend it on historical grounds), our students lose no opportunity to insist on its misleading meaning. In an ultrametric space all balls of positive radius (whether defined by $d(x,a) \leq r$ or by $d(x,a) < r$) are both open and closed. They are *clopen* sets. Also note that in an ultrametric space, *any point* of a ball is a center of this ball. The systematic appearance of totally disconnected spaces in the context of fractals also calls for a renewed view of metric spaces. I propose using a more suggestive notation,

$$B_{<r}(a) = \{x : d(x,a) < r\}, \quad B_{\leq r}(a) = \{x : d(x,a) \leq r\}$$

which has at least the advantage of clarity. In this way I can keep the notation $\overline{A}$ strictly for the closure of a subset A of a topological space X. The algebraic closure of a field K is denoted by K^a.

* * *

Finally, let me thank all the people who helped me during the preparation of this book, read preliminary versions, or corrected mistakes. I would like to mention especially the anonymous referee who noted many mistakes in my first draft, suggested invaluable improvements and exercises; W.H. Schikhof, who helped me to correct many inaccuracies; and A. Gertsch Hamadene, who proofread the whole manuscript. I also received encouragement and help from many friends and collaborators. Among them, it is a pleasure for me to thank

D. Barsky, G. Christol, B. Diarra, A. Escassut, S. Guillod-Griener,
A. Junod, V. Schürch, C. Vonlanthen, M. Zuber.

My wife, Ann, also checked my English and removed many errors.

Cross-references are given by number: (II.3.4) refers to Section (3.4) of Chapter II. Within Chapter II we omit the mention of the chapter, and we simply refer to (3.4). Within a section, lemmas, propositions, and theorems are individually numbered only if several of the same type appear. I have not attempted to track historical priorities and attach names to some results only for convenience. General assumptions are repeated at the head of chapters (or sections) where they are in force.

Figures I.2.5a, I.2.5c, I.2.5d, and I.2.6 are reproduced here (some with minor modifications) with written permission from Marcel Dekker. They first appeared in my contribution to the *Proceedings of the 4th International Conference on p-adic Functional Analysis* (listed in the References).

Alain M. Robert
Neuchâtel, Switzerland, July 1999

Contents

1

p-adic Numbers

The letter p will denote a fixed prime.

The aim of this chapter is the construction of the compact topological ring $\mathbf{Z}_p$ of *p-adic integers* and of its quotient field $\mathbf{Q}_p$, the locally compact field of *p-adic numbers*. This gives us an opportunity to develop a few concepts in *topological algebra*, namely the structures mixing algebra and topology in a coherent way. Two tools play an essential role from the start:

- the p-adic absolute value $|\,.\,|_p = |\,.\,|$ or its additive version, the p-adic valuation $\mathrm{ord}_p = v_p$,
- reduction mod p.

1. The Ring $\mathbf{Z}_p$ of p-adic Integers

We start by a down-to-earth definition of p-adic integers: Other equivalent presentations for them appear below, in (4.7) and (4.8).

1.1. Definition

A p-adic integer *is a formal series* $\sum_{i\geq 0} a_i p^i$ *with integral coefficients* a_i *satisfying*

$$0 \leq a_i \leq p-1.$$

With this definition, a p-adic integer $a = \sum_{i\geq 0} a_i p^i$ can be identified with the sequence $(a_i)_{i\geq 0}$ of its coefficients, and the set of p-adic integers coincides with the Cartesian product

$$X = X_p = \prod_{i \geq 0} \{0, 1, \dots, p-1\} = \{0, 1, \dots, p-1\}^{\mathbf{N}}.$$

In particular, if $a = \sum_{i \geq 0} a_i p^i$, $b = \sum_{i \geq 0} b_i p^i$ (with $a_i,\ b_i \in \{0, 1, \dots, p-1\}$) we have

$$a = b \iff a_i = b_i \text{ for all } i \geq 0.$$

The usefulness of the series representation will be revealed when we introduce algebraic operations on these p-adic integers. Let us already observe that the expansions in base p of natural integers produce p-adic integers (ending with zero coefficients: Finite series are special series), and we obtain a canonical embedding of the set of natural integers $\mathbf{N} = \{0, 1, 2, \dots\}$ into X.

From the definition, we immediately infer that *the set of p-adic integers is not countable.* Indeed, if we take any sequence of p-adic integers, say

$$a = \sum_{i \geq 0} a_i p^i, \quad b = \sum_{i \geq 0} b_i p^i, \quad c = \sum_{i \geq 0} c_i p^i, \quad \dots,$$

we can define a p-adic integer $x = \sum_{i \geq 0} x_i p^i$ by choosing

$$x_0 \neq a_0,\ x_1 \neq b_1,\ x_2 \neq c_2, \quad \dots,$$

thus constructing a p-adic integer different from $a,\ b,\ c, \dots$. This shows that the sequence $a,\ b,\ c, \dots$ does not exhaust the set of p-adic integers. A mapping from the set of natural integers $\mathbf{N}$ to the set of p-adic integers is never surjective.

1.2. Addition of p-adic Integers

Let us define the sum of two p-adic integers a and b by the following procedure. The first component of the sum is $a_0 + b_0$ if this is less than or equal to $p - 1$, or $a_0 + b_0 - p$ otherwise. In the second case, we add a *carry* to the component of p and proceed by addition of the next components. In this way we obtain a series for the sum that has components in the desired range. More succinctly, we can say that *addition is defined componentwise, using the system of carries to keep them in the range* $\{0, 1, \dots, p-1\}$.

An example will show how to proceed. Let

$$a = 1 = 1 + 0p + 0p^2 + \cdots,$$
$$b = (p-1) + (p-1)p + (p-1)p^2 + \cdots.$$

The sum $a + b$ has a first component 0, since $1 + (p-1) = p$. But we have to remember that a carry has to be taken into account for the next component. Hence this next component is also 0, and another carry has to be accounted for in the next place, etc. Eventually, we find that all components vanish, and the result is

$1+b=0$, namely b is an additive inverse of the integer $a=1$ (in the set of p-adic integers), and for this reason written $b=-1$. More generally, if

$$a = \sum_{i \geq 0} a_i p^i,$$

we define

$$b = \sigma(a) = \sum_{i \geq 0} (p-1-a_i)p^i$$

so that $a+b+1=0$. This is best summarized by $a+\sigma(a)+1=0$ or even $\sigma(a)+1=-a$. In particular, all natural integers have an additive inverse in the set of p-adic integers. It is now obvious that the set X of p-adic integers with the precedingly defined addition is an abelian group. The embedding of the monoid $\mathbf{N}$ in X extends to an injective homomorphism $\mathbf{Z} \to X$. Negative integers have the form $-m-1=\sigma(m)$ with all but finitely many components equal to $p-1$. Considering that the rational integers are p-adic integers, from now on we shall denote by $\mathbf{Z}_p$ the group of p-adic integers. (Another natural reason for this notation will appear in (3.6).) The mapping $\sigma : \mathbf{Z}_p \to \mathbf{Z}_p$ obviously satisfies $\sigma^2 = \sigma \circ \sigma =$ id and is therefore an *involution* on the set of p-adic integers. When p is odd, this involution has a fixed point, namely the element $a = \sum_{i \geq 0} \frac{p-1}{2} p^i \in \mathbf{Z}_p$.

1.3. *The Ring of p-adic Integers*

Let us define the product of two p-adic integers by multiplying their expansions componentwise, using the system of carries to keep these components in the desired range $\{0, 1, \ldots, p-1\}$.

This multiplication is defined in such a way that it extends the usual multiplication of natural integers (written in base p). The usual algorithm is simply pursued indefinitely. Again, a couple of examples will explain the procedure. We have found that $-1 = \sum (p-1)p^i$. Now we write

$$-1 = (p-1) \cdot \sum_{i \geq 0} p^i, \quad -(p-1) \sum_{i \geq 0} p^i = 1,$$

$$\sum_{i \geq 0} p^i = \frac{1}{1-p}.$$

Hence $1-p$ is invertible in $\mathbf{Z}_p$ with inverse given as a formal geometric series of ratio p. Since

$$p \cdot \sum_{i \geq 0} a_i p^i = a_0 p + a_1 p^2 + \cdots \neq 1 + 0p + 0p^2 + \cdots,$$

the prime p is not invertible in $\mathbf{Z}_p$ for multiplication. Using multiplication, we can also write the additive inverse of a natural number in the form

$$-m = (-1) \cdot m = \sum (p-1)p^i \cdot \sum_{i \geq 0} m_i p^i,$$

but it is not so easy to deduce the coefficients of $-m$ from this relation. Together with addition and multiplication, $\mathbf{Z}_p$ is a commutative ring. When p is odd, the fixed element under the involution σ is

$$a = \sum_{i\geq 0} \frac{p-1}{2} \cdot p^i = \frac{p-1}{2} \cdot \sum_{i\geq 0} p^i = \frac{p-1}{2} \cdot \frac{1}{1-p} = -\frac{1}{2},$$

but 2 is not an invertible element of $\mathbf{Z}_2$, $-\frac{1}{2} \notin \mathbf{Z}_2$, and the involution $\sigma = \sigma_2$ has no fixed point in $\mathbf{Z}_2$.

1.4. The Order of a p-adic Integer

Let $a = \sum_{i\geq 0} a_i p^i$ be a p-adic integer. If $a \neq 0$, there is a first index $v = v(a) \geq 0$ such that $a_v \neq 0$. This index is the p-adic order $v = v(a) = \mathrm{ord}_p(a)$, and we get a map

$$v = \mathrm{ord}_p : \mathbf{Z}_p - \{0\} \to \mathbf{N}.$$

This terminology comes from a formal analogy between the ring of p-adic integers and the ring of holomorphic functions of a complex variable $z \in \mathbf{C}$. If f is a nonzero holomorphic function in a neighborhood of a point $a \in \mathbf{C}$, we can write its Taylor series near this point

$$f(z) = \sum_{n\geq m} a_n (z-a)^n, \quad (a_m \neq 0, \ |z-a| < \varepsilon).$$

The index m of the first nonzero coefficient is by definition the order (of vanishing) of f at a: this order is 0 if $f(a) \neq 0$ and is positive if f vanishes at a.

Proposition. *The ring $\mathbf{Z}_p$ of p-adic integers is an integral domain.*

PROOF. The commutative ring $\mathbf{Z}_p$ is not $\{0\}$, and we have to show that it has no zero divisor. Let therefore $a = \sum_{i\geq 0} a_i p^i \neq 0$, $b = \sum_{i\geq 0} b_i p^i \neq 0$, and define $v = v(a)$, $w = v(b)$. Then a_v is the first nonzero coefficient of a, $0 < a_v < p$, and similarly b_w is the first nonzero coefficient of b. In particular, p divides neither a_v nor b_w and consequently does not divide their product $a_v b_w$ either. By definition of multiplication, the first nonzero coefficient of the product ab is the coefficient c_{v+w} of p^{v+w}, and this coefficient is defined by

$$0 < c_{v+w} < p, \quad c_{v+w} \equiv a_v b_w \pmod{p}.$$ ■

Corollary of proof. *The order $v : \mathbf{Z}_p - \{0\} \to \mathbf{N}$ satisfies*

$$v(ab) = v(a) + v(b),$$
$$v(a+b) \geq \min(v(a), v(b))$$

if a, b, and $a+b$ are not zero. ■

It is convenient to extend the definition of the order by $v(0) = \infty$ so that the preceding relations are satisfied without restriction on $\mathbf{Z}_p$, with the natural conventions concerning the symbol ∞. The p-adic order is then a mapping $\mathbf{Z}_p \to \mathbf{N} \cup \{\infty\}$ having the two above-listed properties.

1.5. Reduction mod p

Let $\mathbf{F}_p = \mathbf{Z}/p\mathbf{Z}$ be the finite field with p elements. *The mapping*

$$a = \sum_{i \geq 0} a_i p^i \mapsto a_0 \bmod p$$

defines a ring homomorphism $\varepsilon : \mathbf{Z}_p \to \mathbf{F}_p$ *called reduction mod* p. This reduction homomorphism is obviously surjective, with kernel

$$\{a \in \mathbf{Z}_p : a_0 = 0\} = \left\{\Sigma_{i \geq 1} a_i p^i = p \Sigma_{j \geq 0} a_{j+1} p^j\right\} = p\mathbf{Z}_p.$$

Since the quotient is a field, the kernel $p\mathbf{Z}_p$ of ε is a maximal ideal of the ring $\mathbf{Z}_p$. A comment about the notation used here has to be made in order to avoid a paradoxical view of the situation: Far from being p times bigger than $\mathbf{Z}_p$, the set $p\mathbf{Z}_p$ is a subgroup of index p in $\mathbf{Z}_p$ (just as $p\mathbf{Z}$ is a subgroup of index p in $\mathbf{Z}$).

Proposition. *The group* $\mathbf{Z}_p^\times$ *of invertible elements in the ring* $\mathbf{Z}_p$ *consists of the* p*-adic integers of order zero, namely*

$$\mathbf{Z}_p^\times = \{\sum_{i \geq 0} a_i p^i : a_0 \neq 0\}.$$

PROOF. If a p-adic integer a is invertible, so must be its reduction $\varepsilon(a)$ in $\mathbf{F}_p$. This proves the inclusion $\mathbf{Z}_p^\times \subset \{\sum_{i \geq 0} a_i p^i : a_0 \neq 0\}$. Conversely, we have to show that any p-adic integer a of order $v(a) = 0$ is invertible. In this case the reduction $\varepsilon(a) \in \mathbf{F}_p$ is not zero, and hence is invertible in this field. Choose $0 < b_0 < p$ with $a_0 b_0 \equiv 1 \bmod p$ and write $a_0 b_0 = 1 + kp$. Hence, if we write $a = a_0 + p\alpha$, then

$$a \cdot b_0 = 1 + kp + p\alpha b_0 = 1 + p\kappa$$

for some p-adic integer κ. It suffices to show that the p-adic integer $1 + \kappa p$ is invertible, since we can then write

$$a \cdot b_0(1 + \kappa p)^{-1} = 1, \quad a^{-1} = b_0(1 + \kappa p)^{-1}.$$

In other words, it is enough to treat the case $a_0 = 1$, $a = 1 + \kappa p$. Let us observe that we can take

$$(1 + \kappa p)^{-1} = 1 - \kappa p + (\kappa p)^2 - \cdots = 1 + c_1 p + c_2 p^2 + \cdots$$

with integers $c_i \in \{0, 1, \ldots, p - 1\}$. This possibility is assured if we apply the rules for carries suitably. Such a procedure is cumbersome to detail any further, and another, equivalent, definition of the ring $\mathbf{Z}_p$ will be given in (4.7) below, making such verifications easier to handle. ■

Corollary 1. *The ring $\mathbf{Z}_p$ of p-adic integers has a unique maximal ideal, namely*

$$p\mathbf{Z}_p = \mathbf{Z}_p - \mathbf{Z}_p^\times. \qquad ■$$

The statement of the preceding corollary corresponds to a partition $\mathbf{Z}_p = \mathbf{Z}_p^\times \amalg p\mathbf{Z}_p$ (a disjoint union). In fact, one has a partition

$$\mathbf{Z}_p - \{0\} = \coprod_{k\geq 0} p^k\mathbf{Z}_p^\times \quad \text{(disjoint union of } p^k\mathbf{Z}_p^\times = v^{-1}(k)).$$

Corollary 2. *Every nonzero p-adic integer $a \in \mathbf{Z}_p$ has a canonical representation $a = p^v u$, where $v = v(a)$ is the p-adic order of a and $u \in \mathbf{Z}_p^\times$ is a p-adic unit.* ■

Corollary 3. *The rational integers $a \in \mathbf{Z}$ that are invertible in the ring $\mathbf{Z}_p$ are the integers prime to p. The quotients of integers $m/n \in \mathbf{Q}$ $(n \neq 0)$ that are p-adic integers are those that have a denominator n prime to p.* ■

1.6. The Ring of p-adic Integers is a Principal Ideal Domain

The principal ideals of the ring $\mathbf{Z}_p$,

$$(p^k) = p^k\mathbf{Z}_p = \{x \in \mathbf{Z}_p : \mathrm{ord}_p(x) \geq k\},$$

have an intersection equal to $\{0\}$:

$$\mathbf{Z}_p \supset p\mathbf{Z}_p \supset \cdots \supset p^k\mathbf{Z}_p \supset \cdots \supset \bigcap_{k\geq 0} p^k\mathbf{Z}_p = \{0\}.$$

Indeed, any element $a \neq 0$ has an order $v(a) = k$, hence $a \notin (p^{k+1})$. In fact, these principal ideals are the only nonzero ideals of the ring of p-adic integers.

Proposition. *The ring $\mathbf{Z}_p$ is a principal ideal domain. More precisely, its ideals are the principal ideals $\{0\}$ and $p^k\mathbf{Z}_p$ $(k \in \mathbf{N})$.*

Proof. Let $I \neq \{0\}$ be a nonzero ideal of $\mathbf{Z}_p$ and $0 \neq a \in I$ an element of minimal order, say $k = v(a) < \infty$. Write $a = p^k u$ with a p-adic unit u. Hence $p^k = u^{-1}a \in I$ and $(p^k) = p^k\mathbf{Z}_p \subset I$. Conversely, for any $b \in I$ let $w = v(b) \geq k$ and write

$$b = p^w u' = p^k \cdot p^{w-k}u' \in p^k\mathbf{Z}_p.$$

This shows that $I \subset p^k\mathbf{Z}_p$. ■

2. The Compact Space $\mathbf{Z}_p$

2.1. Product Topology on $\mathbf{Z}_p$

The Cartesian product spaces

$$X_p = \prod_{i \geq 0} \{0, 1, 2, \ldots, p-1\} = \{0, 1, 2, \ldots, p-1\}^{\mathbf{N}}$$

will now be considered as topological spaces, with respect to the product topology of the finite discrete sets $\{0, 1, 2, \ldots, p-1\}$. These basic spaces will be studied presently, and we shall give natural models for them (they are homeomorphic for all p). By the Tychonoff theorem, X_p is *compact*. It is also *totally disconnected*: The connected components are points.

Let us recall that the discrete topology can be defined by a metric

$$\delta(a, b) = \begin{cases} 1 & \text{if } a \neq b, \\ 0 & \text{if } a = b, \end{cases}$$

or, using the Kronecker symbol, $\delta(a, b) = 1 - \delta_{ab}$. Several metrics compatible with the product topology on X_p can be deduced from these discrete ones. For $x = (a_0, a_1, \ldots)$, $y = (b_0, b_1, \ldots) \in X_p$, we can define

$$d(x, y) = \sup_{i \geq 0} \frac{\delta(a_i, b_i)}{p^i} = \frac{1}{p^{v(x-y)}},$$

$$d'(x, y) = \sum_{i \geq 0} \frac{\delta(a_i, b_i)}{p^{i+1}}, \text{ and so on.}$$

Although all metrics on a compact metrizable space are *uniformly equivalent*, they are not all equally interesting! For example, we favor metrics that give a faithful image of the coset structure of $\mathbf{Z}_p$: For each integer $k \in \mathbf{N}$, all cosets of $p^k\mathbf{Z}_p$ in $\mathbf{Z}_p$ should be isometric (and in particular have the same diameter).

The *p-adic metric* is the first mentioned above. Unless specified otherwise, we use it and introduce the notation

$$|x| = \begin{cases} d(x, 0) = p^{-v} & \text{if } x \neq 0 \ (v = \operatorname{ord}_p(x)), \\ 0 & \text{if } x = 0 \end{cases}$$

(absolute values will be studied systematically in Chapter II). We recover the p-adic metric from this absolute value by $d(x, y) = |x - y|$. With this metric, *multiplication by p in $\mathbf{Z}_p$ is a contracting map*

$$d(px, py) = \tfrac{1}{p}\, d(x, y)$$

and hence is continuous.

2.2. *The Cantor Set*

In point set topology the Cantor set plays an important role. Let us recall its construction. From the unit interval $C_0 = I = [0, 1]$ one deletes the open middle third. There remains a compact set

$$C_1 = [0, \tfrac{1}{3}] \cup [\tfrac{2}{3}, 1].$$

Deleting again the open middle third of each of the remaining intervals, we obtain a smaller compact set

$$C_2 = [0, \tfrac{1}{9}] \cup [\tfrac{2}{9}, \tfrac{1}{3}] \cup [\tfrac{2}{3}, \tfrac{7}{9}] \cup [\tfrac{8}{9}, 1].$$

Iterating the process, we get a decreasing sequence of nested compact subsets of the unit interval. By definition, the *Cantor set C is the intersection of all C_n.*

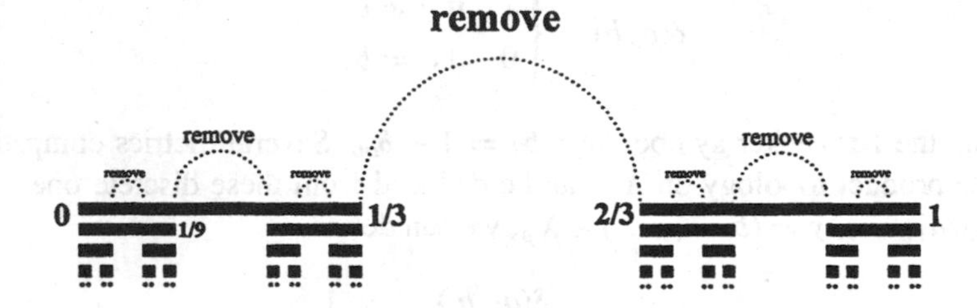

The Cantor set

It is a nonempty compact subset of the unit interval $I = [0, 1]$. The Cantor diagonal process (see 1.1) also shows that this compact set is not countable. If we temporarily adopt a system of numeration in base 3 — hence with *digits* 0, 1, and 2 — the removal of the first middle third amounts to deleting numbers having first digit equal to 1 (keeping first digits 0 and 2). Removing the second, smaller, middle intervals amounts to removing numbers having second digit equal to 1, and so on. Finally, we see that the Cantor set C consists precisely of the numbers $0 \le a \le 1$ that admit an expansion *in base* 3:

$$0.\alpha_1\alpha_2 \ldots = \frac{\alpha_1}{3} + \frac{\alpha_2}{3^2} + \cdots$$

with digits $\alpha_i = 0$ or 2. We obtain these expansions by doubling the elements of arbitrary binary sequences. This leads to considering the bijection

$$\psi : \sum_{i \ge 0} a_i 2^i \mapsto \sum_{i \ge 0} \frac{2a_i}{3^{i+1}}, \quad \mathbf{Z}_2 \to C.$$

The definition of the product topology shows that this mapping is continuous, and hence is a homeomorphism, since the spaces in question are compact.

Binary sequences can also be considered as representing expansions in base 2 of elements in the unit interval. This leads to a surjective mapping

$$\varphi : \sum_{i\geq 0} a_i 2^i \mapsto \sum_{i\geq 0} \frac{a_i}{2^{i+1}}, \quad \mathbf{Z}_2 \to [0, 1].$$

This map is surjective and continuous but is not injective: The numbers $\sum_{i>j} 2^i$ and $2^j \in \mathbf{Z}_2$ have the same image in [0, 1], as is immediately seen (in the decimal system, a decimal expansion having only 9's after place j can be replaced by a decimal expansion with a single 1 in place j). In fact, Card $\varphi^{-1}(t) \leq 2$ for any $t \in [0, 1]$.

We can summarize the situation by a commutative diagram of maps

$$\begin{array}{ccccc} \psi : \mathbf{Z}_2 & \to & C & \subset & [0, 1] \\ \| & & \downarrow & \swarrow & g \\ \varphi : \mathbf{Z}_2 & \to & [0, 1] & & \end{array}$$

The function g identifies contiguous extremities of the Cantor set C and sends them onto points of the interval having two binary expansions (rational numbers of the form $a/2^j$). These constructions will now be generalized.

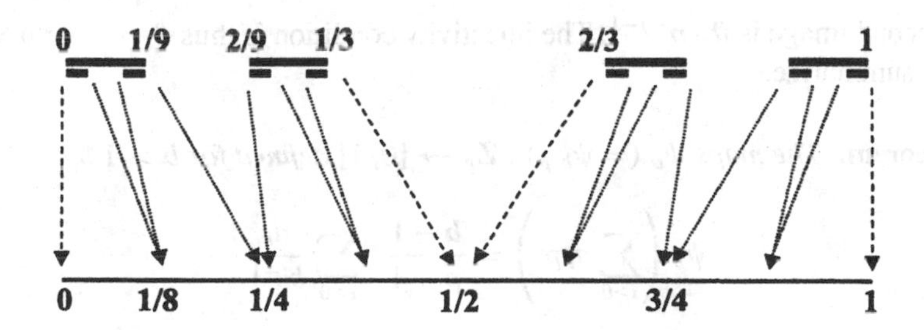

Gluing the extremities of the Cantor set

2.3. *Linear Models of* $\mathbf{Z}_p$

We choose a real number $b > 1$ and use it as numeration base in the unit interval [0, 1]. In other words, we try to write real numbers in this interval in the form $a_0/b + a_1/b^2 + \cdots$ with integral digits $0 \leq a_i < b$. More precisely, fix the prime p and consider the maps $\psi = \psi_b (= \psi_{b,p}) : \mathbf{Z}_p \to [0, 1]$ defined by the infinite series in $\mathbf{R}$

$$\psi\left(\sum_{i\geq 0} a_i p^i\right) = \vartheta \cdot \sum_{i\geq 0} \frac{a_i}{b^{i+1}},$$

with a normalizing constant ϑ chosen so that the maximum of ψ is 1. Since this maximum is attained when all digits a_i are maximal, it is attained at

$-1 = \sum_{i\geq 0}(p-1)p^i \in \mathbf{Z}_p$, and its image must be 1:

$$1 = \vartheta \cdot \sum_{i\geq 0} \frac{p-1}{b^{i+1}} = \vartheta(p-1)\frac{b^{-1}}{1-b^{-1}} = \vartheta \frac{p-1}{b-1},$$

namely

$$\vartheta = \frac{b-1}{p-1}.$$

For $p = 2$ and $b = 3$ we find that $\vartheta = 2$, and we recover the special case studied in the preceding section, where ψ furnished a homeomorphism $\mathbf{Z}_2 \to C \subset [0, 1]$. In general, $\psi = \psi_b$ will be injective if the p-adic integers

$$\sum_{i>j}(p-1)p^i \text{ and } p^j \in \mathbf{Z}_p$$

have distinct images in $[0, 1]$. The first image is

$$\begin{aligned}\vartheta \cdot (p-1) \sum_{i>j} 1/b^{i+1} &= \vartheta(p-1)b^{-j-2}/(1-b^{-1}) \\ &= \vartheta b^{-j-1}(p-1)/(b-1) = b^{-j-1}.\end{aligned}$$

The second image is $\vartheta \cdot b^{-j-1}$. The injectivity condition is thus $\vartheta > 1$, or $b > p$. Let us summarize.

Theorem. *The maps* $\psi_b\ (= \psi_{b,p}) : \mathbf{Z}_p \to [0, 1]$ *defined for* $b > 1$ *by*

$$\psi_b\left(\sum_{i\geq 0} a_i p^i\right) = \frac{b-1}{p-1} \cdot \sum_{i\geq 0} \frac{a_i}{b^{i+1}}$$

are continuous. When $b > p$, ψ_b *is injective and defines a homeomorphism of* $\mathbf{Z}_p$ *onto its image* $\psi_b(\mathbf{Z}_p)$. *When* $b = p$, *we get a surjective map* ψ_p *which is not injective.* ∎

The commutative diagram given in the last section generalizes immediately to our present context.

Comment. When $b > p$, ψ_b gives a linear model of $\mathbf{Z}_p$ in the interval $[0, 1]$; the image is a *fractal subset* A of this interval. The self-similarity dimension d of such a set is "defined" by means of a dilatation producing a union of copies of translates of A. If we denote by $E(A)$ an intuitive — not formally defined — notion of *extent* of A and if λA is a union of m translates of A, this self-similarity dimension d satisfies

$$mE(A) = E(\lambda A) = \lambda^d E(A),$$

and hence $d \cdot \log \lambda = \log m$ and $d = \log m / \log \lambda$. In our case, take $\lambda = b$ so that $m = p$ and the self-similarity dimension of $A = \psi_b(\mathbf{Z}_p)$ in $[0, 1] \subset \mathbf{R}$ is $\log p / \log b < 1$. In this way we obtain a continuous family of fractal models of increasing dimension for $b \searrow p$ degenerating in the limit to a connected interval.

It may be useful to look at symmetric models obtained by replacing the digits $a_i \in \{0, 1, 2, \ldots, p-1\}$ by symmetric ones in $\{-\frac{p-1}{2}, \ldots, \frac{p-1}{2}\}$. Define

$$\nu(k) = k - \frac{p-1}{2} \qquad (0 \leq k \leq p-1).$$

We can choose the normalization constant ϑ of the map

$$\psi' : \sum_{i \geq 0} a_i p^i \mapsto \vartheta \cdot \sum \frac{\nu(a_i)}{b^{i+1}}$$

in order to have

$$\min \psi' = -1, \quad \max \psi' = +1.$$

(When $p = 2$, $\nu(k) = (-1)^{k+1}\frac{1}{2} = \pm\frac{1}{2}$, and the corresponding expansion has fractional digits.) The involution σ induces a change of sign in the image. When $p \neq 2$ it has the origin as fixed point. Here is a picture of centered linear models of $\mathbf{Z}_3$ when $b \searrow 3$.

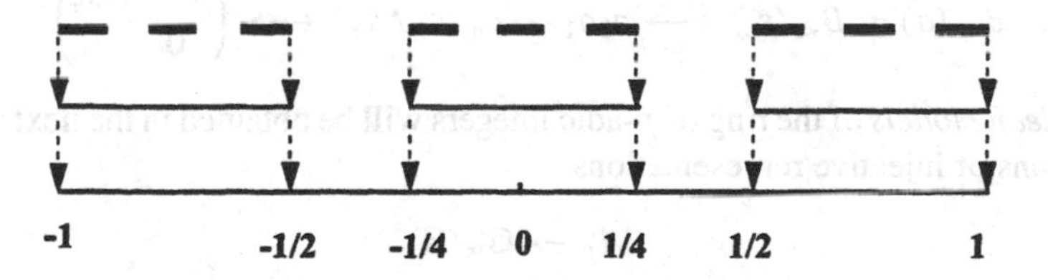

A centered lincar model of $\mathbf{Z}_3$

2.4. *Free Monoids and Balls of $\mathbf{Z}_p$*

Let $B_{<r}(a)$ denote the *ball* defined by $d(x, a) = |x - a| < r$ in $\mathbf{Z}_p$. It is clear that this ball does not change if we replace its radius r by the smallest power p^{-n} that is greater than or equal to r. If the p-adic expansion of a is $a_0 + a_1 p + \cdots + a_n p^n + \cdots = s_n + p^{n+1}\alpha$, the ball does not change either if we replace its center by s_n. This ball is fully determined by the sequence of digits (of variable length giving the radius) $a_0, a_1, \ldots, a_n$, and we associate to it the word

$$a_0 a_1 \cdots a_n \in \mathcal{M}_p$$

in the free monoid generated by $S = \{0, 1, \ldots, p-1\}$.

Conversely, to each (finite) word in the elements of S — say $a_0 a_1 \cdots a_n$ — we associate the ball of center $a = a_0 + a_1 p + \cdots + a_n p^n$ and radius $r = p^{-n}$. We get in this way a bijective map between $\mathcal{M}_p$ and the set of balls of $\mathbf{Z}_p$: Observe that a *ball* $B_{\leq r}(a)$ defined by $d(x, a) \leq r$ is the same as a ball $B_{<r'}(a)$ for some $r' > r$.

The monoid $\mathcal{M}_p$ has several matrix representations

$$\mathcal{M}_p \to \mathrm{Gl}_n(\mathbf{Z}_p).$$

For example, when $n = 2$, we can take

$$s \mapsto T_s = \begin{pmatrix} p & s \\ 0 & 1 \end{pmatrix} \quad (s \in S = \{0, 1, \ldots, p-1\}).$$

Indeed,

$$T_a T_b = \begin{pmatrix} p & a \\ 0 & 1 \end{pmatrix} \begin{pmatrix} p & b \\ 0 & 1 \end{pmatrix} = \begin{pmatrix} p^2 & a + bp \\ 0 & 1 \end{pmatrix},$$

and more generally,

$$T_{a_0} T_{a_1} \cdots T_{a_n} = \begin{pmatrix} p^{n+1} & a_0 + a_1 p + \cdots + a_n p^n \\ 0 & 1 \end{pmatrix}.$$

Observe that in this representation the *length* of a word corresponds to the order of the determinant of the matrix. In terms of balls, the radius appears as the absolute value of the determinant, whereas a center of the ball is read in the upper right-hand corner of the matrix. With the preceding notation

$$B_{<r}(a) = B_{<r}(s_n) \longleftrightarrow a_0 a_1 \cdots a_n \ (\in \mathcal{M}_p) \longleftrightarrow \begin{pmatrix} p^{n+1} & s_n \\ 0 & 1 \end{pmatrix}.$$

Euclidean models of the ring of p-adic integers will be obtained in the next section by means of injective representations

$$\mathcal{M}_p \to \mathrm{Gl}_n(\mathbf{R}).$$

Since $\mathcal{M}_p$ is free, such representations are completely determined by the images of the generators, namely by p matrices $M_0, \ldots, M_{p-1}$.

2.5. *Euclidean Models*

Let V be a Euclidean space, namely a finite-dimensional inner product space over the field $\mathbf{R}$ of real numbers. Select an injective map

$$\nu : S = \{0, 1, 2, \ldots, p-1\} \to V, \quad \nu(S) = \Sigma \subset V,$$

and define the vector mappings (using *vector digits*)

$$\Psi = \Psi_{\nu,b} : \mathbf{Z}_p \to V, \quad \sum_{i \geq 0} a_i p^i \mapsto \vartheta \sum_{i \geq 0} \frac{\nu(a_i)}{b^{i+1}}.$$

Since $\mathbf{Z}_p = \coprod_{a_0 \in S} (a_0 + p\mathbf{Z}_p)$, we have

$$\Psi(\mathbf{Z}_p) = \bigcup_{v \in \Sigma} \left(\vartheta \frac{v}{b} + \frac{1}{b} \Psi(\mathbf{Z}_p) \right).$$

For large enough values of b, the image $F = F_{v,b} = \Psi_{v,b}\mathbf{Z}_p$ will also be a *disjoint* union of self-similar images. In this way we get a construction of spatial models $\Psi(\mathbf{Z}_p)$ by iteration (similar to the construction of the Cantor set as an intersection of compact sets).

More explicitly, let us denote by $\widehat{\Sigma}$ the convex hull of Σ in V. As is known, this is the intersection of all half spaces containing Σ. It is also the intersection of those half spaces containing Σ and having for boundary a hyperplane touching the configuration. Let λ be an affine linear functional on V such that

$$\lambda \leq 1 \text{ on } \Sigma, \quad \lambda(v) = 1 \text{ for some } v \in \Sigma.$$

Choose $\vartheta = b - 1$. Then

$$\lambda\left(\vartheta \sum_{i\geq 0} \frac{\nu(a_i)}{b^{i+1}}\right) \leq \vartheta \sum_{i\geq 0} \frac{1}{b^{i+1}} = 1,$$

so that the image F of Ψ is also contained in the convex hull of Σ: $F \subset \widehat{\Sigma} = K_0$. Moreover, by choice of the constant ϑ,

$$\lambda\left(\vartheta \Sigma_{i\geq 0} \frac{v}{b^{i+1}}\right) = 1.$$

From the self-similarity representation of F we get a better approximation

$$F = \bigcup_{v\in\Sigma}\left(\vartheta\frac{v}{b} + \frac{F}{b}\right) \subset K_1 = \bigcup_{v\in\Sigma}\left(\vartheta\frac{v}{b} + \frac{K_0}{b}\right).$$

Iterating this inclusion in the self-similarity representation of F we get an even better approximation:

$$F = \bigcup_{v\in\Sigma}\left(\vartheta\frac{v}{b} + \frac{F}{b}\right) \subset K_2 = \bigcup_{v\in\Sigma}\left(\vartheta\frac{v}{b} + \frac{1}{b}\bigcup\nolimits_{v\in\Sigma}\left(\vartheta\tfrac{v}{b} + \tfrac{K_0}{b}\right)\right).$$

Eventually, this leads to a representation of the fractal F as the intersection of a decreasing sequence of compact sets K_n. Several pictures will illustrate this construction.

(2.5.1) Take, for example, $p = 3$, $V = \mathbf{R}^3$ with canonical basis $\mathbf{e}_0$, $\mathbf{e}_1$, $\mathbf{e}_2$, and $\nu(k) = \mathbf{e}_k$. Then the corresponding vector maps $\Psi : \mathbf{Z}_3 \to \mathbf{R}^3$ are given by

$$a = \sum_{i\geq 0} a_i 3^i \mapsto \Psi(a) = \vartheta \sum \frac{\mathbf{e}_{a_i}}{b^{i+1}}.$$

Let us choose the constant ϑ such that

$$\Psi(0) = \vartheta \sum \frac{\mathbf{e}_0}{b^{i+1}} = \mathbf{e}_0,$$

namely $\vartheta \sum_{i\geq 0} 1/b^{i+1} = \vartheta/(b-1) = 1$. In this case, the image of Ψ is contained in the plane $x + y + z = 1$. Since the components of the images $\Psi(a)$ are positive, the image of the map Ψ is contained in the unit simplex of $\mathbf{R}^3$ (convex span of the basic vectors). More precisely, the mappings Ψ are injective for $b > 2$, and hence give homeomorphic images — models — of $\mathbf{Z}_3$ in this simplex. When $b = 2$, the image is a *Sierpiński gasket* — hence connected — in this simplex. In general, the image is a fractal having self-similarity dimension $\log 3/\log b$.

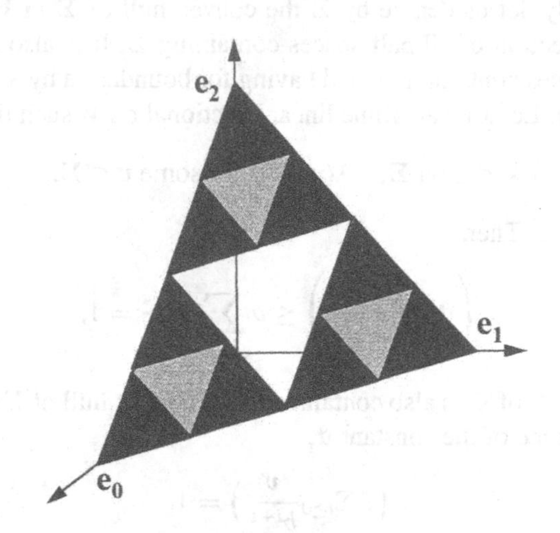

Models of $\mathbf{Z}_3$: Sierpińsky gasket

(2.5.2) Take now $p = 5$, $V = \mathbf{R}^2$, and the map ν defined by $\nu(0) = (0, 0)$, $\nu(1) = (1, 0)$, $\nu(2) = (0, 1)$, $\nu(3) = (-1, 0)$, $\nu(4) = (0, -1)$. With a suitably chosen normalization constant ϑ, the components of an image $\Psi(a) = (x, y)$ will satisfy $-1 \leq x + y \leq 1$ and $-1 \leq x - y \leq 1$. The image of Ψ is a union of the similar subsets $\Psi(k + 5\mathbf{Z}_5)$ $(0 \leq k \leq 4)$. Observe that $\Psi(5\mathbf{Z}_5) = b^{-1}\Psi(\mathbf{Z}_5)$ and that these subsets are disjoint when $b > 3$. In this case, the image is a fractal of self-similarity dimension $\log 5/\log b$. In the limit case $b = 3$ the image is connected.

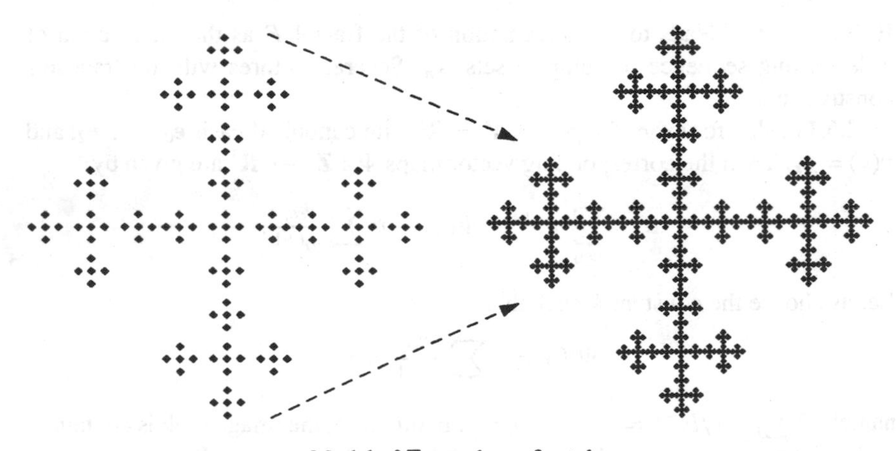

Model of $\mathbf{Z}_5$ as planar fractal

(2.5.3) It is interesting to refine the preceding construction by addition of an extra component. Take $p = 5$ as before but $V = \mathbf{R}^3$ with ν' of the form

$$\nu'(k) = (\nu(k), h_k) \in \mathbf{R}^3,$$

$$h_0 = 0,\ h_1 = h_3 = -h_2 = -h_4 = h > 0.$$

The corresponding vector maps Ψ have images in a tetrahedron bounded by an upper edge parallel to the x-axis and a lower edge parallel to the y-axis (hence two horizontal edges: Choosing h suitably, we get a regular tetrahedron). These edges give linear models of $\mathbf{Z}_2$, and the vertical projection on the horizontal plane (obtained by omitting the third component) is the previous construction. But now, the vector maps Ψ are already injective for $b > 2$, and in the limit case $b = 2$ the image is a well-known connected fractal, parametrized by $\mathbf{Z}_5$. As in (2.2), these vector mappings furnish commutative diagrams

$$\begin{array}{cccccc} \Psi_b: & \mathbf{Z}_5 & \to & \Psi_b(\mathbf{Z}_5) & \to & V \\ & \| & & \downarrow f & \swarrow g & \\ \Phi = \Psi_2: & \mathbf{Z}_5 & \to & \Phi(\mathbf{Z}_5) & & \end{array}$$

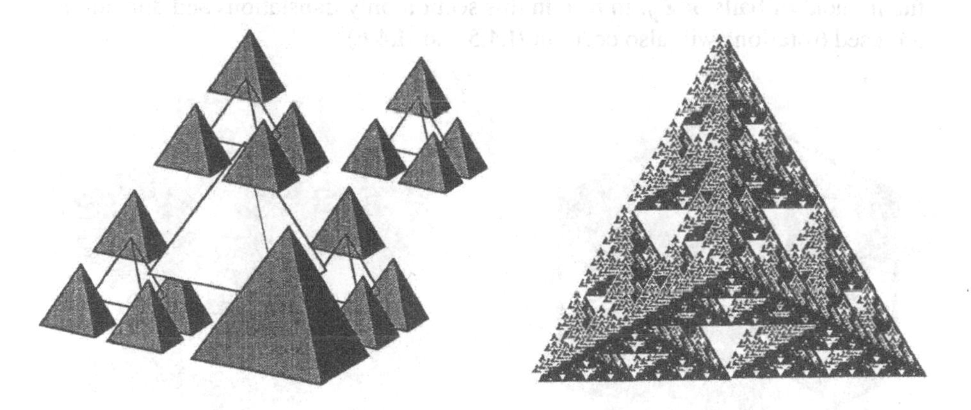

Model of $\mathbf{Z}_5$ as space fractal

(2.5.4) Take $p = 7$, $\nu : \{0, 1, 2, \ldots, 6\} \to \mathbf{R}^3$ given by $\nu(0) = 0$ and

$$\begin{array}{lll} \nu(1) = (1, 0, -1) & \nu(2) = (0, 1, -1) & \nu(3) = (-1, 1, 0) \\ \nu(4) = (-1, 0, 1) & \nu(5) = (0, -1, 1) & \nu(6) = (1, -1, 0). \end{array}$$

With a suitable normalization constant, all the image points will remain in the cube

$$-1 \le x \le 1, \quad -1 \le y \le 1, \quad -1 \le z \le 1.$$

The components of an image also satisfy $x + y + z = 0$, and hence are situated in this plane, intersecting the cube in a regular hexagon. For $b > 3$ we get interesting models of $\mathbf{Z}_7$ in this hexagon. In the limit case $b = 3$, a connected fractal parametrized by $\mathbf{Z}_7$ appears.

(2.5.5) We can give a 3-dimensional model refining the preceding one. Still with $p = 7$, take the canonical basis $\mathbf{e}_1, \mathbf{e}_2, \mathbf{e}_3$ of $\mathbf{R}^3$ and consider the vector map corresponding to the choice $\nu(0) = 0$ and

$$\begin{array}{lll} \nu(1) = \mathbf{e}_1 & \nu(2) = \mathbf{e}_2 & \nu(3) = \mathbf{e}_3 \\ \nu(4) = -\mathbf{e}_1 & \nu(5) = -\mathbf{e}_2 & \nu(6) = -\mathbf{e}_3. \end{array}$$

The image of the corresponding vector map $\Psi : \mathbf{Z}_7 \to \mathbf{R}^3$ is a fractal model contained in the octahedron

$$|x| + |y| + |z| \leq 1$$

(provided that we choose a correct normalization constant ϑ). A suitable projection of this model on a plane brings us back to the preceding planar example (contained in a hexagon).

The preceding constructions are similar to the IFS (iterated function systems) used for representing fractals: They stem from affine Euclidean representations of the monoid of balls of $\mathbf{Z}_p$. In fact, in this section only translations and dilatations are used (rotations will also occur in II.4.5 and II.4.6).

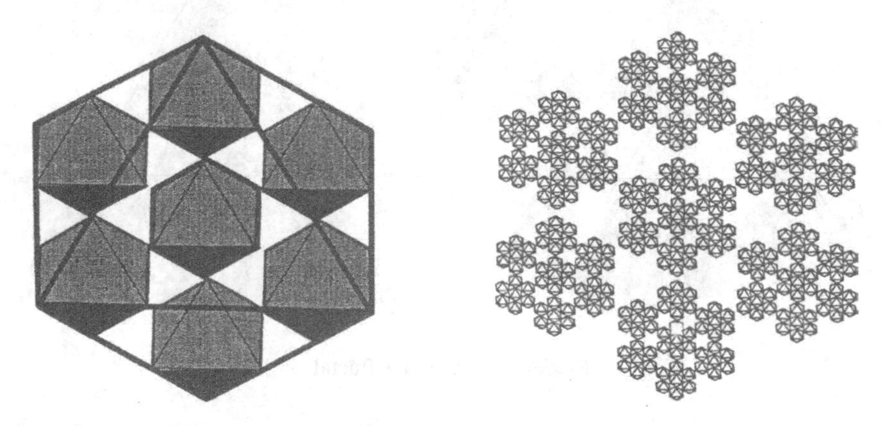

Models of $\mathbf{Z}_7$

2.6. *An Exotic Example*

There is an interesting example connecting different primes. We can add formally (i.e., componentwise) two 2-adic numbers and consider this sum in $\mathbf{Z}_3$. We thus obtain a continuous map

$$\Sigma : \mathbf{Z}_2 \times \mathbf{Z}_2 \to \mathbf{Z}_3, \quad \left(\sum a_i 2^i, \sum b_i 2^i\right) \mapsto \sum (a_i + b_i) 3^i.$$

We can make a commutative diagram

$$\begin{array}{ccc} \mathbf{Z}_2 \times \mathbf{Z}_2 & \xrightarrow{\Sigma} & \mathbf{Z}_3 \\ \downarrow & & \downarrow \\ C \times C & \xrightarrow{+} & C + C \\ \cap & & \cap \\ [0,1]^2 & \xrightarrow{+} & [0,2]. \end{array}$$

Recall that the left vertical map is given by

$$\left(\sum a_i 2^i, \sum b_i 2^i\right) \mapsto \left(\sum \frac{2a_i}{3^{i+1}}, \sum \frac{2b_i}{3^{i+1}}\right)$$

and hence the diagonal composite is

$$\left(\sum a_i 2^i, \sum b_i 2^i\right) \mapsto 2\sum \frac{a_i + b_i}{3^{i+1}}.$$

Consequently, this composite has an image equal to the whole interval $[0, 2]$. Hence addition $C \times C \to [0, 2]$ is also surjective. A good way of viewing the situation is to make a picture of the subset $C \times C$ in the unit square of $\mathbf{R}^2$ and consider addition $(x, y) \mapsto (x + y, 0)$ as a projection on the x-axis. The image of the totally disconnected set $C \times C$ is the whole interval $[0, 2]$.

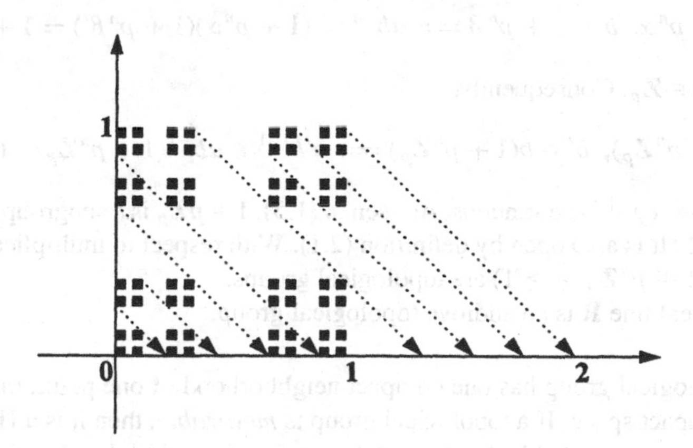

A projection of $C \times C$

3. Topological Algebra

3.1. Topological Groups

Definition. *A topological group is a group G equipped with a topology such that the map $(x, y) \mapsto xy^{-1} : G \times G \to G$ is continuous.*

If G is a topological group, the *inverse map* $x \mapsto x^{-1}$ is continuous (fix $x = e$ in the continuous map $(x, y) \mapsto xy^{-1}$) and hence a homeomorphism of order 2 of G. The translations $x \mapsto ax$ (resp. $x \mapsto xa$) are also homeomorphisms (e.g., the inverse of $x \mapsto ax$ is $x \mapsto a^{-1}x$). A subgroup of a topological group is a topological group for the induced topology.

Examples. (1) With addition, $\mathbf{Z}_p$ is a topological group. We have indeed

$$a' \in a + p^n\mathbf{Z}_p,\ b' \in b + p^n\mathbf{Z}_p \Longrightarrow a' - b' \in a - b + p^n\mathbf{Z}_p$$

for all $n \geq 0$. In other words, using the p-adic metric (2.1), we have

$$|x - a| \leq |p^n| = p^{-n},\ |y - b| \leq |p^n| = p^{-n} \Longrightarrow |(x - y) - (a - b)| \leq p^{-n},$$

proving the continuity of the map $(x, y) \mapsto x - y$ at any point (a, b).

(2) With respect to multiplication, $\mathbf{Z}_p^\times$ is a topological group. There is a fundamental system of neighborhoods of its neutral element 1 consisting of subgroups:

$$1 + p\mathbf{Z}_p \supset 1 + p^2\mathbf{Z}_p \supset \cdots \supset 1 + p^n\mathbf{Z}_p \supset \cdots$$

consists of subgroups: If $\alpha, \beta \in \mathbf{Z}_p$, we see that $(1 + p^n\beta)^{-1} = 1 + p^n\beta'$ for some $\beta' \in \mathbf{Z}_p$ (as in (1.5)), and hence

$$a = 1 + p^n\alpha,\ b = 1 + p^n\beta \Longrightarrow ab^{-1} = (1 + p^n\alpha)(1 + p^n\beta') = 1 + p^n\gamma$$

for some $\gamma \in \mathbf{Z}_p$. Consequently,

$$a' \in a(1 + p^n\mathbf{Z}_p),\ b' \in b(1 + p^n\mathbf{Z}_p) \Longrightarrow a'b'^{-1} \in ab^{-1}(1 + p^n\mathbf{Z}_p) \quad (n \geq 1),$$

and $(x, y) \mapsto xy^{-1}$ is continuous. As seen in (1.5), $1 + p\mathbf{Z}_p$ is a subgroup of index $p - 1$ in $\mathbf{Z}_p^\times$. It is also open by definition (2.1). With respect to multiplication, all subgroups $1 + p^n\mathbf{Z}_p$ $(n \geq 1)$ are topological groups.

(3) The real line $\mathbf{R}$ is an additive topological group.

If a topological group has one compact neighborhood of one point, then it is a locally compact space. If a topological group is *metrizable*, then it is a Hausdorff space and has a countable fundamental system of neighborhoods of the neutral element. Conversely, one can show that these conditions are *sufficient* for metrizability.[1]

Let G be a metrizable topological group. Then there exists a metric d on G that defines the topology of G and is invariant under left translations:

$$d(gx, gy) = d(x, y).$$

[1] Specific references for the text are listed at the end of the book.

A metrizable group G can always be completed, namely, *there exists a complete group $\widehat{G}$ and a homomorphism $j : G \to \widehat{G}$ such that*

- *the image $j(G)$ is dense in $\widehat{G}$,*
- *j is a homeomorphism $G \to j(G)$,*
- *any continuous homomorphism $f : G \to G'$ into a complete group G' can be uniquely factorized as $f = g \circ j : G \to \widehat{G} \to G'$ with a continuous homomorphism $g : \widehat{G} \to G'$.*

3.2. Closed Subgroups of Topological Groups

As already observed, a subgroup of a topological group is automatically a topological group for the induced topology.

Lemma. *Let G be a topological group, H a subgroup of G.*

(a) The closure $\overline{H}$ of H is a subgroup of G.
(b) G is Hausdorff precisely when its neutral element is closed.

Proof. (*a*) Let $\varphi : G \times G \to G$ denote the continuous map $(x, y) \mapsto xy^{-1}$. Since H is a subgroup, we have $\varphi(H \times H) \subset H$ and hence

$$\varphi(\overline{H} \times \overline{H}) = \varphi(\overline{H \times H}) \subset \overline{\varphi(H \times H)} \subset \overline{H}.$$

This proves that $\overline{H}$ is a subgroup.

(*b*) Let us recall that a topological space X is Hausdorff precisely when the diagonal Δ_X is closed in the product space $X \times X$. In any Hausdorff space the points are closed, and thus

$$\begin{aligned} G \text{ Hausdorff} &\Longrightarrow \{e\} \text{ closed} \\ &\Longrightarrow \Delta_G = \varphi^{-1}(e) \text{ closed in } G \times G \\ &\Longrightarrow G \text{ Hausdorff.} \end{aligned}$$

The lemma is completely proved. ∎

Proposition. *Let H be a subgroup of a topological group G. If H contains a neighborhood of the neutral element in G, then H is both open and closed in G.*

Proof. Let V be a neighborhood of the neutral element of G contained in H. Then for each $h \in H$, hV is a neighborhood of h in G contained in H. This proves that H is a neighborhood of all of its elements, and hence is open in G. Consider now the cosets gH of H in G. Since translations are homeomorphisms of G, these cosets are open in G. Any union of such cosets is also open. But H is the complement of the union of all cosets $gH \neq H$. Hence H is closed. ∎

Examples. The subgroups $p^n\mathbf{Z}_p$ $(n \geq 0)$ are open and closed subgroups of the additive group $\mathbf{Z}_p$. The subgroups $1+p^n\mathbf{Z}_p$ $(n \geq 1)$ are open and closed subgroups of the multiplicative group $1 + p\mathbf{Z}_p$.

Let us recall that a subspace Y of a topological space X is called *locally closed* (in X) when each point $y \in Y$ has an open neighborhood V in X such that $Y \cap V$ is closed in V. When this is so, the union of all such open neighborhoods of points of Y is an open set U in which Y is closed. This shows that the locally closed subsets of X are the intersections $U \cap F$ of an open set U and a closed set F of X. In fact, Y is locally closed in X precisely when Y is open in its closure $\overline{Y}$. Locally compact subsets of a Hausdorff space are locally closed (a compact subset is closed in a Hausdorff space). With this concept, the preceding proposition admits the following important generalization.

Theorem. *Let G be a topological group and H a locally closed subgroup. Then H is closed.*

Proof. If H is locally closed in G, then H is open in its closure $\overline{H}$. But this closure is also a topological subgroup of G. Hence (by the preceding proposition) H is closed in $\overline{H}$ (hence $H = \overline{H}$) and also closed in G by transitivity of this notion. ∎

Alternatively, we could replace G by $\overline{H}$, thus reducing the general case to H locally closed and dense in G. This case is particularly simple, since all cosets gH must meet H: $g \in H$ for all $g \in G$, namely $H = G$.

Corollary 1. *Let H be a locally compact subgroup of a Hausdorff topological group G. Then H is closed.* ∎

Corollary 2. *Let Γ be a discrete subgroup of a Hausdorff topological group G. Then Γ is closed.* ∎

The completion $\widehat{G}$ of G is also a topological group. If G is locally compact, it must be closed in its completion, and we have obtained the following corollary.

Corollary 3. *A locally compact metrizable group is complete.* ∎

3.3. Quotients of Topological Groups

As the following statement shows, the use of closed subgroups is well suited for constructing Hausdorff quotients. Let us recall that if H is a subgroup of a group G, then G/H is the set of cosets gH $(g \in G)$. The group G acts by left translations on this set. When H is a normal subgroup of G, this quotient is a group. Let now G be a topological group and

$$\pi : G \to G/H$$

denote the canonical projection. By definition of the quotient topology, the open sets $U' \subset G/H$ are the subsets such that $U = \pi^{-1}(U')$ is open in G. Now, if U is any open set in G, then

$$\pi^{-1}(\pi U) = UH = \bigcup_{h \in H} Uh$$

is open, and this proves that πU is open in G/H. Hence the canonical projection $\pi : G \to G/H$ is a continuous and open map. By complementarity, we also see that the closed sets of G/H are the images of the closed sets of the form $F = FH$ (i.e., $F = \pi^{-1}(F')$ for some complement F' of an open set $U' \subset G/H$). It is convenient to say that a subset $A \subset G$ is *saturated* (with respect to the quotient map π) when $A = AH$, so that the closed sets of G/H are the images of the saturated closed sets of G (but π is not a closed map in general).

Proposition. *Let H be a subgroup of a topological group G. Then the quotient G/H (equipped with the quotient topology) is Hausdorff precisely when H is closed.*

PROOF. Let $\pi : G \to G/H$ denote the canonical projection (continuous by definition of the quotient topology). If the quotient G/H is Hausdorff, then its points are closed and $H = \pi^{-1}(\overline{e})$ is also closed. Assume conversely that H is closed in G. The definition of the quotient topology shows that the canonical projection π is an open mapping. We infer that

$$\pi_2 = \pi \times \pi : G \times G \to G/H \times G/H$$

is also an open map. But $\mathrm{Ker}(\pi_2) = H \times H \subset G \times G$. Hence π_2 induces a topological isomorphism

$$\overline{\pi} : (G \times G)/(H \times H) \to G/H \times G/H.$$

To prove that G/H is Hausdorff, we have to prove that the *diagonal*

$$\Delta = \{(x, x) : x \in G/H\}$$

is closed in the Cartesian product $G/H \times G/H$. Since the map $\overline{\pi}$ is a homeomorphism, it is the same as proving that the inverse image A of this diagonal is closed in $(G \times G)/(H \times H)$. This inverse image is

$$\begin{aligned} A &= \{(g, k) \bmod H \times H : gH = kH\} \\ &= \{(g, k) \bmod H \times H : k^{-1}g \in H\}. \end{aligned}$$

But $R = \{(g, k) : k^{-1}g \in H\} \subset G \times G$ is closed by assumption: It is an inverse image of the closed set H under a continuous map. This closed set R is obviously saturated, i.e., satisfies

$$R = R \cdot (H \times H).$$

This proves that its image $R' = A$ in the same quotient is closed, and the conclusion is attained. ∎

Together with the theorem of the preceding section, this proposition establishes the following diagram of logical equivalences and implications for a topological group G and a subgroup H.

$$\begin{array}{ccc} G/H \text{ finite Hausdorff} & \Longleftrightarrow & H \text{ closed of finite index} \\ \Downarrow & & \Downarrow \\ G/H \text{ discrete} & \Longleftrightarrow & H \text{ open} \\ \Downarrow & & \Downarrow \\ G/H \text{ Hausdorff} & \Longleftrightarrow & H \text{ closed} \end{array}$$

3.4. Closed Subgroups of the Additive Real Line

Let us review a few well-known results concerning the classical real line, viewed as an additive topological group. At first sight, the differences with $\mathbf{Z}_p$ are striking, but a closer look will reveal formal similarities, for example when *compact* and *discrete* are interchanged.

Proposition 1. *The discrete subgroups of* $\mathbf{R}$ *are the subgroups*

$$a\mathbf{Z} \quad (0 \leq a \in \mathbf{R}).$$

Proof. Let $H \neq \{0\}$ be a nontrivial discrete subgroup, hence closed by (3.2). Consider any nonzero h in H, so that $0 < |h| \ (= \pm h) \in H$. The intersection $H \cap [0, |h|]$ is compact and discrete, hence finite, and there is a smallest positive element $a \in H$. Obviously, $\mathbf{Z} \cdot a \subset H$. In fact, this inclusion is an equality. Indeed, if we take any $b \in H$ and assume (without loss of generality) $b > 0$, we can write

$$b = ma + r \quad (m \in \mathbf{N},\ 0 \leq r < a)$$

(take for m the integral part of b/a). Since $r = b - ma \in H$ and $0 \leq r < a$, we must have $r = 0$ by construction. This proves $b = ma \in \mathbf{Z} \cdot a$, and hence the reverse inclusion $H \subset \mathbf{Z} \cdot a$. ∎

Corollary. *The quotient of* $\mathbf{R}$ *by a nontrivial discrete subgroup* $H \neq \{0\}$ *is compact.* ∎

Proposition 2. *Any nondiscrete subgroup of* $\mathbf{R}$ *is dense.*

Proof. Let $H \subset \mathbf{R}$ be a nondiscrete subgroup. Then there exists a sequence of distinct elements $h_n \in H$ with $h_n \to h \in H$. Hence $\varepsilon_n = |h_n - h| \in H$ and $\varepsilon_n \to 0$. Since H is an additive subgroup, we must also have $\mathbf{Z} \cdot \varepsilon_n \subset H$ (for all $n \geq 0$), and the subgroup H is *dense* in $\mathbf{R}$. ∎

Corollary. *(a) The only proper closed subgroups of* $\mathbf{R}$ *are the discrete subgroups* $a\mathbf{Z}$ $(a \in \mathbf{R})$.
(b) The only compact subgroup of $\mathbf{R}$ *is the trivial subgroup* $\{0\}$. ■

Using an isomorphism (of topological groups) between the additive real line and the positive multiplicative line, for example an exponential in base p

$$t \mapsto p^t, \quad \mathbf{R} \to \mathbf{R}_{>0}$$

(the inverse isomorphism is the logarithm to the base p) we deduce parallel results for the closed (resp. discrete) subgroups of the topological group $\mathbf{R}_{>0}$.

Typically, we shall use the fact that the discrete nontrivial subgroups of this group have the form $p^{a\mathbf{Z}}$ $(a > 0)$ or, putting $\theta = p^{-a}$, are the subgroups

$$\theta^{\mathbf{Z}} = \{\theta^m : m \in \mathbf{Z}\}$$

for some $0 < \theta < 1$.

3.5. *Closed Subgroups of the Additive Group of p-adic Integers*

Proposition. *The closed subgroups of the additive group* $\mathbf{Z}_p$ *are ideals: They are*

$$\{0\}, \quad p^m\mathbf{Z}_p \quad (m \in \mathbf{N}).$$

Proof. We first observe that multiplication in $\mathbf{Z}_p$ is separately continuous, since

$$|x'a - xa| = |a||x' - x| \to 0 \quad (x' \to x).$$

Since an abelian group is a $\mathbf{Z}$-module, if $H \subset \mathbf{Z}_p$ is a closed subgroup, then for any $h \in H$,

$$\mathbf{Z}H \subset H \quad \Longrightarrow \quad \mathbf{Z}_p a \subset \overline{\mathbf{Z}a} \subset \overline{H} = H.$$

This proves that a closed subgroup is an ideal of $\mathbf{Z}_p$ (or a $\mathbf{Z}_p$-module). Hence the result follows from (1.6). ■

Corollary 1. *The quotient of* $\mathbf{Z}_p$ *by a closed subgroup* $H \neq \{0\}$ *is discrete.* ■

Corollary 2. *The only discrete subgroup of the additive group* $\mathbf{Z}_p$ *is the trivial subgroup* $\{0\}$.

Proof. Indeed, discrete subgroups are closed: We have a complete list of these (being closed in $\mathbf{Z}_p$ compact, a discrete subgroup is finite hence trivial). Alternatively, if a subgroup H contains a nonzero element h, it contains all multiples of h, and hence $H \supset \mathbf{N} \cdot h$. In particular, $H \ni p^n h \to 0$ $(n \to \infty)$. Since the elements $p^n h$ are distinct, H is not discrete. ■

3.6. Topological Rings

Definition. *A topological ring A is a ring equipped with a topology such that the mappings*

$$(x, y) \mapsto x + y : A \times A \to A,$$
$$(x, y) \mapsto x \cdot y : A \times A \to A$$

are continuous.

The second axiom implies in particular that $y \mapsto -y$ is continuous (fix $x = -1$ in the product). Combined with the first, it shows that

$$(x, y) \mapsto x - y : A \times A \to A$$

is continuous and the additive group of A is a topological group. A topological ring A is a ring with a topology such that A *is an additive topological group and multiplication is continuous on* $A \times A$.

If A is a topological ring, the subgroup $A^\times$ of units is not in general a topological group, since $x \mapsto x^{-1}$ is not necessarily continuous for the induced topology (for an example of this, see the exercises). However, we can consider the embedding

$$x \mapsto (x, x^{-1}) : A^\times \to A \times A,$$

and give $A^\times$ the initial topology: It is finer than the topology induced by A. For this topology, $A^\times$ is a topological group: The continuity of the inverse map, induced by the symmetry $(x, y) \mapsto (y, x)$ of $A \times A$, is now obvious. Still with this topology, the canonical embedding $A^\times \hookrightarrow A$ is continuous, but not a homeomorphism onto its image in general.

Proposition. *With the p-adic metric the ring* $\mathbf{Z}_p$ *is a topological ring. It is a compact, complete, metrizable space.*

Proof. Since we already know that $\mathbf{Z}_p$ is a topological group (3.1), it is enough to check the continuity of multiplication. Fix a and b in $\mathbf{Z}_p$ and consider $x = a + h$, $y = b + k$ in $\mathbf{Z}_p$. Then

$$\begin{aligned} |xy - ab| &= |(a + h)(b + k) - ab| = |ak + hb - hk| \\ &\le \max(|a|, |b|)(|h| + |k|) + |h||k| \to 0 \quad (|h|, |k| \to 0). \end{aligned}$$

This proves the continuity of multiplication at any point $(a, b) \in \mathbf{Z}_p \times \mathbf{Z}_p$. ■

Corollary 1. *The topological group* $\mathbf{Z}_p$ *is a completion of the additive group* $\mathbf{Z}$ *equipped with the induced topology.* ■

To make the completion process explicit, let us observe that if $x = \sum_{i\geq 0} a_i p^i$ is a p-adic number, then

$$x_n = \sum_{0\leq i<n} a_i p^i \in \mathbf{N}$$

defines a Cauchy sequence converging to x.

Corollary 2. *The addition and multiplication of p-adic integers are the only continuous operations on $\mathbf{Z}_p$ extending addition and multiplication of the natural numbers.* ∎

3.7. Topological Fields, Valued Fields

Definition. *A topological field K is a field equipped with a topology such that the mappings*

$$(x, y) \mapsto x + y : \quad K \times K \to K,$$

$$(x, y) \mapsto x \cdot y : \quad K \times K \to K,$$

$$x \mapsto x^{-1} : \quad K^\times \to K^\times$$

are continuous.

Unless explicitly stated otherwise, fields are supposed to be *commutative*. A topological field is a topological ring for which $K^\times = K - \{0\}$ with the induced topology is a topological group. Equivalently, a topological field is a field K equipped with a topology such that

$$(x, y) \mapsto x - y \text{ is continuous on } K \times K,$$

$$(x, y) \mapsto x/y \text{ is continuous on } K^\times \times K^\times.$$

Except for the appendix to Chapter II, we shall be interested only in *valued fields*: Pairs $(K, |\,.\,|)$ where K is a field, and $|\,.\,|$ an *absolute value*, namely a group homomorphism

$$|\,.\,| : K^\times \to \mathbf{R}_{>0}$$

extended by $|0| = 0$ and satisfying the triangle inequality

$$|x + y| \leq |x| + |y| \quad (x, y \in K),$$

or the stronger ultrametric inequality

$$|x + y| \leq \max(|x|, |y|) \quad (x, y \in K).$$

In this case $d(x, y) = |x - y|$ defines an invariant metric (or ultrametric) on K,

$$d(x, y) = d(x - a, y - a) = d(x - y, 0) \quad (a,\ x,\ y \in K).$$

This situation will be systematically considered from (II.1.3) on, and in the appendix of Chapter II we shall show that any *locally compact* topological field can be considered canonically as a valued field.

Proposition 1. *Let K be a valued field. For the topology defined by the metric $d(x, y) = |x - y|$, K is a topological field.*

PROOF. The map $(x, y) \mapsto x - y$ is continuous. Let us check that the map $(x, y) \mapsto xy^{-1}$ is continuous on $K^\times \times K^\times$. We have

$$\frac{x+h}{y+k} - \frac{x}{y} = \frac{hy - kx}{y(y+k)}.$$

Hence if $y \neq 0$ is fixed, $|k| < |y|/2$, and $c = \max(|x|, |y|)$,

$$\left|\frac{x+h}{y+k} - \frac{x}{y}\right| < 2c\frac{|h| + |k|}{|y^2|} \to 0 \quad (|h|, |k| \to 0).$$

This proves that K is a topological field. ∎

Proposition 2. *Let K be a valued field. Then the completion $\widehat{K}$ of K is again a valued field.*

PROOF. The completion $\widehat{K}$ is obviously a topological ring, and inversion is continuous over the subset of invertible elements. We have to show that the completion is a *field*. Let (x_n) be a Cauchy sequence in K that defines a nonzero element of the completion $\widehat{K}$. This means that the sequence $|x_n|$ does not converge to zero. There is a positive $\varepsilon > 0$ together with an index N such that $|x_n| > \varepsilon$ for all $n \geq N$. The sequence $(1/x_n)_{n \geq N}$ is also a Cauchy sequence

$$\left|\frac{1}{x_n} - \frac{1}{x_m}\right| = \left|\frac{x_n - x_m}{x_n x_m}\right| \leq \varepsilon^{-2}|x_n - x_m| \to 0 \quad (n, m \to \infty).$$

The sequence $(1/x_n)_{n \geq N}$ (completed with 1's for $n < N$) defines an inverse of the original sequence (x_n) in the completion $\widehat{K}$. ∎

4. Projective Limits

4.1. *Introduction*

Let $x = \sum_{i \geq 0} a_i p^i$ be a p-adic integer. We have defined its reduction mod p as $\varepsilon(x) = a_0 \bmod p \in \mathbf{F}_p$. We can also consider the finer reduction $a_0 + a_1 p \bmod p^2$ or more generally

$$\varepsilon_n(x) = \sum_{i<n} a_i p^i \bmod p^n \in \mathbf{Z}/p^n\mathbf{Z}.$$

By definition of addition and multiplication of p-adic integers, we get homomorphisms

$$\varepsilon_n : \mathbf{Z}_p \to \mathbf{Z}/p^n\mathbf{Z}.$$

Since $x_n = \sum_{i<n} a_i p^i \to x \ (n \to \infty)$, we would also like to be able to say that the rings $\mathbf{Z}/p^n\mathbf{Z}$ *converge to* $\mathbf{Z}_p$. This convergence relies on the links given by the canonical homomorphisms

$$\varphi_n : \mathbf{Z}/p^{n+1}\mathbf{Z} \to \mathbf{Z}/p^n\mathbf{Z}$$

and the commutative diagram

$$\begin{array}{ccccc} & & \mathbf{Z}/p^{n+1}\mathbf{Z} & & \\ & \varepsilon_{n+1} \nearrow & & \searrow \varphi_n & \\ \mathbf{Z}_p & & \xrightarrow{\varepsilon_n} & & \mathbf{Z}/p^n\mathbf{Z} \end{array}$$

which we interpret by saying that $\mathbf{Z}_p$ is closer to $\mathbf{Z}/p^{n+1}\mathbf{Z}$ than to $\mathbf{Z}/p^n\mathbf{Z}$.

Before proceeding with precise definitions, let us still consider an example emphasizing a similar situation for sets. Consider the finite products $E_n = \prod_{i<n} X_i$ of a sequence $(X_i)_{i\geq 0}$ of sets. We would like to say that these partial products converge to the infinite product $E = \prod_{i\geq 0} X_i$ and thus consider this last product as *limit of the sequence* (E_n). For this purpose, we have to formalize the notion of approximation of E by the E_n. This relation is given by the projections

$$p_n : E \to E_n$$

omitting components of index $i \geq n$. In a sense, these projections are composed of infinitely many arrows — each $\varphi_j : E_{j+1} \to E_j$ omitting a component — as in the chain of maps

$$p_n : E \to \cdots \to E_{n+2} \to E_{n+1} \to E_n.$$

One can consider that any set X, given with a family of maps $f_n : X \to E_n$ which have the same property as above, is an *upper bound* of the sequence (E_n). A limit of the sequence would then be a *least upper bound.* Thus the limit would be an upper bound $(E, (p_n))$ such that every upper bound (X, f_n) is obtained by composition with a map $f : X \to E$ as follows:

$$f_n = p_n \circ f : X \xrightarrow{f} E \to \cdots \to E_{n+2} \to E_{n+1} \to E_n.$$

This factorization plays the role of remainder after division of f_n by all maps $\varphi_j : E_{j+1} \to E_j$ for $j \geq n$:

$$f_n = \varphi_n \circ f_{n+1} = \varphi_n \circ \varphi_{n+1} \circ f_{n+2} = \psi_n \circ f.$$

These preliminary considerations should motivate the following definition.

4.2. Definition

A sequence $(E_n, \varphi_n)_{n\geq 0}$ *of sets and maps* $\varphi_n : E_{n+1} \to E_n$ $(n \geq 0)$ *is called a* projective system. *A set* E *given together with maps* $\psi_n : E \to E_n$ *such that* $\psi_n = \varphi_n \circ \psi_{n+1}$ $(n \geq 0)$ *is called a* projective limit *of the sequence* $(E_n, \varphi_n)_{n\geq 0}$ *if the following condition is satisfied: For each set* X *and maps* $f_n : X \to E_n$ *satisfying* $f_n = \varphi_n \circ f_{n+1}$ $(n \geq 0)$ *there is a unique factorization* f *of* f_n *through the set* E:

$$f_n = \psi_n \circ f : X \to E \to E_n \quad (n \geq 0).$$

The maps $\varphi_n : E_{n+1} \to E_n$ are usually called *transition maps* of the projective system. The whole system, represented by

$$E_0 \leftarrow E_1 \leftarrow \cdots \leftarrow E_n \leftarrow \cdots,$$

is also called an *inverse system.* "The" projective limit $E = \varprojlim E_n$ is also placed at the end of the inverse system:

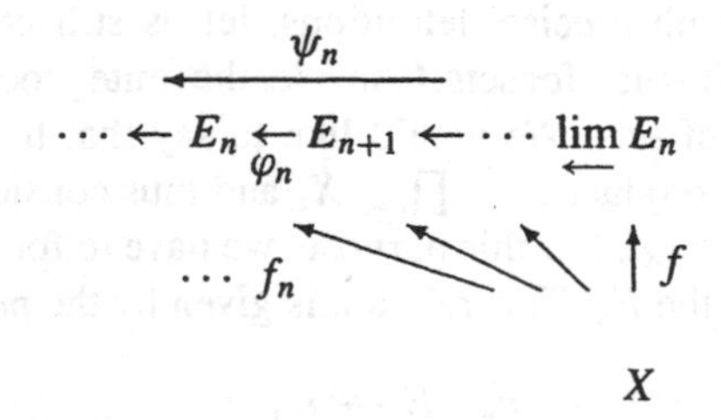

The hypothesis $f_n = \varphi_n \circ f_{n+1}$ can be iterated, and it gives

$$\begin{aligned} f_n &= \varphi_n \circ f_{n+1} = \varphi_n \circ \varphi_{n+1} \circ f_{n+2} \\ &= (\varphi_n \circ \varphi_{n+1} \circ \cdots \circ \varphi_{n+k}) \circ f_{n+k+1} = \psi_n \circ f \end{aligned}$$

for $k \geq 0$. Hence f behaves as a limit of the f_j $(j \to \infty)$ and ψ_n as a limit of composition of transition mappings $\varphi_n \circ \varphi_{n+1} \circ \cdots \circ \varphi_{n+k}$ when $k \to \infty$. The factorization condition is a *universal property* in the sense that it must hold for *all similar data.* Finally, it is obvious that if $(E, (\psi_n)_{n\geq 0})$ is a projective limit of a sequence $(E_n, \varphi_n)_{n\geq 0}$, it will still be a projective limit of any sequence $(E_n, \varphi_n)_{n\geq k}$, since we can always define inductively $\psi_{n-1} = \varphi_{n-1} \circ \psi_n$ for $n < k$. In other words, projective limits do not depend on the first terms of the sequence.

4.3. Existence

Theorem. *For every projective system* $(E_n, \varphi_n)_{n\geq 0}$ *of sets, there is a projective limit* $E = \varprojlim E_n \subset \prod_{n\geq 0} E_n$ *with maps* ψ_n *given by (restriction of) projections. Moreover, if* (E', ψ'_n) *is another projective limit of the same sequence, there is a unique bijection* $f : E' \to E$ *such that* $\psi'_n = \psi_n \circ f$.

PROOF. Let us prove *existence* first. For this purpose, define

$$E = \{(x_n) : \varphi_n(x_{n+1}) = x_n \text{ for all } n \geq 0\} \subset \prod_{n\geq 0} E_n.$$

The elements of E are thus the *coherent sequences* (with respect to the transition maps φ_n) in the product. If $x \in E$, we have by definition

$$\varphi_n(p_{n+1}(x)) = p_n(x);$$

hence for the restrictions ψ_n of the projections p_n to E,

$$\varphi_n \circ \psi_{n+1} = \psi_n.$$

The set E with the maps ψ_n can thus be viewed as an upper bound of the sequence E_n with transition maps φ_n. Let us show that this construction has the required universal property. For this purpose consider any other set E' with maps $\psi'_n : E' \to E_n$ satisfying $\varphi_n \circ \psi'_{n+1} = \psi'_n$, and let us show that there is a unique factorization of ψ'_n by ψ_n. It is clear first that the ψ'_n define a (vector) map

$$(\psi'_n) : E' \to \prod E_n, \quad y \mapsto (\psi'_n(y)).$$

The relations $\psi'_n(y) = \varphi_n(\psi'_{n+1}(y))$ show that the image of the vector map (ψ'_n) is contained in the subset E of coherent sequences. There is thus a unique map $f : E' \to E \subset \prod E_n$ having the required properties $\psi'_n = \psi_n \circ f$, and this one is simply the vector map (ψ'_n) considered as having target E. All that remains is to prove the *uniqueness*. If both $(E, (\psi_n))$ and $(E', (\psi'_n))$ have the universal factorization property, there is also a unique map $f' : E \to E'$ with $\psi_n = \psi'_n \circ f'$. Substituting this expression in $\psi'_n = \psi_n \circ f$, we find that

$$\psi'_n = \psi_n \circ f = \psi'_n \circ f' \circ f,$$

and $f' \circ f$ is a factorization of the identity map $E' \to E'$. Since we are assuming that (E', ψ'_n) has the *unique* factorization property, we must have $f' \circ f = \mathrm{id}_{E'}$. One proves similarly that $f \circ f' = \mathrm{id}_E$. ∎

Corollary. *When all transition maps in a projective system $(E_n, \varphi_n)_{n\geq 0}$ are surjective, then the projective limit $(E, (\psi_n))$ also has surjective projections ψ_n, and in particular, the set E is not empty.*

PROOF. By construction of E in the product $\prod E_n$, it is enough to show that if one component $x_n \in E_n$ is given arbitrarily, then there is a coherent sequence with this component in E_n. It is enough to choose $x_{n+1} \in E_{n+1}$ with $\varphi_n(x_{n+1}) = x_n$ (this is possible by surjectivity of φ_n) and to continue choices accordingly. The (countable!) axiom of choice ensures the possibility of finding a global coherent sequence with prescribed nth component. ∎

4.4. Projective Limits of Topological Spaces

When the projective system $(E_n, \varphi_n)_{n\geq 0}$ is formed of *topological spaces* and *continuous* transition maps, the construction made in the previous section (4.3) immediately shows that the projective limit (E, ψ_n) is a topological space equipped with continuous maps $\psi_n : E \to E_n$ having the universal property with respect to continuous maps. Any topological space X equipped with a family of continuous maps $f_n : X \to E_n$ such that $f_n = \varphi_n \circ f_{n+1}$ $(n \geq 0)$ has the factorization property $f_n = \psi_n \circ f$ with a *continuous function* $f : X \to E$. Indeed, this factorization is simply given in components by the f_n and is continuous by definition of the product topology (and the induced topology on the subset $\varprojlim E_n \subset \prod E_n$). When the topological spaces E_n are Hausdorff spaces, the subspace $\varprojlim E_n$ is closed: It is the intersection of the closed sets defined respectively by the coincidence of the functions p_n and $\varphi \circ p_{n+1}$. For future reference, let us prove a couple of results.

Proposition 1. *A projective limit of nonempty compact spaces is nonempty and compact.*

Proof. Let (K_n, φ_n) be a projective system consisting of compact spaces. The product of the K_n is a compact space (Tychonoff's theorem), and the projective limit is a closed subspace of this compact space. Hence $\varprojlim K_n$ is compact. Define

$$K'_n = \varphi_n(K_{n+1}) \supset K''_n = \varphi_n(\varphi_{n+1}(K_{n+2})) \,(= \varphi_n(K'_{n+1})\,) \supset \cdots .$$

These subsets are compact and nonempty. Their intersection L_n is not empty in the compact space K_n. Moreover, $\varphi_n(L_{n+1}) = L_n$, and the restriction of the maps φ_n to the subsets L_n leads to a projective system having surjective transition mappings. By the corollary in (4.3), this system has a nonempty limit (with surjective projections). Since $\varprojlim L_n \subset \varprojlim K_n$, the proof is complete. ∎

Corollary. *A projective limit of nonempty finite sets is nonempty.* ∎

Proposition 2. *In a projective limit $E = \varprojlim E_n$ of topological spaces, a basis of the topology is furnished by the sets $\psi_n^{-1}(U_n)$, where $n \geq 0$ and U_n is an arbitrary open set in E_n.*

Proof. We take a family $x = (x_i)$ in the projective limit and show that the mentioned open sets containing x form a basis of neighborhoods of this point. If we take two open sets $V_n \subset E_n$ and $V_{n-1} \subset E_{n-1}$, the conjunction of the conditions $x_n \in V_n$ and $x_{n-1} \in V_{n-1}$ means that

$$\psi_n(x) = x_n \in V_n \cap \varphi_{n-1}^{-1}(V_{n-1}).$$

Call U_n the open set $V_n \cap \varphi_{n-1}^{-1}(V_{n-1})$ of E_n. Then the preceding condition is still equivalent to $x \in \psi_n^{-1}(U_n)$. By induction, one can show that a basic open set in the product — say $\prod_{n\le N} V_n \times \prod_{n>N} E_n$ — has an intersection with the projective limit of the form $\psi_N^{-1}(U_N)$ for some open set $U_N \subset E_N$. ∎

Corollary. *The projective limit of the sequence of initial partial products $E_n = \prod_{i<n} X_i$ of a sequence of topological spaces is (homeomorphic to) the topological product $\prod_{i\ge 0} X_i$ of the family.* ∎

Proof. The canonical projections $\prod_{i\ge 0} X_i \to E_n$ furnish a continuous bijective factorization $\prod_{i\ge 0} X_i \to \varprojlim E_n$, which is an open map by definition of the open sets in these two spaces. ∎

Proposition 3. *Let A be a subset of a projective limit $E = \varprojlim E_n$ of topological spaces. Then the closure $\overline{A}$ of A is given by*

$$\overline{A} = \bigcap_{n\ge 0} \psi_n^{-1}(\overline{\psi_n(A)}).$$

Proof. It is clear that A is contained in the above mentioned intersection, and that this intersection is closed. Hence $\overline{A}$ is also contained in the intersection. Conversely, if b lies in the intersection, let us show that b is in the closure of A. Let V be a neighborhood of b. Without loss of generality, we can assume that V is of the form $\psi_n^{-1}(U_n)$ for some open set $U_n \subset E_n$. Hence $\psi_n(b) \subset U_n$. Since by assumption $b \in \psi_n^{-1}(\overline{\psi_n(A)})$, we have $\psi_n(b) \in \overline{\psi_n(A)}$, and the open set U_n containing b must meet $\psi_n(A)$: There is a point $a \in A$ with $\psi_n(a) \in U_n$. This shows that

$$a \in A \cap \psi_n^{-1}(U_n).$$

In particular, this intersection is nonempty, and the given neighborhood of b indeed meets A. ∎

Corollary 1. *If K is a compact subset of a projective limit $E = \varprojlim E_n$, then*

$$K = \bigcap_{n\ge 0} \psi_n^{-1}(\psi_n(K)).$$

Corollary 2. *A subset A of a topological projective limit is dense exactly when all its projections $\psi_n(A)$ are dense.* ∎

4.5. Projective Limits of Topological Groups

It is also clear that if a projective system (G_n, φ_n) is formed of groups G_n and homomorphisms $\varphi_n : G_{n+1} \to G_n$, then the projective limit $G = \varprojlim G_n$ is nonempty since it contains the neutral sequence $(e, e, ...)$. It is even a group having this sequence as neutral element, and the projections $\psi_n : G \to G_n$ are group homomorphisms. The universal factorization property holds in the *category* of groups.

An interesting case is the following. Let G be a group and (H_n) a decreasing sequence of normal subgroups of G. We can then take $G_n = G/H_n$ and (since

$H_{n+1} \subset H_n$), $\varphi_n : G/H_{n+1} \to G/H_n$ the canonical projection homomorphism. The projective limit of this sequence is a subgroup of the product

$$\widehat{G} = \varprojlim G/H_n \subset \prod G/H_n$$

together with the restrictions of projections $\psi_n : \widehat{G} \to G/H_n$. Since the system of quotient maps $f_n : G \to G/H_n$ is always a compatible system, we get a factorization $f : G \to \widehat{G}$ such that $f_n = \psi_n \circ f$. It is easy to determine the kernel of this factorization f:

$$\ker f = f^{-1}\left(\bigcap \ker \psi_n\right) = \bigcap \ker f_n = \bigcap H_n.$$

In fact, we have the following general result.

Proposition. *Let* $G = \varprojlim G_n$ *be a projective limit of groups, and let* $\psi_n : G \to G_n$ *denote the canonical homomorphisms. Then* $\cap \ker \psi_n = \{e\}$ *is reduced to the neutral element and* G *is canonically isomorphic to the projective limit* $\varprojlim (G/\ker \psi_n)$.

PROOF. Let $G' = \bigcap \ker \psi_n$ and consider the embedding $f : G' \to G$ leading to trivial composites $f_n = \psi_n|_{G'} = \psi_n \circ f$. Since the system (G', f_n) obviously admits the trivial factorization $g : G' \to G$ (constant homomorphism with image $e \in G$), we have $f = g$ by uniqueness. This proves that the embedding f is trivial, namely $G' = \{e\}$. Of course, one can also argue that since the projective limit G consists of the coherent sequences in the product $\prod G_n$, with maps ψ_n given by restriction of projections, $\cap \ker \psi_n$ consists only of the trivial sequence. ∎

4.6. Projective Limits of Topological Rings

It would be a tedious task to give a list of all structures for which projective limits can be defined. One can do it for rings, vector spaces, ..., and one can mix structures, for example by looking at topological groups, topological rings, and so on. Just for caution: A projective limit of fields is a ring, not a field in general (because a product of fields is not a field). Coming back to the case of a group G (having no topology at first), in which a decreasing sequence (H_n) of normal subgroups has been chosen, we can consider the projective limit of the system of *discrete* topological groups $G_n = G/H_n$. Let again $\widehat{G} = \varprojlim G/H_n$ and identify G with its image in $\widehat{G}$. Then G is dense in $\widehat{G}$, which can be viewed as a *completion* of G. More precisely, the closure $\widehat{H}_i$ of H_i in $\widehat{G}$ is open and closed in $\widehat{G}$, and these subgroups form a basis of neighborhoods of the identity in $\widehat{G}$. The subgroups H_i make up a basis of neighborhoods of the neutral element in G for a topology, and $\widehat{G}$ is the completion of this topological group. At this point one should recall that a topological group admitting a countable system of neighborhoods of its neutral element is *metrizable*.

Similarly, if A is a commutative ring given with a decreasing sequence (I_n) of ideals and transition homomorphisms $\varphi_n : A/I_{n+1} \to A/I_n$, the projective limit $\widehat{A} = \varprojlim A/I_n$ is a topological ring equipped with continuous homomorphisms (projections) $\psi_n : \widehat{A} \to A/I_n$. By the universal factorization property of this limit, we get a canonical homomorphism $A \to \widehat{A}$ that is injective when $\bigcap I_n = \{e\}$, and in this case $\widehat{A}$ can be identified with the completion of A for the topology of this ring, having the I_n as a fundamental system of neighborhoods of 0.

4.7. *Back to the p-adic Integers*

We apply the preceding considerations to the ring $\mathbf{Z}$ of rational integers and its decreasing sequence of ideals $I_n = p^n\mathbf{Z}$. The inclusions $p^{n+1}\mathbf{Z} \subset p^n\mathbf{Z}$ lead to canonical transition homomorphisms

$$\varphi_n : \mathbf{Z}/p^{n+1}\mathbf{Z} \longrightarrow \mathbf{Z}/p^n\mathbf{Z}.$$

The next theorem gives a second equivalent definition for p-adic integers.

Theorem. *The mapping $\mathbf{Z}_p \to \varprojlim \mathbf{Z}/p^n\mathbf{Z}$ that associates to the p-adic number $x = \sum a_i p^i$ the sequence $(x_n)_{n\geq 1}$ of its partial sums $x_n = \sum_{i<n} a_i p^i \bmod p^n$ is an isomorphism of topological rings.*

PROOF. Since the transition homomorphism φ_n is given by

$$\sum_{i\leq n} a_i p^i \bmod p^{n+1} \mapsto \sum_{i<n} a_i p^i \bmod p^n,$$

the coherent sequences in the product $\prod \mathbf{Z}/p^n\mathbf{Z}$ are simply the sequences (x_n) of partial sums of a formal series $\sum_{i\geq 0} a_i p^i$ $(0 \leq a_i \leq p-1)$, and these are precisely the p-adic integers. The relations

$$x_1 = a_0, \quad x_2 = a_0 + a_1 p, \quad x_3 = a_0 + a_1 p + a_2 p^2, \quad \ldots$$

and conversely

$$a_0 = x_1, \quad a_1 = \frac{x_2 - x_1}{p}, \quad a_2 = \frac{x_3 - x_2}{p^2}, \quad \ldots$$

show that the factorization $\mathbf{Z}_p \to \varprojlim \mathbf{Z}/p^n\mathbf{Z}$ is bijective, and hence an algebraic isomorphism. Since this is a continuous map between two compact spaces, it is a homeomorphism, whence the statement. ∎

One can note that the homomorphisms $\mathbf{Z} \to \mathbf{Z}/p^{n+1}\mathbf{Z} \to \mathbf{Z}/p^n\mathbf{Z}$ furnish a limit homomorphism $\mathbf{Z} \to \varprojlim \mathbf{Z}/p^n\mathbf{Z}$, which can be identified to the canonical embedding $\mathbf{Z} \to \mathbf{Z}_p$. The map

$$\sum_{i<n} a_i p^i \bmod p^n \mapsto \sum_{i<n} a_i p^i \bmod p^n\mathbf{Z}_p$$

obviously defines an isomorphism $\mathbf{Z}/p^n\mathbf{Z} \rightarrow \mathbf{Z}_p/p^n\mathbf{Z}_p$, and in particular,

$$\mathbf{Z}_p/p\mathbf{Z}_p \cong \mathbf{Z}/p\mathbf{Z} = \mathbf{F}_p.$$

More generally, the same argument shows that

$$\mathbf{Z}_p/p^n\mathbf{Z}_p \cong \mathbf{Z}/p^n\mathbf{Z}.$$

On the other hand, the restriction of the reduction homomorphism $\mathbf{Z}_p \rightarrow \mathbf{Z}/p^n\mathbf{Z}$ to the subring

$$\mathbf{Z}_{(p)} = \{a/b : a \in \mathbf{Z},\ 0 \neq b \in \mathbf{N} \text{ and } b \text{ prime to } p\} \subset \mathbf{Q}$$

is already surjective and has kernel $p^n\mathbf{Z}_{(p)}$, hence defines an isomorphism:

$$\mathbf{Z}_{(p)}/p^n\mathbf{Z}_{(p)} \cong \mathbf{Z}/p^n\mathbf{Z}.$$

Starting with the subring $\mathbf{Z}_{(p)} \subset \mathbf{Q}$, we see that $\mathbf{Z}_p$ appears also as a projective limit $\varprojlim \mathbf{Z}_{(p)}/p^n\mathbf{Z}_{(p)}$ and hence as a completion of this ring $\mathbf{Z}_{(p)}$.

Comment. The presentation of the ring $\mathbf{Z}_p$ of p-adic integers as a projective limit of the rings $\mathbf{Z}/p^n\mathbf{Z}$ shows that one can choose any system of representatives for $\mathbf{Z}$ mod $p\mathbf{Z}$ and write a corresponding expansion for any $x \in \mathbf{Z}_p$ in the form $x = \sum s_i p^i$ with *all digits* $s_i \in S$. In particular, when the prime p is odd, it can also be useful to choose the symmetrical system of representatives

$$S = \{-\tfrac{p-1}{2}, \ldots, 0, \ldots, \tfrac{p-1}{2}\}.$$

In practice, we always choose a system of representatives S containing 0 in order to allow *finite* expansions $x = \sum s_i p^i$. For example, if we choose the representative $p \in S$ instead of $0 \in S$, the representations

$$p \cdot 1 + 0 \cdot p + \sum_{i \geq 2} s_i p^i = 0 \cdot 1 + 1 \cdot p + \sum_{i \geq 2} s_i p^i$$

are not permitted, since $0 \notin S$.

4.8. *Formal Power Series and p-adic Integers*

Let us derive yet another presentation of p-adic integers. We denote by $\mathbf{Z}[[X]]$ the ring of formal power series in an indeterminate X with rational integral coefficients. A formal power series is just a sequence $(a_n)_{n\in\mathbf{N}}$ of integers $a_n \in \mathbf{Z}$. Addition is made coefficientwise,

$$(a_n) + (b_n) = (c_n) \text{ with } c_n = a_n + b_n \quad (n \geq 0),$$

and multiplication according to

$$(a_n) \cdot (b_n) = (c_n) \text{ with } c_n = \sum_{i+j=n} a_i b_j \quad (n \geq 0).$$

These composition laws appear naturally if we use the notation $f = f(X) = \sum_{n\geq 0} a_n X^n$ for the sequence $(a_n)_{n\in\mathbf{N}}$. In this way we identify polynomials to formal power series having only finitely many nonzero coefficients: $\mathbf{Z}[X] \subset \mathbf{Z}[[X]]$. We shall use formal power series rings over more general rings of coefficients and shall study their formal properties when needed (VI.1).

Theorem. *The map*

$$\sum a_i X^i \mapsto \sum a_i p^i : \mathbf{Z}[[X]] \to \mathbf{Z}_p$$

is a ring homomorphism. It defines a canonical isomorphism

$$\mathbf{Z}[[X]]/(X-p) \xrightarrow{\sim} \mathbf{Z}_p,$$

where $(X-p)$ denotes the principal ideal generated by the polynomial $X-p$ in the formal power series ring.

PROOF. Let us consider the sequence of homomorphisms

$$f_n : \mathbf{Z}[[X]] \to \mathbf{Z}/p^n\mathbf{Z}, \quad \sum a_i X^i \mapsto \sum_{i<n} a_i p^i \bmod p^n.$$

Since these maps f_n are obviously compatible with the transition homomorphisms φ_n defining the projective limit, we infer that there is a unique homomorphism

$$f : \mathbf{Z}[[X]] \to \varprojlim \mathbf{Z}/p^n\mathbf{Z} = \mathbf{Z}_p$$

compatible with the f_n. If $x = \sum a_i p^i$ is any p-adic integer, then $x = f(\sum a_i X^i)$, and this shows that f is *surjective*. We have to show that the kernel of f is the principal ideal generated by the polynomial $X-p$. In other words, we have to show that if the formal power series $\sum a_i X^i$ is such that $\sum_{i<n} a_i p^i \in p^n\mathbf{Z}$ for every $n > 0$, then this formal power series $\sum a_i X^i$ is divisible by $X-p$. For $n=1$ the condition implies $a_0 \equiv 0 \bmod p$, hence $a_0 = p\alpha_0$ for some integer α_0. Then, for $n=2$ we get

$$a_0 + a_1 p \equiv 0 \bmod p^2 \Longrightarrow \alpha_0 + a_1 \equiv 0 \bmod p,$$

and we infer that there is an integer α_1 such that $\alpha_0 + a_1 = p\alpha_1$. Let us go on:

$$(a_0 + a_1 p) + a_2 p^2 \equiv 0 \bmod p^3 \Longrightarrow \alpha_1 p^2 + a_2 p^2 \equiv 0 \bmod p^3,$$

which gives $\alpha_1 + a_2 = p\alpha_2$ for some integer α_2. Generally, for $n \geq 1$,

$$p^n \alpha_{n-1} + a_n p^n = a_0 + a_1 p + \cdots + a_n p^n \equiv 0 \bmod p^{n+1}$$

furnishes an integer α_n with $\alpha_{n-1} + a_n = p\alpha_n$. All these relations can be summarized by

$$a_0 = p\alpha_0, \quad a_n = p\alpha_n - \alpha_{n-1} \quad (n \geq 1),$$

or still more concisely by

$$a_0 a_1 X + a_2 X^2 + \cdots = (p - X)(\alpha_0 + \alpha_1 X + \alpha_2 X^2 + \cdots),$$

namely

$$\sum a_i X^i = (p - X) \sum \alpha_i X^i.$$

This concludes the proof. ■

5. The Field $\mathbf{Q}_p$ of p-adic Numbers

5.1. The Fraction Field of $\mathbf{Z}_p$

The ring of p-adic integers is an integral domain. Hence we can define the *field of p-adic numbers* as the fraction field of $\mathbf{Z}_p$

$$\mathbf{Q}_p = \operatorname{Frac}(\mathbf{Z}_p).$$

An equivalent definition of $\mathbf{Q}_p$ appears in (5.4).

We have seen that any nonzero p-adic integer $x \in \mathbf{Z}_p$ can be written in the form $x = p^m u$ with a unit u of $\mathbf{Z}_p$ and $m \in \mathbf{N}$ the order of x. The inverse of x in the fraction field will thus be $1/x = p^{-m}u^{-1}$. This shows that this fraction field is generated — multiplicatively, and a fortiori as a ring — by $\mathbf{Z}_p$ and the negative powers of p. We can write

$$\mathbf{Q}_p = \mathbf{Z}_p[1/p].$$

The representation $1/x = p^{-m}u^{-1}$ also shows that $1/x \in p^{-m}\mathbf{Z}_p$ and

$$\mathbf{Q}_p = \bigcup_{m \geq 0} p^{-m}\mathbf{Z}_p$$

is a union over the positive integers m. These considerations also show that a nonzero p-adic number $x \in \mathbf{Q}_p$ can be *uniquely* written as $x = p^m u$ with $m \in \mathbf{Z}$ and a unit $u \in \mathbf{Z}_p^\times$; hence

$$\mathbf{Q}_p^\times = \coprod_{m \in \mathbf{Z}} p^m \mathbf{Z}_p^\times$$

is a disjoint union over the rational integers $m \in \mathbf{Z}$. The definition of the order given in (1.4) for p-adic integers can now be extended to p-adic numbers $x \in \mathbf{Q}_p$. If $0 \neq x = p^m u$ with a unit $u \in \mathbf{Z}_p^\times$, then we define

$$\operatorname{ord}_p(x) = v_p(x) = v_p(p^m u) = m \in \mathbf{Z}.$$

(When the reference to the prime p is not needed, we simply denote this order by $v(x) = \operatorname{ord} x$.) Hence

$$v^{-1}(m) = p^m\mathbf{Z}_p - p^{m+1}\mathbf{Z}_p = p^m\mathbf{Z}_p^\times.$$

We have

$$v(x) \geq 0 \iff x \in \mathbf{Z}_p,$$

and this equivalence is valid even when $x = 0$ with the usual convention $v(0) = +\infty \geq 0$. If $x = a/b$ ($a \in \mathbf{Z}_p$, $0 \neq b \in \mathbf{Z}_p$), then $v(x) = v(a) - v(b) \in \mathbf{Z}$, and the basic relation

$$v(xy) = v(x) + v(y)$$

holds for all $x, y \in \mathbf{Z}_p$ (even when $xy = 0$ with the convention $m + \infty = \infty + \infty = \infty$). The p-adic order is a homomorphism

$$v : \mathbf{Q}_p^\times = \coprod_{m \in \mathbf{Z}} p^m \mathbf{Z}_p^\times \to \mathbf{Z}.$$

Moreover, if $x = p^v u$ is a nonzero p-adic number, with u a p-adic unit, we can write $u = \sum a_i p^i \in \mathbf{Z}_p$ with $a_0 \neq 0$ ($0 \leq a_i \leq p - 1$), and

$$x = \sum a_i p^{i+v} = \sum x_j p^j$$

is a sum starting at the integer $v = \operatorname{ord} x \in \mathbf{Z}$, possibly negative.

As in (1.4), we may compare these expansions to the Laurent expansions of meromorphic functions (in the complex plane, near a pole). The index of the first nonvanishing coefficient is the order of the power series.

By convention, the order of the zero power series is $+\infty$. Hence the relation

$$v(x + y) \geq \min(v(x), v(y))$$

holds in all cases.

Comment. If $\mathbf{Z}_{(p)} \subset \mathbf{Q}$ denotes the subring consisting of rational numbers having denominator prime to p, we have similar formulas

$$\mathbf{Q} = \bigcup_{m \geq 0} p^{-m} \mathbf{Z}_{(p)}, \quad \mathbf{Q}^\times = \coprod_{p \in \mathbf{Z}} p^m \mathbf{Z}_{(p)}^\times,$$

since the group $\mathbf{Z}_{(p)}^\times$ consists of the fractions having both numerator and denominator prime to p.

5.2. *Ultrametric Structure on Q_p*

The map $x \mapsto |x| = 1/p^v$, where $v = \operatorname{ord} x \in \mathbf{Z}$, defines a homomorphism

$$\mathbf{Q}_p^\times \to (\mathbf{R}^\times)_+ = \mathbf{R}_{>0}$$

that we conventionally extend by the definition $|0| = 0$. This map extends the previous *absolute value* on $\mathbf{Z}_p$ and is called *the p-adic absolute value on* $\mathbf{Q}_p$

(cf. (2.1), (3.7); absolute values will be systematically studied in Chapter II, cf. (II.1.3)). This absolute value has the characteristic properties

$$|x| > 0 \text{ if } x \neq 0, \quad |xy| = |x| \cdot |y|, \quad |x + y| \leq \max(|x|, |y|).$$

In particular, we can define a *metric* on $\mathbf{Q}_p$ by

$$d(x, y) = |x - y|.$$

This *distance* satisfies

$$d(x, y) > 0 \text{ if } x \neq y \text{ and } d(y, x) = d(x, y)$$

as well as the triangle inequality in the strong *ultrametric* form

$$d(x, y) \leq \max(d(x, z), d(z, y)) \leq d(x, z) + d(z, y).$$

This metric is invariant on the additive group

$$d(x + z, y + z) = d(x, y)$$

and also satisfies

$$d(zx, zy) = |z| \cdot d(x, y)$$

for all $x, y, z \in \mathbf{Q}_p$. In particular,

$$d(px, py) = \frac{d(x, y)}{p}.$$

From now on we shall always consider $\mathbf{Q}_p$ as a metric field, endowed with this ultrametric distance. By (3.7) $\mathbf{Q}_p$ is a valued field, and hence a topological field.

Theorem. *The field of p-adic numbers $\mathbf{Q}_p$ induces on $\mathbf{Z}_p$ the p-adic topology. It is a locally compact field of characteristic* 0. *It can be identified with the completion of $\mathbf{Z}[1/p] = \{ap^v : a \in \mathbf{Z},\ v \in \mathbf{Z}\}$, or of $\mathbf{Q}$, for the p-adic metric.*

PROOF. With the metric just introduced $\mathbf{Z}_p$ is the unit ball centered at the origin in $\mathbf{Q}_p$: For $x \in \mathbf{Q}_p$ we have equivalences

$$x \in \mathbf{Z}_p \Longleftrightarrow v(x) \geq 0 \Longleftrightarrow |x| \leq 1 \Longleftrightarrow d(x, 0) \leq 1.$$

Similarly, if $k \geq 0$, the ideal $p^k\mathbf{Z}_p$ is the ball defined by $d(x, 0) \leq p^{-k}$. These balls make up a fundamental system of neighborhoods of 0 in $\mathbf{Z}_p$ and $\mathbf{Q}_p$. Since the group $\mathbf{Z}_p$ contains a neighborhood of 0, it is open (and hence closed). In fact, it is a compact neighborhood of 0 in $\mathbf{Q}_p$. This proves that the topological field $\mathbf{Q}_p$ is locally compact, and hence complete (Corollary 3 in (3.2)). Finally, if

$$x = \sum_{i \geq v} x_i p^i \quad (v = \operatorname{ord} x \in \mathbf{Z})$$

is the p-adic expansion of a nonzero element $x \in \mathbf{Q}_p$, the sequence

$$x_n = \sum_{v \leq i < n} x_i p^i$$

of truncated sums is a Cauchy sequence of $\mathbf{Z}[1/p]$ converging to x,

$$x - x_n = \sum_{i \geq n} x_i p^i \in p^n \mathbf{Z}_p,$$

$$d(x, x_n) = |x - x_n| \leq p^{-n} \to 0 \quad (n \to \infty).$$

This proves that $\mathbf{Z}[1/p]$ is dense in $\mathbf{Q}_p$, and this metric space can be viewed as a completion of the ring $\mathbf{Z}[1/p]$ for the induced metric. ∎

5.3. *Characterization of Rational Numbers Among p-adic Ones*

It is easy to recognize rationals among p-adic numbers if we know their expansions. The result is similar to the characterization of rational numbers among real numbers expressed in decimal expansions.

Proposition. *Let* $x = \sum a_i p^i \in \mathbf{Q}_p$ $(i \geq v(x),\ 0 \leq a_i \leq p-1)$. *Then x is a rational number, i.e., $x \in \mathbf{Q}$ precisely when the sequence (a_i) of digits of x is eventually periodic.*

Proof. Multiplying if necessary a p-adic expansion by a power of p, we see that it is enough to consider the case $v(x) \geq 0$, namely $x \in \mathbf{Z}_p$. If the sequence (a_i) is eventually periodic, x is the sum of an integer and a linear combination (with integral coefficients) of series of the form

$$\sum_{j \geq 0} p^{s+jt} = p^s \frac{1}{1 - p^t} \in \mathbf{Q},$$

and hence is a rational number. Conversely, suppose that $x = \sum x_i p^i = a/b$ is the p-adic expansion of a rational number (as we mentioned, we can assume that $x \in \mathbf{Z}_p$; hence the summation is made for $i \geq 0$). Taking a reduced representation, a and b will be relatively prime integers, with b prime to p. Adding a suitably large integer to x, we may assume that x is positive (hence a and b are also positive). Considering the p-adic expansions of these integers, we are able to write an equality

$$\sum_{j \le \beta} b_j p^j \cdot \sum_{i \ge 0} x_i p^i = \sum_{k \le \alpha} a_k p^k.$$

In the left-hand side we have to take into account some carries r_ℓ according to the following identities:

$$b_0 x_\ell + b_1 x_{\ell-1} + \cdots + b_\ell x_0 + r_\ell = a_\ell + r_{\ell+1} p.$$

For $\ell > \max(\alpha, \beta)$, we have more simply

$$b_0 x_\ell + b_1 x_{\ell-1} + \cdots + b_\beta x_{\ell-\beta} + r_\ell = r_{\ell+1} p.$$

It suffices to compute $x_\ell \bmod p$ as a function of $x_{\ell-1}, \ldots, x_{\ell-\beta}$ and r_ℓ, and then to take the representative of this class such that $0 \le x_\ell < p$. This allows the determination of the carry $r_{\ell+1}$ by division by p. In other words, starting with the data

$$(x_{\ell-1}, \ldots, x_{\ell-\beta}, r_\ell) \in (\mathbf{Z}/p\mathbf{Z})^{\beta+1}$$

there is an algorithm (taking into account the fixed values of $b_0, \ldots, b_\beta$) furnishing

$$(x_\ell, x_{\ell-1}, \ldots, x_{\ell-\beta+1}, r_{\ell+1}) \in (\mathbf{Z}/p\mathbf{Z})^{\beta+1}$$

(the values of $x_{\ell-1}, \ldots, x_{\ell-\beta+1}$ are simply copied in a shifted position). Since the set $(\mathbf{Z}/p\mathbf{Z})^{\beta+1}$ is finite, this algorithm will eventually produce a cyclic orbit (as soon as a vector takes a value already attained, it will produce the next vector already attained and start a cycle). ∎

Corollary. *The p-adic integers* $\sum p^{n^2}$ *and* $\sum p^{n!}$ *are not rational.* ∎

5.4. Fractional and Integral Parts of p-adic Numbers

As we have already noticed, any nonzero p-adic number $x \in \mathbf{Q}_p$ can be written as a series $x = \sum_{i \ge m} x_i p^i$ starting at the index $m = v(x) \in \mathbf{Z}$. Let us define

$$[x] = \sum_{i \ge 0} x_i p^i \in \mathbf{Z}_p : \textit{integral part} \text{ of } x,$$

$$\langle x \rangle = \sum_{i < 0} x_i p^i \in \mathbf{Z}[1/p] \subset \mathbf{Q} : \textit{fractional part} \text{ of } x.$$

We thus obtain a decomposition

$$x = [x] + \langle x \rangle : \mathbf{Q}_p = \mathbf{Z}_p + \mathbf{Z}[1/p].$$

If $\langle x \rangle \neq 0$, then $\langle x \rangle = ap^v$ for integers a and $v < 0$. This decomposition *depends on the choice of representatives chosen for digits*; here $0 \le x_i \le p - 1$. With this choice, more can be said of the fractional part as a real number, namely

$$0 \leq \langle x \rangle = \sum_{i<0} x_i p^i = \sum_{1 \leq j \leq -v} \frac{x_{-j}}{p^j} < (p-1) \sum_{j \geq 1} \frac{1}{p^j} = 1.$$

Hence the fractional part of any p-adic number satisfies

$$\langle x \rangle \in [0, 1) \cap \mathbf{Z}[1/p].$$

Let us consider these representatives mod 1, namely in $\mathbf{Z}[1/p]/\mathbf{Z} \subset \mathbf{R}/\mathbf{Z}$. With the normalized exponential, we can embed the circle $\mathbf{R}/\mathbf{Z}$ in the complex numbers:

$$\mathbf{R} \to \mathbf{R}/\mathbf{Z} \to \mathbf{C}^\times : t \mapsto \exp(2\pi i t).$$

This leads us to consider the map (systematically considered by J. Tate, whence the notation)

$$\tau : \mathbf{Q}_p \to \mathbf{C}^\times : x \mapsto \exp(2\pi i \langle x \rangle).$$

For example, if $v(x) = -1$, namely $x = k/p + y$ with $0 < k \leq p-1$ and $y \in \mathbf{Z}_p$, then

$$\tau(x) = \exp(2\pi i k/p) = \zeta^k,$$

where $\zeta = \exp(2\pi i/p)$ is a primitive pth root of unity in $\mathbf{C}$. The image of all elements $x \in \mathbf{Q}_p$ with $v(x) \geq -1$ is the cyclic subgroup of order p in $\mathbf{C}^\times$:

$$p^{-1}\mathbf{Z}_p/\mathbf{Z}_p \cong \tau(p^{-1}\mathbf{Z}_p) = \mu_p \subset \mathbf{C}^\times.$$

It is useful to introduce some notation. The cyclic subgroup of mth roots of unity in $\mathbf{C}$ will be denoted by

$$\mu_m = \{z \in \mathbf{C} : z^m = 1\}.$$

The union of all these cyclic groups is the group of all roots of unity (in $\mathbf{C}$)

$$\mu = \bigcup_{m \geq 1} \mu_m = \{z \in \mathbf{C} : z^m = 1 \text{ for some integer } m \geq 1\}.$$

With respect to the prime p, we have a *direct product* decomposition

$$\mu = \mu_{(p)} \cdot \mu_{p^\infty},$$

where $\mu_{(p)}$ is the group of roots of unity of order prime to p, and μ_{p^∞} the group of roots of unity having order a power of p: pth power roots of unity. Hence μ_{p^∞} is the p-Sylow subgroup of the abelian torsion group μ. It is the union of the increasing sequence of cyclic groups

$$\mu_p \subset \mu_{p^2} \subset \cdots \subset \mu_{p^k} \subset \cdots,$$
$$\mu_{p^\infty} = \bigcup_{k \geq 0} \mu_{p^k} \subset \mathbf{C}^\times.$$

Proposition. *The map $\tau : \mathbf{Q}_p \to \mathbf{C}^\times$, $x \mapsto \exp(2\pi i\langle x\rangle)$ is a homomorphism. It defines an isomorphism $\mathbf{Q}_p/\mathbf{Z}_p \cong \mu_{p^\infty}$ of the additive group $\mathbf{Q}_p/\mathbf{Z}_p$ with the group of pth power roots of unity in the complex field $\mathbf{C}$.*

PROOF. Let us compute the difference

$$\langle x+y\rangle - \langle x\rangle - \langle y\rangle = x + y - [x+y] - (x - [x]) - (y - [y]).$$

It is equal to $[x] + [y] - [x+y] \in \mathbf{Z}_p$, and hence $\langle x+y\rangle - \langle x\rangle - \langle y\rangle \in \mathbf{Z}[1/p] \cap \mathbf{Z}_p = \mathbf{Z}$. This proves that

$$\exp(2\pi i[\langle x+y\rangle - \langle x\rangle - \langle y\rangle]) = 1$$

and $\tau(x+y) = \tau(x) + \tau(y)$. The map τ is a homomorphism. Its kernel is defined by

$$\ker \tau = \{x \in \mathbf{Q}_p : \langle x\rangle \in \mathbf{Z}\}.$$

But $\langle x\rangle \in \mathbf{Z}$ means $x = [x] + \langle x\rangle \in \mathbf{Z}_p$, so that $\ker \tau = \mathbf{Z}_p$. The image of τ consists of the complex numbers of the form

$$\exp(2\pi ik/p^m) = \exp(2\pi i/p^m)^k.$$

Since $\exp(2\pi i/p^m)$ is a root of unity of order p^m, these roots of unity generate — when m varies among natural integers — the subgroup μ_{p^∞}. ■

In particular, we have

$$x \in p^{-k}\mathbf{Z}_p \iff p^k x \in \mathbf{Z}_p \iff \tau(x)^{p^k} = 1 \iff \tau(x) \in \mu_{p^k}.$$

Comment. It is possible to give the factorization of rational numbers into p-integral and p-fractional components independent of the construction of p-adic numbers. Indeed, any rational number has the form

$$x = p^v \frac{a}{b} \quad (v \in \mathbf{Z},\ a \text{ and } b \text{ prime to } p).$$

When $v = -m < 0$, namely when $x \notin \mathbf{Z}_{(p)}$, we can use the Bézout theorem to express the fact that p^m and b are relatively prime,

$$(p^m, b) = 1 = \alpha p^m + \beta b;$$

hence multiplying by x yields

$$x = \frac{a}{p^m b} = \frac{\alpha a}{b} + \frac{\beta a}{p^m} \in \mathbf{Z}_{(p)} + \mathbf{Z}[1/p].$$

This gives an elementary description of the decomposition

$$\mathbf{Q} = \mathbf{Z}_{(p)} + \mathbf{Z}[1/p]$$

induced by the decomposition $\mathbf{Q}_p = \mathbf{Z}_p + \mathbf{Z}[1/p]$.

5.5. *Additive Structure of Q_p and Z_p*

Let us start with the sum formula $\mathbf{Q}_p = \mathbf{Z}_p + \mathbf{Z}[1/p]$ proved in the last section. Observe that this sum is not direct, since

$$\mathbf{Z}_p \cap \mathbf{Z}[1/p] = \mathbf{Z}.$$

The various embeddings that we have obtained are gathered in the following commutative diagrams giving the additive (resp. multiplicative) structure of $\mathbf{Q}_p$ (resp. $\mathbf{Q}_p^\times$).

$$\begin{array}{ccccc} & & \mathbf{Q} & & \\ & \nearrow & & \nwarrow & \\ \mathbf{Z}_{(p)} & & & & \mathbf{Z}[1/p] \\ & \nwarrow & & \nearrow & \\ & & \mathbf{Z} & & \end{array} \qquad \begin{array}{ccccc} & & \mathbf{Q}_p & & \\ & \nearrow & & \nwarrow & \\ \mathbf{Z}_p & & & & \mathbf{Z}[1/p] \\ & \nwarrow & & \nearrow & \\ & & \mathbf{Z} & & \end{array}$$

$$\begin{array}{ccccc} & & \mathbf{Q}^\times & & \\ & \nearrow & & \nwarrow & \\ \mathbf{Z}_{(p)}^\times & & & & p^{\mathbf{Z}} \\ & \nwarrow & & \nearrow & \\ & & (1) & & \end{array} \qquad \begin{array}{ccccc} & & \mathbf{Q}_p^\times & & \\ & \nearrow & & \nwarrow & \\ \mathbf{Z}_p^\times & & & & p^{\mathbf{Z}} \\ & \nwarrow & & \nearrow & \\ & & (1) & & \end{array}$$

If we embed $\mathbf{Z}$ in the direct sum $\mathbf{Z}_p \oplus \mathbf{Z}[1/p]$ by means of $m \mapsto (m, -m)$ and call Γ the image, then the addition homomorphisms

$$\mathbf{Z}_{(p)} \oplus \mathbf{Z}[1/p] \to \mathbf{Z}_{(p)} + \mathbf{Z}[1/p] = \mathbf{Q},$$

$$\mathbf{Z}_p \oplus \mathbf{Z}[1/p] \to \mathbf{Z}_p + \mathbf{Z}[1/p] = \mathbf{Q}_p$$

have kernel Γ and furnish isomorphisms

$$\left(\mathbf{Z}_{(p)} \oplus \mathbf{Z}[1/p]\right) / \Gamma \cong \mathbf{Q},$$

$$\left(\mathbf{Z}_p \oplus \mathbf{Z}[1/p]\right) / \Gamma \cong \mathbf{Q}_p.$$

Thus we have the following diagrams with vertical short exact sequences.

$$\begin{array}{ccccc} & & \mathbf{Z} & & \\ & & \downarrow & & \\ \mathbf{Z}_{(p)} & \rightarrow & \mathbf{Z}_{(p)} \oplus \mathbf{Z}[1/p] & \leftarrow & \mathbf{Z}[1/p], \\ & & \downarrow & & \\ & & \mathbf{Q} & & \end{array}$$

$$\begin{array}{ccccc} & & \mathbf{Z} & & \\ & & \downarrow & & \\ \mathbf{Z}_p & \rightarrow & \mathbf{Z}_p \oplus \mathbf{Z}[1/p] & \leftarrow & \mathbf{Z}[1/p]. \\ & & \downarrow & & \\ & & \mathbf{Q}_p & & \end{array}$$

Here is another pair of diagrams describing the inclusion relations between the various abelian groups of numbers that we have considered:

$$\begin{array}{ccccc} \mathbf{Z}[1/p] & \hookrightarrow & \mathbf{Q} & \hookrightarrow & \mathbf{Q}_p \\ \cup & & \cup & & \cup \\ \mathbf{Z} & \hookrightarrow & \mathbf{Z}_{(p)} & \hookrightarrow & \mathbf{Z}_p, \end{array} \qquad \begin{array}{ccccc} p^{\mathbf{Z}} & \hookrightarrow & \mathbf{Q}^\times & \hookrightarrow & \mathbf{Q}_p^\times \\ \cup & & \cup & & \cup \\ (1) & \hookrightarrow & \mathbf{Z}_{(p)}^\times & \hookrightarrow & \mathbf{Z}_p^\times. \end{array}$$

Comment. The subgroup $\mathbf{Z}_p$ of $\mathbf{Q}_p$ admits no direct complement. Indeed, for any subgroup Γ of $\mathbf{Q}_p$

$$\Gamma \cap \mathbf{Z}_p = \{0\} \Longrightarrow \Gamma \text{ discrete in } \mathbf{Q}_p \Longrightarrow \Gamma = \{0\}.$$

In a sense, the subgroup $\mathbf{Z}[1/p]$ is the best *near supplement* that one can take, and we have unique sum decompositions with two components:

$$x \in \mathbf{Z}_p, \quad y \in [0,1) \cap \mathbf{Z}[1/p].$$

But this system of representatives $[0, 1) \cap \mathbf{Z}[1/p]$ is not a subgroup.

5.6. Euclidean Models of Q_p

It is easy to give Euclidean models of the fields $\mathbf{Q}_p$ extending the models of $\mathbf{Z}_p$ given in (2.5) if we only observe that the inclusions of additive topological groups

$$\frac{1}{p}\mathbf{Z}_p \supset \mathbf{Z}_p \text{ and } \mathbf{Z}_p \supset p\mathbf{Z}_p$$

are similar. In other words, a dilatation of ratio p of the Euclidean model of $\mathbf{Z}_p$ gives a model of $(1/p)\mathbf{Z}_p$. Iteration gives a model of

$$\mathbf{Q}_p = \bigcup_{m \geq 0} p^{-m}\mathbf{Z}_p.$$

An illustration shows a piece of $\mathbf{Q}_7$, with central portion $\mathbf{Z}_7$.

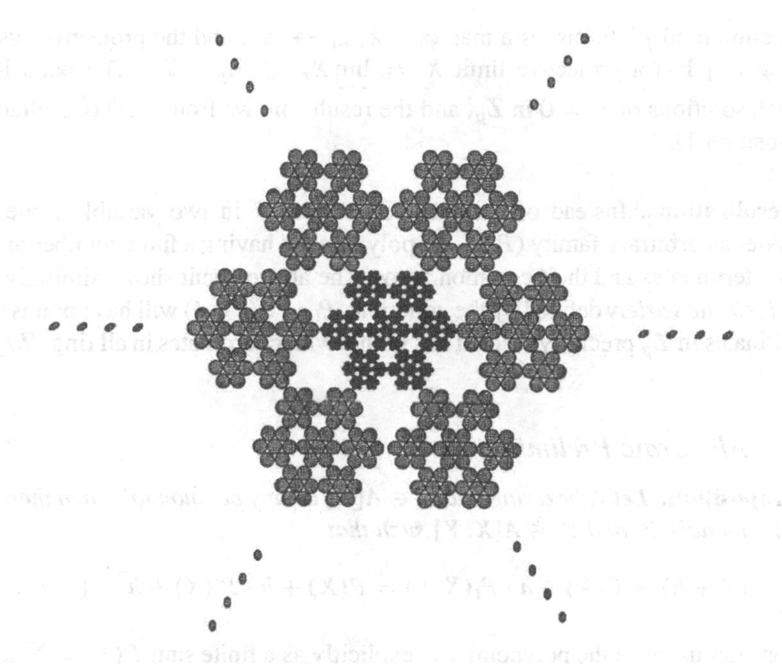

A piece of $\mathbf{Q}_7$ with $\mathbf{Z}_7$ as central portion

6. Hensel's Philosophy

6.1. *First Principle*

Let us explain the first principle in a particular case. Let $P(X, Y) \in \mathbf{Z}[X, Y]$ be a polynomial with integral coefficients. When speaking of solutions of the implicit equation $P = 0$ in a ring A, we mean a pair $(x, y) \in A \times A = A^2$ such that $P(x, y) = 0$.

Proposition. *The following properties are equivalent:*

(*i*) $P = 0$ *admits a solution in* $\mathbf{Z}_p$.
(*ii*) *For each* $n \geq 0$, $P = 0$ *admits a solution in* $\mathbf{Z}/p^n\mathbf{Z}$.
(*iii*) *For each* $n \geq 0$, *there are integers* a_n, b_n *such that*

$$P(a_n, b_n) \equiv 0 \bmod p^n.$$

PROOF. (*iii*) is a simple reformulation of (*ii*). Now for $x = \sum_{i\geq 0} a_i p^i \in \mathbf{Z}_p$, define $x_n = \sum_{i<n} a_i p^i \bmod p^n \in \mathbf{Z}/p^n\mathbf{Z}$. Then if $(x, y) \in \mathbf{Z}_p \times \mathbf{Z}_p$, then

$$P(x_n, y_n) = P(x, y) \bmod p^n\mathbf{Z}_p \in \mathbf{Z}_p/p^n\mathbf{Z}_p \ (= \mathbf{Z}/p^n\mathbf{Z}),$$

and hence $(i) \Rightarrow (ii)$. Conversely, to prove $(ii) \Rightarrow (i)$ let us consider the finite sets

$$X_n = \{(x, y) \in \mathbf{Z}/p^n\mathbf{Z} \times \mathbf{Z}/p^n\mathbf{Z} : P(x, y) = 0\}.$$

Reduction mod p^n furnishes a map $\varphi_n : X_{n+1} \to X_n$, and the projective system $(X_n, \varphi_n)_{n\geq 1}$ has a projective limit $X = \varprojlim X_n \subset \mathbf{Z}_p \times \mathbf{Z}_p$. The pairs in X furnish solutions of $P = 0$ in $\mathbf{Z}_p$, and the result follows from (4.4) (Corollary of Proposition 1). ■

Generalizations. Instead of a single polynomial P in two variables, one can consider an arbitrary family $(P_i)_{i\in I}$ of polynomials having a finite number $m \geq 2$ of indeterminates and their common zeros. The above result shows similarly that the *algebraic variety* defined by the equations $P_i = 0$ $(i \in I)$ will have points with coordinates in $\mathbf{Z}_p$ precisely when it has points with coordinates in all rings $\mathbf{Z}/p^n\mathbf{Z}$ $(n \geq 1)$.

6.2. *Algebraic Preliminaries*

Proposition. *Let A be a ring and $P \in A[X]$ be any polynomial. Then there are polynomials P_1 and $P_2 \in A[X, Y]$ such that*

$$P(X+h) = P(X) + h \cdot P_1(X, h) = P(X) + h \cdot P'(X) + h^2 P_2(X, h).$$

PROOF. Let us write the polynomial P explicitly as a finite sum $P(X) = \sum a_n X^n$ with some coefficients $a_n \in A$. Then

$$\begin{aligned} P(X+h) &= \sum a_n (X+h)^n = \sum a_n (X^n + nX^{n-1}h + h^2(\cdots)) \\ &= \sum a_n X^n + h \sum n a_n X^{n-1} + h^2 \cdot P_2(X, h); \end{aligned}$$

hence the result. ■

6.3. *Second Principle*

The idea for improving approximate solutions will now be given in its simplest form. Take a polynomial $P \in \mathbf{Z}[X]$ and an integer x such that $P(x) \equiv 0 \bmod p$. We can look for a better approximation $\hat{x}$ of $P(X) = 0$ in the form of an integer such that $P(\hat{x}) \equiv 0 \bmod p^2$. Without loss of generality, we may assume that x is an integer a_0 between 0 and $p-1$. We are looking for an integer $\hat{x} = a_0 + a_1 p$ (again with $0 \leq a_1 < p$) such that $P(\hat{x}) \equiv 0 \bmod p^2$. But we have just seen that we can write

$$P(a_0 + a_1 p) = P(a_0) + P'(a_0) \cdot a_1 p + (a_1 p)^2 \cdot b$$

for some integer b. By assumption, $P(a_0) = pt$, and the desired congruence holds mod p^2 if $t + P'(a_0) \cdot a_1 \equiv 0 \bmod p$. We can suppose $t \not\equiv 0$ (there is nothing to prove otherwise). When $P'(a_0) \not\equiv 0 \bmod p$ we can take $a_1 \equiv -t/P'(a_0) \bmod p$ and

$$\hat{x} = a_0 + a_1 p = a_0 - \frac{pt}{P'(a_0)} = x - \frac{P(a_0)}{P'(a_0)}$$

exactly as in the classical *Newton approximation method.* With this choice, we have

$$P(\hat{x}) = P(a_0 + a_1 p) \equiv 0 \bmod p^2.$$

We shall occasionally use the notation

$$N_P(x) = x - \frac{P(x)}{P'(x)}$$

for the Newton map. It is obvious that $\hat{x} = N_P(x)$ can be *far* from x when $P'(x)$ is *small.*

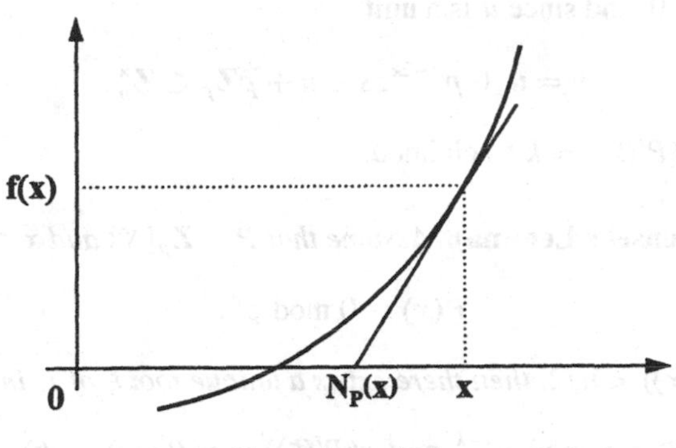

Newton's method

6.4. The Newtonian Algorithm

In this section we show that even when the derivative vanishes mod p, we can still construct a better approximation of a root of $P = 0$, but we have to be less demanding concerning its location.

Proposition. *Let $P \in \mathbf{Z}_p[X]$ and $x \in \mathbf{Z}_p$ be such that $P(x) \equiv 0 \bmod p^n$. If $k = v(P'(x)) < n/2$, then $\hat{x} = N_P(x) = x - P(x)/P'(x)$ satisfies*

(1) $P(\hat{x}) \equiv 0 \bmod p^{n+1}$ *(a definite improvement),*
(2) $\hat{x} \equiv x \bmod p^{n-k}$ *(a controlled loss),*
(3) $v(P'(\hat{x})) = v(P'(x))$ $(= k)$ *(an invitation to iteration).*

Proof. Put $P(x) = p^n y$ for some $y \in \mathbf{Z}_p$, and $P'(x) = p^k u$ for some unit $u \in \mathbf{Z}_p^\times$. By definition of $\hat{x}$,

$$\hat{x} - x = -\frac{P(x)}{P'(x)} = -p^{n-k} y u^{-1} \in p^{n-k}\mathbf{Z}_p.$$

On the other hand, still by choice of $\hat{x}$, the first two terms of the Taylor expansion of the polynomial P at the point x cancel each other:

$$P(\hat{x}) = P(x) - \frac{P(x)}{P'(x)} P'(x) + (\hat{x} - x)^2 \cdot t.$$

By (6.2) the t in the last term belongs to $\mathbf{Z}_p$. Hence

$$P(\hat{x}) = (\hat{x} - x)^2 \cdot t \in p^{2n-2k}\,\mathbf{Z}_p = p^n \cdot p^{n-2k}\,\mathbf{Z}_p \subset p^{n+1}\mathbf{Z}_p$$

(recall that $2k < n$). It only remains to compute the order of $P'(\hat{x})$. For this, we use a first-order Taylor expansion of P' at the point x (6.2):

$$\begin{aligned} P'(\hat{x}) &= P'(x + (\hat{x} - x)) = P'(x) + (\hat{x} - x) \cdot s \\ &= p^k u + p^{n-k} z \cdot s = p^k(u + p^{n-2k} zs) \;=\; p^k v. \end{aligned}$$

Since $n - 2k > 0$, and since u is a unit,

$$v = u + p^{n-2k} zs \in u + p\mathbf{Z}_p \subset \mathbf{Z}_p^\times,$$

which proves $v(P'(\hat{x})) = k$ as claimed. ∎

Theorem (Hensel's Lemma). *Assume that $P \in \mathbf{Z}_p[X]$ and $x \in \mathbf{Z}_p$ satisfies*

$$P(x) \equiv 0 \bmod p^n.$$

If $k = v(P'(x)) < n/2$, then there exists a unique root ξ of P in $\mathbf{Z}_p$ such that

$$\xi \equiv x \bmod p^{n-k} \text{ and } v(P'(\xi)) = v(P'(x)) \; (= k).$$

Proof. *Existence.* Let $x_0 = x$ and construct an improved root $x_1 \in \mathbf{Z}_p$,

$$x_1 \equiv x_0 \bmod p^{n-k} \text{ and } P(x_1) \equiv 0 \bmod p^{n+1}, \quad v(P'(x_1)) = v(P'(x_0)) \; (=k).$$

Similarly, we can find an improvement x_2 of the approximate root x_1 in the form of a p-adic integer satisfying

$$x_2 \equiv x_1 \bmod p^{n+1-k} \text{ and } P(x_2) \equiv 0 \bmod p^{n+2}.$$

Iterating the construction, we get a Cauchy sequence $(x_n)_{n \geq 0}$ having a p-adic limit ξ satisfying $P(\xi) = 0$ and $\xi \equiv x \bmod p^{n-k}$.

Uniqueness. Let ξ and η be two roots of P satisfying the required conditions: In particular,

$$\eta \equiv \xi \bmod p^{n-k},$$

and since $n > 2k$, we have $n - k \geq k + 1$, and a fortiori

$$\eta \equiv \xi \bmod p^{k+1}.$$

Now,

$$\underbrace{P(\eta)}_{=0} = \underbrace{P(\xi)}_{=0} + P'(\xi)(\eta - \xi) + (\eta - \xi)^2 a$$

for some p-adic integer a. Hence

$$(\eta - \xi)\Big(P'(\xi) + (\eta - \xi)a\Big) = 0.$$

But

$$\underbrace{P'(\xi)}_{\text{order } k} + \underbrace{(\eta - \xi)a}_{\text{order} \geq k+1} \neq 0,$$

so that the only possibility is $\eta - \xi = 0$, and uniqueness follows. ∎

Note that the uniqueness part of the proof shows that ξ is the unique root satisfying the a priori weaker congruence $\xi \equiv x \pmod{p^{k+1}}$.

6.5. *First Application: Invertible Elements in* $\mathbf{Z}_p$

Let us consider the first-degree polynomial $P(X) = aX - 1$, where $a \neq 0$ is a p-adic integer. In order to be able to find an approximate root mod p, we have to assume that $a \notin p\mathbf{Z}_p$ (in the p-adic expansion of a, the constant term $a_0 \neq 0$). When this is the case, $P'(X) = a$ and $k = v(P'(x)) = 0$, and any root mod p can be improved to a root mod p^n ($n \geq 2$). Eventually, we find a genuine root in $\mathbf{Z}_p$, which means that a is invertible in this ring. Thus we have another "proof" of the implication

$$a \in \mathbf{Z}_p - p\mathbf{Z}_p \Longrightarrow a \in \mathbf{Z}_p^\times.$$

However, this proof is deceptive, since Newton's method assumes a priori that we know how to *divide*: In the first step we are led to replacing x by

$$\hat{x} = x - \frac{P(x)}{P'(x)} = x - \frac{ax - 1}{a} = \frac{1}{a} \; (!)$$

Numerically, it is better to apply Newton's method to the *rational function* $f(X) = 1/X - a$, for which $f'(X) = -1/X^2$. Hence

$$\hat{x} = N_f(x) = x - \frac{f(x)}{f'(x)} = x + x^2 f(x) = 2x - ax^2.$$

With this function, Newton's method uses a *polynomial*, and no division is required to evaluate the successive approximations of the inverse.

6.6. *Second Application: Square Roots in* $\mathbf{Q}_p$

Consider now the quadratic polynomials $P(X) = X^2 - a$, where a is a p-adic integer. It is obvious that such an equation can have a root x in $\mathbf{Z}_p$ only if $v(a) = v(x^2) = 2v(x)$ is even. Then if we divide a by a suitable even power p^{2m} of p, we are brought back to the case $v(a) = 0$, namely $a \in \mathbf{Z}_p^\times$. Since $P'(x) = 2x$, we see that the case $p = 2$ has to be treated separately.

Case p odd. Hensel's lemma will apply as soon as we can find an approximate root mod p. But we know that in the cyclic group $\mathbf{F}_p^\times$ squares make up a subgroup of index two. The *quadratic residue symbol* of Legendre distinguishes them:

$$\left(\frac{a}{p}\right) = \begin{cases} +1 & \text{if } a \text{ is a square mod } p, \\ -1 & \text{if } a \text{ is not a square mod } p. \end{cases}$$

Let us choose an integer $1 < a < p$ that is not a square mod p. Then the three numbers a, p, ap have no square root in $\mathbf{Q}_p$. They make a full set of representatives for the classes mod squares

$$\mathbf{Q}_p^\times/(\mathbf{Q}_p^\times)^2 \cong (p^{\mathbf{Z}}/p^{2\mathbf{Z}}) \times (\mathbf{Z}_p^\times/(\mathbf{Z}_p^\times)^2) \cong \mathbf{Z}/2\mathbf{Z} \times \mathbf{Z}/2\mathbf{Z}.$$

Since every quadratic extension of $\mathbf{Q}_p$ is generated by a square root of an element (every quadratic extension of a field of characteristic 0 is generated by a square root), we see that we obtain all quadratic extensions of the field $\mathbf{Q}_p$ for $p \geq 3$ — up to isomorphism — in the form of the three distinct fields

$$\mathbf{Q}_p(\sqrt{a}),\ \mathbf{Q}_p(\sqrt{p}),\ \mathbf{Q}_p(\sqrt{ap}).$$

Case $p = 2$. Observe that $\mathbf{Z}_2^\times = 1 + 2\mathbf{Z}_2$, since the only possibility for the nonzero constant digit is 1. Now we have

$$a \in \mathbf{Z}_2^\times \textit{ is a square} \iff a \in 1 + 8\mathbf{Z}_2.$$

PROOF. If $a = b^2 \in \mathbf{Z}_2^\times$ for some $b = 1 + b_1 2 + b_2 2^2 + \cdots = 1 + 2c$, then $b^2 = 1 + 4(c + c^2)$, and since $c \equiv c^2 \bmod 2\mathbf{Z}_2$, we have $b^2 \in 1 + 8\mathbf{Z}_2$ as claimed. Conversely, if $a \equiv 1 \bmod 8\mathbf{Z}_2$, we can apply Hensel's lemma to the resolution of the equation $X^2 - a = 0$, starting with the approximate solution $x = 1$. By assumption, this is an approximate solution mod 2^3 ($n = 3 > 2k = 2$ is suitable). We get an improved solution $\hat{x}$,

$$\hat{x}^2 \equiv a \bmod 2^3 \quad \text{but } \hat{x} \equiv x \bmod 2^2 \text{ only},$$

since $n - k = 3 - 1 = 2$. By iteration, we get an exact root $\xi \equiv 1 \bmod 4$ satisfying $x^2 = a$ in $\mathbf{Z}_2$. ∎

We have

$$\mathbf{Q}_2^\times/(\mathbf{Q}_2^\times)^2 \cong (2^{\mathbf{Z}}/2^{2\mathbf{Z}}) \times (\mathbf{Z}_2^\times/(\mathbf{Z}_2^\times)^2).$$

Since

$$\mathbf{Z}_2^\times = 1 + 2\mathbf{Z}_2 = \{\pm 1\} \cdot (1 + 4\mathbf{Z}_2),$$

we also have

$$\mathbf{Z}_2^\times/(\mathbf{Z}_2^\times)^2 \cong \{\pm 1\} \times (1 + 4\mathbf{Z}_2)/(1 + 8\mathbf{Z}_2),$$

so that finally

$$\mathbf{Q}_2^\times/(\mathbf{Q}_2^\times)^2 \cong \mathbf{Z}/2\mathbf{Z} \times \mathbf{Z}/2\mathbf{Z} \times \mathbf{Z}/2\mathbf{Z}.$$

There are — up to isomorphism — seven quadratic extensions of the field $\mathbf{Q}_2$. They are obtained by adjoining roots of elements in the nontrivial classes of $\mathbf{Q}_2^\times/(\mathbf{Q}_2^\times)^2$. If we choose the elements

$$-1, \ \pm(1+4) = \pm 5, \ \pm 2, \ \pm 2 \cdot 5,$$

we get the seven nonisomorphic quadratic extensions

$$\mathbf{Q}_2(\sqrt{-1}), \ \mathbf{Q}_2(\sqrt{\pm 5}), \ \mathbf{Q}_2(\sqrt{\pm 2}), \ \mathbf{Q}_2(\sqrt{\pm 10}).$$

Examples. (1) Since $3^2 \equiv 1 \bmod 8$, $x = 3$ is an approximate root of $x^2 - 1 = 0$. Newton's method leads to the improvement $\hat{x} = 7$, which is an improved solution mod 16, but we only have $7 \equiv 3 \bmod 4$ as the theory predicts (and there is no exact root $\xi \equiv 3 \bmod 4$, since the only roots are $\xi = \pm 1$).
(2) Since $a = -7 = 1 - 8 \equiv 1 \bmod 8$, we obtain

$$\sqrt{-7} \in \mathbf{Z}_2^\times \subset \mathbf{Q}_2.$$

(3) The preceding considerations prove that the equations

$$X^2 + 1 = 0 \text{ and } X^2 - 3 = 0$$

have no solution in $\mathbf{Q}_2$. The polynomials $X^2 + 1$ and $X^2 - 3$ are irreducible in $\mathbf{Q}_2[X]$.

We shall determine later the structure of the multiplicative group $1 + 4\mathbf{Z}_2$.

6.7. *Third Application: nth Roots of Unity in $\mathbf{Z}_p$*

Let ξ be any root of unity in $\mathbf{Q}_p$, say $\xi^n = 1$. Then $nv(\xi) = v(1) = 0$ and $v(\xi) = 0$. This proves that all roots of unity in $\mathbf{Q}_p$ lie in $\mathbf{Z}_p^\times \subset \mathbf{Q}_p^\times$. In particular, each root of unity has a well-defined reduction mod p, $\varepsilon(\xi) \in \mathbf{F}_p^\times$. Let us show that the group $\mathbf{Z}_p^\times$ contains roots of unity in each class mod $p\mathbf{Z}_p$, i.e. above each element of $\mathbf{F}_p^\times$.

The polynomial $P(X) = X^{p-1} - 1$ has derivative $P'(X) = (p-1)X^{p-2}$. For any unit $x \in \mathbf{Z}_p^\times$, $k = v(P'(x)) = 0$, and the simplest case (6.3) of the approximation method applies. Since the polynomial $X^{p-1} - 1$ has $p - 1$ distinct roots in the field $\mathbf{F}_p$, namely all elements of $\mathbf{F}_p^\times$, Hensel's lemma furnishes $p - 1$ distinct roots in $\mathbf{Z}_p^\times$. This shows that the field $\mathbf{Q}_p$ of p-adic numbers always contains a cyclic subgroup of order $p - 1$,

$$\mu_{p-1} \subset \mathbf{Z}_p^\times \subset \mathbf{Q}_p^\times,$$

consisting of roots of unity.

Proposition 1. *When p is an odd prime, the group of roots of unity in the field* $\mathbf{Q}_p$ *is* μ_{p-1}.

PROOF. We have to prove that the reduction homomorphism $\varepsilon : \mu(\mathbf{Q}_p) \to \mathbf{F}_p^\times$ is bijective. It is surjective by Hensel's lemma. So assume that $\zeta = 1 + pt \in \ker \varepsilon$ ($t \in \mathbf{Z}_p$) is a root of unity, say ζ has order $n \geq 1$,

$$\zeta^n = (1 + pt)^n = 1.$$

Hence $npt + \binom{n}{2}p^2t^2 + \cdots + p^nt^n = 0$, or

$$t\left(n + \binom{n}{2}pt + \cdots + p^{n-1}t^{n-1}\right) = 0.$$

This shows that $t = 0$ (when $p \nmid n$) or $p \mid n$. In the second case, replace ζ by ζ^p and n by n/p: Starting the same computation, we see that $t = 0$ or $p^2 \mid n$ (original n), and so on. Finally, we are reduced to the case $n = p$. In this case, the above equation is simply

$$t\left(p + \binom{p}{2}pt + \cdots + p^{p-1}t^{p-1}\right) = 0,$$

and since $p \geq 3$,

$$p + \binom{p}{2}pt + \cdots + p^{p-1}t^{p-1} = p + p^2(\cdots) \neq 0.$$

This proves that $t = 0$ in all cases and $\zeta = 1$. ∎

When p is odd, $p - 1$ is even and -1 belongs to μ_{p-1}. The number -1 will have a square root in $\mathbf{Q}_p$ precisely when $(p - 1)/2$ is still even, namely when $p \equiv 1 \bmod 4$. We have

$$\sqrt{-1} \in \mathbf{Q}_p \iff 4 \mid p - 1 \iff p \equiv 1 \bmod 4.$$

A number $i = \sqrt{-1}$ can thus be found in $\mathbf{Q}_5$, $\mathbf{Q}_{13}$, ...

Proposition 2. *The group of roots of unity in the field* $\mathbf{Q}_2$ *is* $\mu_2 = \{\pm 1\}$.

PROOF. We have

$$-1 = 1 + 2 + 2^2 + \cdots \in 1 + 2\mathbf{Z}_2$$

and

$$\{\pm 1\} = \mu_2 \subset \mathbf{Z}_2^\times = 1 + 2\mathbf{Z}_2.$$

On the other hand, $\mathbf{F}_2^\times = \{1\}$, and the only roots of unity in $\mathbf{Z}_2^\times$ have order a power of 2. But -1 is not a square of $\mathbf{Z}_2^\times$ (6.6), and there is no fourth root of 1 in $\mathbf{Q}_2$.

To summarize, we give a TABLE.

Field	Units	Squares	Roots of unity	Number of quadratic extensions
$\mathbf{Q}_2$	$\mathbf{Z}_2^\times = 1 + 2\mathbf{Z}_2$	$1 + 8\mathbf{Z}_2$ index 4 in $\mathbf{Z}_2^\times$	$\mu_2 = \{\pm 1\}$	7
$\mathbf{Q}_p$ p odd prime	$\mathbf{Z}_p^\times \supset 1 + p\mathbf{Z}_p$ index p-1	index 2 in $\mathbf{Z}_p^\times$	μ_{p-1}	3

6.8. Fourth Application: Field Automorphisms of Q_p

It is possible to determine all automorphisms of the field $\mathbf{Q}_p$ (over the prime field $\mathbf{Q}$). For this purpose, we need a lemma.

Lemma. *Let $x \in \mathbf{Q}_p^\times$. Then the following properties are equivalent:*

(i) x is a unit: $x \in \mathbf{Z}_p^\times$.
(ii) x^{p-1} possesses nth roots for infinitely many values of n.

PROOF. If x is a unit, then $x \not\equiv 0 \bmod p\mathbf{Z}_p$ and $x^{p-1} \equiv 1 \bmod p\mathbf{Z}_p$. Let us put $a = x^{p-1}$ and consider the equation $P(X) = X^n - a = 0$. It has an approximate root 1 mod p, and when n is not a multiple of p, $P'(1) = n$ does not vanish mod p. By Hensel's lemma, there is an exact solution of this equation, namely there exists an element $\xi \in \mathbf{Z}_p$ such that $\xi^n = a = x^{p-1}$. This proves $(i) \Rightarrow (ii)$.
Conversely, if $x^{p-1} = y_n^n$, we have

$$(p-1)v(x) = n \cdot v(y_n),$$

and n divides $(p-1)v(x)$. This can happen for infinitely many values of n only if $v(x) = 0$; hence x is a unit (we are assuming $x \neq 0$ from the outset). ■

Theorem. *The only field automorphism of $\mathbf{Q}_p$ is the identity.*

PROOF. Let φ be an automorphism of the field $\mathbf{Q}_p$. By the algebraic characterization of units of $\mathbf{Q}_p^\times$, the automorphism φ must preserve units. Hence if $x \in \mathbf{Q}_p^\times$ is written in the form $x = p^n u$ (where $n = v(x)$ and $u \in \mathbf{Z}_p^\times$ is a p-adic unit), we shall have

$$\varphi(x) = \varphi(p^n u) = \varphi(p^n)\varphi(u) = p^n \varphi(u)$$

and $v(\varphi(x)) = n = v(x)$. This shows that the algebraic automorphisms of the field $\mathbf{Q}_p$ preserve the p-adic order: They are automatically *continuous*. Now, if $y \in \mathbf{Q}_p$ is an arbitrary element, we can take a sequence of rational numbers $r_n \in \mathbf{Q}$ with

$r_n \to y$. For example, we can take these rational numbers by truncating the p-adic expansion of y. Now, since the automorphism φ is trivial on rational numbers,

$$\varphi(y) = \varphi(\lim_{n\to\infty} y_n) = \lim_{n\to\infty} \varphi(y_n) = \lim_{n\to\infty} y_n = y. \qquad \blacksquare$$

Note. The preceding theorem is similar to the following well-known result:

The only algebraic automorphism of the real field **R** *is the identity.*

Indeed, if φ is a field automorphism of **R**, we have $\varphi(x^2) = \varphi(x)^2$ for all x, and hence $\varphi(y) \geq 0$ for all $y \geq 0$ (write $y = x^2$), and then also

$$\varphi(u) \leq \varphi(v) \text{ for all } u \leq v$$

(put $y = v - u$). This means that these algebraic automorphisms automatically preserve the order relation $\leq$. Since they must be trivial on the prime field **Q**, they must be trivial. In detail: If $t \in \mathbf{R}$ and $a, b \in \mathbf{Q}$, then

$$a \leq t \leq b \Longrightarrow a = \varphi(a) \leq \varphi(t) \leq \varphi(b) = b.$$

Thus we see that

$$|\varphi(t) - t| \leq b - a$$

is arbitrarily small; hence $\varphi(t) - t = 0$.

Comment. Let us stress that in both the p-adic and the real cases, we are considering purely algebraic automorphisms over the prime field **Q**: The proofs show that they are automatically *continuous*, and hence trivial. But there are infinitely many automorphisms of the complex field **C**: Only two of them are continuous, namely the identity and the complex conjugation. For example, the nontrivial automorphism

$$a + b\sqrt{2} \mapsto a - b\sqrt{2} \quad (a, b \in \mathbf{Q})$$

of the field $\mathbf{Q}(\sqrt{2})$ extends to any algebraically closed extension of this field; in particular it extends to **C**. This extension is a *discontinuous* automorphism of **C**.

Appendix to Chapter 1: The p-adic Solenoid

The fields **R** of real numbers and $\mathbf{Q}_p$ of p-adic numbers can be linked in an interesting topological group, the *solenoid*. We present a couple of constructions and properties of this mathematical structure.

A.1. Definition and First Properties

The canonical group homomorphisms

$$\varphi_n : \mathbf{R}/p^{n+1}\mathbf{Z} \to \mathbf{R}/p^n\mathbf{Z}, \quad x \bmod p^{n+1}\mathbf{Z} \mapsto x \bmod p^n\mathbf{Z} \quad (n \geq 0)$$

make up a projective system $(\mathbf{R}/p^n\mathbf{Z}, \varphi_n)_{n\geq 0}$ of topological groups.

Definition. *The* p-adic solenoid $\mathbf{S}_p$ *is the projective limit* $\mathbf{S}_p = \varprojlim \mathbf{R}/p^n\mathbf{Z}$ *of the projective system* $(\mathbf{R}/p^n\mathbf{Z}, \varphi_n)$.

By definition, the solenoid $\mathbf{S}_p$ is a compact abelian group equipped with canonical projections

$$\psi_n : \mathbf{S}_p \to \mathbf{R}/p^n\mathbf{Z} \quad (n \geq 0)$$

that are continuous surjective homomorphisms. In particular,

$$\psi = \psi_0 : \mathbf{S}_p \to \mathbf{R}/\mathbf{Z}$$

is continuous and surjective, and the solenoid can be viewed as a *covering* of the circle. The kernel of this covering is obviously $\ker \psi = \varprojlim \mathbf{Z}/p^n\mathbf{Z} = \mathbf{Z}_p$, and we have the following short exact sequence of continuous homomorphisms,

$$0 \to \mathbf{Z}_p \to \mathbf{S}_p \to \mathbf{R}/\mathbf{Z} \to 0,$$

presenting the circle as a quotient of the solenoid, or the solenoid as a *covering of the circle with fiber* $\mathbf{Z}_p$. Also observe that

$$p^n\mathbf{Z}_p = \ker(\psi_n) \subset \mathbf{Z}_p = \ker(\psi) \subset \mathbf{S}_p.$$

Alternatively, one could define the solenoid as the projective limit of the system having transition homomorphisms

$$\varphi'_n : \mathbf{R}/\mathbf{Z} \to \mathbf{R}/\mathbf{Z}, \quad x \bmod \mathbf{Z} \mapsto px \bmod \mathbf{Z} \quad (n \geq 1).$$

A.2. Torsion of the Solenoid

We recall the following well-known fact:

> *For each positive integer* $m \geq 1$ *there is a unique cyclic subgroup of order* m *in the circle: It is* $m^{-1}\mathbf{Z}/\mathbf{Z} \subset \mathbf{R}/\mathbf{Z}$.

Proposition 1. *For each positive integer* $m \geq 1$ *prime to* p *the solenoid* $\mathbf{S}_p$ *has a unique cyclic subgroup* C_m *of order* m.

Proof. Let us denote temporarily by C_m^n the cyclic subgroup of order m of the circle $\mathbf{R}/p^n\mathbf{Z}$ (it is the subgroup $m^{-1}\mathbf{Z}/p^n\mathbf{Z}$). Since the transition maps

$$\varphi_n : \mathbf{R}/p^{n+1}\mathbf{Z} \longrightarrow \mathbf{R}/p^n\mathbf{Z}$$

have a kernel of order p prime to m (by assumption), they induce isomorphisms $C_m^{n+1} \to C_m^n$. The projective limit of this constant sequence is the cyclic subgroup $C_m \subset \mathbf{S}_p$. To prove uniqueness, let us consider any homomorphism $\sigma : \mathbf{Z}/m\mathbf{Z} \to \mathbf{S}_p$. The composite

$$\psi_n \circ \sigma : \mathbf{Z}/m\mathbf{Z} \to \mathbf{S}_p \to \mathbf{R}/p^n\mathbf{Z}$$

has an image in the unique cyclic subgroup C_m^n of the circle $\mathbf{R}/p^n\mathbf{Z}$. Hence σ has an image in C_m, and this concludes the proof. ∎

Observe that this unique cyclic subgroup C_m of order m (prime to p) of $\mathbf{S}_p$ has a projection $\psi(C_m)$ in the circle given by

$$\psi(C_m) = m^{-1}\mathbf{Z}/\mathbf{Z} \subset \mathbf{R}/\mathbf{Z}.$$

Since $\psi^{-1}(m^{-1}\mathbf{Z}/\mathbf{Z}) \cong C_m \times \mathbf{Z}_p$, the cyclic group C_m is the maximal finite subgroup contained in $\psi^{-1}(m^{-1}\mathbf{Z}/\mathbf{Z})$.

Proposition 2. *The p-adic solenoid $\mathbf{S}_p$ has no p-torsion.*

Proof. Let $\sigma : \mathbf{Z}/p\mathbf{Z} \to \mathbf{S}_p$ be any homomorphism of a cyclic group of order p into the solenoid. I claim that all composites

$$\varphi_n \circ \psi_{n+1} \circ \sigma : \mathbf{Z}/p\mathbf{Z} \to \mathbf{S}_p \to \mathbf{R}/p^{n+1}\mathbf{Z} \to \mathbf{R}/p^n\mathbf{Z}$$

are trivial. Indeed, the composite

$$\psi_{n+1} \circ \sigma : \mathbf{Z}/p\mathbf{Z} \to \mathbf{S}_p \to \mathbf{R}/p^{n+1}\mathbf{Z}$$

must have an image in the unique cyclic subgroup of order p of the circle $\mathbf{R}/p^{n+1}\mathbf{Z}$, and this subgroup is precisely the kernel of the connecting homomorphism φ_n and $\psi_n \circ \sigma = \varphi_n(\psi_{n+1} \circ \sigma)$. Consequently, there is no element of order p in $\mathbf{S}_p$ (and a fortiori no element of order p^k for $k \geq 1$ in $\mathbf{S}_p$). ∎

A.3. Embeddings of R and $\mathbf{Q}_p$ in the Solenoid

Theorem. *The p-adic solenoid contains a dense subgroup isomorphic to* **R**. *It also contains a dense subgroup isomorphic to* $\mathbf{Q}_p$.

Proof. The projection maps $f_n : \mathbf{R} \to \mathbf{R}/p^n\mathbf{Z}$ are compatible with the transition maps of the projective system defining the solenoid

$$f_n = \varphi_n \circ f_{n+1} : \mathbf{R} \to \mathbf{R}/p^{n+1}\mathbf{Z} \to \mathbf{R}/p^n\mathbf{Z}.$$

Hence there is a unique factorization $f : \mathbf{R} \to \mathbf{S}_p$ such that

$$f_n = \psi_n \circ f : \mathbf{R} \to \mathbf{S}_p \to \mathbf{R}/p^n\mathbf{Z}.$$

If $x \neq 0 \in \mathbf{R}$, as soon as $p^n > x$ we have $f_n(x) \neq 0 \in \mathbf{R}/p^n\mathbf{Z}$ and consequently $f(x) \neq 0 \in \mathbf{S}_p$. This shows that the homomorphism f is injective (this also follows from (I.4.5), since $\bigcap_{n\geq 1} \ker f_n = \bigcap_{n\geq 1} p^n\mathbf{Z} = \{0\}$). The density of the image of f follows from the density of the images of the f_n (I.4.4, Proposition 3) (in fact, all f_n are surjective). Consider now the subgroups

$$H_k = \psi^{-1}(p^{-k}\mathbf{Z}/\mathbf{Z}) \subset \mathbf{S}_p \quad (k \geq 0).$$

We have $H_0 = \mathbf{Z}_p$ by definition, and this is a subgroup of index p^k of H_k:

$$H_k = \varprojlim_n \, p^{-k}\mathbf{Z}/p^n\mathbf{Z} \cong p^{-k}\mathbf{Z}_p \quad (k \geq 1).$$

Hence

$$\mathbf{Q}_p \cong \psi^{-1}(\mathbf{Z}[1/p]/\mathbf{Z}) = \bigcup \psi^{-1}(p^{-k}\mathbf{Z}/\mathbf{Z}) = \bigcup H_k \subset \mathbf{S}_p.$$

The density of this subgroup of $\mathbf{S}_p$ follows from the density of all images

$$\psi_n(\mathbf{Q}_p) = \mathbf{Z}[1/p]/p^n\mathbf{Z} \subset \mathbf{R}/p^n\mathbf{Z}$$

(I.4.4, Proposition 3). ∎

Corollary. *The solenoid is a (compact and) connected space.*

PROOF. Recall that for any subspace A of a topological space X we have

$$A \text{ connected}, \ A \subset B \subset \overline{A} \Longrightarrow B \text{ connected}.$$

In our context, take for A the connected subspace $f(\mathbf{R}) \subset \mathbf{S}_p$, which is dense in the solenoid. The conclusion follows. ∎

Let us summarize the various homomorphisms connected to the solenoid in a commutative diagram.

$$\begin{array}{ccccc}
\mathbf{Z} & \hookrightarrow & \mathbf{Z}_p & = & \mathbf{Z}_p \\
\downarrow & & \downarrow & & \downarrow \\
\mathbf{R} & \hookrightarrow & \mathbf{S}_p & \leftarrow & \mathbf{Q}_p \\
\downarrow & & \downarrow & & \downarrow \\
\mathbf{R}/\mathbf{Z} & = & \mathbf{R}/\mathbf{Z} & \leftarrow & \mathbf{Q}_p/\mathbf{Z}_p
\end{array}$$

A.4. *The Solenoid as a Quotient*

The sequence of continuous homomorphisms

$$f_n : \mathbf{R} \times \mathbf{Q}_p \to \mathbf{R}/p^n\mathbf{Z}, \ (t, x) \longmapsto t + \sum_{i<n} a_i p^i \bmod p^n\mathbf{Z}$$

(if $x = \sum_{\nu \leq i < \infty} a_i p^i$, $\nu = \mathrm{ord}_p(x)$) is compatible with the sequence of connecting homomorphisms defining the projective limit $\mathbf{S}_p$. Hence there is a unique factorization consisting of a continuous homomorphism

$$f : \mathbf{R}\times \mathbf{Q}_p \to \mathbf{S}_p,\ (t,x)\mapsto t+x$$

having composites $\psi_n \circ f = f_n$. Alternatively, the two injective continuous homomorphisms $j_1 : \mathbf{R} \to \mathbf{S}_p$, $j_2 : \mathbf{Q}_p \to \mathbf{S}_p$ furnish a unique continuous homomorphism

$$j_1 + j_2 : \mathbf{R} \oplus \mathbf{Q}_p \to \mathbf{S}_p,$$

which coincides with the preceding one (we are identifying the product and the direct sum). This homomorphism f will therefore be called the *sum homomorphism.*

Lemma. *The kernel of the homomorphism f defined above is the subgroup*

$$\ker f = \Gamma = \{(a, -a) : a \in \mathbf{Z}[1/p]\} \subset \mathbf{R} \times \mathbf{Q}_p.$$

It is a discrete subgroup of the product $\mathbf{R} \times \mathbf{Q}_p$.

PROOF. If $f(t,x) = 0$, we have in particular $f_0(t,x) = \psi_0 \circ f(t,x) = 0 \in \mathbf{R}/\mathbf{Z}$, namely $t + \sum_{i\le 0} a_i p^i \in \mathbf{Z}$, $t \in -\sum_{i\le 0} a_i p^i + \mathbf{Z} \subset \mathbf{Z}[1/p]$. Similarly, $f_n(t,x) = 0$ gives

$$t + \sum_{i<n} a_i p^i \in p^n\mathbf{Z} \quad (n \ge 1).$$

This proves that the p-adic expansion of the element $t \in \mathbf{Z}[1/p]$ is given by $t = -\lim \sum_{i<n} a_i p^i$ in $\mathbf{Q}_p$. Hence $t = -x \in \mathbf{Q}_p$. Conversely, it is obvious that $\Gamma \subset \ker f$. Let us show that the (closed) subgroup Γ is *discrete*. For this it is enough to show that a suitable neighborhood of 0 in $\mathbf{R} \times \mathbf{Q}_p$ contains only the neutral element of Γ. Consider the open set

$$(-1,1) \times \mathbf{Z}_p \subset \mathbf{R} \times \mathbf{Q}_p.$$

If a pair $(a, -a)$ is in $\Gamma \cap (-1,1) \times \mathbf{Z}_p$, then the p-adic expansion of $a \in \mathbf{Z}[1/p]$ must be of the form $\sum_{i\ge 0} a_i p^i$. But we have seen (I.5.4) that in the p-adic field $\mathbf{Q}_p$, the intersection $\mathbf{Z}[1/p] \cap \mathbf{Z}_p = \mathbf{Z}$ contains only the rational integers. In particular, $a \in \mathbf{Z} \cap (-1,1) = \{0\}$. Hence

$$\Gamma \cap ((-1,1) \times \mathbf{Z}_p) = \{0\} \subset \mathbf{R} \times \mathbf{Q}_p,$$

and the proof is concluded. ∎

Theorem. *The sum homomorphism $f : \mathbf{R} \times \mathbf{Q}_p \to \mathbf{S}_p$ furnishes an isomorphism $f' : (\mathbf{R} \times \mathbf{Q}_p)/\Gamma_p \cong \mathbf{S}_p$ both algebraically and topologically.*

PROOF. Since all maps f_n are surjective, the map f has a dense image (I.5.4). Moreover, using the integral and fractional parts introduced there,

$$f(t,x) = f(t + \langle x\rangle, x - \langle x\rangle) = f(s,y),$$

where $s \in \mathbf{R}$ and $y = x - \langle x \rangle = [x] \in \mathbf{Z}_p$. Going one step further, we have

$$f(s, y) = f(s - [s], y + [s]) = f(u, z),$$

where $u = s - [s] \in [0, 1)$ and $z = y + [s] \in \mathbf{Z}_p$. This proves

$$\operatorname{Im} f = f(\mathbf{R} \times \mathbf{Q}_p) = f([0, 1) \times \mathbf{Z}_p).$$

A fortiori, the image of f is equal to $f([0, 1] \times \mathbf{Z}_p)$, and hence is compact and closed. Consequently, f is surjective (and f' is bijective). In fact, the preceding equalities also show that the Hausdorff quotient (recall that the subgroup Γ_p is discrete and closed) is also the image of the compact set $\Omega = [0, 1] \times \mathbf{Z}_p$ and hence is compact. The continuous bijection

$$f' : (\mathbf{R} \times \mathbf{Q}_p)/\Gamma_p \to \mathbf{S}_p$$

between two compact spaces is automatically a homeomorphism. ∎

Corollary 1. *The solenoid can also be viewed as a quotient of* $\mathbf{R} \times \mathbf{Z}_p$ *by the discrete subgroup* $\Delta_{\mathbf{Z}} = \{(m, -m) : m \in \mathbf{Z}\}$

$$f' : (\mathbf{R} \times \mathbf{Z}_p)/\Delta_{\mathbf{Z}} \cong \mathbf{S}_p.$$

Proof. Since the restriction of the sum homomorphism $f : \mathbf{R} \times \mathbf{Q}_p \to \mathbf{S}_p$ to the subgroup $\mathbf{R} \times \mathbf{Z}_p$ is already surjective, this restriction gives a (topological and algebraic) isomorphism

$$f' : (\mathbf{R} \times \mathbf{Z}_p)/\ker f' \cong \mathbf{Z}_p.$$

But

$$\ker f' = (\ker f) \cap (\mathbf{R} \times \mathbf{Z}_p) = \Delta_{\mathbf{Z}} = \{(m, -m) : m \in \mathbf{Z}\}.$$ ∎

These presentations of the solenoid can be gathered in commutative diagrams of homomorphisms:

$$\begin{array}{ccccccccc} & & \mathbf{Z}[1/p] & & & & \mathbf{Z} & & \\ & \swarrow & & \searrow & & \swarrow & & \searrow & \\ \mathbf{R} & & & & \mathbf{Q}_p \quad \mathbf{R} & & & & \mathbf{Z}_p \\ & \searrow & & \swarrow & & \searrow & & \swarrow & \\ & & \mathbf{S}_p & & & & \mathbf{S}_p & & \end{array}$$

Corollary 2. *The solenoid can also be viewed as a quotient of the topological space* $[0, 1] \times \mathbf{Z}_p$ *by the equivalence relation identifying* $(1, x)$ *to* $(0, x + 1)$ $(x \in \mathbf{Z}_p)$.

PROOF. This follows immediately from the previous corollary, since the restriction of the sum homomorphism to $[0, 1] \times \mathbf{Z}_p$ is already surjective, whereas its restriction to $[0, 1) \times \mathbf{Z}_p$ is bijective. ∎

Comment. This last corollary gives a good topological model of the solenoid: One has to *glue the two extremities of the cylinder* $[0, 1] \times \mathbf{Z}_p$ having basis $\mathbf{Z}_p$ *by a twist representing the unit shift of* $\mathbf{Z}_p$. This gives a model for the solenoid as a *very twisted rope*! On the other hand, it is clear that instead of the subgroup $\Gamma \cong \mathbf{Z}[1/p]$ consisting of the elements $(a, -a)$ $(a \in \mathbf{Z}[1/p])$ we could equally well have taken the diagonal subgroup Δ, image of

$$a \mapsto (a, a) : \mathbf{Z}[1/p] \to \mathbf{R} \times \mathbf{Q}_p,$$

the isomorphism $(\mathbf{R} \times \mathbf{Q}_p)/\Delta \cong \mathbf{S}_p$ now being given by *subtraction*.

A.5. Closed Subgroups of the Solenoid

Lemma. *Let* $\sigma : C_{p^m} \to C_{p^{m-1}}$ *be a surjective homomorphism between two cyclic groups of orders* p^m *and* p^{m-1}. *Then the only subgroup* $H \subset C_{p^m}$ *not contained in the kernel of* σ *is* $H = C_{p^m}$.

PROOF. Recall that any subgroup of a cyclic group is cyclic and that the number of generators of $C_n \cong \mathbf{Z}/n\mathbf{Z}$ is given by the Euler φ-function $\varphi(n)$. In particular, if $n = p^m$ is a power of p, the number of generators is

$$\varphi(p^m) = p^{m-1}(p-1) = p^m - p^{m-1}.$$

Consequently, all elements not in the kernel of a surjective homomorphism of a cyclic group of order p^m onto a cyclic group of order p^{m-1} are generators of the cyclic group of order p^m (the kernel has order p^{m-1}). ∎

Proposition. *For each integer* $k \geq 0$, *there is exactly one subgroup* $H_k \subset S_p$ *having a projection of order* p^k *in the circle:* $\psi(H_k) = p^{-k}\mathbf{Z}/\mathbf{Z} \subset \mathbf{R}/\mathbf{Z}$. *This subgroup is* $H_k = \psi^{-1}(p^{-k}\mathbf{Z}/\mathbf{Z}) \subset \mathbf{S}_p$.

PROOF. We can apply the lemma to each surjective homomorphism

$$p^{-k}\mathbf{Z}/p^{n+1}\mathbf{Z} \to p^{-k}\mathbf{Z}/p^n\mathbf{Z}$$

in the sequence of connecting homomorphisms defining the solenoid as a projective limit. The projective limit of these cyclic groups is $p^{-k}\mathbf{Z}_p$. ∎

As a preliminary observation to the following theorem, let us assume that the solenoid contains a cyclic subgroup H of some finite order $m > 1$. Taking a generator x of H and n large enough so that $\psi_n(x) \neq 0$, we see that the restriction of this homomorphism ψ_n to H must be injective. A fortiori, the restriction of ψ_{n+1} (and all ψ_N for $N > n$) to H must be injective. The restriction of $\varphi_n : \mathbf{R}/p^{n+1}\mathbf{Z} \to \mathbf{R}/p^n\mathbf{Z}$

to $\psi_{n+1}(H)$ must be injective. Hence $H \cong \psi_{n+1}(H)$ has no element of order p and m is prime to p.

Theorem. *The closed subgroups of the solenoid* $\mathbf{S}_p$ *are*

(1) C_m, *the cyclic subgroup of order m relatively prime to p* ($m \geq 1$),
(2) $C_m \times p^k\mathbf{Z}_p$, *where m is prime to p and* $k \in \mathbf{Z}$,
(3) $\mathbf{S}_p$ *itself (connected).*

PROOF. Let H be a closed subgroup of the solenoid $\mathbf{S}_p$. Since H is compact, its image $\psi(H)$ is a closed subgroup of the circle $\mathbf{R}/\mathbf{Z}$. The only possibilities are

$$\psi(H) = n^{-1}\mathbf{Z}/\mathbf{Z} \text{ cyclic of order } n \geq 1,$$

or

$$\psi(H) = \mathbf{R}/\mathbf{Z} \text{ is the whole circle.}$$

(1) The easiest case is the second one,

$$\psi(H) = \mathbf{R}/\mathbf{Z} \text{ is the whole circle,}$$

in which case $\psi_n(H) \subset \mathbf{R}/p^n\mathbf{Z}$ must be a closed subgroup of *finite index*. Hence it must be open in this circle. By connectivity, $\psi_n(H) \subset \mathbf{R}/p^n\mathbf{Z}$. Since this must hold for all $n \geq 1$, we conclude that

$$H = \overline{H} = \bigcap_{n\geq 1} f_n^{-1}(\overline{(f_n H)}) = \mathbf{S}_p$$

and $H = \mathbf{S}_p$ in this case.

(2) If $\psi(H) = \{0\}$, then $H \subset \psi^{-1}(0) = \mathbf{Z}_p \subset \mathbf{S}_p$, and we have shown in (3.5) that the only possibilities are

$$H = \{0\}, \quad p^k\mathbf{Z}_p \text{ for some integer } k \geq 0.$$

These possibilities occur in the list for $C_m = \{0\}$ ($m = 1$).

(3) We can now assume that $\psi(H) = a^{-1}\mathbf{Z}/\mathbf{Z}$ is cyclic and not trivial. Write $a = p^k \cdot m$ with $k \geq 0$ and m prime to p. By the Chinese remainder theorem (or the p-Sylow decomposition theorem) this cyclic group is a direct product of the cyclic subgroups $m^{-1}\mathbf{Z}/\mathbf{Z}$ and $p^{-k}\mathbf{Z}/\mathbf{Z}$. If $k \geq 1$, the above lemma shows that $\psi_{n+1}(H)$ must contain an element of order p^{k+1}. As in the proposition, we see that H contains $\psi^{-1}(p^{-k}\mathbf{Z}/\mathbf{Z}) = p^{-k}\mathbf{Z}_p \subset \mathbf{S}_p$, and finally $H = C_m \times p^{-k}\mathbf{Z}_p$. If $k = 0$, two possibilities occur: Either $\psi_n(H)$ is cyclic of order m for all n, or there is a first n such that this group $\psi_n(H)$ contains an element of order p. In the first case $H = C_m$, while $H = C_m \times p^n\mathbf{Z}_p$ in the second. ■

A.6. *Topological Properties of the Solenoid*

We have seen in (I.A.4) that the solenoid $\mathbf{S}_p$ can be viewed as a quotient of the cylinder $[0, 1] \times \mathbf{Z}_p$, and an image of $[0, 1) \times \mathbf{Z}_p$. This leads to considering the second projection of this product as a (discontinuous) map $(t, x) \mapsto x$. This map has continuous restrictions to all subspaces $[0, \eta] \times \mathbf{Z}_p$ $(0 < \eta < 1)$. It furnishes continuous *retractions* of these subspaces onto the neutral $\mathbf{Z}_p$-fiber of the solenoid.

Recall that we have a continuous surjective homomorphism $\psi : \mathbf{S}_p \to \mathbf{R}/\mathbf{Z}$ leading to a presentation of the solenoid by the short exact sequence of continuous homomorphisms

$$0 \to \mathbf{Z}_p \to \mathbf{S}_p \to \mathbf{R}/\mathbf{Z} \to 0.$$

The subspaces $\psi^{-1}([0, \eta])$ $(0 < \eta < 1)$ have *continuous retractions on the fiber* $\mathbf{Z}_p$, simply since $\psi^{-1}([0, \eta])$ is homeomorphic to $[0, \eta] \times \mathbf{Z}_p$. The following statement is then an immediate consequence of these observations.

Proposition 1. *Let U be any proper subset of the circle $\mathbf{R}/\mathbf{Z}$. Then the subspace $\psi^{-1}(U) \subset \mathbf{S}_p$ of the solenoid is homeomorphic to $U \times \mathbf{Z}_p$. The map*

$$(t, x) = (t - [t], x + [t]) \mapsto (0, x + [t])$$

furnishes by restriction a continuous retraction of $\psi^{-1}([0, \eta]) \subset \mathbf{S}_p$ onto the neutral fiber $\mathbf{Z}_p \subset \mathbf{S}_p$ $(0 < \eta < 1)$. ∎

The solenoid has still another important topological property that we explain and prove now.

Definition. *A compact and connected topological space K is called* indecomposable *when the only partition of K in two compact and connected subsets is the trivial one.*

Proposition 2. *The solenoid $\mathbf{S}_p$ is an indecomposable compact connected topological space.*

PROOF. Let us take two compact connected subsets A and B covering $\mathbf{S}_p$. We have to show that if $A \neq \mathbf{S}_p$, then $B = \mathbf{S}_p$. Thus we assume $A \neq \mathbf{S}_p$ from now on: $B \neq \emptyset$. Since we have

$$K = \bigcap_{n \geq 1} \psi_n^{-1}(\psi_n(K))$$

for every compact set K, the assumption $A \neq \mathbf{S}_p$ leads to $\psi_n(A) \neq \mathbf{R}/p^n\mathbf{Z}$ for some integer $n = n_0$ and hence also for all integers $n \geq n_0$ (the transition maps φ_m are surjective). It will suffice to show $\psi_n(B) = \mathbf{R}/p^n\mathbf{Z}$ for all $n \geq n_0$. Take such an n and an element $b \in B$. Then

$$\varphi_n^{-1}(b) \subset \mathbf{R}/p^{n+1}\mathbf{Z}$$

has cardinality $p \geq 2$, and the restriction of φ_n to the connected set

$$C = \varphi_n^{-1}\psi_n(B) = \psi_{n+1}(B)$$

is not injective. The proof will be complete as soon as the following statement (in which the situation and notation are simplified) is established.

Let $a > 1$ be any integer, $\varphi : \mathbf{R}/a\mathbf{Z} \to \mathbf{R}/\mathbf{Z}$ the canonical projection, and C a connected subset of $\mathbf{R}/a\mathbf{Z}$ containing two distinct points $s \neq t$ with $\varphi(s) = \varphi(t)$. Then $\varphi(C) = \mathbf{R}/\mathbf{Z}$.

In terms of the restriction $\varphi|_C$ of the map φ to C, we have to prove

$$\varphi|_C \text{ not injective} \implies \varphi|_C \text{ surjective}$$

under the stated assumptions. It is obviously enough to do so when $C \neq \mathbf{R}/a\mathbf{Z}$. In this case, take a point $P \notin C \subset \mathbf{R}/a\mathbf{Z}$ and consider a *stereographic projection* from the point P of the circle $\mathbf{R}/a\mathbf{Z}$ onto a line $\mathbf{R}$. This is a homeomorphism

$$f : \mathbf{R}/a\mathbf{Z} - \{P\} \xrightarrow{\sim} \mathbf{R}.$$

The image $f(C)$ of the subset C is a connected subset of the real line containing the images of two different congruent points mod $\mathbf{Z}$. Since any connected set in the real line is an interval, this proves that $f(C)$ contains the whole interval J linking these two different congruent points. Hence C contains a whole arc I of the circle having image $\varphi(I) = \mathbf{R}/\mathbf{Z}$. ■

Exercises for Chapter 1

1. Compute the squares of the following numbers

$$6,\ 76,\ 376,\ 9376,\ \ldots.$$

Show that one can continue the sequence in a unique way: For example, the number

$$743\,74008\,17871\,09376$$

appears in the 18th position. Define the *limit*

$$\alpha := \sum_{i \geq 0} a_i 10^i = \cdots a_6 a_5 a_4 a_3 a_2 a_1 a_0 = \cdots 109376$$

as a 10-adic integer: $\alpha \in \mathbf{Z}_{10}$. Give the 10-adic expansion of -1.

Observe that by definition $\alpha^2 = \alpha$, and find the four solutions $0, 1, \alpha, \beta$ of $x^2 = x$ in $\mathbf{Z}_{10}$. What are $\alpha + \beta, \alpha\beta$?

Prove that $\mathbf{Z}_{10} \cong \mathbf{Z}_5 \times \mathbf{Z}_2$. (*Hint.* Consider the map $x \mapsto (\alpha x, \beta x)$.)

2. (*a*) Give the 5-adic expansion of the integers $15, -1, -3$. The integers $2, 3, 4$ are invertible in $\mathbf{Z}_5$: Give the 5-adic expansions of the inverses. Give the expansion of $\frac{1}{3}$ in $\mathbf{Z}_7$.

   (*b*) What is the p-adic expansion of $\frac{1}{2}$ if the prime p is odd?

   (*c*) If f is a positive integer, give the expansion of $1/(1 - p^f)$ in $\mathbf{Z}_p$.

   (*d*) More generally, find the expansion of $1/m$ in $\mathbf{Z}_p$ when the integer m is not divisible by p. (*Hint.* Let f be the multiplicative order of p mod m so that $p^f - 1 = nm$. Then use $1/m = -n/(1 - p^f)$.)

3. (*a*) Show $x \in p^n\mathbf{Z}_p \iff -x \in p^n\mathbf{Z}_p$ and so $\mathrm{ord}_p(-x) = \mathrm{ord}_p(x)$.
   (*b*) Check as in (1.5) that if $\alpha \in \mathbf{Z}_p$, then $(1+p^n\alpha)^{-1} = 1 + p^n\alpha'$ for some $\alpha' \in \mathbf{Z}_p$.
   (*c*) Using the p-adic metric, reformulate (*b*) in the form
   if $0 < r < 1$, then

$$|x-1| < r,\ |y-1| < r \Longrightarrow |xy-1| < r.$$

   (*d*) Let σ denote the involution introduced in (1.2). Show that $\sigma(B_{<r}(a)) = B_{<r}(\sigma(a))$.

4. Show that there is a square root of 2 in $\mathbf{Z}_7$. (Compute the first coefficients in $a = a_0 + a_1 7 + a_2 7^2 + \cdots$ iteratively using $a^2 = 2$; do not be surprised if no regular pattern appears: The same happens for the computation of the decimal expansion of $\sqrt{2}$ in $\mathbf{R}$; cf. also (I.5.3).)

5. (*a*) Solve the equation $x^2 = 1$ in all $\mathbf{Z}/2^n\mathbf{Z}$ $(n \geq 1)$. Guess the result by making a small table with the first values $n \leq 4$ or 5.
   (*Hint.* Consider separately the cases $n = 1,\ 2,\ \geq 3$. When $n \geq 3$, observe that if $x^2 = 1$, then x is the class of an odd integer $2k+1$ $(0 \leq k < 2^{n-1})$, and $4k(k+1)$ has to be divisible by 2^n. In (VII.1.7) we show that the unit group in $\mathbf{Z}/2^n\mathbf{Z}$ is a product of two cyclic groups $(n \geq 3)$, from which the result also follows.)
   (*b*) Solve the equation $x^2 = 1$ in $\mathbf{Z}_2$.

6. (*a*) Let N be a positive integer. Show that the subset $\{N, N+1, N+2, \ldots\}$ is dense in $\mathbf{Z}_p$.
   (*b*) For which values of a and $b \in \mathbf{Z}_p$ is the subset $a + b\mathbf{N}$ dense in $\mathbf{Z}_p$?
   (*c*) Show that the subset $\{-1, -2, -3, \ldots\}$ is dense in $\mathbf{Z}_p$.

7. Let $j_p : \mathbf{Q} \to \mathbf{Q}_p$ denote the canonical injection.
   (*a*) Determine the subring $j_p^{-1}(\mathbf{Z}_p)$ of the field $\mathbf{Q}$ (this subring is simply written $\mathbf{Q} \cap \mathbf{Z}_p = \mathbf{Z}_{(p)}$). What is $j_p^{-1}(\mathbf{Z}_p) \cap \mathbf{Z}[1/p]$?
   (*b*) Show that

$$\bigcap_{p \text{ prime}} j_p^{-1}(\mathbf{Z}_p) = \mathbf{Z}$$

   (this equality is sometimes simply written $\bigcap_p (\mathbf{Q} \cap \mathbf{Z}_p) = \mathbf{Z}$).

8. Let X be a nonempty set and $E = X^{\mathbf{N}}$ the set of sequences in X. For two different sequence $a = (a_n)$, $b = (b_n)$ let us put

$$d(a,b) = \frac{1}{\min\{n : a_n \neq b_n\}} = \frac{1}{\nu}.$$

   (*a*) Show that d defines an ultrametric distance on E.
   (*b*) Show that E is complete for the preceding metric.

9. The distance between two subsets A, B of a metric space is defined by $d(A,B) = \inf_{a \in A, b \in B} d(a,b)$. Show that if the metric d is ultrametric, then

$$d(B_{\leq r}(a), B_{\leq r}(b)) = \begin{cases} d(a,b) & \text{if } r < d(a,b), \\ 0 & \text{if } r \geq d(a,b). \end{cases}$$

   More generally, the distance of two disjoint balls B, B' is equal to the constant value of $d(x,x')$ for $x \in B$, $x' \in B'$.

10. Let K be a (commutative) field and let $K[[X]]$ be the ring of formal power series $f(X) = \sum_{n \geq 0} a_n X^n$. Choose $0 < \theta < 1$ and for $g(X) = \sum_{n \geq 0} b_n X^n \neq f(X)$, define

$$d(f(X), g(X)) = \theta^{\min\{n : a_n \neq b_n\}}.$$

Show that d defines an ultrametric distance on $K[[X]]$ for which this space is complete. Show that the space of polynomials $K[X]$ is dense in $K[[X]]$, and hence this is a completion of the space of polynomials. The ball $\{f(X) : d(f(X), 0) \leq \theta^n\}$ is the ideal $(X^n) = X^n K[[X]]$. The fraction field $K((X)) = K[[X]][X^{-1}]$ consists of the Laurent series $\sum_{i \geq \nu} a_n X^n$ ($\nu \in \mathbf{Z}$). It is a completion of the ring $K[X, X^{-1}]$.

11. Let $\theta > 1$ and for any nonzero polynomial $f \in \mathbf{R}[X]$ define $|f| = \theta^{\deg f}$. Extend this definition by $|0| = 0$ and $|f/g| = |f|/|g|$ for a rational fraction $f/g \in \mathbf{R}(X)$. Show that this defines an ultrametric absolute value on the field $\mathbf{R}(X)$.

12. Let E be a compact metric space and $f : \mathbf{Z}_2 \to E$ be a continuous surjective map. For each ball $B \subset \mathbf{Z}_2$ of positive radius, let $A_B = f(B)$ be the compact image of B in the space E. Observe that

$$A_B = A_{B'} \cup A_{B''} \text{ if } B = B' \sqcup B'',$$

$$\bigcap_{B \ni x} A_B = \{f(x)\}.$$

Conversely, recall that $\mathcal{M}_2$ denotes the free monoid generated by two letters, say 0 and 1, and $\mathcal{P}(E)$ denotes the set of parts (power set) of E. For any map $\varphi : \mathcal{M}_2 \to \mathcal{P}(E)$ having the properties
(*a*) $\varphi(\emptyset) = E, \quad \varphi(w) = \varphi(w0) \cup \varphi(w1) \quad (w \in \mathcal{M}_2)$,
(*b*) $\delta(\varphi(w_n)) \to 0$ when the w_n are the initial segments of an infinite word,
(*c*) $\bigcap \varphi(w_n) \neq \emptyset$ when the w_n are the initial segments of an infinite word,
show that there exists a continuous *surjective* map

$$f : \mathbf{Z}_2 \to E \text{ such that } f(B_w) = \varphi(w).$$

13. Let E be a compact metric space. Show that there exists a continuous surjective map $f : \mathbf{Z}_2 \to E$. In other words, the metric space is a topological quotient of the space $\mathbf{Z}_2$. (*Hint.* Let $(K_i)_{1 \leq i \leq k}$ be a covering of E by closed sets of diameter ≤ 1. If $k > 1$ call $A_0 = K_0 \cup \cdots \cup K_\ell$ and $A_1 = K_{\ell+1} \cup \cdots \cup K_k$ with, e.g., $\ell = [k/2]$. If $\ell > 1$, start again and define similarly shorter unions A_{00}, A_{01} such that $A_0 = A_{00} \cup A_{01}$. This leads to finitely many words w_i so that $K_i = A_{w_i}$. Proceeding similarly for each of them, show how to define a map $\varphi : \mathcal{M}_2 \to E$ having the properties listed in the previous exercise.)
Conclude that all spaces $\mathbf{Z}_p$ are homeomorphic to $\mathbf{Z}_2$.
Give an explicit continuous surjective map $\mathbf{Z}_2 \to \overline{\{1/n : n \geq 1\}}$

14. Let E be a compact metric space. Show the equivalence
(i) there is a continuous surjective map $f : [0, 1] \to E$,
(ii) E is path-connected.
(*Hint.* Use the previous exercise to construct a continuous surjective map $f_0 : C \to E$, where C is the Cantor subset of the unit interval, and extend f_0 through the missing intervals — this is possible if the space E is path-connected.)

In particular, for every compact, convex subset K of a (real or complex) Hilbert space, there is a continuous surjective $f:[0,1] \to E$ ("space-filling curve" or Peano curve).

15. Let E be a compact metric space with the following properties:
 (a) E is totally disconnected.
 (b) E has no isolated point (hence is not a singleton set!).
 Show that E is homeomorphic to $\mathbf{Z}_2$.

16. Show that the planar fractal image of $\mathbf{Z}_5$ is path-connected when it is connected (cf. picture in text).

17. Construct a planar model of $\mathbf{Q}_7$ using $\nu : \{0, 1, \dots 6\} \to \mathbf{C}$ defined by

$$\nu(0) = 0, \quad \nu(j) = e^{j \cdot 2\pi i/6} \quad (1 \le j \le 6).$$

Observe the appearance of the von Koch curve in the image of

$$X_7 = \left\{ \sum_{n \ge 0} a_n 7^n : 1 \le a_n \le 6 \right\} \subset \mathbf{Q}_7.$$

18. (a) Give an example of a discrete subset of $[0, 1] \subset \mathbf{R}$ that is not closed.
 (b) Prove that if A is a discrete subset of a Hausdorff topological space X, then A is open in $\overline{A}$ (the same is true for any locally compact subset in a Hausdorff space).
 (c) Let G be a topological group that is Hausdorff, and Γ a discrete subgroup. Prove directly that Γ is closed in G (cf. I.3.2).
 (d) Let G be a group having more than one element, let G_d denote the topological group G with the discrete topology, and let G_0 denote the topological group G with the topology having only $\emptyset$ and G as open sets (not Hausdorff!). Prove that $\Gamma = G_d \times \{e\}$ is a discrete subgroup of the topological group $G_d \times G_0$. What is its closure?

19. (a) Let H be a normal subgroup of a topological group G. Prove that the subgroup $\overline{H}$ is also normal.
 (b) For any topological group G, the quotient $G/\overline{\{e\}}$ is a Hausdorff topological group.
 (c) Let H be a closed subgroup of a locally compact (topological) group G. Prove that the space G/H is locally compact.
 (d) Let G be a locally compact totally discontinuous group, so that the connected component of the neutral element in G is $\{e\}$. Prove that any neighborhood of the neutral element contains a clopen subgroup. (*Hint.* Start with a compact neighborhood K of e. There is a clopen neighborhood U of e contained in K. Since U is compact and disjoint from the closed set $F = G - U$, there is a symmetric neighborhood W of e such that $UW \cap FW = \emptyset$ and hence

$$UW \subset (FW)^c \subset F^c = U.$$

By induction $W^n \subset UW^n \subset U$. The subgroup generated by W is open and contained in U.)

20. Here is an example of a topological ring A that does not induce on its units $A^\times$ a topology compatible with the group structure (cf. (I.3.5)). Let H be a complex Hilbert space with orthonormal basis $(e_i)_{i \ge 0}$. Hence the elements of H are the series $x = \sum_{i \ge 0} x_i e_i$ such

that $x_i \in \mathbf{C}$ and $\sum_{i\geq 0} |x_i|^2 < \infty$. Consider the sequence of continuous operators T_n in H defined by

$$T_n : e_i \mapsto \begin{cases} e_i & \text{if } i \neq n, \\ e_n/n & \text{if } i = n. \end{cases}$$

Prove that for every $x \in H$, $\|T_n x - x\|^2 \to 0$, and hence $T_n x \to x$ and $T_n \to I$ for the strong topology on the ring A of bounded operators on H. But $T_n^{-1} \not\to I$ for the strong topology (consider the vector $x = \sum_{n\geq 1} e_n/n$).

21. Let K be an ultrametric field.
    (*a*) Show that if K is locally compact, then all balls of K are compact (and conversely).
    (*b*) Two balls of K having the same radius $r > 0$ are homeomorphic.
    (*Hint.* Consider separately the cases $|K^\times|$ discrete or dense; remember that all spheres are clopen, and if necessary, use a bijection $(0, r] \cap |K^\times| \xrightarrow{\sim} (0, r) \cap |K^\times|$.)

22. Let G be a group and

$$G = G_0 \supset G_1 \supset G_2 \supset \cdots \supset G_n \supset \cdots$$

be a decreasing sequence of normal subgroups of G. Show that there is a unique group topology on G for which $(G_n)_{n\geq 0}$ is a fundamental system of neighborhoods of e. For this topology, the G_n are clopen subgroups and

$$G \text{ Hausdorff} \iff \bigcap_{n\geq 0} G_n = \{e\}.$$

When this is the case, show that G is metrizable. (*Hint.* Note that G/G_n is discrete and metrizable. One can embed G in the countable metrizable product $\prod G/G_n$.)

23. Let $A = \mathrm{M}_2(\mathbf{Z}_p)$ be the noncommutative ring of 2×2 matrices having coefficients in $\mathbf{Z}_p$. Show that A is a topological ring (for the product topology). The units in A constitute a group $A^\times = \mathrm{Gl}_2(\mathbf{Z}_p)$:

$$g \in \mathrm{Gl}_2(\mathbf{Z}_p) \iff g \in \mathrm{M}_2(\mathbf{Z}_p) \text{ and } \det g \in \mathbf{Z}_p^\times.$$

Show that $\mathrm{Gl}_2(\mathbf{Z}_p)$ is a topological group with the topology induced from A. Let $G_n \subset G$ denote the normal subgroup consisting of matrices $g = (g_{ij})$ congruent to the identity matrix mod p^n,

$$g_{ij} \equiv \delta_{ij} \bmod p^n \mathbf{Z}_p$$

($\delta_{ij} = 1$ if $i = j$ and $= 0$ if $i \neq j$ is the Kronecker symbol). Show that the G_n form a fundamental system of neighborhoods of the identity in $\mathrm{Gl}_2(\mathbf{Z}_p)$.

24. Let $(A_n)_{n\geq 0}$ be a decreasing sequence of subsets of a set E. Consider the canonical inclusions $A_{n+1} \subset A_n$ as transition homomorphisms. Show that the intersection $A = \bigcap_{n\geq 0} A_n$ together with the inclusions $A \to A_n$ has the universal property characterizing the projective limit $\varprojlim A_n$ and hence may be identified with it: $\varprojlim A_n = \bigcap A_n$.

25. Let $(X_n, \varphi_n)_{n\geq 0}$ and $(Y_n, \psi_n)_{n\geq 0}$ be two projective systems. One can consider canonically $(X_n \times Y_n, \varphi_n \times \psi_n)_{n\geq 0}$ as a projective system. Prove

$$\varprojlim (X_n \times Y_n) \cong \varprojlim X_n \times \varprojlim Y_n.$$

26. Let $a \in \mathbf{Z}$ be a rational integer. Show that $X^2 + X + a = 0$ has a root in $\mathbf{Q}_2$ if and only if a is even.

27. (*a*) In which fields $\mathbf{Q}_p$ does one find the golden ratio (root of $x^2 = x + 1$)?
    (*b*) How many solutions of $X^4 + X^2 + 1 = 0$ are in $\mathbf{Q}_7$? (Either make a list of solutions mod 7, or consider $Y = X^2$ and solve in two steps.)

28. (*a*) Show that if $a \in 1 + p\mathbf{Z}_p$ and the integer n is prime to p, then there is an nth root of a in $\mathbf{Q}_p$.
    (*b*) Give an example of $a \in 1 + p\mathbf{Z}_p$ having no pth root in $\mathbf{Q}_p$.
    (*c*) Show that if $a \in 1 + p^3\mathbf{Z}_p$, then a has a pth root in $\mathbf{Q}_p$.

29. Let n be a positive integer, $\nu = \mathrm{ord}_p n$; hence $n = p^\nu n'$ and $(p, n') = 1$. For integers $a,\ b \in \mathbf{Z}$, prove

$$a \equiv b \pmod{n\mathbf{Z}_{(p)}} \Longleftrightarrow a \equiv b \pmod{p^\nu \mathbf{Z}}.$$

(*Hint.* Observe that $n\mathbf{Z}_{(p)} = p^\nu \mathbf{Z}_{(p)}$ and $n\mathbf{Z}_{(p)} \cap \mathbf{Z} = p^\nu \mathbf{Z}$.)

30. Let p and q be distinct primes.
    (*a*) Prove that the fields $\mathbf{Q}_p$ and $\mathbf{Q}_q$ are not isomorphic.
    (*b*) Prove that the fields $\mathbf{Q}_p$ and $\mathbf{R}$ are not isomorphic.
    (*c*) Prove that the fields $\mathbf{Q}_p(\mu_{q-1})$ and $\mathbf{Q}_q(\mu_{p-1})$ are not isomorphic.
    (*Hint.* Look at roots of unity. Observe that for each prime p, the field $\mathbf{Q}_p$ has an algebraic extension of degree 4, which is not the case of the field $\mathbf{R}$. For part (*c*), use the lemma in (6.8).)

31. Let p and q be distinct primes. What is the projective limit

$$\varprojlim \mathbf{R}^2/(p^n\mathbf{Z} \times q^n\mathbf{Z})?$$

2

Finite Extensions of the Field of p-adic Numbers

The field $\mathbf{Q}_p$ is not algebraically closed: It admits algebraic extensions of arbitrarily large degrees. These extensions are the p-adic fields to be studied here. Each one is a finite-dimensional, hence locally compact, normed space over $\mathbf{Q}_p$. A main result is the following: The p-adic absolute value on $\mathbf{Q}_p$ has a *unique extension* to any finite algebraic extension K of $\mathbf{Q}_p$.

1. Ultrametric Spaces

1.1. Ultrametric Distances

Let (X, d) be a *metric space*. Thus X is equipped with a *distance function* $d : X \times X \longrightarrow \mathbf{R}_{\geq 0}$ satisfying the characteristic properties

$$d(x, y) > 0 \Longleftrightarrow x \neq y,$$
$$d(y, x) = d(x, y),$$
$$d(x, y) \leq d(x, z) + d(z, y)$$

for all x, y, and $z \in X$. For $r \geq 0$ and $a \in X$ we define[1]

$$B_{\leq r}(a) = \{x \in X : d(x, a) \leq r\}$$
$$= \textit{dressed ball} \text{ of radius } r \text{ and center } a,$$

[1] Let me use this unconventional terminology in this section only. From (II.2) on, I shall rely on the reader for a proper distinction between "open" and "closed" balls.

$$\begin{aligned} B_{<r}(a) &= \{x \in X : d(x,a) < r\} \\ &= \textit{stripped ball} \text{ of radius } r \text{ and center } a. \end{aligned}$$

Hence $B_{<r}(a)$ is empty if $r = 0$, and the stripped balls form a basis of a topology on X: In particular, *all stripped balls are open.*

Definition. *An* ultrametric distance *on a space X is a distance (or metric) satisfying the strong inequality*

$$d(x,y) \leq \max(d(x,z), d(z,y)) \quad (\leq d(x,z) + d(z,y))$$

for all x, y, and $z \in X$. An ultrametric space (X, d) is a metric space in which the distance satisfies this strong inequality.

The following results are valid in ultrametric spaces.

Lemma 1. (a) *Any point of a ball is a center of the ball.*
(b) *If two balls have a common point, one is contained in the other.*
(c) *The diameter of a ball is less than or equal to its radius.*

PROOF. (a) If $b \in B_{<r}(a)$, then $d(a,b) < r$ and

$$x \in B_{<r}(a) \Longleftrightarrow d(x,a) < r \overset{d(a,b)<r}{\Longleftrightarrow} d(x,b) < r \Longleftrightarrow x \in B_{<r}(b)$$

proving $B_{<r}(a) = B_{<r}(b)$. The case of a dressed ball is similar.
(b) Take, for example, a common point c of the balls $B_{<r}(a)$ and $B_{\leq r'}(b)$. By the previous part, we have

$$B_{<r}(a) = B_{<r}(c) \text{ and } B_{\leq r'}(b) = B_{\leq r'}(c).$$

Now, it is clear that $B_{<r}(c) \subset B_{\leq r'}(c)$ if $r \leq r'$, while $B_{\leq r'}(c) \subset B_{<r}(c)$ if $r' < r$. All other cases are treated similarly. Part (c) is obvious. ∎

It is immediately seen by induction that ultrametric distances also satisfy the strong inequality for finite sequences $x_1, x_2, \ldots, x_n \in X$:

$$d(x_1, x_n) \leq \max(d(x_1,x_2), d(x_2,x_3), \ldots, d(x_{n-1},x_n)).$$

Consider a cycle containing $n \geq 3$ distinct points: x_i $(1 \leq i \leq n)$, $x_{n+1} = x_1$. We may assume $d(x_1, x_n) = \max_{i \leq n} d(x_i, x_{i+1})$: Renumber these points if necessary, and observe that $d(x_n, x_{n+1}) = d(x_n, x_1) = d(x_1, x_n)$. Since

$$d(x_1,x_n) \leq \max(d(x_1,x_2), \ldots, d(x_{n-1},x_n))$$

by the ultrametric inequality, it follows that

$$d(x_1,x_n) = d(x_i,x_{i+1})$$

for at least one index $1 \leq i \leq n-1$. In other words, the cycle has at least two pairs of consecutive points with *equal maximal distance*. In particular, in a set a, b, c of cardinality 3, at least two pairs have the same (maximal) length. A picturesque way of formulating this property is this:

> *In an ultrametric space, all triangles are* isosceles (*or* equilateral), *with at most one short side.*

Here is an image of the situation. Let x be the earth and y, z be two stars in a galaxy not containing the earth, so that $d(x, y) > d(y, z)$. Then we consider that $d(x, y) = d(x, z)$ (this is the distance of the galaxy containing y, z to the earth). In other words, ultrametric distances behave as *orders of magnitude*.

Let us denote by $S_r(a) = \{x \in X : d(x, a) = r\}$ the sphere of center a and radius $r > 0$. Then if a ball B does not contain the point a, it lies on the sphere $S_r(a)$, where $r = d(a, B)$

$$\text{if } B = B_{<s}(b), \text{ then } r = d(a, b) \geq s \text{ and } B \subset S_r(a),$$

and similarly,

$$\text{if } B = B_{\leq s}(b), \text{ then } r = d(a, b) > s \text{ and } B \subset S_r(a).$$

Let us reformulate these properties in the form of another lemma.

Lemma 2. (*a*) *If* $d(x, z) > d(z, y)$, *then* $d(x, y) = d(x, z)$.
(*b*) *If* $d(x, z) \neq d(z, y)$, *then* $d(x, y) = \max(d(x, z), d(z, y))$.
(*c*) *If* $x \in S_r(a)$, *then* $B_{<r}(x) \subset S_r(a)$ *and*

$$S_r(a) = \bigcup_{x \in S_r(a)} B_{<r}(x). \qquad ■$$

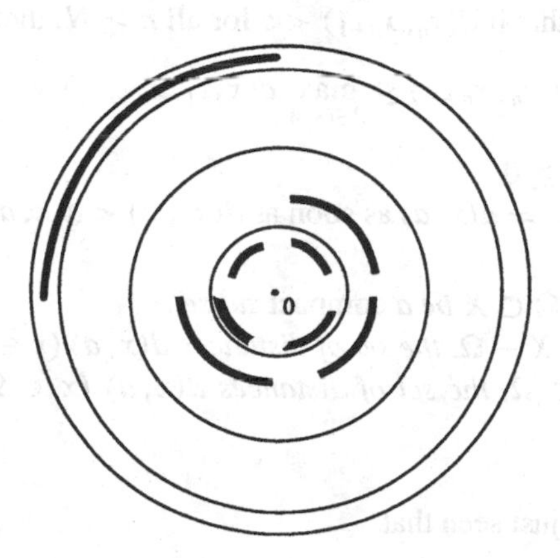

Balls within a ball

The stripped balls are open in any metric space: By definition, they make up a basis of the topology. Similarly, the dressed balls are closed in any metric space. In an ultrametric space we have some other peculiarities.

Lemma 3. *(a) The spheres $S_r(a)$ $(r > 0)$ are both open and closed.*
(b) The dressed balls of positive radius are open.
(c) The stripped balls are closed.
(d) Let B and B' be two disjoint balls.
Then $d(B, B') = d(x, x')$ for any $x \in B, x' \in B'$.

Proof. (a) The spheres are closed in all metric spaces, since the distance function $x \mapsto d(x, a)$ is continuous. A sphere of positive radius is open in an ultrametric space by part (c) of the previous lemma.

(b) If $r > 0$, then $B_{\leq r}(a) = B_{<r}(a) \cup S_r(a)$ is open.

(c) If $r > 0$, the sphere $S_r(a)$ is open; hence $B_{<r}(a) = B_{\leq r}(a) - S_r(a)$ is closed. If $r = 0$, $B_{<r}(a) = \emptyset$ is closed.

(d) Take four points: $x, y \in B$ and $x', y' \in B'$. The 4-cycle of points x, x', y', y has two pairs with maximal distance: They can only be $d(x, x') = d(y, y')$, since we assume that the balls are disjoint. All pairs of points $x \in B, x' \in B'$ are at the same distance, and $d(B, B') := \inf_{x \in B, x' \in B'} d(x, x')$ is this common value. ∎

Due to the frequent appearance of simultaneously open and closed sets in ultrametric spaces, it is useful to introduce a definition.

Definition. *An open and closed set will be called a* clopen *set.*

Lemma 4. *(a) A sequence $(x_n)_{n \geq 0}$ with $d(x_n, x_{n+1}) \to 0$ $(n \to \infty)$ is a Cauchy sequence.*
(b) If $x_n \to x \neq a$, then $d(x_n, a) = d(x, a)$ for all large indices n.

Proof. (a) Observe that if $d(x_n, x_{n+1}) < \varepsilon$ for all $n \geq N$, then also

$$d(x_n, x_{n+m}) \leq \max_{0 \leq i < m} d(x_{n+i}, x_{n+i+1}) < \varepsilon$$

for all $n \geq N$ and $m \geq 0$.

(b) In fact, $d(x_n, a) = d(x, a)$ as soon as $d(x_n, x) < d(x, a)$. ∎

Proposition. *Let $\Omega \subset X$ be a* compact *subset.*
(a) For every $a \in X - \Omega$, the set of distances $d(x, a)$ $(x \in \Omega)$ is finite.
(b) For every $a \in \Omega$, the set of distances $d(x, a)$ $(x \in \Omega - \{a\})$ is discrete in $\mathbf{R}_{>0}$.

Proof. (a) We have just seen that

$$d(x, y) < d(x, a) \Longrightarrow d(y, a) = d(x, a);$$

hence the function $f : x \mapsto d(x, a)$, $\Omega \to \mathbf{R}_{>0}$ is locally constant and continuous. Its range is finite: The sets $f^{-1}(c)$ (for $c \in f(\Omega)$) form an open partition of the compact set Ω.

(*b*) The map $f : x \mapsto d(x, a)$, $\Omega - \{a\} \to \mathbf{R}_{>0}$ is locally constant as before. For $\varepsilon > 0$, its restriction to the compact subset $\Omega - B_{<\varepsilon}(a)$ has finite range. This proves that all sets

$$[\varepsilon, \infty) \cap \{d(x, a) : x \in \Omega,\ x \neq a\}$$

are *finite*. Hence $f(\Omega - \{a\})$ is discrete in $\subset \mathbf{R}_{>0}$. ∎

Let us summarize.

Properties of ultrametric distances.

(*a*) *Any point of a ball is a possible center of the ball* $b \in B_{\leq r}(a) \Longrightarrow B_{\leq r}(b) = B_{\leq r}(a)$ *(and similarly for stripped balls).*
(*b*) *If two balls have a common point, then one is contained in the other.*
(*c*) *A sequence* $(x_n)_{n \in \mathbf{N}}$ *is a Cauchy sequence precisely when* $d(x_n, x_{n+1}) \to 0 \quad (n \to \infty)$.
(*d*) *In a compact ultrametric space* X, *for each* $a \in X$, *the set of nonzero distances* $\{d(x, a) : a \neq x \in X\}$ *is discrete in* $\mathbf{R}_{>0}$.

1.2. Ultrametric Principles in Abelian Groups

Let G be an additive (abelian) group equipped with an invariant metric d, namely a metric satisfying

$$d(x + z, y + z) = d(x, y) \quad (x, y \text{ and } z \in G).$$

For $x \in G$, define

$$|x| = d(x, 0).$$

Then

$$|-x| = d(-x, 0) = d(0, x) = d(x, 0) = |x|$$

and

$$\begin{aligned} |x + y| = d(x + y, 0) &\leq d(x + y, y) + d(y, 0) \\ &\leq d(x, 0) + d(y, 0) = |x| + |y|. \end{aligned}$$

This shows that $x \mapsto -x$ and $(x, y) \mapsto x + y$ are continuous and G is a *topological group* when equipped with the metric d. We shall say that G is a *valued group* when such a metric d has been chosen.

Assuming that this metric satisfies the *ultrametric inequality*, we shall have similarly

$$\begin{aligned} |x + y| = d(x + y, 0) &\leq \max(d(x + y, y), d(y, 0)) \\ &\leq \max(d(x, 0), d(y, 0)) = \max(|x|, |y|). \end{aligned}$$

In particular, all nonempty balls centered at the neutral element $0 \in G$ are *subgroups* of G. These subgroups are

$$B_{\le r}(0) = \{x \in G : |x| \le r\} \quad (r \ge 0),$$
$$B_{<r}(0) = \{x \in G : |x| < r\} \quad (r > 0).$$

Instead of applying (1.1) to see that the balls $B_{\le r}(0)$ and $B_{<r}(0)$ are open and closed when $r > 0$, one can observe that these subgroups are neighborhoods of the origin and use (I.3.2) to reach the same conclusion.

Conversely, if we are given a function $G \to \mathbf{R}_{\ge 0} : x \mapsto |x|$ satisfying

$$|x| > 0 \ \text{ for } x \ne 0, \quad |-x| = |x|,$$
$$|x+y| \le |x| + |y| \quad (\text{resp.} \le \max(|x|, |y|))$$

then we can define an invariant metric (resp. ultrametric) on G by

$$d(x, y) = |x - y|.$$

The characteristic properties of distances are immediately verified (see the specific references at the end of the volume). A pair $(G, |\,.\,|)$ consisting of an abelian group G and a function $G \to \mathbf{R}_{\ge 0} : x \mapsto |x|$ satisfying the preceding properties, with the ultrametric inequality

$$|x+y| \le \max(|x|, |y|) \quad (x, y \in G),$$

will be called *an abelian ultrametric group.*

The study of convergence for series in a complete abelian group is simpler in ultrametric analysis than in classical analysis. Let $(a_i)_{i \ge 0}$ be a sequence and define

$$s_n = \sum_{i<n} a_i .$$

If this sequence of partial sums s_n has a limit s, then

$$a_n = s_{n+1} - s_n \to s - s = 0.$$

This *necessary* condition for convergence of the series $\sum_{i \ge 0} a_i$ is *sufficient in any complete ultrametric group*. Indeed, if $s_{n+1} - s_n = a_n \to 0$, the sequence (s_n) is a Cauchy sequence and hence converges. Moreover, reordering the terms of a convergent series, and grouping terms, alters neither its convergence nor its sum.

Proposition. *Let $(a_i)_{i \in \mathbf{N}}$ be a sequence in a complete ultrametric abelian group. Assume that $a_i \to 0$, so that the series $\sum_{i \ge 0} a_i$ converges: Let s be its sum. Then*

(a) for any bijection $\sigma : \mathbf{N} \to \mathbf{N}$ we have $s = \sum_{i \ge 0} a_{\sigma(i)}$,
(b) for any partition $\mathbf{N} = \coprod_j I_j$ we have $s = \sum_j \left(\sum_{i \in I_j} a_i\right)$.

PROOF. (*a*) For $\varepsilon > 0$, define the finite set

$$I(\varepsilon) = \{i : |a_i| > \varepsilon\}$$

and the corresponding sum

$$s(\varepsilon) = \sum_{i \in I(\varepsilon)} a_i.$$

For any finite set $J \supset I(\varepsilon)$,

$$\left| \sum_{J-I(\varepsilon)} a_i \right| \leq \max_{i \notin I(\varepsilon)} |a_i| \leq \varepsilon.$$

This proves that the family (a_i) is *summable*. This notion is independent of the order on **N**. Explicitly, for $\varepsilon > 0$, $\eta > 0$ we have

$$|s(\varepsilon) - s(\eta)| \leq \max(\varepsilon, \eta),$$

since $s(\varepsilon) - s(\eta)$ is a finite sum of terms having absolute values between ε and η. In particular, $(s(1/n))_{n>0}$ is a Cauchy sequence, and we call s its limit. If $\varepsilon > 0$, letting $n \to \infty$ in

$$|s(\varepsilon) - s(1/n)| \leq \max(\varepsilon, 1/n)$$

we get

$$|s(\varepsilon) - s| \leq \varepsilon.$$

Hence we can say that $s(\varepsilon) \to s$ when $\varepsilon \to 0$. Now, if $a'_i = a_{\sigma(i)}$ is a rearrangement of the terms of the series and $s'_n = \sum_{i<n} a'_i$, the inequality

$$|s - s'_n| \leq \varepsilon$$

holds when $\{\sigma(i) : i < n\}$ contains the finite set $I(\varepsilon)$, hence for all sufficiently large n.

(*b*) Let $s_j = \sum_{i \in I_j} a_i$, so that we have to prove $s = \sum_j s_j$. Take any $\varepsilon > 0$ and define the finite sets

$$I_j(\varepsilon) = I_j \cap I(\varepsilon).$$

Obviously, the nonempty $I_j(\varepsilon)$ make a partition of the finite set $I(\varepsilon)$, and

$$s(\varepsilon) = \sum_{I(\varepsilon)} a_i = \sum_j \left(\sum_{i \in I_j(\varepsilon)} a_i \right).$$

Finally,

$$|s - \sum_j s_j| \le \max\left(|s - s(\varepsilon)|, \left|\sum_j \left(\sum_{i \in I_j(\varepsilon)} a_i\right) - \sum_j s_j\right|\right)$$

$$\le \max\left(|s - s(\varepsilon)|, \max_j \left|\sum_{i \in I_j(\varepsilon)} a_i - s_j\right|\right) \le \varepsilon.$$

Since this is true for all $\varepsilon > 0$, the conclusion follows. ■

Corollary. *Let $(a_{ij})_{i\ge 0, j\ge 0}$ be a double sequence such that for any $\varepsilon > 0$ the set of pairs (i, j) with $|a_{ij}| > \varepsilon$ is finite. Then this double family is summable and*

$$\sum_{i\ge 0}\left(\sum_{j\ge 0} a_{ij}\right) = \sum_{j\ge 0}\left(\sum_{i\ge 0} a_{ij}\right).$$

PROOF. The family $(a_{ij})_{i\ge 0, j\ge 0}$ is summable over the countable set $\mathbf{N} \times \mathbf{N}$ by hypothesis, and the sum of the corresponding series $\sum_{i,j} a_{ij}$ can be computed in any order. It can also be computed using the two groupings mentioned. ■

Comments (1) Summable families over arbitrary index sets will be considered later (cf. (IV.4.1)). The above proposition will be generalized correspondingly.

(2) In classical analysis, there is a distinction between conditionally convergent and absolutely convergent — or commutatively convergent, or summable — series (of real or complex numbers): This distinction disappears in non-Archimedean analysis, since *the sum of a convergent series* can be computed in any order, any grouping. But *in both contexts a grouping in a divergent series may produce a convergent one*: Think of $a_i = (-1)^i$, $|a_i| = 1 \not\to 0$; here is a grouping that leads to a convergent series

$$(1-1) + (1-1) + \cdots = 0 + 0 + \cdots = 0,$$

and here is another grouping,

$$1 + (-1+1) + (-1+1) + \cdots = 1 + 0 + 0 + \cdots = 1$$

leading to a different sum. Or think of the divergent series $\sum_{n\ge 0} a_n$ where all $a_n = 1$. A suitable grouping of its terms leads to a convergent series:

$$1 + \underbrace{(1 + \cdots + 1)}_{p \text{ terms}} + \underbrace{(1 + \cdots + 1)}_{p^2 \text{ terms}} + \cdots = 1 + p + p^2 + \cdots = \frac{1}{1-p}.$$

Basic Principles of Ultrametric Analysis in an Abelian Group

(1) *The strongest wins*

$$|x| > |y| \Longrightarrow |x+y| = |x|.$$

(2) *Equilibrium: All triangles are isosceles (or equilateral)*

$$a+b+c=0, |c| < |b| \Longrightarrow |a| = |b|.$$

(3) *Competitivity*

$$a_1 + a_2 + \cdots + a_n = 0 \Longrightarrow$$
$$\text{there is } i \neq j \text{ such that } |a_i| = |a_j| = \max |a_k|.$$

(4) *A dream realized*

$$(a_n)_{n\geq 0} \text{ is a Cauchy sequence} \Longleftrightarrow d(a_n, a_{n+1}) \to 0.$$

(5) *Another dream come true (in a* complete *group)*

$$\sum\nolimits_{n\geq 0} a_n \text{ converges} \Longleftrightarrow a_n \to 0.$$

When $\sum_{n\geq 0} a_n$ *converges,* $\sum_{n\geq 0} |a_n|$ *may diverge but* $\left|\sum_{n\geq 0} a_n\right| \leq \sup |a_n|$ *and the infinite version of* (3) *is valid.*

(6) *Stationarity of the absolute value*

$$a_n \to a \neq 0 \Longrightarrow \text{there is } N \text{ with } |a_n| = |a| \text{ for } n \geq N.$$

1.3. Absolute Values on Fields

Definition 1. *An* absolute value *on a field* K *is a homomorphism*

$$f : K^\times \to \mathbf{R}_{>0}$$

extended by $f(0) = 0$ *and such that* $f(x+y) \leq f(x) + f(y) \quad (x, y \in K)$.

The trivial homomorphism $f(x) = 1$ $(x \in K^\times)$ defines the *trivial absolute value* on K. We shall usually denote by $f(x) = |x|$ an absolute value, and by definition, such a function will always have the *characteristic properties*

$$\begin{aligned} |x| &\geq 0, \\ |x| = 0 &\Longleftrightarrow x = 0, \\ |xy| &= |x| \cdot |y|, \\ |x+y| &\leq |x| + |y| \end{aligned}$$

for all $x, y \in K$. The pair $(K, |\,.\,|)$ is a *valued field* (I.3.7).

If $x^n = 1 \in K$, then $|x|^n = |x^n| = 1$ and $|x| = 1$. In particular, $|-1| = |1| = 1$. Also, $|2| = |1+1| \le 1+1 = 2$, and by induction

$$|n| \le n \quad (n \in \mathbf{N})$$

(here $n = n \cdot 1_K \in K$ in the left-hand side of the inequality, whereas $n \in \mathbf{R}_{>0}$ in the right-hand side). Also, quite generally,

$$\left|\frac{x}{y}\right| = \frac{|x|}{|y|} \quad (x \in K, y \in K^\times).$$

By induction

$$|x_1 + x_2 + \cdots + x_n| \le |x_1| + |x_2| + \cdots + |x_n|$$

for every positive integer n.

Definition 2. *An* ultrametric field *is a pair* $(K, |\,.\,|)$ *consisting of a field* K *and an* ultrametric absolute value *on* K, *namely an absolute value satisfying the strong triangle inequality*

$$|x + y| \le \max(|x|, |y|) \le |x| + |y| \quad (x, y \in K).$$

As before, induction shows that

$$|x_1 + x_2 + \cdots + x_n| \le \max(|x_1|, |x_2|, \ldots, |x_n|).$$

In this case, we have $|2| = |1+1| \le 1$ and by induction

$$|n| \le 1 \quad (n \in \mathbf{N}).$$

Hence ultrametric fields have the non-Archimedean property

$$|nx| \le |x| \quad (n \in \mathbf{N}).$$

The following lemma is obvious (cf. (1.2)).

Lemma. *All balls containing* 0 *in an ultrametric field* K *are additive subgroups. The dressed unit ball* $B_{\le 1}(0)$ *is a subring of* K. *The balls* $B_{\le r}(0)$ $(r < 1)$ *are ideals of* $B_{\le 1}(0)$. *The balls* $B_{<r}(0)$ $(r \le 1)$ *are ideals of* $B_{\le 1}(0)$. ∎

Proposition. *Let* $x \mapsto |x|$ *be an absolute value on a field* K. *Then:*

(1) $d(x, y) = |x - y|$ *defines a metric on* K.
(2) *For each exponent* $0 < \alpha \le 1$, $x \mapsto |x|^\alpha$ *still defines an absolute value on the field* K.
(3) *If* $x \mapsto |x|$ *is an ultrametric absolute value, then for each positive exponent* $\alpha > 0$, $x \mapsto |x|^\alpha$ *still defines an ultrametric absolute value on the field* K.

PROOF. All statements are obvious except perhaps the triangle inequality, which is nevertheless a simple exercise. ■

The trivial absolute value defines the *discrete metric*: $d(x, y) = 1$ if $x \neq y$.

1.4. Ultrametric Fields: The Representation Theorem

Let K be an ultrametric field. We use the general notation

$$A = \{x \in K : |x| \leq 1\}\text{: dressed unit ball,}$$
$$M = \{x \in K : |x| < 1\}\text{: stripped unit ball.}$$

Hence

$$A = A^\times \sqcup M$$

is a disjoint union, where $A^\times$, the multiplicative group of invertible elements in A, is the unit sphere $|x| = 1$.

Proposition. *The subset A is a maximal subring of K, and M is the unique maximal ideal of the ring A.*

PROOF. Indeed, if A' is any subring strictly containing A, it will contain an element y such that $|y| = r > 1$ together with all its powers y^n. Hence $B_{\leq r^n} = y^n A \subset A'$, and since $r^n = |y^n| \to \infty$, we see that $K = \bigcup_{n\geq 1} y^n A = A'$. Moreover, any ideal not contained in M contains a unit, and hence coincides with the whole ring A. This shows that M is the unique maximal ideal of A. ■

Definition 1. *A subring A of a field K such that*

$$\textit{for every } x \in K^\times,\ x \in A \textit{ or } 1/x \in A$$

is called a valuation ring *of K. A commutative ring A having a single maximal ideal is called a* local ring.

The unit ball in an ultrametric field is a local ring and a valuation ring.

Definition 2. *If K is an ultrametric field, its* residue field *is the quotient $k = A/M$ of its dressed unit ball, the maximal subring of K, by its unique maximal ideal.*

The residue field parametrizes the stripped balls of unit radius in the dressed unit ball of K: If $S \subset A$ is a set of representatives for the classes mod M, then

$$A = B_{\leq 1}(0) = \coprod_{x\in S} B_{<1}(x).$$

Theorem. *Let K be a complete ultrametric field, A its maximal subring defined by $|x| \leq 1$. Choose an element ξ with $|\xi| < 1$ together with a set of* *representatives $S \subset A$ containing 0 for the classes $A/\xi A$. Then each nonzero element $x \in K^\times$ is a sum*

$$x = \sum_{i \geq m} a_i \xi^i \quad (m \in \mathbf{Z},\ a_i \in S,\ a_m \neq 0)$$

with $m \geq 0$ precisely when $x \in A$. The map $x \mapsto (s_n)$ where $s_n = \sum_{m \leq i < n} a_i \xi^i$ defines an isomorphism $A \cong \varprojlim A/\xi^n A$.

Proof. The conditions $|\xi| < 1$, $\xi \in A$ is not a unit, and $\xi \in M$ are all equivalent. Starting with $x \in A$, there is a unique $a_0 \in S$ with $x - a_0 \in \xi A$,

$$x = a_0 + \xi x_1 \quad (x_1 \in A).$$

Repeating the procedure for x_1, and so on, we get by induction

$$x = a_0 + a_1 \xi + \cdots + a_{n-1} \xi^{n-1} + x_n \xi^n$$

with $a_i \in S$ and $x_n \in A$. In the notation of the statement of the theorem, we can write $x = s_n + x_n \xi^n$. Since $|x_n \xi^n| \leq |\xi^n| = |\xi|^n \to 0$, the sequence $(s_n)_{n \geq 0}$ is a Cauchy sequence, and the series $\sum_{n \geq 0} a_i \xi^i$ converges to the element $x \in A$. Since for any $x \in K$ there is an integer k such that $|\xi^k x| \leq 1$, namely such that $\xi^k x \in A$, the preceding expansion can be derived for this element, and we obtain a series expansion for x starting at the index $i = m = -k$. ■

Observe that even when K is not complete, each $x \in K^\times$ has a series representation as indicated in the theorem, but an arbitrary series

$$\sum_{i \geq m} a_i \xi^i \quad (m \in \mathbf{Z},\ a_i \in S,\ a_m \neq 0)$$

will — in general — converge only in the completion of K. In other words, even when K is not complete, we get an *injection*

$$A \hookrightarrow \widehat{A} = \varprojlim A/\xi^n A.$$

1.5. General Form of Hensel's Lemma

Theorem (Hensel's Lemma). *Let K be a complete ultrametric field with maximal subring A and $f \in A[X]$. Assume that $x \in A$ satisfies*

$$|f(x)| < |f'(x)|^2.$$

Then there is a root $\xi \in A$ of f such that $|\xi - x| = |f(x)/f'(x)| < |f'(x)|$. This is the only root of f in the stripped ball of center x and radius $|f'(x)|$.

Proof. In spite of the similarity with (I.6.4) (particular case $K = \mathbf{Q}_p$), we give a complete proof with absolute values (instead of congruences). The idea is again to use Newton's method iteratively. Since the polynomials f and f' have coefficients in the ring A, we have $|f(x)| \leq 1$ and $0 < |f'(x)| \leq 1$.

First step: Estimates concerning the distance of $\widehat{x} = x - f(x)/f'(x)$ *to* x.

The assumption is $c := |f(x)/f'(x)^2| < 1$. We have

$$\widehat{x} - x = -\frac{f(x)}{f'(x)} = -\frac{f(x)}{f'(x)^2} \cdot f'(x),$$
$$|\widehat{x} - x| = c|f'(x)|.$$

Similarly

$$(\widehat{x} - x)^2 = \left(\frac{f(x)}{f'(x)}\right)^2 = \frac{f(x)}{f'(x)^2} \cdot f(x),$$
$$|\widehat{x} - x|^2 = c|f(x)|.$$

The second-order expansion (I.6.2) of f at the point x gives

$$f(\widehat{x}) = \underbrace{f(x) + (\widehat{x} - x)f'(x)}_{=0\text{: Newton's choice}} + (\widehat{x} - x)^2 r \quad (r \in A : \ |r| \leq 1),$$
$$|f(\widehat{x})| \leq |\widehat{x} - x|^2 = c|f(x)| < |f(x)|,$$

and $\widehat{x}$ is an improved approximation to a root. The first-order expansion (I.6.2) of f' at the point x gives

$$f'(\widehat{x}) = f'(x) + (\widehat{x} - x)s \quad (s \in A : \ |s| \leq 1),$$
$$|f'(\widehat{x}) - f'(x)| \leq |\widehat{x} - x| = c|f'(x)| < |f'(x)|.$$

It shows that

$$|f'(\widehat{x})| = |\underbrace{f'(x)}_{\text{strongest}} + (f'(\widehat{x}) - f'(x))| = |f'(x)|.$$

The invitation to iteration is clear.

Second step: Further iterations.

Let now $\widehat{\widehat{x}} = N_f(\widehat{x})$

$$\widehat{c} := \left|\frac{f(\widehat{x})}{f'(\widehat{x})^2}\right| \leq \frac{c|f(x)|}{|f'(x)|^2} = c^2.$$

This iteration furnishes

$$|f(\widehat{\widehat{x}})| \leq \widehat{c}|f(\widehat{x})| \leq \widehat{c}c|f(x)| \leq c^{2+1}|f(x)|,$$

and since $|f(x)| = c|f'(x)|^2$ by definition, we obtain

$$|f(\widehat{\widehat{x}})| \leq c^4|f'(x)|^2.$$

We can construct the sequence $x_0 = x$, $x_1 = \widehat{x}$, $x_2 = \widehat{\widehat{x}}$, ... inductively with $x_{i+1} = \widehat{x_i}$. Define also $c_1 = \widehat{c}$, $c_{i+1} = \widehat{c_i}$. The preceding estimates show that

$$|f(x_i)| \leq c_{i-1} \cdots c_1 c |f(x_0)| \leq c^{2^i-1} |f(x_0)| = c^{2^i} |f'(x_0)|^2 \to 0 \quad (i \to \infty),$$
$$|x_2 - x_1| = |\widehat{\widehat{x}} - \widehat{x}| \leq c^2 |f'(x_0)| < c|f'(x_0)| = |x_1 - x_0|,$$

and by induction,

$$|x_{i+1} - x_i| \leq c^{2^i} |f'(x_0)| < c|f'(x_0)| = |x_1 - x_0| \quad (i \geq 1).$$

In particular, $|x_i - x_0| = |x_1 - x_0| = |\widehat{x} - x| = c|f'(x_0)|$ is constant for $i \geq 1$ (these x_i are closer to each other than to x_0).

Third step: *The limit root* ξ.

The sequence $(x_i)_{i\geq 0}$ is a Cauchy sequence, so it converges in the complete field K. Since all iterates x_i belong to the closed subring A, we have

$$\xi = \lim_{n\to\infty} x_i \in A,$$
$$|\xi - x_0| = |x_1 - x_0| = |\widehat{x} - x| = c|f'(x)| < |f'(x)|,$$
$$f(\xi) = f(\lim_{n\to\infty} x_i) = \lim_{n\to\infty} f(x_i) = 0.$$

Fourth step: *Uniqueness of the root* ξ.

Let ξ be as before and η have the required properties, say $\eta = \xi + h$. Hence $|h| = |\eta - \xi| < |f'(x)| = |f'(\xi)|$. The second-order expansion (I.6.2) of f at the point ξ gives

$$0 = f(\eta) = \underbrace{f(\xi)}_{=0} + hf'(\xi) + h^2 t \quad (t \in A : |t| \leq 1),$$
$$0 = h(\underbrace{f'(\xi)}_{\text{strongest}} + ht) = h\underbrace{(f'(\xi) + ht)}_{\neq 0};$$

hence $h = 0$, i.e., $\eta = \xi$. ■

Observe that when the absolute value is trivial, it takes only the values 0 and 1, the assumption reduces to

$$0 = |f(x)| < |f'(x)|^2 = 1,$$

and the statement is trivially correct.

1.6. Characterization of Ultrametric Absolute Values

Theorem. *Let $x \mapsto |x|$ be an absolute value on a field K. Then the following properties are equivalent:*

(*i*) *$|n| \leq 1$ for all natural integers $n \in \mathbf{N}$.*
(*ii*) *The absolute value is bounded on $\mathbf{N} \cdot 1_K$.*

(iii) $|1+x| \leq 1$ *for every* $x \in K$ *such that* $|x| \leq 1$.
(iv) $x \mapsto |x|$ *is an ultrametric absolute value.*
(v) $\{x \in K : |x| \leq 1\}$ *is a subring of* K.

PROOF. We proceed according to the following scheme of implications:

$$(i) \Rightarrow (ii) \Rightarrow (iii) \Rightarrow (iv) \Rightarrow (v) \Rightarrow (iii)$$

and

$$(iv) \Rightarrow (i).$$

Among these implications, several are trivial, namely,

$$(i) \Rightarrow (ii), \quad (iv) \Rightarrow (v) \Rightarrow (iii) \quad \text{and} \quad (iv) \Rightarrow (i).$$

It only remains to prove two implications. For *(ii)* $\Rightarrow$ *(iii)* we can assume $|n| \leq M$ for all integers n and use the binomial formula to compute

$$|1+x|^n = |(1+x)^n| = \left|\sum \binom{n}{i} x^i\right| \leq \sum M|x|^i.$$

When $|x| \leq 1$, we obtain

$$|1+x|^n \leq (n+1)M,$$
$$|1+x| \leq (n+1)^{1/n} \cdot M^{1/n}$$

for all integers $n \geq 1$. Since $(n+1)^{1/n} \to 1$ as well as $M^{1/n} \to 1$ for $n \to \infty$, we infer $|1+x| \leq 1$. To prove *(iii)* $\Rightarrow$ *(iv)*, we can — without loss of generality — assume that $|x| \geq |y|$ in $|x+y|$, $|x| > 0$ and estimate this quantity as follows:

$$|x+y| = |x| \cdot |1+y/x| \leq |x| = \max(|x|, |y|). \qquad \blacksquare$$

Corollary. *Any absolute value on a field of characteristic* $p \neq 0$ *is ultrametric.*

PROOF. Indeed, any absolute value is bounded on the image of **N** in a field of characteristic p, since this image is a *finite* prime field. The second condition of the theorem is automatically satisfied. ■

The absolute values that are *not bounded* on the prime field of K (necessarily of characteristic zero) are sometimes called *Archimedean absolute values*: They have the property that

if $x \neq 0$, then for each y there is an $n \in \mathbf{N}$ such that $|nx| > |y|$.

1.7. Equivalent Absolute Values

Distinct absolute values can define the same topology on a field K. It is not always useful to distinguish them.

Theorem. *Let $|\,.\,|_1$ and $|\,.\,|_2$ be two absolute values on a field K. Then the following conditions are equivalent:*

(i) There is an $\alpha > 0$ with $|\,.\,|_2 = |\,.\,|_1^\alpha$.
(ii) $|\,.\,|_1$ and $|\,.\,|_2$ define the same topology on K.
(iii) The stripped unit balls for $|\,.\,|_1$ and $|\,.\,|_2$ coincide.

We say that $|\,.\,|_1$ and $|\,.\,|_2$ are equivalent absolute values *when these conditions are satisfied.*

PROOF. (*i*) $\Rightarrow$ (*ii*) Since $|x-a|_2 < r \iff |x-a|_1 < r^{1/\alpha}$, the stripped balls are the same for the two topologies. Hence the topologies defined by $|\,.\,|_1$ and $|\,.\,|_2$ are the same.

(*ii*) $\Rightarrow$ (*iii*) Let us observe that

$$|x|_1 < 1 \iff x^n \to 0 \text{ (for the topology defined by } |\,.\,|_1)$$

and similarly for $|\,.\,|_2$. By assumption we obtain

$$|x|_1 < 1 \iff |x|_2 < 1.$$

(*iii*) $\Rightarrow$ (*i*) Let us assume $|x|_1 < 1 \iff |x|_2 < 1$. Since $|1/x|_1 = 1/|x|_1$ and similarly for $|\,.\,|_2$, we see that

$$|x|_1 > 1 \iff |x|_2 > 1$$

and consequently

$$|x|_1 = 1 \iff |x|_2 = 1.$$

If $|\,.\,|_1$ is trivial, $|x|_1 = 1$ for all $x \in K^\times$, and the same is true for $|\,.\,|_2$, so that we can take $\alpha = 1$ in the statement of (*i*). Otherwise, we can find $x_0 \in K^\times$ with $|x_0|_1 \neq 1$, and replacing x_0 by $1/x_0$ if necessary, we can assume $|x_0|_1 < 1$. Define

$$\alpha = \frac{\log|x_0|_2}{log|x_0|_1},$$

so that $|x_0|_2 = |x_0|_1^\alpha$ by definition. Take then any element $x \in K^\times$ with $|x|_1 < 1$ and consider the rational numbers $r > 0$ such that $|x|_1^r < |x_0|_1$. These rational numbers $r = m/n$ are those for which

$$|x|_1^m < |x_0|_1^n, \quad \left|\frac{x^m}{x_0^n}\right|_1 < 1.$$

By assumption, these are the same as those for which

$$\left|\frac{x^m}{x_0^n}\right|_2 < 1,$$

namely $|x|_2^m < |x_0|_2^n$ or $|x|_2^r < |x_0|_2$. On the other hand, these rational numbers are precisely those for which

$$r \log|x|_1 < \log|x_0|_1 \quad (\text{resp. } r \log|x|_2 < \log|x_0|_2)$$

or

$$r > \log|x_0|_1/\log|x|_1 \quad (\text{resp. } r > \log|x_0|_2/\log|x|_2)$$

(all logarithms in question are negative). This proves

$$\log|x_0|_1/\log|x|_1 = \log|x_0|_2/\log|x|_2,$$
$$\log|x|_2/\log|x|_1 = \log|x_0|_2/\log|x_0|_1 = \alpha.$$

Hence $|x|_2 = |x|_1^\alpha$, as was to be shown. ■

2. Absolute Values on the Field Q

2.1. *Ultrametric Absolute Values on* Q

Let us recall that if p is a prime number, we can define an absolute value on the field $\mathbf{Q}$ of rational numbers by the following procedure. If $x = p^m a/b$ with $a, b, m \in \mathbf{Z}$, $b \neq 0$, and p prime to a and b, we put

$$|x|_p = p^{-m}.$$

In other words, we put $|p|_p = 1/p < 1$ and $|n|_p = 1$ for any integer n prime to p, and extend it multiplicatively for products. Since

$$\mathbf{Q}^\times = p^{\mathbf{Z}} \times \mathbf{Z}_{(p)}^\times = \coprod_{m \in \mathbf{Z}} p^m \mathbf{Z}_{(p)}^\times,$$

this defines the absolute value uniquely. This absolute value is an ultrametric absolute value on $\mathbf{Q}$.

Theorem (Ostrowski). *Let $x \mapsto |x|$ be a nontrivial ultrametric absolute value on the field* $\mathbf{Q}$. *Then there exists a prime p and a real number $\alpha > 0$ such that*

$$|x| = |x|_p^\alpha \quad (x \in \mathbf{Q}).$$

PROOF. Since the integers generate $\mathbf{Q}$ (by multiplication and quotients), the absolute value must be nontrivial on $\mathbf{N}$. As we have seen, any ultrametric absolute value satisfies $|n| \leq 1$ $(n \in \mathbf{N})$. Hence there must exist a positive integer n with $|n| < 1$. The smallest such integer is a prime p because in any factorization $n = n_1 \cdot n_2$, we have $|n_1| \cdot |n_2| = |n| < 1$, and consequently one factor n_i must satisfy $|n_i| < 1$. Let us call this prime p so that by definition

$$|n| = 1 \quad \text{for } 1 \leq n < p$$

but $0 < |p| < 1$. I claim that for every integer $m \in \mathbf{Z}$ prime to p, we have $|m| = 1$. Indeed, if m is prime to p, the Bézout theorem asserts that there are integers u and v with $up + vm = 1$. Hence

$$1 = |1| = |up + vm| \leq \max(|up|, |vm|) \leq 1.$$

Since by assumption $|up| = |u||p| < |u| \leq 1$, the maximum must be $|vm| = 1$ and hence $|m| = 1$ (we know a priori that $|v| \leq 1$ and $|m| \leq 1$). There is now a unique positive real number α such that

$$|p| = (1/p)^\alpha$$

(indeed, take $\alpha = (\log|p|)/(\log(1/p))$ — a quotient of two negative numbers — independent from the basis of logarithms chosen). Then if the rational number x is written in the form $x = p^v a/b \in \mathbf{Q}^\times$ with p prime to a and b (i.e., $a/b \in \mathbf{Z}_{(p)}^\times$), we shall have

$$|x| = |p|^v = (1/p)^{v\alpha} = |x|_p^\alpha$$

and the theorem is completely proved. ∎

2.2. *Generalized Absolute Values*

Observe that if $|\,.\,|$ is an absolute value and $\alpha > 0$, then $|\,.\,|^\alpha$ is not an absolute value in general. For example if $|\,.\,|$ is the usual absolute value on $\mathbf{Q}$ and $\alpha = 2$, then $f(x) = |x|^2$ does not satisfy the triangle inequality

$$4 = f(2) = f(1+1) > f(1) + f(1) = 2.$$

But it satisfies

$$f(x+y) = |x+y|^2 \leq (|x| + |y|)^2 \leq (2\max\{|x|, |y|\})^2 = 4\max(f(x), f(y)).$$

This is one reason for considering generalized absolute values.

Definition. *A* generalized absolute value *on a field K is a homomorphism $f : K^\times \to \mathbf{R}_{>0}$ extended by $f(0) = 0$ for which there exists a constant $C > 0$ such that*

$$f(x+y) \leq C\max(f(x), f(y)) \quad (x, y \in K).$$

Observations. (1) For any generalized absolute value f and *any* $\alpha > 0$ (not only for $0 < \alpha \leq 1$), f^α is also a generalized absolute value: Replace C by C^α.

(2) The ultrametric absolute values are those for which the above inequality holds with $C = 1$. Moreover, if f is a (usual) absolute value, then

$$f(x+y) \leq f(x) + f(y) \leq 2\max(f(x), f(y)),$$

and (usual) absolute values are generalized absolute values: The above inequality holds with $C = 2$. Let us prove a converse.

Theorem. *Let f be a generalized absolute value on a field K for which*

$$f(x+y) \le 2\max(f(x), f(y)) \quad (x, y \in K).$$

Then f is a usual absolute value: It satisfies the identity

$$f(x+y) \le f(x) + f(y) \quad (x, y \in K).$$

PROOF. Iterating the defining inequality for generalized absolute values, we find that

$$\begin{aligned} f(a_1+a_2+a_3+a_4) &\le C\max(f(a_1+a_2), f(a_3+a_4)) \\ &\le C^2 \max_{1\le i\le 4} f(a_i). \end{aligned}$$

More generally, by induction if $n = 2^r$, then

$$f(a_1+\cdots+a_n) \le C^r \max f(a_i).$$

Since we are assuming that the constant $C = 2$ can be taken in the preceding inequalities, we have

$$f(a_1+\cdots+a_n) \le 2^r \max f(a_i) = n \max f(a_i).$$

Now, if n is not a power of 2, say $2^{r-1} < n < 2^r$, we can complete the sum by taking coefficients $a_i = 0$ for $n < i \le 2^r$ and still write

$$f(a_1+\cdots+a_n) \le 2^r \max f(a_i) < 2n \max f(a_i).$$

We shall have to use two particular cases of this general inequality:

(1) $f(n) \le 2n$ (take $a_i = 1$ for $1 \le i \le n$),
(2) $f\left(\sum_{1\le i\le n} a_i\right) \le 2n \max f(a_i) \le 2n \sum_{1\le i\le n} f(a_i)$.

To estimate $f(a+b)$, we shall estimate $f((a+b)^n)$ thanks to the binomial formula (the nth power of $a+b$ is a sum of $n+1$ monomials)

$$\begin{aligned} f((a+b)^n) &= f\left(\sum \binom{n}{i} a^i b^{n-i}\right) \\ &\le 2(n+1)\sum f\left(\binom{n}{i}\right) \cdot f(a)^i f(b)^{n-i} \\ &\le 2(n+1)\sum 2\binom{n}{i} f(a)^i f(b)^{n-i} \\ &= 4(n+1)(f(a)+f(b))^n. \end{aligned}$$

Let us extract nth roots:

$$f(a+b) \le 4^{1/n}(n+1)^{1/n} \cdot (f(a)+f(b)) \to f(a)+f(b) \quad (n \to \infty). \qquad \blacksquare$$

2.3. Ultrametric Among Generalized Absolute Values

We can give a generalization of (1.6).

Theorem. *Let f be a generalized absolute value on a field K. If f is bounded on the image of the natural numbers $\mathbf{N}$ in K, then it is an ultrametric absolute value.*

PROOF. Let $n = 2^r$ be a power of 2 and consider a sum of n terms a_i. As in (2.2), we see by induction that

$$f(a_1 + \cdots + a_n) \le C^r \max f(a_i).$$

Take now $x \in K$ and consider the element

$$(1+x)^{n-1} = \sum_{0 \le i < n} \binom{n-1}{i} x^i.$$

Since this sum has n elements, we have

$$(f(1+x))^{n-1} = f((1+x)^{n-1}) \le C^r \max_i \left[f\left(\binom{n-1}{i}\right) \cdot f(x^i) \right].$$

If f is bounded on the image of $\mathbf{N}$ in K, say $f(k) \le A$ for all $k \in \mathbf{N}$, we shall have

$$(f(1+x))^{n-1} \le C^r A \max(1, f(x)^{n-1})$$

and

$$f(1+x) \le C^{r/(n-1)} A^{1/(n-1)} \max(1, f(x)).$$

Letting again $n \to \infty$, we obtain

$$f(1+x) \le \max(1, f(x)).$$

If now $a \ne 0$ and $b \in K$, then $f(a) \ne 0$ and

$$\begin{aligned} f(a+b) &= f(a) f(1 + b/a) \\ &\le f(a) \max(1, f(b/a)) = \max(f(a), f(b)). \end{aligned}$$

∎

2.4. Generalized Absolute Values on the Rational Field

The ultrametric absolute values on the rational field $\mathbf{Q}$ have been determined in (2.1). Here, we treat the generalized absolute values.

Theorem. *Any nontrivial generalized absolute value on the rational field $\mathbf{Q}$ is either a power of the usual absolute value or a power of the p-adic absolute value.*

Proof. Take any nontrivial generalized absolute value f and assume that

$$f(x+y) \le C \cdot \max(f(x), f(y)).$$

If $C \le 1$, then f is ultrametric, and we conclude by (2.1). Assume now $C \ge 1$. By induction — regardless of the size and number of addends — we can prove

$$f(a_0 + \cdots + a_r) \le C^r \cdot \max f(a_i).$$

Let us fix an integer $n \ge 2$ and put $A = A_n = \max(f(1), \ldots, f(n)) \ge 1$. Now, any integer $m \ge 2$ can be expanded in base n, say

$$m = \sum_{0 \le i \le r} m_i n^i \quad (0 \le m_i < n,\ m_r \ne 0).$$

Hence

$$\begin{aligned} f(m) &\le C^r \cdot \max f(m_i) f(n^i) \\ &\le C^r A_n \cdot \max f(n)^i = C^r A_n \max(1, f(n)^r). \end{aligned}$$

But $m_r \ne 0$, $n^r \le m$, and thus $r \le \log m / \log n$, so that we can write

$$\begin{aligned} f(m) &\le A_n C^{\log m/\log n} \cdot \max(1, f(n)^{\log m/\log n}), \\ f(m)^{1/\log m} &\le A_n^{1/\log m} C^{1/\log n} \cdot \max(1, f(n)^{1/\log n}). \end{aligned}$$

Let us replace m by m^k (keeping n fixed), so that the left-hand side is unchanged, and let $k \to \infty$, whence $A_n^{1/(k \log m)} \to 1$. We obtain

$$f(m)^{1/\log m} \le C^{1/\log n} \cdot \max(1, f(n)^{1/\log n}).$$

In other words, we have obtained an inequality in which the constant A_n does not appear. We can now replace n by n^k, and since $C^{1/k} \to 1$, we have simply

$$f(m)^{1/\log m} \le \max(1, f(n)^{1/\log n}).$$

First case: *There is an integer $n \ge 2$ with $f(n) \le 1$.*

We can use such an integer n in the inequality just found and deduce

$$f(m) \le 1 \text{ for every integer } m \ge 2.$$

Hence f is an ultrametric absolute value by (2.3). Finally, Ostrowski's theorem (2.1) applies: f is a power of the p-adic absolute value

$$f(x) = |x|_p^\alpha \quad (x \in \mathbf{Q})$$

for some real α (determined by the condition $f(p) = |p|_p^\alpha = (1/p)^\alpha$).

Second case: *We have $f(n) > 1$ for every integer $n \geq 2$.*
The general inequality

$$f(m)^{1/\log m} \leq \max(1, f(n)^{1/\log n})$$

is now simply

$$f(m)^{1/\log m} \leq f(n)^{1/\log n}.$$

Since we can permute the roles of n and m, we must even have

$$f(m)^{1/\log m} = f(n)^{1/\log n}.$$

Hence $f(n)^{1/\log n} = e^{\alpha}$ is independent from n. This leads to

$$f(n) = e^{\alpha \log n} = n^{\alpha}$$

for all integers $n \geq 1$, and with the usual absolute value

$$f(n) = |n|^{\alpha} \quad (n \in \mathbf{Z}).$$

By the multiplicativity property, we also have

$$f(x) = |x|^{\alpha} \quad (x \in \mathbf{Q}).$$

Since $0 < \alpha < \infty$, the map f is a power of the usual absolute value, and the theorem is completely proved. ∎

Comment. The preceding result shows that for a generalized absolute value f on the field $\mathbf{Q}$, the only possibilities are

- *f is trivial,*
- *$|p| < 1$ for some prime p and f is a power of the p-adic absolute value,*
- *$|n| \geq 1$ for all positive integers n and f is a power of the usual Archimedean absolute value.*

Observe that the two nontrivial cases can also be classified according to the value of $|2|$: If $|2| < 1$, f is a power of the 2-adic absolute value; if $|2| = 1$, f is a power of the p-adic absolute value for some odd prime p; if $|2| > 1$, f is a power of the usual Archimedean absolute value.

3. Finite-Dimensional Vector Spaces

3.1. Normed Spaces over $\mathbf{Q}_p$

Let V be a vector space over the field $\mathbf{Q}_p$. A *norm* on V is a mapping

$$\|.\| : V - \{0\} \to \mathbf{R}_{>0}$$

extended by $\|0\| = 0$ and satisfying the following characteristic properties:

$$\|ax\| = |a|\|x\| \quad (a \in \mathbf{Q}_p, x \in V),$$
$$\|x + y\| \leq \max(\|x\|, \|y\|) \quad (x, y \in V).$$

In particular, the norms that we are considering are *ultrametric norms*. A *normed space over* $\mathbf{Q}_p$ is simply a vector space over this field equipped with a norm. A norm defines an invariant (ultra-)metric on the underlying additive group of V. Hence a norm defines a topology on V, which becomes an additive topological group in which scalar multiplications

$$x \mapsto ax : V \to V \quad (a \in \mathbf{Q}_p^\times)$$

are continuous homeomorphisms.

Examples. (1) Let $V = \mathbf{Q}_p$ with norm $\|x\| = c|x|$ where $c > 0$ is a fixed, arbitrarily chosen positive real constant. This example shows that $\{\|v\| : v \in V\}$ can be different from the set of absolute values of scalars, i.e. the absolute values of the elements of the field $\mathbf{Q}_p$. (This is a difference from real and complex normed spaces).

(2) Let $V = \mathbf{Q}_p^n$ for some positive integer n. Then for $x = (x_i)_{1 \leq i \leq n} \in V$ we can put $\|x\|_\infty = \sup_{1 \leq i \leq n} |x_i| = \max_{1 \leq i \leq n} |x_i|$. This defines an ultrametric norm on V.

Two norms $x \mapsto \|x\|$ and $x \mapsto \|x\|'$ are called *equivalent* when they induce (uniformly) equivalent metrics on V, namely when there exist two constants $0 < c \leq C < \infty$ with

$$c\|x\| \leq \|x\|' \leq C\|x\|.$$

This happens precisely when the topologies defined by these two norms are the same (exercise).

Theorem. *Let V be a finite-dimensional vector space over $\mathbf{Q}_p$. Then all norms on V are equivalent.*

PROOF. Let $n = \dim V$ and choose a basis $(e_i)_{1 \leq i \leq n}$ of V. Hence

$$x = (x_i) \mapsto v = \sum x_i e_i = \varphi(x)$$

defines an algebraic isomorphism $\varphi : \mathbf{Q}_p^n \xrightarrow{\sim} V$. On the space $\mathbf{Q}_p^n$ we consider the sup norm given in the above example. We have to show that the isomorphism φ is bicontinuous. But

$$\left\|\sum x_i e_i\right\| \leq \max\|x_i e_i\| = \max|x_i|\|e_i\| \leq \max\|e_i\| \cdot \max|x_i| = C\|x\|_\infty,$$

where $C = \max \|e_i\|$. (Note that the strong triangle inequality is not really necessary here since it would be enough to observe that $\|\sum x_i e_i\| \le \sum \|x_i e_i\| \le \sum \|e_i\| \cdot \max |x_i| = C'\|x\|_\infty$.) This proves that $\|\varphi(x)\| \le C\|x\|_\infty$ and φ is continuous. Finally, we show that φ is an open map. Let $B = B_{\le 1} = \{x \in \mathbf{Q}_p^n : \|x\|_\infty \le 1\}$ be the unit ball in $\mathbf{Q}_p^n$. We have to show that $\varphi(B)$ contains an open ball of positive radius centered at 0 in V. Denote by S_1 the unit sphere

$$S_1 = \{x \in \mathbf{Q}_p^n : \|x\|_\infty = 1\}$$

in $\mathbf{Q}_p^n$. Then S_1 is a closed subset of the compact set $B_{\le 1}$, and hence is compact. This implies that $\varphi(S_1)$ is also compact. This image does not contain the origin of V (remember that φ is bijective). Hence the distance from 0 to $\varphi(S_1)$ is positive, and the minimum is attained for some point $\varphi(x_0)$:

$$x \in S_1 \Longrightarrow \|\varphi(x)\| \ge \|\varphi(x_0)\| = \varepsilon > 0.$$

If $v \in V - \{0\}$ has norm $\|v\| < \varepsilon$, the norm of all multiples λv where $|\lambda| \le 1$ will also satisfy $\|\lambda v\| < \varepsilon$. Hence in particular, if $\|v\| < \varepsilon$, then

$$\lambda \in K,\ |\lambda| \le 1 \Longrightarrow \lambda v \notin \varphi(S_1).$$

Since (e_i) is a basis, we can write

$$v = \sum v_i e_i = \varphi((v_i)).$$

Without loss of generality we may assume that the largest component is the last one:

$$0 \ne |v_n| = \max |v_i| = \|(v_i)\|_\infty.$$

With $\lambda = 1/v_n$ we have $\lambda v = \varphi((v_i/v_n)) = \varphi(w) \in \varphi(S_1)$. The remark made before proves that this scalar λ satisfies $|\lambda| > 1$, so that

$$\|(v_i)\|_\infty = |v_n| = \frac{1}{|\lambda|} < 1.$$

This shows that $v = \varphi((v_i))$ with $\|(v_i)\|_\infty < 1$: $v \in \varphi(B)$, where $B = B_{\le 1}(0, \mathbf{Q}_p^n)$. Consequently,

$$B_{<\varepsilon}(V) \subset \varphi(B). \qquad \blacksquare$$

Corollary 1. *Let V and W be two finite-dimensional normed vector spaces over $\mathbf{Q}_p$ and $\alpha : V \to W$ a linear map. Then α is continuous.* ■

Corollary 2. *Any algebraic isomorphism of a finite-dimensional normed vector space over $\mathbf{Q}_p$ is bicontinuous.* ■

Corollary 3. *Let V be a finite-dimensional vector space over $\mathbf{Q}_p$. A subset $S \subset V$ that is bounded with respect to one norm on V is bounded with respect to any other norm on V.* ∎

Remark. Observe that the proof could be simplified if we knew that all norms of elements of V were absolute values of scalars, namely if $\|V\| = |\mathbf{Q}_p|$. But this equality is in general not satisfied.

3.2. Locally Compact Vector Spaces over $\mathbf{Q}_p$

There are not many compact normed spaces over $\mathbf{Q}_p$. In fact, any nonzero element x of a vector space generates a line, and the norm is an unbounded continuous function on this line because

$$\|\lambda x\| = |\lambda|\|x\| \quad (\lambda \in \mathbf{Q}_p).$$

This shows that the only compact normed space is the trivial normed space $\{0\}$. Let us turn to locally compact normed spaces over $\mathbf{Q}_p$.

Theorem. *If V is a locally compact normed space over $\mathbf{Q}_p$, then its dimension is finite.*

Proof. Let us select a compact neighborhood Ω of 0 in V. Also choose a scalar $a \in \mathbf{Q}_p$ with $0 < |a| < 1$ (for example $a = p$ with $|a| = 1/p$ will do). The *interiors* of the translates $x + a\Omega$ $(x \in V)$ cover the whole space. A fortiori there is a finite covering of the compact set Ω of the form

$$\Omega \subset \bigcup_{\text{finite}} (a_i + a\Omega) \quad (\text{for some } a_i \in V).$$

Consider the finite-dimensional subspace L generated by the elements a_i. By (3.1), this finite-dimensional subspace is isomorphic to a normed space $\mathbf{Q}_p^d$, and hence is complete. Consequently, this subspace L is *closed*, and in the Hausdorff quotient V/L (I.3.3) the image A of the set Ω is a compact neighborhood of 0 and satisfies

$$A \subset aA \quad (\text{or } a^{-1}A \subset A),$$

whence $a^{-n}A \subset A$ by induction. Since $|a^{-n}| \to \infty$, we see that

$$A \subset V/L \subset \bigcup_{n \geq 1} a^{-n}A \subset A.$$

In particular V/L is compact: $V/L = 0$, $V = L$ is a finite-dimensional space. ∎

Corollary. *In a locally compact normed vector space over $\mathbf{Q}_p$, the compact subsets are the closed bounded sets.*

PROOF. The compact subsets of any metric space are closed and bounded (by continuity of the distance function). Conversely, if V is a locally compact normed vector space over $\mathbf{Q}_p$, it has finite dimension and its norm is equivalent to the sup norm of this space (3.1). But in $\mathbf{Q}_p^n$ any bounded set is contained in a (compact) product of balls of $\mathbf{Q}_p$. Hence the closed bounded sets are compact subsets of $\mathbf{Q}_p^n$. ■

3.3. Uniqueness of Extension of Absolute Values

Let K be a finite (hence algebraic) extension of the field $\mathbf{Q}_p$. We can consider K as a finite-dimensional vector space over $\mathbf{Q}_p$. Each absolute value on K that extends the p-adic absolute value of $\mathbf{Q}_p$ is a norm on this vector space, and we can apply the results of (3.1).

Proposition. *There is at most one absolute value on K that extends the p-adic absolute value of $\mathbf{Q}_p$.*

PROOF. Let $|\,.\,|$ and $|\,.\,|'$ be two absolute values on K that extend the absolute value of $\mathbf{Q}_p$. These two norms must be equivalent, and there exist constants $0 < c \leq C < \infty$ such that

$$c|x| \leq |x|' \leq C|x| \quad (x \in K).$$

Replace x by x^n in the preceding inequalities:

$$c|x^n| \leq |x^n|' \leq C|x^n|.$$

Since $|\,.\,|$ and $|\,.\,|'$ are absolute values, they are multiplicative, and the preceding inequality is simply

$$c|x|^n \leq |x|'^n \leq C|x|^n,$$

or

$$c^{1/n}|x| \leq |x|' \leq C^{1/n}|x|.$$

Letting $n \to \infty$, we have $c^{1/n} \to 1$ and $C^{1/n} \to 1$. This proves $|x| = |x|'$. ■

Application. Let K be a Galois extension of $\mathbf{Q}_p$ and *assume that the p-adic absolute value of $\mathbf{Q}_p$ extends to K*. Then for each automorphism σ of $K/\mathbf{Q}_p$ we can consider the absolute value $|x|' = |\sigma x|$. By the preceding proposition, this absolute value must coincide with the original one. Let $G = \mathrm{Gal}\,(K/\mathbf{Q}_p)$ and for each $x \in K$, consider the element

$$N(x) = \prod_{\sigma \in G} \sigma x \in \mathbf{Q}_p.$$

We must have

$$|N(x)| = \left|\prod_{\sigma\in G} \sigma x\right| = \prod_{\sigma\in G} |\sigma x| = |x|^{\#(G)}.$$

Hence with $d = \#(G) = [K : \mathbf{Q}_p] = \dim_{\mathbf{Q}_p}(K)$,

$$|x| = |N(x)|^{1/d}.$$

Since $N(x) \in \mathbf{Q}_p$, this formula gives an explicit expression for the extension of the absolute value of $\mathbf{Q}_p$ (provided that one exists!). This observation can be used to prove the *existence* of an extension of the absolute value of $\mathbf{Q}_p$.

3.4. Existence of Extension of Absolute Values

Let again K be a finite extension of degree d of the field $\mathbf{Q}_p$. The relative *norm* (as defined in field theory, not to be confused with a vector space norm!) is a multiplicative homomorphism (3.3)

$$N = N_{K/\mathbf{Q}_p} : K^\times \to \mathbf{Q}_p^\times, \quad x \mapsto N(x),$$

which coincides with the dth power on $\mathbf{Q}_p^\times$. It can be defined either by embedding K in a Galois extension L and taking a product over the d distinct embeddings $K \hookrightarrow L$ or by using the determinant of the $\mathbf{Q}_p$-linear map $y \mapsto xy$ of the $\mathbf{Q}_p$-vector space K.

Theorem. *Let K be a finite extension of degree d of the field $\mathbf{Q}_p$ of p-adic numbers. For each $x \in K$, let ℓ_x denote the $\mathbf{Q}_p$-linear operator $y \mapsto xy$ in K. Then*

$$f(x) = |N(x)|^{1/d} = |\det \ell_x|^{1/d}$$

defines an absolute value on K that extends the p-adic one. This is the unique absolute value on K having this property.

Proof. If $a \in \mathbf{Q}_p$, it is obvious that $N(a) = a^d$ whence $|N(a)|^{1/d} = |a|$, and the proposed formula is an extension of the p-adic value. The multiplicativity $f(xy) = f(x) \cdot f(y)$ is a consequence of the multiplicativity of the determinant (or of the norm). We still have to check the ultrametric inequality. For this crucial point we use the local compactness of K. Let us choose any norm $x \mapsto \|x\|$ on K with $\|K\| = |\mathbf{Q}_p|$. For example, pick a basis $e_1, \ldots, e_d$ of K over $\mathbf{Q}_p$ and use the sup norm on components in this basis. Since the continuous function f does not vanish on the compact set $\|x\| = 1$, it is both bounded above and below on this set, say

$$0 < \varepsilon \le f(x) \le A < \infty \quad (\|x\| = 1).$$

For $x \in K^\times$ choose $\lambda \in \mathbf{Q}_p$ with $\|x\| = |\lambda|$. Hence the vector x/λ has norm 1,

$$\varepsilon \le f(x/\lambda) \le A \quad (x \neq 0),$$

and since $f(x/\lambda) = f(x)/|\lambda|$,

$$\varepsilon|\lambda| \le f(x) \le A|\lambda| \quad (x \neq 0),$$
$$\varepsilon\|x\| \le f(x) \le A\|x\| \quad (x \neq 0).$$

Thus with $a = \varepsilon^{-1}$ we have both $\|x\| \le af(x)$ and $f(x) \le A\|x\|$. Suppose now $f(x) \le 1$ (hence $\|x\| \le a$). We infer

$$\begin{aligned} f(1+x) &\le A\|1+x\| \le A \max(\|1\|, \|x\|) \\ &\le A \max(\|1\|, a) = C = C \max(f(1), f(x)). \end{aligned}$$

If more generally $f(y) \ge f(x)$, we can divide by y and apply the preceding inequality to x/y, since $f(x/y) = f(x)/f(y) \le 1$. Finally, multiplying both sides by $f(y)$, we obtain the general inequality

$$f(x+y) \le C \max(f(x), f(y)).$$

This proves that f is a generalized absolute value. Since f extends the p-adic absolute value, it is bounded on $\mathbf{N} \subset \mathbf{Q}_p \subset K$ and is an ultrametric absolute value by (2.3).

The uniqueness of the extension has already been proved in (3.3). ■

3.5. *Locally Compact Ultrametric Fields*

In *locally compact ultrametric fields* K, we shall use adapted notation

$$R = B_{\le 1} \supset P = B_{<1}$$

instead of

$$A = B_{\le 1} \supset M = B_{<1},$$

which will still be used in the general — not necessarily locally compact — case.

We are going to prove the following general result.

Theorem. *Let K be a field equipped with a nontrivial ultrametric absolute value and consider the corresponding (ultra-)metric space. Then K is locally compact precisely when the following three conditions are satisfied:*

(1) *K is a complete metric space.*
(2) *The residue field $k = R/P$ is finite.*
(3) *$|K^\times|$ is a discrete subgroup of $\mathbf{R}_{>0}$,*
hence of the form $\theta^{\mathbf{Z}}$ for some $0 < \theta < 1$.

PROOF. Assume first that the field K is locally compact. Hence there is a compact neighborhood of 0 in K. This neighborhood contains a ball $B_{\leq\varepsilon}(0)$, where $\varepsilon > 0$. This ball $B_{\leq\varepsilon}$ is compact. Using dilatations, we see that all balls $B_{\leq r}(0)$ of K are compact. Any Cauchy sequence in K is bounded, hence contained in a compact ball: It must converge in K. This shows that K is complete (recall more generally that every locally compact topological group is complete (I.3.2)). Now the residue field parametrizes the open unit balls contained in the unit ball $B_{\leq 1}(0)$. If this last one is compact, the preceding partition in open sets must be finite, which proves (2). Finally, since the open unit ball $B_{<1}(0)$ is closed in the compact ball $B_{\leq 1}(0)$, the continuous function $x \mapsto |x|$ must attain a maximal value over the compact set $B_{<1}(0)$. Call $\theta < 1$ this maximal absolute value. The only possible nonzero absolute values are now the integral powers of θ. Indeed, a multiplicative subgroup of $\mathbf{R}_{>0}$ is either discrete or dense (I.3.4). (Alternatively, one could use the last property of ultrametric distances mentioned in (1.1) for the compact sets $B_{\leq s} - B_{<r}, 0 < r < s$.)

Conversely, assume that the three conditions are satisfied and choose an element $\pi \in K$ with largest possible absolute value less than 1: $\pi \in P \subset R \Longrightarrow \pi R \subset P$. The reverse inclusion also holds:

$$x \in P \Longrightarrow |x| \leq |\pi| \Longrightarrow x = \pi \cdot x/\pi \quad (x/\pi \in R) \Longrightarrow x \in \pi R.$$

This proves that $P = (\pi) = \pi R$ is principal. By the representation theorem (1.4), the complete ring R is topologically isomorphic to the projective limit $\widehat{R} = \varprojlim R/\pi^n R$ of the finite rings $R/\pi^n R$:

$$R = B_{\leq 1}(0) \text{ is isomorphic to } \widehat{R} \text{ compact.}$$

The field K is locally compact, since it has a compact neighborhood of 0. ∎

4. Structure of p-adic Fields

4.1. *Degree and Residue Degree*

Let K be a finite extension of the field $\mathbf{Q}_p$ of p-adic numbers. Hence K is locally compact and complete. Let us choose an element π of maximal absolute value smaller than 1, say $0 < |\pi| = \theta < 1$, and come back to the usual notation for the ring

$$R = \{x \in K : |x| \leq 1\}$$

and its maximal ideal $P = \pi R$. The residue field $k = R/P$ is finite, hence a finite extension of $\mathbf{F}_p = \mathbf{Z}_p/p\mathbf{Z}_p$. If $f = [k : \mathbf{F}_p] = \dim_{\mathbf{F}_p}(k)$, then

$$k \cong \mathbf{F}_q, \quad q = \#(k) = (\#(\mathbf{F}_p))^f = p^f,$$

since there is — up to isomorphism — only one finite field having q elements. Since the integer p belongs to P, we have

$$1/p = |p| = \theta^e, \quad |\pi| = |p|^{1/e}$$

for some integer $e \geq 1$.

Definitions. *The* residue degree *of the finite extension K of $\mathbf{Q}_p$ is the integer*

$$f = [k : \mathbf{F}_p] = \dim_{\mathbf{F}_p}(k).$$

The ramification index *of K over $\mathbf{Q}_p$ is the integer*

$$e = [\,|K^\times| : |\mathbf{Q}_p^\times|\,] = [\,|K^\times| : p^{\mathbf{Z}}\,] = \#(|K^\times|/p^{\mathbf{Z}}).$$

Warning. I hope that the degree f will not occur next to a polynomial $f(X)$ or a function f, or if it does, let me rely on the reader to distinguish them (using P for a polynomial could similarly lead to a confusion with the maximal ideal $P = \pi R$ in a finite extension K of $\mathbf{Q}_p$, and here π is not 3.14159...!) In the same vein, k will usually denote a residue field and here, we try to avoid its use as a summation index.

Let a_1 and $a_2 \in \mathbf{Q}_p^\times$, x_1 and $x_2 \in K^\times$ be such that

$$|a_1 x_1| = |a_2 x_2| \quad (\neq 0).$$

Then $|x_1| = |a_2/a_1| \cdot |x_2| \in p^{\mathbf{Z}}\,|x_2|$, and the absolute values of x_1 and x_2 belong to the same coset mod $p^{\mathbf{Z}}$. Consequently, in a finite sum

$$\sum a_i x_i \quad (a_i \in \mathbf{Q}_p^\times,\ x_i \in K^\times)$$

of nonvanishing terms, if the $|x_i|$ belong to distinct cosets mod $p^{\mathbf{Z}}$, we cannot have a competition of absolute values, and necessarily $\sum a_i x_i \neq 0$. This argument shows that $n = [K : \mathbf{Q}_p] = \dim(K) \geq e$. One can also see directly that $n \geq f$ (exercise!). Let us prove that $n \geq ef$ (we even prove $n = ef$ below).

Proposition. *In the situation described in this section, we have $ef \leq n$.*

PROOF. Let us choose a family $(s_i)_{1 \leq i \leq f}$ in R such that the images $\tilde{s}_i \in k$ make up a basis over the prime field $\mathbf{F}_p$. I claim that the elements

$$(s_i \pi^j)_{1 \leq i \leq f,\, 0 \leq j < e}$$

are independent over $\mathbf{Q}_p$. Consider indeed a nontrivial linear combination

$$\sum c_{ij} s_i \pi^j = \sum x_j \pi^j \quad (c_{ij} \in K),$$

where $x_j = \sum_i c_{ij} s_i$. Then for each j there is an index $\ell = \ell(j)$ such that

$$|c_{\ell j}| \geq |c_{ij}| \quad \text{for all } i,$$

and $x_j/c_{\ell j} = \sum_i (c_{ij}/c_{\ell j}) s_i = \sum_i \gamma_i s_i$ is a nontrivial linear combination with coefficients in R (and $\gamma_\ell = 1$). Consider this relation mod P: Define $\widetilde{\gamma_i} = \gamma_i \bmod P$. Since $(\widetilde{s_i})_i$ is a basis of the residue field $k = R/P$ considered as a vector space over its prime field, we have

$$0 \neq \sum_i \widetilde{\gamma_i} \widetilde{s_i} \in R/P$$

simply because $\widetilde{\gamma_\ell} = 1$. Hence

$$\sum_i \gamma_i s_i \notin P, \quad \left|\sum_i \gamma_i s_i\right| = 1$$

and $|x_j| = |c_{\ell j}| \in |\mathbf{Q}_p^\times|$ is a power of p. There can be no competition among the absolute values of the distinct terms $x_j \pi^j$, and this proves

$$\sum c_{ij} s_i \pi^j = \sum x_j \pi^j \neq 0.$$

This proves the expected linear independence, and hence the inequality stated in the proposition. ∎

Theorem. *For each finite extension K of $\mathbf{Q}_p$, we have*

$$ef = [K : \mathbf{Q}_p] = n.$$

Proof. By the above proposition it is enough to prove the existence of a set of generators of K over $\mathbf{Q}_p$ containing ef elements. We shall show that the family

$$(s_i \pi^j)_{1 \leq i \leq f,\, 0 \leq j < e}$$

of the Proposition generates the $\mathbf{Q}_p$-vector space K. For this purpose we use the representation theorem (1.4) for the complete field K and the element $\xi = p \in P \subset R$. In this case R/pR is finite with representatives

$$S = \left\{ \sum_{1 \leq i \leq f, 0 \leq j < e} c_{ij} s_i \pi^j : 0 \leq c_{ij} \leq p - 1 \right\}.$$

Hence one can write any element $x \in R$ as a series

$$x = \sum_{\ell \geq 0} c_\ell p^\ell \quad (c_\ell \in S).$$

If we write explicit expressions for the coefficients

$$c_\ell = \sum_{1 \leq i \leq f, 0 \leq j < e} c_{ij\ell} s_i \pi^j \in S,$$

we obtain

$$x = \sum_{\ell \geq 0} \sum_{1 \leq i \leq f, 0 \leq j < e} c_{ij\ell} s_i \pi^j p^\ell,$$

and if we sum in a different order (only ℓ can take infinitely many values, and $p^\ell \to 0$: The family in question is summable by the Proposition in (1.2)), then

$$x = \sum_{1 \leq i \leq f, 0 \leq j < e} \left(\sum_\ell c_{ij\ell} p^\ell \right) \cdot s_i \pi^j.$$

But $c_{ij} = \sum_\ell c_{ij\ell} p^\ell \in \mathbf{Z}_p$ and $x = \sum_{ij} c_{ij} s_i \pi^j$. This proves that the ef elements $s_i \pi^j$ $(1 \leq i \leq f,\ 0 \leq j \leq p-1)$ constitute a spanning set of the field K considered as a vector space over $\mathbf{Q}_p$. Together with the proposition, this concludes the proof of the theorem. ∎

A finite extension K of $\mathbf{Q}_p$ is said to be

unramified when $e = 1$, i.e., when $[K : \mathbf{Q}_p] = f$,
totally ramified when $f = 1$, i.e., when $[K : \mathbf{Q}_p] = e$,
tamely ramified when p does not divide e,
wildly ramified when e is a power of p.

In other words, an extension $K/\mathbf{Q}_p$ is

unramified when p is a generator of the maximal ideal $P \subset R$,
totally ramified when the residue field does not grow.

Comment. Let us come back to the analogy between p-adic numbers and functions of a complex variable already mentioned in (I.1.4) and (I.5.1), since it is also responsible for the preceding terminology. Let us explain this in its simplest form.

Let $\xi \neq 0$ be a meromorphic function defined in a neighborhood of 0 in $\mathbf{C}$. It is known that there is a representation

$$\xi(z) = \sum_{n \geq m} a_n z^n \quad (a_m \neq 0)$$

valid in a punctured disc $0 < |z| < \varepsilon$. The smallest index $m \in \mathbf{Z}$ such that $a_m \neq 0$ is the *order* of ξ at the origin. This integer is positive when ξ vanishes and is negative when ξ has a pole at the origin. In this way, we see an analogy between the field L of meromorphic functions defined in a (variable) neighborhood of the origin in $\mathbf{C}$ and the p-adic field $\mathbf{Q}_p$ consisting of the formal expansions $x = \sum_{n \geq m} a_n p^n$ $(m \in \mathbf{Z})$. The functions that are holomorphic at the origin make a maximal subring L_0 of L comparable to $\mathbf{Z}_p$ in $\mathbf{Q}_p$. The *local* construction of the field L is also a justification for calling $\mathbf{Q}_p$ a *local field*.

Now take an integer $e > 1$ and consider the change of variable

$$u \mapsto z = u^e : \mathbf{C} \to \mathbf{C}.$$

This is a canonical example of a *ramified covering of degree e* at the origin, in a topological sense: The inverse image of any $z \neq 0$ consists of e distinct preimages, while $u = 0$ is the only preimage of $z = 0$. If $\xi = \sum_{n \geq m} a_n z^n$ $(0 < |z| < \varepsilon)$ is as before a meromorphic function in a neighborhood of the origin, we can make the change of variable $z = u^e$ and obtain a new expansion

$$\eta(u) = \xi(u^e) = \sum_{n \geq m} a_n u^{en}.$$

In this way, the field L is embedded in the field L' consisting of convergent Laurent series in the variable u. There is no function $\sqrt[e]{z}$ defined in a neighborhood of $z = 0$ in $\mathbf{C}$, so that the field $L' = L(z^{1/e})$ is a proper extension of the field of convergent Laurent series L in the variable z. This extension L' is totally ramified over L, with degree e: It is obviously comparable to the extension $\mathbf{Q}_p(\pi)$ of $\mathbf{Q}_p$ if $\pi = p^{1/e}$. Observe that with meromorphic functions it is traditional to work with the order-of-vanishing function $\mathrm{ord}_0(\xi) = m$, instead of a corresponding ultrametric absolute value $|\xi|_0 = \theta^m$ (for a choice $0 < \theta = |z|_0 < 1$; there is no canonical choice for θ here).

The rational field $\mathbf{Q}$ can similarly be compared to the field of rational functions $\mathbf{C}(z)$, the completions $\mathbf{Q}_p$ (letting now the prime p vary) corresponding to the fields of meromorphic functions near a variable point $a \in \mathbf{C}$ instead of the origin.

4.2. *Totally Ramified Extensions*

Let us recall the following well-known irreducibility result stated over $\mathbf{Z}_p$ rather than over $\mathbf{Z}$.

Theorem 1 (Eisenstein). *Let $f(X) \in \mathbf{Z}_p[X]$ be a monic polynomial of degree $n \geq 1$ with $f(X) \equiv X^n \bmod p$, $f(0) \not\equiv 0 \bmod p^2$. In other words,*

$$f(X) = X^n + a_{n-1}X^{n-1} + \cdots + a_0,$$
$$\mathrm{ord}\,(a_i) \geq 1 \;\; (0 \leq i \leq n-1), \qquad \mathrm{ord}\,(a_0) = 1.$$

Then f is irreducible in the rings $\mathbf{Z}_p[X]$ and $\mathbf{Q}_p[X]$.

Proof. Take a factorization $f = g \cdot h$ in $\mathbf{Z}_p[X]$ — or in $\mathbf{Q}_p[X]$; this is the same by an elementary lemma attributed to Gauss — say

$$g = b_\ell X^\ell + \cdots + b_0, \quad h = c_m X^m + \cdots + c_0.$$

Hence

$$\ell + m = n, \quad b_\ell c_m = 1, \quad b_0 c_0 = a_0.$$

Since a_0 is not divisible by p^2, p can divide only one of the two coefficients b_0 and c_0. Without loss of generality we can assume that p divides c_0 but p does not divide b_0. Consider now all these polynomials mod p. By assumption $\widetilde{f} = X^n$ is a monomial, so that its factorization $\widetilde{f} = \widetilde{g} \cdot \widetilde{h}$ must be a product of monomials and $\widetilde{h} = \widetilde{c_0}$ is a constant. Considering that $b_\ell c_m = 1$, the only possibility now is $m = 0$ and a trivial factorization. ∎

The preceding argument mod p can be made directly on the coefficients. Let $r \geq 1$ be the smallest power of X in h having a coefficient not divisible by p:

$$p \text{ does not divide } c_r \text{ but } p \text{ divides } c_{r-1}, c_{r-2}, \ldots, c_0.$$

The coefficient of X^r in the product of g and h is

$$a_r = b_0 c_r + b_1 c_{r-1} + b_2 c_{r-2} + \cdots = b_0 c_r + p(\cdots).$$

Since $b_0 c_r$ is not divisible by p, the preceding equality shows that p does not divide a_r either. By assumption, this shows that $r = n$. Summing up,

$$n = m + \ell \geq m \geq r = n$$

implies $m = n$ and $\ell = 0$. The factorization $g \cdot h$ of f is necessarily trivial.

The same proof shows the following more general result.

Let A be a factorial ring with fraction field K, and π a prime of A. Any polynomial

$$f = a_n X^n + a_{n-1} X^{n-1} + \cdots + a_0 \in A[X] \quad (\textit{of degree } n \geq 1)$$

with a_n not divisible by π, a_i divisible by π for $0 \leq i \leq n-1$, a_0 not divisible by π^2, is irreducible in the rings $K[X]$ and $A[X]$.

Definition. *A monic polynomial $f(X) \in \mathbf{Z}_p[X]$ of degree $n \geq 1$ satisfying the conditions of the theorem, namely*

$$f(X) \equiv X^n \bmod p, \quad f(0) \not\equiv 0 \bmod p^2,$$

is called an Eisenstein polynomial.

Theorem 2. *Let K be a finite, totally ramified extension of $\mathbf{Q}_p$. Then K is generated by a root of an Eisenstein polynomial.*

Proof. The maximal ideal P of the subring $R = B_{\leq 1}$ of K is principal and generated by an element π with $|\pi|^e = |p|$. Since $n = [K : \mathbf{Q}_p] = e$ by assumption, the linearly independent powers $(\pi^i)_{0 \leq i < e}$ generate K and $K = \mathbf{Q}_p[\pi]$. The irreducible polynomial of this element can be factored (in a Galois extension of $\mathbf{Q}_p$ containing K) as

$$f(X) = \prod_\sigma (X - \pi^\sigma) = X^e + \sum_{0 < i < e} a_i X^i \pm \prod_\sigma \pi^\sigma.$$

The constant term has absolute value $|\Pi_\sigma \pi^\sigma| = |\pi|^e = |p|$ (by (3.3) all automorphisms σ are *isometric*), whereas the intermediate coefficients a_i satisfy $|a_i| < 1$

(each is divisible by one π^σ at least, and $a_i \in \mathbf{Z}_p$). Hence these intermediate coefficients are in $p\mathbf{Z}_p$ as required: f is an Eisenstein polynomial. ■

Examples. (1) In the field $\mathbf{Q}_2$, -1 is not a square (I.6.6), and we can construct the quadratic extension $K = \mathbf{Q}_2(i) = \mathbf{Q}_2[X]/(X^2+1)$. Since

$$(i+1)^2 = i^2 + 2i + 1 = 2i,$$

the element $i+1$ is a square root of $2i$. With the (unique) extension of the 2-adic absolute value we have

$$|i+1|^2 = |2i| = |2| = \tfrac{1}{2}, \quad |i+1| = \sqrt{\tfrac{1}{2}},$$

so that $i+1$ is a generator of the maximal ideal P of the maximal subring R of the field K: $P = (i+1)R$. The quadratic extension K is totally ramified of index $e = 2$, hence *wildly ramified.* Let $x = i+1$. Then $x - 1 = i$ and $(x-1)^2 = -1$ shows that x is a root of the polynomial

$$X^2 - 2X + 2 = (X-1)^2 + 1.$$

This is an Eisenstein polynomial (relative to the prime 2), and $K = \mathbf{Q}_2(i)$ is also obtained as a splitting field of this Eisenstein polynomial.

(2) For $p \neq 2$ let us add a primitive pth root of unity to $\mathbf{Q}_p$. In other words, we are adding to $\mathbf{Q}_p$ a root of $\zeta^p - 1 = 0$ with $\zeta \neq 1$. Hence ζ is a root of

$$\Phi_p(X) = (X^p - 1)/(X-1) = X^{p-1} + \cdots + X + 1.$$

This is the *pth cyclotomic polynomial*: It is irreducible, since the change of variable $X - 1 = Y$ produces

$$\Phi_p(X) = [(Y+1)^p - 1]/Y = Y^{p-1} + p(\cdots) + p,$$

an Eisenstein polynomial. Hence we obtain an extension of $\mathbf{Q}_p$ of degree $p-1$ prime to p. We shall prove that it is totally ramified. Since the powers ζ^i are also roots of the same equation when i is not a multiple of p, the powers ζ^i $(1 \le i \le p-1)$ form a complete set of conjugates of ζ, and

$$\Phi_p(X) = \prod_{1 \le i \le p-1} (X - \zeta^i).$$

Obviously, $\Phi_p(1) = 1 + \cdots + 1 = p$, so that

$$p = \Phi_p(1) = \prod_{1 \le i \le p-1} (1 - \zeta^i).$$

But all absolute values $|1 - \zeta^i|$ are equal by (3.3), since these elements are conjugate. The preceding inequality leads to

$$|p| = \prod_{1 \le i \le p-1} |1 - \zeta^i| = |1 - \zeta|^{p-1}.$$

This proves that $\pi = 1 - \zeta$ is a generator of P in R: The extension $K = \mathbf{Q}_p(\zeta)$ is ramified with degree $n = e = p - 1$, hence totally and tamely ramified.

In the course of the preceding deduction we have used the uniqueness of extension of valuations again. However, in the present context, it is obvious that

$$1 - \zeta^i = (1 - \zeta)(1 + \cdots + \zeta^{i-1})$$

implies $|1 - \zeta^i| \le |1 - \zeta|$. But the roles of ζ and ζ^i may be reversed: ζ^i is also a generator of the cyclic group μ_p of order p when $1 \le i \le p - 1$, so that ζ is a power $(\zeta^i)^j$ of ζ^i (take j such that $ij \equiv 1 \bmod p$). This furnishes the equality $|1 - \zeta^i| = |1 - \zeta|$. By the way, this proves that

$$\left|1 + \zeta + \cdots + \zeta^{i-1}\right| = \left|\frac{1 - \zeta^i}{1 - \zeta}\right| = 1$$

and $1 + \zeta + \cdots + \zeta^{i-1}$ are *units* of the maximal subring $R \subset K = \mathbf{Q}_p(\zeta)$. These are the so-called *cyclotomic units* of K. Since $\zeta \equiv 1 \bmod P$, we have $1 + \zeta + \cdots + \zeta^{i-1} \equiv i \bmod P$.

4.3. Roots of Unity and Unramified Extensions

Let K be a (commutative) field of characteristic 0 and let $\mu(K)$ be the multiplicative group consisting of the roots of unity in K. Since every element of this group has finite order (by definition), we can apply the Sylow decomposition theorem (or the Chinese remainder theorem) and write a direct product $\mu(K) = \mu_{p^\infty}(K) \cdot \mu_{(p)}(K)$, where elements in $\mu_{p^\infty}(K)$ have a pth power order and elements in $\mu_{(p)}(K)$ have order prime to p. We shall prove that when K is a finite extension of $\mathbf{Q}_p$, the group $\mu_{(p)}(K)$ is finite, and compute its order. (In the next section we show the finiteness of the group $\mu_{p^\infty}(K)$.)

In any valued field, all roots of unity are on the unit sphere:

$$\zeta^m = 1 \Longrightarrow |\zeta|^m = |\zeta^m| = |1| = 1 \Longrightarrow |\zeta| = 1.$$

In the case of an ultrametric extension of $\mathbf{Q}_p$,

$$K \supset A = B_{\le 1} \supset M = B_{<1},$$

we see that $\mu = \mu(K) \subset A^\times \subset K^\times$. By reduction mod M,

$$\varepsilon : A \to A/M = k$$

we obtain $\varepsilon(\mu) \subset k^\times$. To explain the effect of reduction mod M on roots of unity let us give a lemma.

Proposition 1. *Let K be any ultrametric extension of $\mathbf{Q}_p$. Then*

$$\mu_{p^\infty}(K) = \mu(K) \cap (1 + M).$$

PROOF. First, if $\zeta \in \mu(K)$ has order a power of p, denote by $\tilde{\zeta} = \varepsilon(\zeta) \in k$ its reduction. Then

$$\zeta^{p^f} = 1 \Longrightarrow \tilde{\zeta}^{p^f} = \tilde{1} \in k \Longrightarrow \tilde{\zeta} = \tilde{1} \Longrightarrow \zeta \in 1 + M,$$

since the field k has characteristic p. Conversely, if $\zeta \in 1 + M$ has order $n > 1$, write $\zeta = 1 + \xi$ with $0 \neq |\xi| < 1$. Then

$$1 = (1 + \xi)^n = 1 + n\xi + \cdots + \xi^n = 1 + \xi(n + \xi\alpha)$$

implies $n + \xi\alpha = 0$, and

$$|n| = |\xi\alpha| \leq |\xi| < 1$$

implies $p \mid n$. If $n \neq p$, we can replace ζ by ζ^p, which has order $n/p > 1$, and iterate the procedure. Eventually, we see that n is a power of p. ■

Corollary 1. *The restriction of the reduction map ε to $\mu(K)$ has kernel $\mu_{p^\infty}(K)$. It is injective on $\mu_{(p)}(K)$: The distance between two distinct roots of unity of order prime to p is 1.* ■

Corollary 2. *If the residue degree $f = f(K/\mathbf{Q}_p)$ is finite, then the group $\mu_{(p)}(K)$ of roots of unity having order prime to p in K is finite and*

$$\#\left(\mu_{(p)}(K)\right) \leq p^f - 1.$$ ■

When $K/\mathbf{Q}_p$ is finite, the next proposition shows that the order of $\mu_{(p)}(K)$ is exactly $p^f - 1$.

Proposition 2. *Assume that the extension K of $\mathbf{Q}_p$ is* complete *with residue field k algebraic over $\mathbf{F}_p$. Then we have a split exact sequence*

$$(1) \to \mu_{p^\infty}(K) \to \mu(K) \to k^\times \to (1).$$

If the residue field is finite, say $f = [k : \mathbf{F}_p] < \infty$, then the cyclic group $\mu_{(p)}(K)$ has order $p^f - 1$.

PROOF. Let $\varepsilon : \mu \to k^\times$ be the group homomorphism obtained by restriction of the reduction (ring) homomorphism $A \to A/M$. It will be enough to show that ε induces an isomorphism $\mu_{(p)}(K) \cong k^\times$. By the preceding proposition, the reduction map induces an isomorphism of $\mu_{(p)}(K)$ into $k^\times$. We have to prove that it is surjective. Let $\alpha \in k^\times$ and replace k by the finite field $\mathbf{F}_p(\alpha) \cong \mathbf{F}_q$ so that

α is a root of unity of order prime to p, dividing $m = q - 1 = \#(k^\times)$. Choose an element $a \in A$ in the coset α (mod M) and consider the solutions x of the following problem:

$$X^m - 1 = 0 \text{ with } x \equiv a \pmod{M} \quad (\text{i.e., } \varepsilon(x) = \alpha).$$

Since m is prime to p, and K is complete, Hensel's lemma (1.4) can be applied, and this furnishes an element x in $K^\times$ with $x^m = 1$; hence $x \in \mu_{(p)}(K)$ and $\varepsilon(x) = \varepsilon(a) = \alpha$.

This proves that — when the residue field k is algebraic — the restriction of the reduction mod M is an isomorphism $\mu_{(p)}(K) \xrightarrow{\sim} k^\times$. ∎

Application. Let K be a locally compact (i.e., finite) extension of $\mathbf{Q}_p$ and adopt the usual notation corresponding to this case:

$$\begin{aligned} K \supset R = B_{\leq 1}(K) \supset P = \pi R, \\ k = R/P, \quad f = [k : \mathbf{F}_p], \quad q = p^f = \#(k). \end{aligned}$$

Then we have canonical isomorphisms

$$\begin{aligned} \mu_{(p)}(K) \times (1 + P) &\xrightarrow{\sim} R^\times \quad \text{(multiplication)}, \\ \mu_{(p)}(K) &\xrightarrow{\sim} k^\times \quad \text{(reduction mod } P). \end{aligned}$$

With a choice of π, we also have an isomorphism

$$\pi^{\mathbf{Z}} \times \mu_{(p)}(K) \times (1 + P) \xrightarrow{\sim} K^\times \quad \text{(multiplication)}.$$

We infer that if p^s is the highest power of p such that K has a root of unity of order p^s, then

$$\mu_{p^\infty}(K) = \mu(K) \cap (1 + P) \text{ has order } p^s.$$

The p-adic logarithm will furnish a way of analyzing more precisely the structure of the abelian group $1 + P$ (cf. V.4.5).

It is useful to *relativize* the definitions of ramification index and residue degree as follows. Let $K \subset L$ be two finite extensions of the p-adic field $\mathbf{Q}_p$ and denote by R the maximal subring of K, P its maximal ideal, $k = R/P$ (residue field of K) as before. Introduce the maximal subring R_L of L, the maximal ideal P_L of R_L, and $k_L = R_L/P_L$ (residue field of L). We can define

$$\begin{aligned} e &= e(L/K) = [\,|L^\times| : |K^\times|\,], \\ f &= f(L/K) = [k_L : k] = \dim_k k_L, \\ n &= [L : K] = \dim_K(L). \end{aligned}$$

Then $n = ef$ simply because this relation holds for both index and degree over $\mathbf{Q}_p$:

$$n' = e'f' \quad (\text{where } n' = [L : \mathbf{Q}_p], \ldots)$$
$$n'' = e''f'' \quad (\text{where } n'' = [K : \mathbf{Q}_p], \ldots),$$

and we can divide these relations,

$$n = \frac{n'}{n''} = \frac{e'}{e''} \cdot \frac{f'}{f''}.$$

Theorem. *Let $K \subset L$ be two finite extensions of $\mathbf{Q}_p$. Then there is a unique maximal intermediate extension $K \subset K_{ur} \subset L$ that is unramified over K.*

Proof. If the residue field k_L of L has order q_L, we have seen that $L^\times$ contains a cyclic subgroup $\mu_{(p)}(L)$ of order $q_L - 1$ consisting of the roots of unity having order prime to p in L. More precisely, if $q = \#(k)$ and $f = f(L/K)$ is the residue degree of the extension, then $q_L = q^f$. The unramified extensions of K contained in L correspond one-to-one to the extensions of $k = \mathbf{F}_q$ in k_L. This correspondance is order-preserving, hence the uniqueness of a maximal unramified extension. Explicitly,

$$K_{\text{ur}} = K(\mu_{(p)}(L)) = K(\mu_{q_L - 1}) \subset L. \qquad \blacksquare$$

4.4. Ramification and Roots of Unity

Let us keep the notation introduced in the preceding section for the group of roots of unity in the extension K of $\mathbf{Q}_p$.

Theorem. *Let ζ be a root of unity in the field K having order p^t ($t \geq 1$). Then $|\zeta - 1| = |p|^{1/\varphi(p^t)} < 1$, where $\varphi(p^t) = p^{t-1}(p-1)$ denotes the Euler φ-function.*

Proof. (1) Case $t = 1$, the root ζ has order p. In this case $\zeta^p = 1$ but $\zeta \neq 1$ and $\zeta = 1 + \xi$ ($|\xi| < 1$) is a root of the polynomial $(X^p - 1)/(X - 1)$:

$$0 = \frac{(1+\xi)^p - 1}{\xi} = \frac{1}{\xi}(p\xi + p\xi^2 x + \xi^p) \quad (|x| \leq 1).$$

Hence

$$p(1 + \xi x) + \xi^{p-1} = 0,$$

and since $|\xi| < 1$ and $|x| \leq 1$, we have $|1 + px| = 1$,

$$|\xi^{p-1}| = |-p(1 + px)| = |p|,$$
$$|\zeta - 1| = |\xi| = |p|^{1/(p-1)} < 1.$$

Since this absolute value occurs frequently in p-adic analysis, let us introduce a special notation for it:

$$\tfrac{1}{p} = |p| \leq |p|^{\frac{1}{p-1}} := r_p < 1,$$

so that

$$r_2 = \tfrac{1}{2} = |2|, \quad r_p > \tfrac{1}{p} \quad (p \text{ odd prime}).$$

(2) General case: The order of ζ is precisely p^{t+1} ($t + 1 \geq 1$). Then ζ^{p^t} has order p, and by the special case already treated,

$$|\zeta^{p^t} - 1| = r_p < 1.$$

Let us write $\zeta = 1 + \eta$ with $|\eta| < 1$, so that

$$\zeta^{p^t} - 1 = (1+\eta)^{p^t} - 1 = \eta^{p^t} + p\eta y$$

with $|y| \leq 1$. Since

$$|p\eta y| < |p| \leq r_p = |1 - \zeta^{p^t}|,$$

we see that $r_p = |1 - \zeta^{p^t}| = |\eta^{p^t}|$ and finally $|\eta| = r_p^{1/p^t}$ as expected. ■

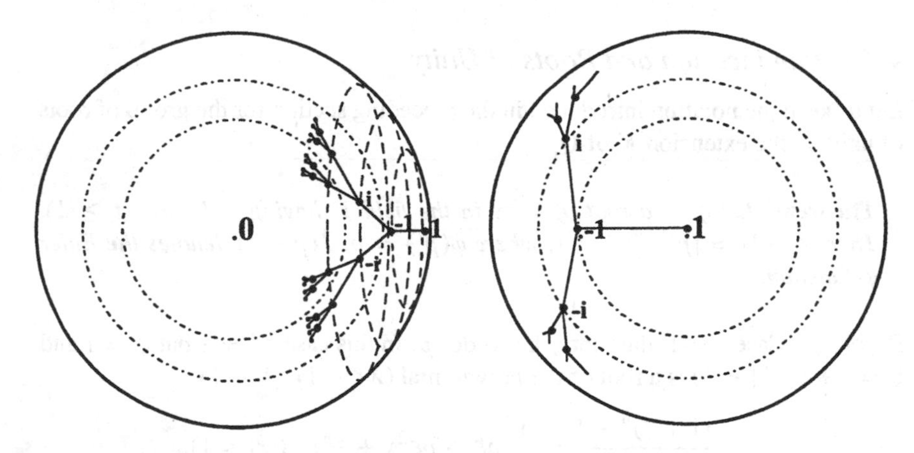

Location of the 2^nth roots of unity on the unit sphere

The appearance of the Euler φ-function is even more natural if we proceed as in (4.2). Let us give this deduction as a reminder of the properties of the cyclotomic polynomials. Recall that

$$\Phi_p(X) = \frac{X^p - 1}{X - 1}$$

denotes the pth cyclotomic polynomial (of degree $p - 1$).

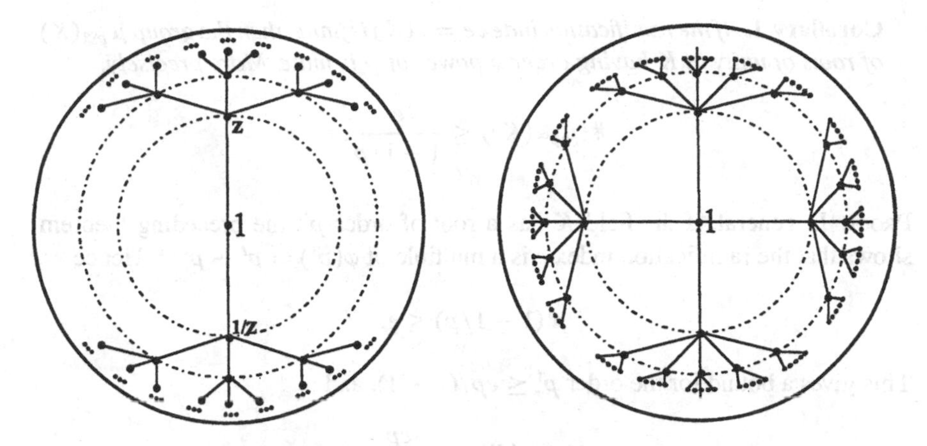

Location of the p^nth roots of unity on the unit sphere ($p = 3$ and 5)

Then, it is well known that the p^tth cylotomic polynomial (of degree $\varphi(p^t) = p^{t-1}(p-1)$) is given by

$$\Phi_{p^t}(X) = \Phi_p(X^{p^{t-1}}) = \frac{X^{p^t} - 1}{X^{p^{t-1}} - 1}$$
$$= X^{(p-1)p^{t-1}} + \cdots + X^{p^{t-1}} + 1.$$

If ζ is a root of unity of order p^t, then the other roots of unity having the same order are the powers ζ^j of ζ, where the integer j is prime to p, hence the preceding cyclotomic polynomial has a factorization

$$\Phi_{p^t}(X) = X^{(p-1)p^{t-1}} + \cdots + X^{p^{t-1}} + 1$$
$$= \prod_{1 \le j \le p^t - 1, p \nmid j} (X - \zeta^j)$$

with a product restricted to the integers j prime to p: There are $\varphi(p^t)$ linear factors in this product. On the other hand, substituting $X = 1$, we get

$$p = \prod_{1 \le j \le p^t - 1, p \nmid j} (1 - \zeta^j).$$

But $\zeta \equiv 1 \pmod{P}$ and

$$\frac{1 - \zeta^j}{1 - \zeta} = 1 + \zeta + \cdots + \zeta^{j-1} \equiv j \pmod{P}.$$

When p is prime to j, we infer $|1 - \zeta^j| = |1 - \zeta|$, and all factors in the above product have the same absolute value,

$$|p| = |1 - \zeta|^{\varphi(p^t)}, \quad |\zeta - 1| = |p|^{1/\varphi(p^t)}.$$

Corollary 1. *If the ramification index $e = e(K)$ is finite, then the group $\mu_{p^\infty}(K)$ of roots of unity in K having order a power of p is finite. More precisely,*

$$\#(\mu_{p^\infty}(K)) \leq \frac{e}{1 - 1/p}.$$

PROOF. In general, if the field K has a root of order p^t, the preceding theorem shows that the ramification index e is a multiple of $\varphi(p^t) = p^t - p^{t-1}$. Hence

$$p^t(1 - 1/p) \leq e.$$

This gives a bound for the order $p^t \leq ep/(p-1)$, and

$$\#(\mu_{p^\infty}(K)) \leq \frac{ep}{p-1}. \qquad \blacksquare$$

Observe that the result of this corollary is valid for any valued field K of characteristic 0, provided that its absolute value extends the p-adic one on $\mathbf{Q}$.

In particular, if $e = 1$, we have $\#(\mu_{p^\infty}(K)) \leq p/(p-1)$,

$$\#(\mu_{p^\infty}(K)) = 1 \text{ if } p \geq 3,$$

whereas $\#(\mu_{2^\infty}(K)) \leq 2$ if $p = 2$. This proves again a result obtained in (I.6.7).

Corollary 2. *The group of roots of unity in $\mathbf{Q}_p$ is precisely*

$$\mu(\mathbf{Q}_p) = \mu_{(p)}(\mathbf{Q}_p) = \mu_{p-1} \quad p \text{ odd prime},$$
$$\mu(\mathbf{Q}_2) = \mu_2(\mathbf{Q}_2) = \{\pm 1\}. \qquad \blacksquare$$

Example. Let K be the extension generated over $\mathbf{Q}_p$ by a primitive pth root of unity and K' the extension of K generated by a primitive root of unity of order p^2. Both extensions are totally ramified. The degrees of these cyclotomic extensions are determined by the previous theory, and a diagram summarizes the situation.

$$\begin{array}{ccc} & K' = \mathbf{Q}_p(\zeta_{p^2}) & \\ \text{degree } p & | & \text{wild} \\ & K = \mathbf{Q}_p(\zeta_p) & \\ \text{degree } p-1 & | & \text{tame} \\ & \mathbf{Q}_p & \end{array}$$

The element $\pi = \zeta_p - 1$ has absolute value $|\pi| = |p|^{1/(p-1)}$ generating the group of values $|K^\times|$: $P = \pi R \subset R \subset K = \mathbf{Q}_p(\zeta_p)$. Similarly, the element $\pi' = \zeta_{p^2} - 1$ has absolute value $|\pi'| = |p|^{1/p(p-1)}$ generating the group of values $|K'^\times|$:

$$P' = \pi' R' \subset R' \subset K' = \mathbf{Q}_p(\zeta_{p^2}).$$

4.5. Example 1: The Field of Gaussian 2-adic Numbers

The ring of Gaussian integers $\mathbf{Z}[i]$ is a square lattice generated by 1 and $i = \sqrt{-1}$ in the complex field:

$$\mathbf{Z}[i] = \mathbf{Z} \oplus i\mathbf{Z} \subset \mathbf{C}.$$

It is known that this ring is a *principal ideal domain.* We can also embed it in an algebraic extension of the 2-adic field $\mathbf{Q}_2$. Since we have seen that -1 has no root in $\mathbf{Q}_2$, the extension $K = \mathbf{Q}_2(i)$ has degree 2 over $\mathbf{Q}_2$. Observe that $(1+i)^2 = 2i$; hence $|1+i| = |2|^{1/2}$, and this extension is *totally and wildly ramified:* $e = 2$. The general notation gives in this case

$$K = \mathbf{Q}_2(i) \supset R = \mathbf{Z}_2[i] \supset P = (1+i)R.$$

We shall consider the generator

$$\pi = i - 1 = i(1+i)$$

of the maximal ideal P,

$$\pi^2 = -2i, \quad |\pi| = |2|^{1/2}.$$

Since the residue field of K is

$$k = R/P = \mathbf{F}_2,$$

we can consider representations with "digits" in the representative system

$$S = \{0, 1\} \subset \mathbf{Q}_2 \subset \mathbf{Q}_2(i).$$

Expansions in base $b = \pi$ of nonzero elements of $K = \mathbf{Q}_2(i)$ have the form:

$$\textstyle\sum_{i \geq \nu} a_i b^i \ (a_i \in S,\ \nu \in \mathbf{Z},\ a_\nu \neq 0),$$

while elements of $\mathbf{Z}_2[i]$ have expansions

$$\textstyle\sum_{i \geq 0} a_i b^i,\ (a_i \in S).$$

A parametrization of $\mathbf{Z}_2[i]$ is given by the set of binary sequences, hence a bijective map

$$\Phi : S^{\mathbf{N}} \to \mathbf{Z}_2[i], \quad (a_i) \mapsto \sum a_i b^i,$$

or equivalently,

$$\Phi : \mathcal{P}(\mathbf{N}) \to \mathbf{Z}_2[i], \quad J \mapsto \sum_J b^i.$$

Proposition. *The elements of* $\mathbf{Z}_2[i]$ *admitting a finite expansion* $\sum_{0 \leq i \leq n} a_i b^i$, *(where* $a_i \in S$, $n \in \mathbf{N}$*) in base* b *are precisely the Gaussian* *integers, and we have*

$$\mathbf{Z}[i] = \left\{ \sum_J b^j : J \text{ a finite subset of } \mathbf{N} \right\}.$$

PROOF. Since $\mathbf{Z}[i]$ is a ring containing 1 and b, it certainly contains all polynomials in b. We have to prove the converse inclusion, namely:

Every Gaussian integer admits a finite representation in base b.

Let $F = \Phi(S^{(\mathbf{N})}) \subset \mathbf{Z}[i]$ be the image of the finite binary sequences. It will be enough to prove that this image is a *subgroup* of $\mathbf{Z}[i]$, since it contains the generators 1 and b. In other words, we have to prove

$$F + F \subset F \text{ and } -F \subset F.$$

Starting with

$$b = i - 1, \quad b + 1 = i,$$
$$b^2 = -2i, \qquad b^4 = -4,$$

we infer successively

$$b^2(b+1) = 2,$$
$$b^2(b+1) + 1 = 3,$$
$$b^4 + b^2(b+1) + 1 = 3 - 4 = -1.$$

We have obtained the expansions

$$2 = b^2 + b^3,$$
$$-1 = 1 + b^2 + b^3 + b^4,$$

and more generally,

$$2b^i = b^{i+2} + b^{i+3},$$
$$-b^i = b^i + b^{i+2} + b^{i+3} + b^{i+4}.$$

These expansions give reduction algorithms to prove that for finite subsets J and K of $\mathbf{N}$,

$$\sum_J b^j + \sum_K b^k \in F,$$
$$-\sum_J b^j \in F. \qquad \blacksquare$$

4.6. *Example 2: The Hexagonal Field of 3-adic Numbers*

Here, we consider the quadratic extension $K = \mathbf{Q}_3(\sqrt{-3})$ of the field $\mathbf{Q}_3$ of 3-adic numbers. Since it is obtained by adjunction of the root of a generator of the maximal ideal $3\mathbf{Z}_3$ of $\mathbf{Z}_3$, it is totally and tamely ramified with index $e = 2$. This quadratic extension contains $\zeta = (1+\sqrt{-3})/2$, which is a root of unity of order 6: One can check in succession

$$\zeta^2 = \zeta - 1,\ \zeta^3 = \zeta^2 - \zeta = (\zeta - 1) - \zeta = -1.$$

Also observe that if we add a root η of unity of order 3 to $\mathbf{Q}_3$, we obtain a totally ramified extension of degree 2, for which $\eta - 1$ is a generator of the maximal ideal. In fact, we can take $\eta = \zeta^2$ and check (with the 3-adic absolute value)

$$(\eta - 1)^2 = \zeta^4 - 2\zeta^2 + 1 = -\zeta - 2(\zeta - 1) + 1 = 3(1 - \zeta) = -3\zeta^2,$$

and since $\eta + 1 = \zeta^2 + 1 = \zeta$, it follows that $|\eta + 1| = 1$ and

$$|\eta^2 - 1|^2 = |\eta - 1|^2 = |-3\zeta^2|, \quad |\eta - 1| = |3|^{1/2} = \sqrt{1/3}.$$

We shall now take the generator $b = \sqrt{-3}$ and consider expansions $\sum_i a_i b^i$ having coefficients a_i in a fixed set of representatives (containing 0) of the residue field

$$k = R/P = R/bR = \mathbf{Z}_3/3\mathbf{Z}_3 = \mathbf{F}_3.$$

We could take $\{0, 1, -1\}$ as a set of representatives. However, we shall take $S = \{0, 1, \zeta\}$: Indeed by definition $2\zeta = 1 + b$, so that

$$-1 = -2\zeta + b = \zeta + b - 3\zeta = \zeta + b + \zeta b^2 \equiv \zeta \pmod{b}$$

and we can replace the representative -1 by ζ. It is easy to check that

$$\begin{aligned} 2\zeta &= 1 + b, \\ 2 &= \zeta + b + b^2 + \zeta b^3, \\ 1 + \zeta &= \zeta b + \zeta b^2 + b^3 + \zeta b^4. \end{aligned}$$

These relations show how to compute sums. Finally, a picture *shows*(!) how the image of the finite ternary sequences

$$F = \Phi(S^{(\mathbf{N})}) \subset \mathbf{Z}[\zeta]$$

fills in the whole lattice $\mathbf{Z}[\zeta]$.

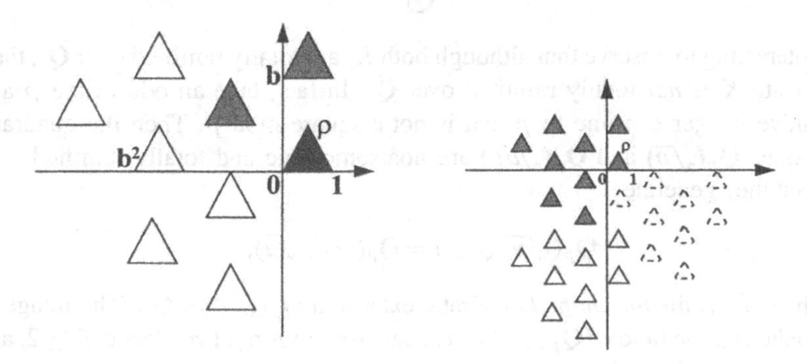

Finite sums $\sum_{0 \le i < 3^n} a_i b^i$

As in the preceding section, we have obtained unique representations for the elements of the *hexagonal lattice* $\mathbf{Z}[\zeta]$, which is the ring of integers in $\mathbf{Q}(\sqrt{-3})$.

Proposition. *Let* $b = \sqrt{-3}$, $\zeta = (1+\sqrt{-3})/2$, *and* $S = \{0, 1, \zeta\}$. *Then the finite sums* $\sum_j a_j b^j$ $(a_j \in S)$ *fill up the hexagonal lattice* $\mathbf{Z}[\zeta]$ *in* $\mathbf{C}$ *(or in* $K = \mathbf{Q}_3(\sqrt{-3})$*).* ■

4.7. Example 3: A Composite of Totally Ramified Extensions

Let us consider the following quadratic extensions of $\mathbf{Q}_3$:

$$K_1 = \mathbf{Q}_3(\sqrt{-3}), \quad K_2 = \mathbf{Q}_3(\sqrt{3}).$$

They are both totally (tamely) ramified, since $|\sqrt{-3}| = |\sqrt{3}| = |3|^{1/2}$. Hence $n = e = 2$, $f = 1$ for both. Let $K = K_1 \cdot K_2$ denote the composite (in a common extension). Obviously, $\sqrt{-1} = \sqrt{-3}/\sqrt{3} \in K$, and the cyclic group of roots of unity in K contains μ_4. But the residue field of $\mathbf{Q}_3$ is $\mathbf{F}_3 = \mathbf{Z}_3/3\mathbf{Z}_3$; it contains only the roots of unity ± 1. Hence the residue field of K contains the quadratic extension $\mathbf{F}_9$ and its cyclic group of units μ_8. On the other hand, as we have seen in the preceding example,

$$K_1 = \mathbf{Q}_3(\sqrt{-3}) \supset \mathbf{Q}(\sqrt{-3}) \supset \mathbf{Z}[\zeta]$$

where $\zeta = \zeta_6 = (1+\sqrt{-3})/2$ is a root of unity of order 6. Altogether, K contains $\mu_8 \cdot \mu_3 = \mu_{24}$ (Chinese remainder theorem). Both the residue degree and the ramification index of K must be greater than 1. The only possibility is $e(K) = 2$, $f(K) = 2$ (and $n(K) = 4$).

$$\begin{array}{ccc} & \mathbf{Q}_3(\sqrt{3}, \sqrt{-3}) = \mathbf{Q}_3(\sqrt{3}, \sqrt{-1}) & \\ \nearrow & & \nwarrow \\ K_1 & & K_2 \\ \nwarrow & & \nearrow \\ & \mathbf{Q}_3 & \end{array}$$

It is interesting to observe that although both K_i are totally ramified over $\mathbf{Q}_3$, their composite K is *not* totally ramified over $\mathbf{Q}_3$. In fact, take an odd prime p and a positive integer a prime to p that is not a square mod p. Then the quadratic extensions $\mathbf{Q}_p(\sqrt{p})$ and $\mathbf{Q}_p(\sqrt{ap})$ are nonisomorphic and totally ramified over $\mathbf{Q}_p$. But they generate

$$\mathbf{Q}_p(\sqrt{p}, \sqrt{ap}) = \mathbf{Q}_p(\sqrt{p}, \sqrt{a}),$$

which contains the *unramified* quadratic extension $\mathbf{Q}_p(\sqrt{a})$ of $\mathbf{Q}_p$. The image of $\sqrt{a}$ in the residue field of $\mathbf{Q}_p(\sqrt{a})$ is a square root of a mod p. Hence $f \geq 2$, and since $ef = n = 2$, we have $e = 1$.

Appendix to Chapter 2: Classification of Locally Compact Fields

In this appendix we shall give an approach to the classification of locally compact (commutative) fields of characteristic 0. This contains our main case of interest, namely that of ultrametric fields. For this purpose we shall take for granted the existence of a Haar measure on such a field: On any locally compact group G there exists a positive Radon measure μ on G — or equivalently a regular Borel measure μ on G — that is left invariant. Thus we view this measure either as a (1) positive *continuous* linear functional

$$\mu : C_c(G;\mathbf{R}) \to \mathbf{R}, \quad f \mapsto \mu(f)$$

on the space of compactly supported continuous functions on G, invariant under left translations

$$\mu(f) = \int_G f(x)d\mu(x) = \int_G f(gx)d\mu(x) \quad (g \in G),$$

or as a

(2) σ-additive function on a suitable σ-algebra of subsets containing the relatively compact open sets U of G. We also write $\mu(U)$ for the measure of the subset U.

If U is a relatively compact open subset of K, we denote by vol (U) the measure of U. By left invariance of this measure, we have vol (U) = vol (gU) for any $g \in G$. The Radon measure can be extended as a linear form on a vector space of functions containing the characteristic functions of relatively compact open sets $U \subset G$, and if we denote by φ_U the characteristic function of U, the two points of views are linked by the relation vol $(U) = \mu(\varphi_U)$. By abuse of notation, we shall also write vol $(U) = \mu(U)$.

The uniqueness of Haar measures will play an essential role and will be admitted here without proof:

> *Let μ and ν be two Haar measures on a locally compact group G; then there exists a positive constant α such that $\mu = \alpha\nu$.*

For a general classification of locally compact fields, not necessarily commutative and in any characteristic, the reader can consult the references given at the end of this volume.

A.1. Haar Measures

Let K be a locally compact commutative field (the general definition of topological fields was given in (I.3.7)) and let us choose and fix a Haar measure μ on the additive group K. By invariance, we have vol (U) = vol $(U + a)$ for any $a \in K$.

For any automorphism α of the field K, the invariant measure $\alpha(\mu)$ defined by $\alpha(\mu)(U) = \mu(\alpha U)$ (for all U in the suitable σ-algebra) is proportional to μ, say $\alpha(\mu) = m(\alpha) \cdot \mu$. Since two Haar measures are proportional, this scalar $m(\alpha)$ is independent of the choice of Haar measure. Now take in particular for automorphism α an automorphism of the form $\alpha : x \mapsto ax$ where $a \neq 0 \in K$. In this case we shall simply denote by $m(a)$ the resulting scalar. By definition

$$\text{vol}\,(aU) = m(a) \cdot \text{vol}\,(U) \quad (a \in K^\times).$$

The associativity of multiplication in K gives immediately

$$m(ab) = m(a)m(b) \quad (a, b \in K^\times).$$

Hence m is a homomorphism $K^\times \to (\mathbf{R}^\times)_{>0}$, $m(1) = 1$ and $m(a^{-1}) = m(a)^{-1}$. This homomorphism m is the *modulus* of K. It is conventionally extended by $m(0) = 0$. We shall eventually show that it is a *generalized absolute value* on K.

A.2. Continuity of the Modulus

Take a compact neighborhood V of 0 in K and choose $a \in K$. Since aV is compact and the Haar measure is regular, for each $\varepsilon > 0$ we can find an open set $U \supset aV$ with

$$\text{vol}\,(U) \leq \text{vol}\,(aV) + \varepsilon.$$

By continuity of multiplication in K, there is a neighborhood W of a such that $U \supset WV$. Thus for $x \in W$

$$\text{vol}\,(xV) \leq \text{vol}\,(U) \leq \text{vol}\,(aV) + \varepsilon,$$
$$m(x) \leq m(a) + \varepsilon/\text{vol}\,(V).$$

Since $m(x) \geq 0$ and $m(0) = 0$, this inequality proves that m is continuous at the point 0. It also proves that m is *upper semicontinuous* at each point $a \in K$. But for $a \neq 0$ we can write $m(a) = 1/m(a^{-1})$, whence m is also *lower semicontinuous* at such points. This proves the continuity of the modulus on K.

A.3. Closed Balls are Compact

For $r \geq 0$ we denote by $B_r = \{x \in K : m(x) \leq r\}$ a *closed ball* in K. Fix again a compact neighborhood V of 0 in K. We shall prove

B_r is contained in a compact set of the form yV.

As a *first step,* we construct a sequence $(\pi_n)_{n \geq 0} \subset V$ with $\pi_n \to 0$. Since

$$0 \cdot V = \{0\} \subset V,$$

there is a neighborhood U of 0 in K for which we still have $UV \subset V$ (take V_0 an open neighborhood of 0 in V and choose U such that UV is contained in V_0). We can find an element $\pi \in U \cap V$ with $0 < m(\pi) < 1$. Hence $\pi^2 \in UV \subset V$, $\pi^3 = \pi \cdot \pi^2 \in UV \subset V$, and by induction, $\pi^n \in V$ $(n \geq 1)$. But V is compact, so that the sequence (π^n) must have a cluster value π' in V. By continuity of m, $m(\pi')$ must be a cluster value of the sequence $(m(\pi^n))$. Since $m(\pi^n) = m(\pi)^n \to 0$, the only possibility is $m(\pi') = 0$ and $\pi' = 0$. This proves that the sequence

(π^n) has only one cluster value in the compact set V: It must converge, and $\pi^n \to 0$. Finally, observe that since $\pi \in U$ and $UV \subset V$, we have $\pi V \subset V$ and $V \subset \pi^{-1}V$. We see by induction that the sequence of compact sets $\pi^{-n}V$ increases monotonically.

Second step: We show that $B_r \subset \pi^{-N}V$ for some large $N \geq 1$. Since we already know that B_r is closed and $\pi^{-N}V$ is compact, this will indeed show that B_r is compact. Let $a \in B_r$. By the first part $\pi^n a \to 0$, and there is a first integer n such that $\pi^n a \in V$. If $a \notin V$, this first positive n is such that $\pi^n a \in V$ but $\pi^{n-1}a \notin V$. In other words, $\pi^n a \in V - \pi V$. The set $V - \pi V$ is relatively compact (in V compact) and

$$0 \notin \Omega := \overline{V - \pi V}.$$

We can define $r' = \inf_\Omega m(x) > 0$ and choose $N \geq 1$ such that $m(\pi)^N \cdot r \leq r'$. Hence

$$m(\pi)^N \cdot r \leq r' \leq m(\pi^n a) = m(\pi)^n m(a) \leq m(\pi)^n r \quad (a \in B_r).$$

This shows that $m(\pi)^N \leq m(\pi)^n$ and hence $n \leq N$. Thus we have $a \in \pi^{-n}V \subset \pi^{-N}V$ for all $a \in B_r$: The ball B_r is contained in the compact set $\pi^{-N}V$. ■

Corollary 1. *The balls B_r $(r > 0)$ make up a fundamental system of neighborhoods of 0 in K. In particular,*

$$a^n \to 0 \text{ in } K \iff m(a) < 1.$$

Proof. If V is any compact neighborhood of 0 in K, put $r = \max_V m(x)$ in order to have $V \subset B_r$. Since 0 is not in the closure of $B_r - V$, the minimum r' of $m(x)$ on the closure Ω of $B_r - V$ is positive; for $0 < r'' < r'$ it is clear that $B_{r''} \subset V$. ■

Corollary 2. *Any discrete subfield of K is finite.*

Proof. Let F be a discrete subfield of K. Choose any $a \in K$ with $m(a) > 1$. Then we have $m(a^{-n}) = m(a)^{-n} \to 0$, whence $a^{-n} \to 0$, and since F is discrete it shows $a \notin F$. This proves $F \subset B_1$. But we know that F is closed (I.3.2). Thus F is compact and discrete, hence finite. ■

Remark. If the field K has characteristic 0 but is not assumed to be commutative, we see here that its center is a locally compact nondiscrete (commutative) field. Indeed, this center is *closed* and contains the rational field $\mathbf{Q}$ by assumption, hence is not finite. It is locally compact and not discrete by Corollary 2.

A.4. The Modulus is a Strict Homomorphism

We claim that $\Gamma = m(K^\times)$ is closed in $\mathbf{R}_{>0}$ and $m : K^\times \to \Gamma$ is an open map. For $r > 0$, the compact set $m(B_r)$ is simply $m(B_r) = \{0\} \cup (\Gamma \cap [0, r])$. In particular, if $0 < \varepsilon < r < \infty$, $\Gamma \cap [\varepsilon, r]$ is closed in $\mathbf{R}_{>0}$. Since the *interiors* of the intervals $[\varepsilon, r]$ cover $\mathbf{R}_{>0}$, we can conclude that Γ is closed in this topological space. If V is a neighborhood of 1 in $K^\times$, we have now to prove that $m(V)$ is a neighborhood of 1 in Γ. It is enough to show that for every sequence (γ_n) in Γ such that $\gamma_n \to 1$, there is a subsequence γ_{n_i} in $m(V)$ (a subset A is not a neighborhood of 1 in Γ when there is a sequence $\gamma_n \to 1$ in Γ and $\gamma_n \notin A$). Let us write $\gamma_n = m(x_n)$ for some elements $x_n \in V$. Since V is compact, the sequence (x_n) must have — at least — one cluster point $x \in V$. By continuity of m, $m(x)$ must be a cluster point of $m(x_n) = \gamma_n \to 1$. This proves $m(x) = 1$, namely $x \in N := \ker(m) \subset K^\times$. But VN is a neighborhood of $x \in N$. By definition of a cluster point, for each n_0 there must be an integer $n > n_0$ with $x_n \in VN$ and hence $\gamma_n = m(x_n) \in m(VN) = m(V)$. This proves the existence of the subsequence of (γ_n) in $m(V)$ as desired.

Corollary. *If the field K is locally compact and nondiscrete, the subgroup $m(K^\times)$ is either $\mathbf{R}_{>0}$ or of the form $\{\theta^n : n \in \mathbf{Z}\} = \theta^{\mathbf{Z}}$ for some $0 < \theta < 1$. When $C = \max\{m(1 + x) : x \in B_1\} = 1$, the second case occurs.*

PROOF. Since $1+B_1$ is a neighborhood of 1 in $K^\times$, its image must be a neighborhood of 1 in Γ. When $C = 1$, this neighborhood is contained in $(0, 1]$ and its image under $t \mapsto t^{-1}$ is a neighborhood of 1 in Γ contained in $[1, \infty)$. The intersection of these two neighborhoods of 1 in Γ is reduced to the single point $\{1\}$, thus proving that Γ is discrete in this case. ∎

In an obvious sense, the modulus m defines the topology of K: Any neighborhood of an element $x \in K$ has the form $x + V$ for some neighborhood V of 0 in K, and $m(V)$ contains a neighborhood of $0 \in \Gamma$, namely,

$$\text{there is an } \varepsilon > 0 \text{ such that} \quad m(x) < \varepsilon \Longrightarrow x \in V,$$

which implies that the given neighborhood $x + V$ contains $x + B_\varepsilon$.

A.5. Classification

Let us recall the result obtained above (Corollary 2 in A.3): In a nondiscrete locally compact field, any discrete subfield is finite. Now the discussion of cases can be made according to the value of the constant

$$C = \max_{x \in B_1} m(1 + x) \geq 1.$$

It is obvious that

$$m(a + b) \leq C \cdot \max(m(a), m(b)),$$

since if $0 \neq m(a) \geq m(b)$, we can divide by a so that $x = b/a \in B_1$ and $m(1+b/a) \leq C$, $m(a+b) \leq C \cdot m(a) = C \cdot \max(m(a), m(b))$. Hence m defines a generalized absolute value (II.2.2) on K. In every case, a suitable power of C will be less than or equal to 2, and a power of m is a metric defining the topology of K. This shows that *any locally compact field is metrizable.*

First case: $C > 1$. In this case, the field K is not ultrametric; hence it is automatically of characteristic 0 and contains the field $\mathbf{Q}$. If K is not discrete, $\mathbf{Q}$ is not discrete either (because infinite, by the result just recalled), and the metric induced by K on $\mathbf{Q}$ must be equivalent to the usual Archimedean metric. The completion $\mathbf{R}$ of $\mathbf{Q}$ for this metric must also be contained in K. Hence K is a real vector space. Being locally compact, it must be finite-dimensional. One can show that the only possible cases are $K = \mathbf{R}, \mathbf{C}$ (or $\mathbf{H}$: Hamilton quaternions if it is not commutative).

Second case: $C = 1$. Then K is ultrametric. If we assume K to be of characteristic 0, it contains the field $\mathbf{Q}$, and as before, the induced metric on $\mathbf{Q}$ is not trivial. By the classification of ultrametric absolute values on $\mathbf{Q}$ we infer that K must induce a p-adic metric on $\mathbf{Q}$ and contain a completion $\mathbf{Q}_p$. Since K is assumed to be locally compact, its degree over $\mathbf{Q}_p$ is finite (II.3.2). We leave out the positive characteristic case (interested readers can find a complete discussion in the specific references given at the end of this book).

It is easy to see that contrary to the real case, there are extensions of $\mathbf{Q}_p$ of arbitrarily large degree (cf. (III.1.3)).

A.6. *Finite-Dimensional Topological Vector Spaces*

In order to approach the structure of locally compact fields (having no a priori norm), we have to give a few general definitions and results concerning topological vector spaces. Instead of limiting ourselves to the field of scalars $\mathbf{Q}_p$, let us treat the case of arbitrary valued fields: This general context has the advantage of emphasizing the individual properties needed to establish each result. Thus we shall consider in this section that K is any ultrametric valued field (II.1.3), nondiscrete: $|K^\times| \neq \{1\}$. In particular, K is a metric space.

Definition. *A* topological vector space *over K is a vector space V (over K) equipped with a Hausdorff topology for which*

the additive group V is a topological group,
the multiplication $(a, v) \mapsto a \cdot v : K \times V \to V$ is continuous.

Let U be a neighborhood of 0 in such a topological vector space. By continuity of multiplication at (0, 0), there is $\varepsilon > 0$ and a neighborhood $U_0 \subset U$ of 0 such that

$$U_1 := \{av : a \in K,\ |a| \leq \varepsilon,\ v \in U_0\} \subset U.$$

This neighborhood $U_1 \subset U$ of 0 has the property

$$a \in K, \ |a| \leq 1 \Longrightarrow aU_1 \subset U_1.$$

Definition. *A nonempty subset U in a topological vector space V is* balanced *when*

$$a \in K, \ |a| \leq 1 \Longrightarrow aU \subset U.$$

The balanced neighborhoods of 0 in a topological vector space play the role of the balls in normed spaces. We have just proved that in a topological vector space *there is a fundamental system of neighborhoods of* 0 *consisting of balanced ones.*

Theorem 1. *A one-dimensional topological vector space V over K is isomorphic as a topological vector space to K. More precisely, for each $0 \neq v \in V$, the map $a \mapsto av : K \to V$ is a bijective linear homeomorphism.*

Proof. Fix $0 \neq v \in V$. The one-to-one linear map $a \mapsto av : K \to V$ is continuous, since V is a topological vector space over K. We have to show the continuity of the inverse, namely

$$\forall \varepsilon > 0 \, \exists \, U \text{ neighborhood of } 0 \text{ in } V \text{ such that } av \in U \Longrightarrow |a| < \varepsilon.$$

We proceed as follows. If $\varepsilon > 0$ is chosen, we take $b \in K$ with $0 < |b| \leq \varepsilon$ and a balanced neighborhood U of 0 in V such that $U \not\ni bv \neq 0$ (this is possible, since we assume that V is Hausdorff). Now, if $av \in U$, then

$$bv = \frac{b}{a} \cdot \underbrace{av}_{\in U} \notin U \underset{U \text{ balanced}}{\Longrightarrow} \left|\frac{b}{a}\right| > 1 \Longrightarrow |a| < |b| \leq \varepsilon. \qquad \blacksquare$$

Lemma. *A linear form $\varphi : V \to K$ on a topological vector space V is continuous precisely when its kernel is closed in V.*

Proof. If the linear form φ is continuous, its kernel is closed. Conversely, assume that the kernel of φ is closed. We may assume $\varphi \neq 0$ and take $v_0 \in V$ with $\varphi(v_0) \neq 0$. Replace v_0 by $v_0/\varphi(v_0)$, so that $\varphi(v_0) = 1$. The linear variety

$$\{\varphi = 1\} = v_0 + \ker \varphi$$

is closed and does not contain the origin. Hence there is a balanced neighborhood U of 0 that does not meet this closed subset:

$$(v_0 + \ker \varphi) \cap U = \emptyset.$$

I claim that $\varphi(U) \subset B_{<1}$, so that φ is bounded, continuous at the origin, and continuous. Now, if $v \in U$ and $\varphi(v) \neq 0$, consider the scalar $a = 1/\varphi(v)$. We have

$$\varphi(av) = 1 \Longrightarrow av \notin U \underset{U \text{ balanced}}{\Longrightarrow} |a| > 1.$$

This proves $|\varphi(v)| < 1$, as expected. ■

Theorem 2. *Assume that the field K is complete. Then a finite-dimensional topological vector space V over K is isomorphic as a topological vector space to a Cartesian product K^d. More precisely, for any basis (e_i) of V, the linear map*

$$(\lambda_i) \mapsto \sum_i \lambda_i e_i : K^d \to V$$

is an isomorphism of topological vector spaces.

PROOF. We proceed by induction on the dimension of the vector space V: The dimension-1 case is covered by the first theorem. Assume that the statement is true up to dimension $d-1$. If $\dim_K V = d$, select a basis $e_1, \ldots, e_d$ of V and consider the linear span W of the first $d-1$ e_i. By the induction assumption, the space W is isomorphic to K^{d-1} and hence complete and closed in V. The linear form

$$\varphi : \sum_i \lambda_i e_i \mapsto \lambda_d, \ V \to K$$

is continuous, since its kernel $\ker(\varphi) = W$ is closed. The one-to-one linear map

$$K^d = K^{d-1} \times K \overset{\approx}{\to} W \times K e_d \overset{\text{sum}}{\longrightarrow} V$$

is continuous. Its inverse is

$$x \mapsto (x - \varphi(x)e_d \, , \, \varphi(x)e_d)$$

and hence is also continuous. ■

A.7. *Locally Compact Vector Spaces Revisited*

We have seen in (3.2) that locally compact *normed* spaces V over $\mathbf{Q}_p$ are finite-dimensional. Using the existence of Haar measures, we can now prove the same statement without the assumption that the topology is derived from a norm.

Theorem. *Any locally compact vector space over $\mathbf{Q}_p$ is finite-dimensional.*

PROOF (WEIL). The proof is based on (A.6): A finite-dimensional subspace of a locally compact vector space V over $\mathbf{Q}_p$ is isomorphic as a topological vector space to a finite product $\mathbf{Q}_p^d$, hence is complete, and hence is closed in V and *locally compact*. Let now V be any locally compact vector space over $\mathbf{Q}_p$. In particular, it is a locally compact abelian group, and we can choose a Haar measure μ on V. We can define a *modulus homomorphism*

$$m_V : \mathbf{Q}_p^\times \to \mathbf{R}_{>0}$$

as for locally compact fields (A.1). For $0 \neq a \in \mathbf{Q}_p$, the map $U \mapsto \operatorname{vol}(aU)$ is also a Haar measure on V, and by uniqueness, there is a unique positive scalar $m_V(a) > 0$ such that $\operatorname{vol}(aU) = m_V(a) \cdot \operatorname{vol}(U)$ (for all relatively compact open sets $U \subset V$). For example, If $W = \mathbf{Q}_p^d$ has dimension d over $\mathbf{Q}_p$, then $m_W(a) = |a|^d$. Since $p^n \to 0$ in $\mathbf{Q}_p$, we have

$$m_V(p)^n = m_V(p^n) \to 0,$$

and this proves $m_V(p) < 1$ for all locally compact $\mathbf{Q}_p$-vector spaces V. Select now a d-dimensional vector subspace W of V. Integrating in succession over W and $F = V/W$,

$$f \mapsto \int_F d\mu_F(y) \int_W f(x+y) d\mu_W(x),$$

we get an invariant Radon measure on V, which we may take for μ_V (or we can change the choice of Haar measure on F to obtain this equality). Hence

$$\int_V f(x) d\mu_V(x) = \int_F d\mu_F(y) \int_W f(x+y) d\mu_W(x)$$

for all continuous functions f with compact support on G. We see that

$$\begin{aligned} m_V(a) &= m_W(a) \cdot m_F(a) = |a|^d \cdot m_F(a), \\ m_V(p) &= |p|^d \cdot m_F(p) \leq |p|^d, \\ \log m_V(p) &\leq d \cdot \log |p|, \end{aligned}$$

and by division by $\log |p| < 0$,

$$d \leq \log m_V(p) / \log |p|.$$

This shows that the dimension d of finite-dimensional subspaces of V is bounded, and this implies that V itself is finite-dimensional. ∎

A.8. Final Comments on Regularity of Haar Measures

Let us consider the Haar measure on the locally compact group $G = \mathbf{R} \times \mathbf{R}_d$ where the first copy of $\mathbf{R}$ has the usual topology and the second copy the discrete topology. The usual Lebesgue measure μ_L is a Haar measure on $\mathbf{R}$, and we can take for Haar measure of $\mathbf{R}_d$ the counting measure

$$\mu_d(A) = \#(A) \quad (A \subset \mathbf{R}_d).$$

The product of these two Haar measures is a Haar measure on the product $\mathbf{R} \times \mathbf{R}_d$. The subset $A = \{0\} \times \mathbf{R}_d$ has the discrete topology, and

$$\mu(A) = \inf_{U \text{ open} \supset A} \mu(U) = \infty,$$

simply since each open set $U \supset A$ contains an uncountable family of open intervals of positive length. However, a compact set $K \subset A$ is finite (because discrete), hence $\mu(K) = 0$, and $\sup_{K \text{ compact} \subset A} \mu(K) = 0$ is different from $\mu(A) = \infty$. In general, inner regularity holds only for subsets having $\mu(A) < \infty$ (and in a suitable algebra containing the Borel subsets). This pathology disappears in locally compact spaces that are *countable at infinity*. This last property holds for all locally compact fields: we have seen this in characteristic zero in (A.5).

Exercises for Chapter 2

1. Let X be an ultrametric space. Show that the spheres of radius $r > 0$ in X are the complements of one open ball of maximal radius r in a closed ball of radius r.

2. Let X be an ultrametric space.
   (*a*) Fix a positive radius $r > 0$. Show that the condition $d(x, y) \leq r$ is an equivalence relation $x \sim y$ between elements of X. The equivalence classes are the closed balls of radius r, and the quotient space is the *uniformly discrete* metric space of closed balls of fixed radius r (the inequality $d(x, y) < r$ also defines an equivalence relation, for which the equivalence classes are the open balls of radius r).
   (*b*) Fix $a \in X$ and assume that $\{d(x, a) : x \in X\}$ is dense in $\mathbf{R}_{\geq 0}$. Show that the ordered set of closed balls containing the point a (with respect to inclusion) is isomorphic to the half line $[0, \infty) \subset \mathbf{R}$.
   (*c*) Assume that for each $x \in X$, $\{d(x, y) : y \in X\}$ is dense in $\mathbf{R}_{\geq 0}$. Define T_X as the ordered set of closed balls in X (with respect to inclusion). Prove that this is a *tree*. Recall that we denote by $\delta(A)$ the diameter of a bounded subset of a metric space, so that $\delta B_{\leq r} = r$. We have two natural maps

$$\begin{array}{ccccc} X \times \mathbf{R}_{\geq 0} & \to & T_X & (a, r) & \mapsto & B = B_{\leq r}(a) \\ & & \downarrow \delta & & & \downarrow \delta \\ & & \mathbf{R}_{\geq 0} & & & \delta(B) = r \end{array}$$

   For $r > 0$, the fiber $\delta^{-1}(r)$ is the uniformly discrete metric space consisting of closed balls of fixed radius r. If X is separable, this fiber is countable. For any subset $A \subset X$ define $T_X(A)$ as the subset consisting of the (dressed) balls B meeting A. Prove that this is a subtree of T_X. Take for A successively sets containing only one, two, or three elements: What are the possible configurations?
   (*d*) The metric space $\mathbf{Z}_p$ can be embedded in an ultrametric space X satisfying the condition required in (*c*) (cf. Chapter III). Sketch $T_X(\mathbf{Z}_p)$ and show that the picture does not depend on the choice of ambient space X.

3. Let $|\,.\,|$ be an absolute value on a field K.
   (*a*) Prove the triangle inequality

$$|x + y|^\alpha \leq |x|^\alpha + |y|^\alpha \quad (x, y \in K,\ 0 < \alpha \leq 1).$$

(*b*) When the absolute value is ultrametric, prove the same result for all $\alpha > 0$.
(*c*) If $a = a_0 + \sum_{1\le i\le n} a_i$ and $|a_i| \le |a|$ for $1 \le i \le n$, prove $|a| = \max_{0\le i\le n} |a_i|$.

4. As corollary of the proof of Theorem 1 of (II.1.4) we see that (with the notation of the theorem): If $A/\xi A$ is finite, then $A/\xi^n A$ is also finite and $\#(A/\xi^n A) = [\#(A/\xi A)]^n$. More generally, show that in any integral domain A,

$$\#(A/(ab)) = \#(A/(a)) \cdot \#(A/(b))$$

if $ab \neq 0$. (*Hint.* Observe that multiplication by b on $(a) = aA$ leads to an isomorphism of the A-modules $A/(a)$ and $Ab/(ab)$. Then use the isomorphism $A/abA \cong A/aA \times aA/abA$.)

5. (*a*) Let $P(X) = X^2 - 2X + 1 \in \mathbf{Z}[X]$. This polynomial has the root $x = 1$. Find explicitly the sequence of iterates given by Newton's method starting at an element $x \neq 1$: Does this sequence converge in $\mathbf{Q}_p$?
(*b*) Let A be the maximal subring of an ultrametric field as in (1.4), and let $P(X) \in A[X]$ be a polynomial having a *simple* root $x = \xi$.
Show that for any x in the open ball of center ξ and radius $|P'(\xi)| \neq 0$ Newton's method furnishes a sequence of iterates that converges to ξ.

6. Prove directly the following: If $a_n \to 0$ and $b_n \to 0$ in an ultrametric field, then $c_n = \sum_{0\le i\le n} a_i b_{n-i} \to 0$ and

$$\sum_{n\ge 0} a_n \cdot \sum_{n\ge 0} b_n = \sum_{n\ge 0} c_n.$$

[*Hint.* The assumption implies that the two sequences are bounded, say

$$|a_i| \le C, \ |b_i| \le C \quad \text{for all } i \ge 0,$$

and for each given $\varepsilon > 0$ there exists $N = N_\varepsilon$ such that

$$|a_i| \le \varepsilon, \ |b_i| \le \varepsilon \quad (i \ge N).$$

For $i + j \ge 2N$, we have $|a_i b_j| \le \varepsilon C$, since one index at least is greater or equal to N.]

7. Show that two norms on a vector space define the same topology when there exist two constants c, C such that

$$c\|x\| \le \|x\|' \le C\|x\|.$$

(The unit ball for one norm must contain a ball for the other norm; observe that this condition is independent from the ultrametricity.)

8. Let K'/K be a finite extension of ultrametric fields. Show directly that the residue field k' of K' has finite degree over the residue field of K and

$$f = [k' : k] \le n = [K' : K]$$

(cf. 4.1 and 4.3).

9. Let K be a valued field that is an extension of $\mathbf{Q}_p$, and let $\xi \in K$. Suppose that there exist integers $a_0(j), a_1(j), \dots, a_{n-1}(j) \in \mathbf{Z}$ $(j \ge 1)$ such that

$$|\xi^n + a_{n-1}(j)\xi^{n-1} + \cdots + a_0(j)| \to 0 \quad (j \to \infty).$$

(*a*) Show that $|\xi| \leq 1$. (If you cannot, glimpse at the proof of Proposition 3 in (III.2.1)).
(*b*) Prove that ξ is algebraic of degree less than or equal to n over $\mathbf{Q}_p$.
(*Hint.* Consider the nonempty sets $X_m \subset (\mathbf{Z}/p^m\mathbf{Z})^n$ consisting of the families $(a_0 \bmod p^m, \ldots, a_{n-1} \bmod p^m)$ such that $|\xi^n + a_{n-1}\xi^{n-1} + \cdots + a_0| \leq |p^m| = 1/p^m$. Then any element of $\varprojlim X_m \neq \emptyset$ gives a polynomial dependence relation for ξ over $\mathbf{Q}_p$.)

10. Let $s < t$ and ζ a root of unity of order p^s, ζ' a root of unity of order p^t, both in $\mathbf{Q}_p^a$. What is the distance $|\zeta - \zeta'|$?

11. Let K be an ultrametric extension of $\mathbf{Q}_p$. Prove that if the group $\mu(K)$ of roots of unity in K is infinite, then this field K is not locally compact. (*Hint.* Can you find a convergent subsequence?)

12. Show that the quadratic extensions $\mathbf{Q}_5(\sqrt{2})$ and $\mathbf{Q}_5(\sqrt{3})$ of $\mathbf{Q}_5$ in $\mathbf{Q}_5^a$ coincide, by two methods:
(a) Use the fact that 6 has a square root in $\mathbf{Q}_5$.
(b) $X^2 - 2$ and $X^2 - 3$ are irreducible over $\mathbf{F}_5$ (hence $\mu_{24} \subset \mathbf{Q}_5(\sqrt{2})$, $\mu_{24} \subset \mathbf{Q}_5(\sqrt{3})$).

13. Consider the following quadratic extensions of $\mathbf{Q}_7$ in $\mathbf{Q}_7^a$

$$\mathbf{Q}_7(\sqrt{-1}),\ \mathbf{Q}_7(\sqrt{2}),\ \mathbf{Q}_7(\sqrt{3}),\ \mathbf{Q}_7(\sqrt{5}),\ \mathbf{Q}_7(\sqrt{6}).$$

By (I.6.6), they cannot be distinct: Give identities. What is the degree of $\mathbf{Q}_7(\sqrt{2}, \sqrt{3})$ over $\mathbf{Q}_7$? (What is the degree of $\mathbf{Q}(\sqrt{2}, \sqrt{3})$ over $\mathbf{Q}$?)

Exercises for Appendix to Chapter 2

1. Let U be a neighborhood of 0 in a topological vector space V over a valued field K. Show that

$$\bigcap_{\lambda \in K, |\lambda| \geq 1} \lambda U$$

is a balanced neighborhood of 0 contained in U.

2. Let K be a nondiscrete ultrametric field. Assume that K is not complete and consider the topological vector space $\widehat{K}$ over K. If $a, b \in \widehat{K}$ are linearly independent over K, the two-dimensional subspace $Ka + Kb$ is not isomorphic, as a topological vector space, to K^2. (*Hint.* The one-dimensional subspaces of K^2 are not dense in this space!)

3. Let K be a finite extension of $\mathbf{Q}_p$ (hence locally compact). A *character* of K is a continuous homomorphism $\chi : K \to \mathrm{U}(1) = \{z \in \mathbf{C}^\times : |z| = 1\}$.
(a) Prove that such a character χ is locally constant and takes its values in μ_{p^∞}.
(b) If ψ is a fixed nontrivial character, consider the characters $\psi_a(x) = \psi(ax)$ $(a \in K)$. Show that $a \mapsto \psi_a$ is an injective homomorphism $f : K \to K^\sharp$ where $K^\sharp$ is the (multiplicative) group of characters of K. (For a nontrivial character on K, one can take the composite of the trace $T_{K/\mathbf{Q}_p}$ and the Tate homomorphism τ (I.5.4).)
(c) Define a topology on $K^\sharp$ having for neighborhoods of a given character χ the subsets

$$V_{\varepsilon,A}(\chi) = \{\chi' \in K^\sharp : |\chi'(x) - \chi(x)| \leq \varepsilon\}$$

($\varepsilon > 0$, A a compact subset of K). Show that the above-defined homomorphism $f : a \mapsto \psi_a$ is continuous.

(d) Show that the inverse homomorphism $\psi_a \mapsto a$ is continuous on $f(K)$. Conclude that this image is locally compact, and hence closed in $K^\sharp$. (*Hint.* Use Corollary 1 in (I.3.2).)

Comment. For any locally compact abelian group G, one can define its Pontryagin dual

$$G^\sharp = \{\chi : G \to \mathrm{U}(1) \text{ a continuous homomorphism}\}$$

and show that $G^\sharp$ is again a locally compact abelian group with $(G^\sharp)^\sharp$ canonically isomorphic to G. When $G = K$ is the additive group of a locally compact field, one can show (as above) that K and $K^\sharp$ are isomorphic. This generalizes the known situation for the field $\mathbf{R}$.

3

Construction of Universal p-adic Fields

In order to be able to define K-valued functions by means of series (mainly power series), we have to assume that K is *complete*. It turns out that the algebraic closure $\mathbf{Q}_p^a$ is not complete, so we shall consider its completion $\mathbf{C}_p$: This field turns out to be algebraically closed and is a natural domain for the study of "analytic functions." However, this field is not spherically complete (2.4), and spherical completeness is an indispensable condition for the validity of the analogue of the Hahn-Banach theorem (Ingleton's theorem (IV.4.7); spherical completeness also appears in (VI.3.6)). This is a reason for enlarging $\mathbf{Q}_p^a$ in a more radical way than just completion, and we shall construct a spherically complete, algebraically closed field Ω_p (containing $\mathbf{Q}_p^a$ and $\mathbf{C}_p$) having still another convenient property, namely $|\Omega_p| = \mathbf{R}_{\geq 0}$. This ensures that all spheres of positive radius in Ω_p are nonempty: $B_{<r}(a) \neq B_{\leq r}(a)$ for all $r \geq 0$. In fact, we shall define the big ultrametric extension Ω_p first — using an ultraproduct — and prove all its properties (this method is due to B. Diarra) and then define $\mathbf{C}_p$ as the topological closure of $\mathbf{Q}_p^a$ in $\mathbf{C}_p$. This simplifies the proof that $\mathbf{C}_p$ is algebraically closed. By a *universal* p-adic field we mean a *complete, algebraically closed* extension of $\mathbf{Q}_p$.

In this chapter $\mathbf{Q}_p^a$ *denotes a fixed algebraic closure of* $\mathbf{Q}_p$.

1. The Algebraic Closure $\mathbf{Q}_p^a$ of $\mathbf{Q}_p$

1.1. Extension of the Absolute Value

There is a canonical absolute value on $\mathbf{Q}_p^a$. Indeed, the absolute value of $\mathbf{Q}_p$ extends uniquely to $\mathbf{Q}_p^a$, as the following observation shows. If K_1 and K_2 are two finite extensions of $\mathbf{Q}_p$ in $\mathbf{Q}_p^a$, the two extensions (II.3.4) of the absolute value of $\mathbf{Q}_p$ to these fields must agree on their intersection $K_1 \cap K_2$ by uniqueness (II.3.3). Hence all the extensions of the absolute value of $\mathbf{Q}_p$ to finite subextensions of $\mathbf{Q}_p^a$ define a unique extension of this absolute value to $\mathbf{Q}_p^a$. As a consequence, this algebraic closure is an ultrametric field, and we set

$$A^a := \text{the maximal subring of } \mathbf{Q}_p^a : |x| \leq 1,$$

$$M^a := \text{the maximal ideal of } A^a : |x| < 1,$$

$$k^a := A^a/M^a \text{ the residue field of } \mathbf{Q}_p^a.$$

We shall see below that $\mathbf{Q}_p^a$ is *not complete*, hence *not locally compact*. Moreover, the residue field k^a is infinite, and $|(\mathbf{Q}_p^a)^\times|$ is a dense subgroup of $\mathbf{R}_{>0}$. Hence *none* of the conditions of (II.3.5) for local compactness are satisfied!

1.2. Maximal Unramified Subextension

We have seen in (II.4.3) that every finite extension K of $\mathbf{Q}_p$ contains a maximal unramified subextension: Since K is complete, the group $\mu_{(p)}(K)$ of roots of unity in K having order prime to p is isomorphic to the cyclic group $k^\times$ of order $q - 1 = p^f - 1$, where f is the residue degree of K:

$$K_{\text{ur}} = \mathbf{Q}_p(\zeta_{q-1}) = \mathbf{Q}_p(\mu_{q-1}) \subset K.$$

It is not difficult to generalize this result to the algebraically closed extension $\mathbf{Q}_p^a$.

Proposition. *The residue field k^a of the algebraic closure $\mathbf{Q}_p^a$ of $\mathbf{Q}_p$ is an algebraic closure of the prime field $\mathbf{F}_p$.*

PROOF. Since any algebraic element $x \in \mathbf{Q}_p^a$ generates a finite-dimensional extension K of $\mathbf{Q}_p$, the residue field of K is also finite-dimensional over $\mathbf{F}_p$. This proves that the residue field of $\mathbf{Q}_p^a$ is algebraic over $\mathbf{F}_p$. Conversely, if $\xi \neq 0$ is algebraic over $\mathbf{F}_p$, it belongs to the cyclic group $\mathbf{F}_p(\xi)^\times$ and hence is a root of unity of order m prime to p. Now consider the cyclotomic extension $\mathbf{Q}_p(\mu_m)$ obtained by adjoining to $\mathbf{Q}_p$ all roots of unity of order m. If $\zeta \neq \eta$ are two such roots, then $|\zeta - \eta| = 1$ and the reductions of ζ and η are distinct (cf. II.4.3). Hence the residue field of $\mathbf{Q}_p(\mu_m)$ contains m distinct mth roots of unity and contains ξ. ∎

We shall denote by $\mathbf{F}_p^a = \mathbf{F}_{p^\infty} = \bigcup_{f \geq 1} \mathbf{F}_{p^f}$ an algebraic closure of $\mathbf{F}_p$ and by $(\mathbf{Q}_p^a)_{ur} = \mathbf{Q}_p(\mu_{(p)}) \subset \mathbf{Q}_p^a$ the extension generated by all roots of unity having order prime to p. This is the *maximal unramified* extension of $\mathbf{Q}_p$ in $\mathbf{Q}_p^a$.

Corollary. *The residue field of the maximal unramified extension of $\mathbf{Q}_p$ in $\mathbf{Q}_p^a$ is an algebraic closure of the prime field $\mathbf{F}_p$.* ∎

1.3. Ramified Extensions

One can give another reason for the fact that the extension $\mathbf{Q}_p^a$ has infinite degree over $\mathbf{Q}_p$. Choose algebraic numbers $\pi_e = p^{1/e}$ $(e \geq 2)$. We have

$$|\pi_e^e| = |p| = 1/p, \quad |\pi_e| = (1/p)^{1/e},$$

and consequently the ramification index of $\mathbf{Q}_p(\pi_e)$ is greater than or equal to e. This proves that $\mathbf{Q}_p$ *has algebraic extensions of arbitrarily large degree.* Indeed, the polynomial $X^e - p$ is an Eisenstein polynomial, and hence is irreducible (II.4.2) in $\mathbf{Z}_p[X]$ or $\mathbf{Q}_p[X]$: This defines an extension of degree e of $\mathbf{Q}_p$. More generally, if K is any finite extension of $\mathbf{Q}_p$ (contained in $\mathbf{Q}_p^a$), it is locally compact, and we can choose a generator π for the maximal ideal P of R. The polynomial $X^e - \pi$ is an Eisenstein polynomial, hence is irreducible in $R[X]$ and $K[X]$, whence K is not algebraically closed. These simple observations show that

$$|(\mathbf{Q}_p^a)^\times| \supset p^{\mathbf{Q}} = \{p^\nu : \nu \in \mathbf{Q}\} = \bigcup_{e \geq 1} p^{(1/e)\mathbf{Z}}.$$

Proposition. *The absolute values of algebraic numbers over $\mathbf{Q}_p$ are fractional powers of p: $|(\mathbf{Q}_p^a)^\times| = p^{\mathbf{Q}}$.*

PROOF. If $x \in \mathbf{Q}_p^a - \mathbf{Q}_p$ is any algebraic number not in $\mathbf{Q}_p$, it satisfies a nontrivial polynomial equation

$$\sum_{0 \leq i \leq n} a_i x^i = 0 \quad (a_i \in \mathbf{Q}_p)$$

of degree $n \geq 2$. By the principle of competition, there are two distinct indices, say $i > j$, with

$$|a_i x^i| = |a_j x^j| \neq 0.$$

Hence

$$|x|^{i-j} = |a_j / a_i| \in p^{\mathbf{Z}},$$

and $|x| \in p^{(1/e)\mathbf{Z}}$ $(e = i - j \geq 1)$. ∎

1.4. The Algebraic Closure $\mathbf{Q}_p^a$ is not Complete

A complete metric space X is a *Baire space*: A countable union of closed subsets X_n in X having no interior point cannot have an interior point. In particular, such a countable union cannot be equal to X. Recall that *locally compact spaces and complete metric spaces are Baire spaces.*

Theorem. *The algebraic closure $\mathbf{Q}_p^a$ of $\mathbf{Q}_p$ is* not *a Baire space.*

Proof. Let us define the sequence of subsets

$$X_n = \{x \in \mathbf{Q}_p^a : \deg x = [\mathbf{Q}_p(x) : \mathbf{Q}_p] = n\} \subset \mathbf{Q}_p^a \quad (n \geq 1)$$

so that $\mathbf{Q}_p^a = \cup_{n\geq 1} X_n$. It is also obvious that $\lambda X_n \subset X_n$ ($\lambda \in \mathbf{Q}_p$), $X_m + X_n \subset X_{mn}$, and in particular,

$$X_n + X_n \subset X_{n^2}.$$

(*a*) These subsets are *closed*. If $x \neq 0$ is in the closure of X_n, say $x = \lim x_i$ with a sequence (x_i) in X_n, then for each x_i let $f_i(X) \in \mathbf{Q}_p[X]$ be a polynomial of least degree with x_i as a root and coefficients scaled so they lie in $\mathbf{Z}_p$ and at least one of them is in $\mathbf{Z}_p^\times$. Extracting if necessary a subsequence of (f_i), we can assume that it converges (in norm, coefficientwise), say $f_i \to f$, so $f \in \mathbf{Z}_p[X]$ has degree less than or equal to n and at least one coefficient in $\mathbf{Z}_p^\times$, so $f(X) \neq 0$. By the ultrametric property, the convergence $f_i \to f$ is uniform on all bounded sets of $\mathbf{Q}_p^a$. Since the convergent sequence (x_i) is bounded, we have

$$f(x) - f_i(x_i) = \underbrace{f(x) - f(x_i)}_{\to 0} + \underbrace{f(x_i) - f_i(x_i)}_{\to 0} \to 0.$$

This implies $f(x) = \lim f_i(x_i) = 0$ and $x \in X_n$.

(*b*) The subsets X_n have *no interior point*. Since for any closed ball B of positive radius in $\mathbf{Q}_p^a$ we have $\mathbf{Q}_p^a = \mathbf{Q}_p \cdot B$, such a ball cannot be contained in a subset X_n, and no translate can be contained in X_n. ■

Corollary. *The space $\mathbf{Q}_p^a$ is neither complete nor locally compact.* ■

1.5. Krasner's Lemma

Theorem 1 (Krasner's Lemma). *Let $K \subset \mathbf{Q}_p^a$ be a finite extension of $\mathbf{Q}_p$ and let $a \in \mathbf{Q}_p^a$ (so that a is algebraic over $\mathbf{Q}_p$). Denote by a^σ the conjugates of a over K and put $r = \min_{a^\sigma \neq a} |a^\sigma - a|$. Then every element $b \in B_{<r}(a; \mathbf{Q}_p^a)$ generates (over K) an extension containing $K(a)$.*

Proof. Take any algebraic element b such that $a \notin K(b)$. Since we are in characteristic 0, Galois theory asserts that there is a conjugate $a^\sigma \neq a$ of a over $K(b)$ (the automorphism σ fixes $K(b)$ elementwise) and we can estimate the distance of a to b as follows:

$$|b - a^\sigma| = |(b-a)^\sigma| = |b - a|,$$
$$|a - a^\sigma| \leq \max(|a - b|, |b - a^\sigma|) = |b - a|.$$

This shows that

$$|b - a| \geq |a - a^\sigma| \geq r.$$

Hence if $b \in B_{<r}(a)$, namely $|b - a| < r$, we have

$$a \in K(b), \quad K(a) \subset K(b).$$ ■

Examples. (*a*) Take $K = \mathbf{Q}_2$ and $a = \sqrt{-1} = i$. Then $i^\sigma = -i$ and

$$r = |i - i^\sigma| = |2i| = |2| = \tfrac{1}{2}.$$

Hence for $b \in \mathbf{Q}_2^a$,

$$|b - i| < \tfrac{1}{2} \Longrightarrow i \in \mathbf{Q}_2(b).$$

(*b*) Take $K = \mathbf{Q}_3$ and $a = \sqrt{-3}$. Then $a^\sigma = -\sqrt{-3}$ and

$$r = |a - a^\sigma| = |2\sqrt{-3}| = |\sqrt{-3}| = |3|^{1/2} = \sqrt{\tfrac{1}{3}}.$$

Hence for $b \in \mathbf{Q}_3^a$,

$$|b - \sqrt{-3}| < \sqrt{\tfrac{1}{3}} \Longrightarrow \sqrt{-3} \in \mathbf{Q}_3(b).$$

Recall that the norm of a polynomial $f(X) = \sum_{i \leq n} a_n X^n$ is the sup norm on the coefficients $\|f\| = \max_{i \leq n} |a_n|$.

Theorem 2 (Continuity of Roots of Equations). *Let K be a finite extension of the p-adic field $\mathbf{Q}_p$ and fix an algebraic element $a \in \mathbf{Q}_p^a$ of degree n over K corresponding to a monic irreducible polynomial $f \in K[X]$ (of degree n). There is a positive ε such that any monic polynomial $g \in K[X]$ of degree n with $\|g - f\| < \varepsilon$ has a root $b \in K(a)$ also generating this field: $K(b) = K(a)$.*

Proof. Let us factorize the polynomial g in the algebraic closure $\mathbf{Q}_p^a$ of K, say $g(X) = \Pi(X - b_i)$, and evaluate it at the root a of f:

$$\prod (a - b_i) = g(a) = g(a) - f(a).$$

With $M = \max_{0 \leq i \leq n}(|a|^i) = \max(1, |a|^n)$ we can estimate

$$\prod |a - b_i| = |g(a) - f(a)| \leq \|g - f\| \cdot M,$$

hence for one index i at least,

$$|a - b_i| \leq \|g - f\|^{1/n} \cdot M^{1/n}.$$

By the preceding theorem, if $\varepsilon > 0$ is chosen small enough, then $\|g - f\| < \varepsilon$ will imply $K(b_i) \supset K(a)$ for some i. But the degree of b_i is less than or equal to n, since it is a root of the nth degree polynomial $g \in K[X]$. This proves $K(b_i) = K(a)$. ∎

Corollary 1. *Let $f \in K[X]$ be a monic irreducible polynomial, $a \in \mathbf{Q}_p^a$ a root of f, and $(g_i)_{i \in \mathbf{N}}$ a sequence of monic polynomials with coefficients in K of the same degree as f. If $g_i \to f$ (coefficientwise), there is a sequence (x_i) of roots of these polynomials such that $x_i \in K(a)$ for large i and $x_i \to a$.*

PROOF. As soon as $\|g_i - f\| < \varepsilon$ is small enough, the above result is applicable and shows that $|a - x_i|$ is small for at least one root x_i of g_i. More precisely, the inequality

$$|a - x_i| \leq \|g_i - f\|^{1/n} \cdot M^{1/n}$$

shows that $|a - x_i| \to 0$, and the convergence $x_i \to a$ in $K(a)$ follows. ∎

Corollary 2. *The algebraic closure $\mathbf{Q}_p^a$ of $\mathbf{Q}_p$ is a separable metric space.*

PROOF. Take $a \in \mathbf{Q}_p^a$ and let f be its minimal polynomial over $\mathbf{Q}_p$. Since $\mathbf{Q}$ is dense in $\mathbf{Q}_p$, we can find monic polynomials $g \in \mathbf{Q}[X]$ as close to f as we want. If we choose a sequence $g_n \to f$, the continuity principle for the roots shows that a is a limit of roots x_n of the polynomials g_n. This shows that the algebraic closure of $\mathbf{Q}$ is dense in $\mathbf{Q}_p^a$. But this algebraic closure is a countable field since the set of polynomials of fixed degree with coefficients in the countable field $\mathbf{Q}$ is countable. ∎

1.6. A Finiteness Result

In the last two sections of this chapter, let us prove a couple of theorems easily obtained with the techniques developed above. (We shall not need them in the sequel.)

Theorem. *Let K be a finite extension of $\mathbf{Q}_p$ and $n \geq 1$ an integer. Then there are only finitely many extensions of K of degree n in $\mathbf{Q}_p^a$.*

PROOF. (1) Let F be an extension of degree n of K and let e be its relative ramification index, f its residue degree: $n = ef$. The cyclic subgroup consisting of roots of unity in F having order prime to p is isomorphic to the cyclic group of nonzero elements in the residue field of F (II.4.3). These roots generate the maximal unramified subextension F_{ur} of K in F,

$$[F_{\mathrm{ur}} : K] = f$$

(II.4.4), and the extension F/F_{ur} is totally ramified of degree e. If the residue degree f is given, there is only one unramified extension of degree f of K in $\mathbf{Q}_p^a$. Hence the announced result will be established as soon as the same finiteness property for totally ramified extensions is established.

(2) Let us show that there are only finitely many totally ramified extensions of given degree $n = e$ of K. Fix such an extension F and let $K \supset R \supset P = \pi R$ (conventional notation). By (II.4.2, II.4.4) it is generated by an element having minimal polynomial

$$X^n + a_{n-1}X^{n-1} + \cdots + u_0\pi,$$

which is an Eisenstein polynomial. Its coefficients a_i belong to P, and $u_0 \in R^\times$ is a unit: $u_0 \in K$ and $|u_0| = 1$. The Cartesian product

$$P^{n-1} \times R^\times$$

is compact, and by continuity of the roots of equations (1.5), each element of this product has an open neighborhood corresponding to polynomials having their roots generating the same extension F in $\mathbf{Q}_p^a$. This completes the finiteness proof. ∎

1.7. *Structure of Totally and Tamely Ramified Extensions*

It is possible to improve the result (II.4.2) concerning the generation of totally ramified extensions.

Theorem. *Let $K \subset L \subset \mathbf{Q}_p^a$ be finite extensions of $\mathbf{Q}_p$. Assume that L/K is totally and tamely ramified of degree e. Then there exists a generator π of the maximal ideal P of $R \subset K$ such that L is generated by an eth root of π in $\mathbf{Q}_p^a$.*

PROOF. By assumption $e = [L : K]$ is prime to p. The proof will be accomplished in three steps.

(1) Consider arbitrary generators π of $P \subset R \subset K$ and π_L of $P_L \subset R_L \subset L$. Since L/K is totally ramified of degree e, $|\pi_L|^e = |\pi|$ and $\pi_L^e/\pi = u$ is a unit in R_L. Since the residue degree of L/K is 1, the residue fields are the same, and there is a unit ζ of R (one can take a root of unity in K) such that $\zeta \equiv u \pmod{P}_L$. Let us write

$$\pi_L^e = \pi \cdot u, \quad u = \zeta + \pi_L v \quad (v \in R_L).$$

Hence

$$\pi_L^e = \pi \cdot (\zeta + \pi_L v) = \zeta\pi + \pi\pi_L v.$$

The element $\zeta\pi$ is also a generator of the ideal P of R. We are going to show that L is generated by a root of the equation $X^e - \zeta\pi$. Let us replace the generator π by $\pi' = \zeta\pi$ and simply denote it by π again. Thus we assume from now on that the generators π_L and π are linked by a relation

$$\pi_L^e = \pi + \pi\pi_L v \quad (v \in R_L).$$

(2) The polynomial $f = X^e - \pi$ is an Eisenstein polynomial (II.4.2) of $R[X]$. Hence it is irreducible over $K[X]$. We have

$$f(\pi_L) = \pi_L^e - \pi = \pi\pi_L v, \quad |f(\pi_L)| = |\pi\pi_L v| < |\pi|.$$

Let us factor f in $\mathbf{Q}_p^a$:

$$f(X) = X^e - \pi = \prod_{1 \le i \le e} (X - \alpha_i)$$

(where $\prod \alpha_i = \pm\pi$). Since f is irreducible, the roots α_i are conjugate and have the same absolute value in $\mathbf{Q}_p^a$, say $|\alpha_i| = c$ independent of i. Hence

$$c^e = \prod |\alpha_i| = |\pi|,$$

$$|\alpha_i| = c = |\pi|^{1/e} = |\pi_L|,$$

and

$$|\pi_L - \alpha_i| \le \max(|\alpha_i|, |\pi_L|) = |\pi_L|.$$

If we come back to the polynomial f, then

$$\left|\prod_{1\le i\le e} (\pi_L - \alpha_i)\right| = |f(\pi_L)| < |\pi| = |\pi_L|^e$$

shows that at least one of the factors is smaller than $|\pi_L|$. Without loss of generality we may assume

$$|\pi_L - \alpha_1| < |\pi_L|.$$

(3) The roots of $f(X) = X^e - \pi = 0$ are the $\alpha_i = \zeta_i\alpha$, where $\zeta_i^e = 1$. Since e is prime to p, we have $|\zeta_i - 1| = 1$ when $\zeta_i \ne 1$ by Proposition 1 in (II.4.3). This proves

$$|\alpha_i - \alpha| = |\alpha| = c = |\pi_L| \quad (i \ne 1),$$

$$|\pi_L - \alpha| < |\pi_L| = |\alpha - \alpha_i| \quad (i \ne 1).$$

By Krasner's lemma, we infer that

$$K(\alpha) \subset K(\pi_L),$$

and since the element α has degree e, this inclusion is an equality. ∎

Example. If we add a primitive pth root ζ_p of unity to $\mathbf{Q}_p$, we obtain a totally ramified extension K of degree $p-1$. Hence $K/\mathbf{Q}_p$ is tamely ramified and can be generated by a $(p-1)$-th root of the generator $-p$ of $p\mathbf{Z}_p$.

For $p = 3$, we have seen in (II.4.6) that $b = \sqrt{-3}$ works:

$$\mathbf{Q}_3(\zeta_3) = \mathbf{Q}_3(\sqrt{-3}).$$

2. Definition of a Universal p-adic Field

2.1. *More Results on Ultrametric Fields*

Let us start with a couple of general results concerning (nondiscrete) ultrametric fields.

Proposition 1. *Let K be an ultrametric field and $\widehat{K}$ its completion. Then $\widehat{K}$ is still an ultrametric field and*

(a) $|K| = |\widehat{K}|$,
(b) *K and $\widehat{K}$ have the same residue field.*

PROOF. Let A be the ring of Cauchy sequences in K. The ideal I of A consisting of Cauchy sequences $a = (a_n)$ with $a_n \to 0$ (also called *null* Cauchy sequences) is a maximal ideal: If $a_n \not\to 0$, then $a_n \neq 0$ except for finitely many indices n and a is invertible in the quotient A/I. We can define $\widehat{K} = A/I$ with a canonical injection $K \hookrightarrow \widehat{K}$ given by constant sequences. If $a = (a_n) \in A - I$ is a Cauchy sequence that is not null, the sequence $(|a_n|)$ is stationary (stationarity principle), and we define an absolute value on $\widehat{K}$ by

$$|a| = \lim_{n\to\infty} |a_n| \in |K^\times| \subset \mathbf{R}_{>0} \text{ for } a \neq 0 \quad \text{and } |0| = 0.$$

Obviously, the canonical injection $K \hookrightarrow \widehat{K}$ is an isometric embedding, and we view K as a subfield of $\widehat{K}$: The absolute value of $\widehat{K}$ extends the absolute value of K. The residue field k of K parametrizes the open unit balls $B_{<1}(a)$ ($a = 0$ or $|a| = 1$) contained in the closed unit ball: $k^\times$ parametrizes the open unit balls contained in the unit sphere $S_1 = \{x \in K : |x| = 1\}$. Any Cauchy sequence of the closed unit ball has all its final terms in an open unit ball; hence it corresponds to a fixed element in the residue field k. ■

An extension L of an ultrametric field K having same residue field $k_L = k$ and the same absolute values $|L| = |K|$ is called an *immediate extension* of K. Hence the completion of K is an immediate extension of K.

Proposition 2. *Let K be a nondiscrete ultrametric field and put*

$$A = \{x \in K : |x| \leq 1\} : \textit{maximal subring of } K$$
$$M = \{x \in K : |x| < 1\} : \textit{maximal ideal of } A.$$

Then, either M is principal, or $M = M^2$ and the ring A is not Noetherian.

PROOF. By hypothesis $\Gamma = |K^\times| \neq \{1\}$, and either $\Gamma \cap (0, 1)$ has a maximal element θ or it has a sequence tending to 1. In the first case we can choose $\pi \in M$ with $|\pi| = \theta$, and $M = \pi A$ is principal. In the second case, for each $x \in M$, namely $|x| < 1$, we can find an element y such that $|x| < |y| < 1$, so that

$$x = y \cdot (x/y) \in M^2.$$

Since y and x/y belong to M, this shows that $x \in M^2$, and we have proved $M = M^2$. In this last case, the subgroup $\Gamma = |K^\times|$ is dense in $\mathbf{R}_{>0}$, and all the ideals

$$I_r = B_{\leq r} = B_{\leq r}(0; K) = \{x \in K : |x| \leq r\}$$

for $r \in \Gamma \cap (0, 1)$ are distinct: The ring A is not Noetherian. ■

Proposition 3. *With the same notation as before:*

(a) *If K is algebraically closed, so is the residue field k.*
(b) *If L is an algebraic extension of K, the residue field k_L of L is also an algebraic extension of the residue field k of K.*

PROOF. In any ultrametric field, $|\xi| > 1$, $|a_i| \leq 1$ $(i < n)$ implies

$$|\xi^n| > |\xi^i| \geq |a_i \xi^i| \quad (i < n),$$

$$|\xi|^n > |\sum_{i<n} a_i \xi^i|,$$

and hence

$$|\xi^n + \sum_{i<n} a_i \xi^i| = |\xi|^n > 1,$$

$$\xi^n + \sum_{i<n} a_i \xi^i \neq 0.$$

This proves that any root of a monic polynomial having coefficients $|a_i| \leq 1$ belongs to the closed unit ball $|x| \leq 1$.

(a) Let $X^n + \sum_{i<n} \alpha_i X^i \in k[X]$ be a monic polynomial of degree $n \geq 1$. Choose liftings $a_i \in A$ of the coefficients, i.e., $\alpha_i = a_i \pmod{M}$, and consider the monic polynomial

$$X^n + \sum_{i<n} a_i X^i \in A[X].$$

Since the field K is algebraically closed, this polynomial has a root $x \in K$. By the preliminary observation, $x \in A$ and $x \bmod M$ is a root of the reduced polynomial $X^n + \sum_{i<n} \alpha_i X^i$. This proves that k is algebraically closed.

(b) Let $0 \neq \xi \in k_L$ and choose a representative $x \in A_L - M_L$ of the coset $\xi \neq 0$: $|x| = 1$. By assumption, this element is algebraic over K, and hence x satisfies a nontrivial polynomial equation

$$\sum_{i \leq n} a_i x^n = 0 \quad (n \geq 1,\ a_i \in K).$$

By the principle of competitivity, at least two monomials have maximal competing absolute values

$$|a_i| = |a_i x^i| = |a_j x^j| = |a_j| \quad \text{for some } i < j.$$

Dividing by a_i, we obtain a polynomial equation with coefficients $|a'_k| \leq 1, a'_k \in A$ and at least two of them not in M. By reduction mod M we get a nontrivial polynomial equation satisfied by ξ. ■

2.2. Construction of a Universal Field Ω_p

Let R be the normed ring $\ell^\infty(\mathbf{Q}_p^a)$ consisting of bounded sequences $\alpha = (\alpha_i)_{i\in\mathbf{N}}$ of $\mathbf{Q}_p^a$ with the sup norm

$$\|\alpha\| = \sup_{i\in\mathbf{N}} |\alpha_i|.$$

Let us also choose and fix an ultrafilter $\mathcal{U}$ on $\mathbf{N}$ containing the subsets $[n, \infty)$ ($n \in \mathbf{N}$). (Readers not familiar with ultrafilters can find all definitions and properties used here in the Appendix to this Chapter.) Recall that for each subset $A \subset \mathbf{N}$ either $A \in \mathcal{U}$ or $A^c = \mathbf{N} - A \in \mathcal{U}$. On the other hand (here is the reason for choosing an ultrafilter), *each bounded sequence of real numbers has a limit along* $\mathcal{U}$, and we put

$$\varphi(\alpha) = \lim_{\mathcal{U}} |\alpha_i| \geq 0.$$

Proposition 1. *The subset $\mathcal{J} = \varphi^{-1}(0)$ is a maximal ideal of the ring R, and the field $\Omega_p = R/\mathcal{J}$ is an extension of the field $\mathbf{Q}_p^a$.*

Proof. Let us show that each element $\alpha \notin \mathcal{J}$ is invertible mod $\mathcal{J}$. But if $\alpha = (\alpha_n)$ is not in the ideal $\mathcal{J}$, the limit $r = \varphi(\alpha) > 0$ does not vanish, so we can find a subset $A \in \mathcal{U}$ such that $r/2 < |\alpha_i| < 2r$ ($i \in A$). Define a sequence $\beta = (\beta_i)$ by

$$\beta_i = \frac{1}{\alpha_i} \text{ for } i \in A \quad \text{and} \quad \beta_i = 0 \text{ for } i \notin A.$$

Since $|\beta_i| < 2/r$ ($i \in A$), the sequence β is bounded $\beta \leq 2/r$ and $\beta \in R$. By construction $1 - \alpha\beta$ vanishes on the set A, hence $1 - \alpha\beta \in \mathcal{J}$. This shows that α mod $\mathcal{J}$ is invertible in the quotient Ω_p. Consequently, the quotient is a field, and $\mathcal{J}$ a maximal ideal of R. Finally, constant sequences provide an embedding $\mathbf{Q}_p^a \to \Omega_p$. ∎

The map φ defines an absolute value on the field Ω_p: For $a = (\alpha \bmod \mathcal{J})$ we put

$$|a| = |a|_\Omega = \varphi(\alpha) = \lim_{\mathcal{U}} |\alpha_i|.$$

This absolute value extends the absolute value on $\mathbf{Q}_p^a$ (considered as a subfield of Ω_p through constant sequence).

Proposition 2. *The absolute value $|\,.\,|_\Omega$ coincides with the quotient norm of $R/\mathcal{J}$, namely for $a = (\alpha \bmod \mathcal{J})$,*

$$|a|_\Omega = \|\alpha \bmod \mathcal{J}\|_{R/\mathcal{J}} := \inf_{\beta\in\mathcal{J}} \|\alpha - \beta\|.$$

Proof. We have $\lim_{\mathcal{U}} |\gamma_i| \le \sup |\gamma_i|$ $(\gamma \in R)$, and hence

$$\lim_{\mathcal{U}} |\alpha_i| = \lim_{\mathcal{U}} |\alpha_i - \beta_i| \le \sup |\alpha_i - \beta_i| \quad (\beta \in \mathcal{J}),$$
$$|a|_\Omega \le \|\alpha - \beta\| \quad (\beta \in \mathcal{J}).$$

This proves

$$|a|_\Omega \le \|a\|_{R/\mathcal{J}}.$$

Conversely, if $a = \alpha \bmod \mathcal{J}$, then for any subset $A \in \mathcal{U}$ we can define the sequence $\beta = (\beta_i)$ as $\beta_i = 0$ $(i \in A)$ and $\beta_i = \alpha_i$ $(i \notin A)$ so that $\beta \in \mathcal{J}$ and $\|\alpha - \beta\| = \sup_{i\in A} |\alpha_i|$ and

$$\|a\|_{R/\mathcal{J}} \le \inf_{A\in\mathcal{U}} \sup_{i\in A} |\alpha_i| = \lim \sup |\alpha_i| = \lim_{\mathcal{U}} |\alpha_i| = |a|_\Omega. \qquad \blacksquare$$

From now on we shall simply write $|a| = |a|_\Omega$ for either the absolute value on the field Ω_p or the quotient norm in $R/\mathcal{J}$.

Proposition 3. *We have*

$$|\Omega_p^\times| = \mathbf{R}_{>0}.$$

Proof. This is a simple consequence of the fact that $|\mathbf{Q}_p^a|$ is dense in $\mathbf{R}_{\ge 0}$. Indeed, each positive real number r is limit of a sequence (r_n) of elements $r_n \in |\mathbf{Q}_p^a|$, say $r_n = |\alpha_n|$ $(\alpha_n \in \mathbf{Q}_p^a)$, so that the sequence α is bounded and defines an element a in the quotient Ω_p with $|a| = r$. $\blacksquare$

Comment. This construction of Ω_p is reminiscent of nonstandard analysis. Let $X = \mathbf{Q}_p^a$ and in the Cartesian product $X^{\mathbf{N}}$ introduce the equivalence relation

$$(x_n) \sim (y_n) \iff \{n \in \mathbf{N} : x_n = y_n\} \in \mathcal{U}.$$

The quotient $^*X := X^{\mathbf{N}}/\sim$ is an ultrapower of X (as systematically used in nonstandard analysis, in the construction of superstructures). The subset ^bX consisting of classes of bounded sequences is the set of *limited elements* in this ultraproduct $^*X$, and the classes of sequences tending to zero (along $\mathcal{U}$) are the *infinitesimal elements* $^iX \subset {}^bX$. The quotient $^bX/{}^iX = R/\mathcal{J} = \Omega_p$ has more simply been obtained in one step.

2.3. *The Field Ω_p is Algebraically Closed*

Let $f \in \Omega_p[X]$ be a monic polynomial of degree $n \ge 1$, say

$$f(X) = X^n + a_{n-1}X^{n-1} + \cdots + a_0 \quad (a_k \in \Omega_p).$$

We show that this polynomial f has a root in the field Ω_p. Select representative families for the coefficients:

$$a_k = (\alpha_{ki})_i \bmod \mathcal{J}.$$

We can consider the polynomials

$$f_i(X) = X^n + \sum_{k<n} \alpha_{ki} X^i \in \mathbf{Q}_p^a[X].$$

Since the field $\mathbf{Q}_p^a$ is algebraically closed, each of these has all its roots in $\mathbf{Q}_p^a$. More precisely, the product of the roots of f_i is (up to sign) the constant term α_{0i} of this polynomial, so that we can choose at least one root ξ_i with $\xi_i \le |\alpha_{0i}|^{1/n}$. The sequence $\xi = (\xi_i)$ is bounded $\|\xi\| \le \|a_0\|^{1/n}$, $\xi \in R$, and the class x of ξ is a root of f in Ω_p. ■

2.4. *Spherically Complete Ultrametric Spaces*

Consider a decreasing sequence $(B_{\le r_n}(a_n))_{n\ge 0}$ of closed balls in an ultrametric space X:

$$d(a_i, a_n) \le r_n \text{ for all pairs } i \ge n.$$

When $r_n \searrow 0$, the sequence of centers is a Cauchy sequence; hence it converges if we assume that the space X is complete. The limit of this sequence belongs to every $B_{\le r_n}(a_n)$ (these balls are closed). In particular, this shows that the intersection of the sequence is not empty.

At first, it seems surprising that even in a complete space, a nested sequence of closed balls may have an empty intersection *when the decreasing sequence of radii has a positive limit.* Consider, however, the following situation. In the discrete space $\mathbf{N}$ with the ultrametric distance $d(n, m) = 1 - \delta_{mn}$, the decreasing sequence of closed sets $F_n = [n, \infty)$ has an empty intersection (they all have diameter equal to 1). This space is complete (it is uniformly discrete), and a small modification of the metric (cf. the exercises) transforms these sets F_n into closed balls of strictly decreasing radii.

Definition. *An ultrametric space X is called* spherically complete *when all decreasing sequences of closed balls have a nonempty intersection.*

A spherically complete space X is complete: If (x_n) is any Cauchy sequence of X, consider the decreasing sequence (r_n) where $r_n = \sup_{m>n} |x_m - x_n|$ (which converges to 0). Then $(B_{\le r_n}(x_n))$ is a decreasing sequence of closed balls having for intersection a limit of the sequence.

Comment. Any extension of an ultrametric field K which has the same residue field and the same value group (in $\mathbf{R}^\times$) is called an *immediate extension* of K. It can be proved that each ultrametric field admits an immediate extension that is spherically complete. For example, there is a spherically complete extension of

$\mathbf{Q}_p^a$ that has residue field $\mathbf{F}_{p^\infty}$ and value group $p^{\mathbf{Q}}$. In fact, spherically complete extensions are *maximal elements* among extensions having prescribed residue field and value group.

2.5. *The Field Ω_p is Spherically Complete*

Let us consider any decreasing sequence $(B_n)_{n\geq 0}$ of closed balls $B_n = B_{\leq r_n}(a_n)$ in the field Ω_p. The ultrametric inequality shows that such a sequence of balls decreases if

$$|a_{n+1} - a_n| \leq r_n \text{ and } (r_n) \text{ decreases.}$$

Take liftings $\alpha_n \in R$ of the centers $a_n \in R/\mathcal{J}$ in the following way. Since

$$|a_{n+1} - a_n| \leq r_n < r_{n-1}$$

and since the absolute value is the quotient norm, we can proceed by induction and, once α_n has been chosen, successively choose the next lifting α_{n+1} still satisfying $\|\alpha_{n+1} - \alpha_n\| < r_{n-1}$. Then $\|\alpha_k - \alpha_n\| < r_{n-1}$ for all $k \geq n$. The ith component will a fortiori satisfy $|\alpha_{ki} - \alpha_{ni}| < r_{n-1}$ ($k \geq n$). Consider now the *diagonal sequence* $\xi = (\xi_i)$ in R defined by $\xi_i = \alpha_{ii}$. Then

$$\|\xi - \alpha_n\| \leq \sup_{i\geq n} |\xi_i - \alpha_{ni}| \leq r_{n-1}$$

because the interval $[n, \infty)$ of $\mathbf{N}$ belongs to the ultrafilter $\mathcal{U}$, whence for $x = \xi$ mod $\mathcal{J}$,

$$|x - a_n| \leq \|\xi - \alpha_n\| \leq r_{n-1},$$
$$|x - a_{n-1}| \leq \max(|x - a_n|, |a_n - a_{n-1}|) \leq r_{n-1},$$

namely $x \in B_{n-1}$. Since this happens for all integers $n > 0$, we infer $x \in \bigcap B_n$, and the intersection of the given decreasing sequence of balls is not empty.

The field Ω_p is spherically complete, hence complete.

3. The Completion $\mathbf{C}_p$ of the Field $\mathbf{Q}_p^a$

3.1. *Definition of $\mathbf{C}_p$*

Let us define

$$\mathbf{C}_p = \overline{\mathbf{Q}_p^a} = \text{closure of } \mathbf{Q}_p^a \text{ in } \Omega_p.$$

Hence $\mathbf{C}_p$ is a completion of $\mathbf{Q}_p^a$:

$$\mathbf{C}_p = \widehat{\mathbf{Q}_p^a}.$$

Proposition. *The field $\mathbf{C}_p$ is a separable metric space.*

PROOF. The algebraic closure $\mathbf{Q}_p^a$ of $\mathbf{Q}_p$ is a separable metric space (by Corollary 2 in (1.5)) and is dense in $\mathbf{C}_p$. Any countable dense subset of $\mathbf{Q}_p^a$ is automatically dense in $\mathbf{C}_p$: For example $\mathbf{Q}^a$ is dense in $\mathbf{C}_p$. ■

The universal field $\mathbf{C}_p$ is not locally compact: $|\mathbf{C}_p^\times| = p^{\mathbf{Q}} = \{p^\nu : \nu \in \mathbf{Q}\}$ is dense in $\mathbf{R}_{>0}$. We shall use the following notation

$$\mathbf{A}_p = \{x \in \mathbf{C}_p : |x| \le 1\}\text{: maximal subring of } \mathbf{C}_p,$$

$$\mathbf{M}_p = \{x \in \mathbf{C}_p : |x| < 1\}\text{: maximal ideal of } \mathbf{A}_p.$$

Hence $\mathbf{M}_p = \mathbf{M}_p^2$, and $\mathbf{A}_p$ is not a Noetherian ring (2.1).

3.2. *Finite-Dimensional Vector Spaces over a Complete Ultrametric Field*

Let us formulate and prove a generalization of (II.3.1) (cf. Theorem 2 in (II.A.6) for the most general version).

Theorem 1. *Let K be a complete (nondiscrete) ultrametric field and V a finite-dimensional vector space over K. Then all norms on V are equivalent.*

PROOF. We use induction on the dimension n of V. Since the property is obvious for $n = 1$, it is enough to establish it in dimension n assuming that it holds in dimension $n-1$. Choose a basis $(\mathbf{e}_i)_{1\le i\le n}$ of V and consider the vector space isomorphism $\varphi : K^n \to V$ sending the canonical basis of K^n onto the chosen basis of V. Considering that K^n is equipped with the sup norm, we have to show that for any given norm $\|\,.\,\|$ on V, the mapping φ is *bicontinuous*. First, for $\mathbf{x} = (x_i) \in K^n$, we have

$$\|x_1\mathbf{e}_1 + \cdots + x_n\mathbf{e}_n\| \le \sum |x_i|\|\mathbf{e}_i\| \le \max|x_i| \cdot \Sigma\|\mathbf{e}_i\|,$$

$$\|\varphi(\mathbf{x})\| \le C\|\mathbf{x}\| \quad (C = \Sigma\|\mathbf{e}_i\|),$$

which proves the continuity of the map φ. Conversely, let F be the subspace of V generated by the last $n-1$ basis vectors. Since the dimension of F is $n-1$, the induction hypothesis shows that on this subspace, the given norm is equivalent to the sup norm of the components. In particular, F is complete and closed in V. Since $\mathbf{e} = \mathbf{e}_1 \notin F$, we can define

$$d(\mathbf{e}, F) = \inf_{\mathbf{y}\in F} \|\mathbf{e} - \mathbf{y}\| > 0$$

and put $\gamma = d(\mathbf{e}, F)/\|\mathbf{e}\| \le 1$. By the induction hypothesis, there is also a constant c_F such that

$$\|\mathbf{y}\| \ge c_F \cdot \max_{2\le i\le n} |x_i| \quad (\mathbf{y} = \Sigma_{2\le i\le n}\, x_i\mathbf{e}_i \in F).$$

For each $\mathbf{v} = \varphi(\mathbf{x}) \in E - F$, say $\mathbf{v} = \xi\mathbf{e} + \mathbf{y}$ ($\xi \neq 0$, $\mathbf{y} \in F$), we can write

$$\mathbf{v} = \xi(\mathbf{e} + \mathbf{y}/\xi)$$

with

$$\begin{aligned}\|\mathbf{v}\| &= |\xi| \cdot \|\mathbf{e} + \mathbf{y}/\xi\| = |\xi| \cdot \|\mathbf{e} - \mathbf{y}'\| \\ &\geq |\xi| \cdot d(\mathbf{e}, F) = |\xi| \cdot \gamma \|\mathbf{e}\| = \gamma \cdot \|\xi\mathbf{e}\|,\end{aligned}$$

and hence

$$\|\mathbf{y}\| = \|\mathbf{v} - \xi\mathbf{e}\| \leq \max(\|\mathbf{v}\|, \|\xi\mathbf{e}\|) \leq \max(\|\mathbf{v}\|, \gamma^{-1}\|\mathbf{v}\|) = \|\mathbf{v}\|/\gamma$$

(since $\gamma \leq 1$). This shows that $\|\mathbf{v}\| \geq \gamma\|\mathbf{y}\|$. We have thus proved

$$\|\mathbf{v}\| \geq \gamma\|\xi\mathbf{e}\|, \quad \|\mathbf{v}\| \geq \gamma\|\mathbf{y}\|,$$
$$\|\mathbf{v}\| \geq \gamma \cdot \max(\|\xi\mathbf{e}\|, \|\mathbf{y}\|),$$

and since $\|\mathbf{y}\| \geq c_F \max_{i\geq 2} |x_i|$, we have

$$\begin{aligned}\|\varphi(\mathbf{x})\| = \|\mathbf{v}\| &\geq \gamma \max(|\xi|\|\mathbf{e}\|, c_F \max_{i\geq 2} |x_i|) \\ &\geq c \max_{i\geq 1} |x_i| = c \cdot \|\mathbf{x}\|,\end{aligned}$$

with $x_1 = \xi$ and $c = c_V = \gamma \min(c_F, \|\mathbf{e}\|)$. ∎

Corollary. *If K is a complete (nondiscrete) ultrametric field and L is a finite extension of K, there is at most one extension of the absolute value of K to L. Any K-automorphism of L is isometric.*

PROOF. Same as in (II.3.3). ∎

We can now give Krasner's lemma (1.5) in a more general form.

Theorem 2. *Let Ω be any algebraically closed extension of $\mathbf{Q}_p$ and $K \subset \Omega$ any complete subfield. Select an algebraic element a ($\in \Omega$) over K and denote by a^σ its conjugates over K. Let $r = \min_{a^\sigma \neq a} |a^\sigma - a|$. Then every algebraic element b over K, $b \in B_{<r}(a)$, generates with K an extension containing $K(a)$.*

PROOF. We can proceed as in (1.5), since we now have uniqueness of the extension of absolute values for finite extensions of K. For any algebraic element b such that $a \notin K(b)$, a has a conjugate $a^\sigma \neq a$ over $K(b)$ (the automorphism σ leaves all elements of $K(b)$ fixed), and

$$|b - a^\sigma| = |(b - a)^\sigma| = |b - a|,$$
$$|a - a^\sigma| \leq \max(|a - b|, |b - a^\sigma|) = |b - a|.$$

Hence

$$|b-a| \geq |a-a^{\sigma}| \geq r.$$

Taking the contrapositive, $|b-a| < r \Longrightarrow a \in K(b)$ and $K(a) \subset K(b)$. ∎

3.3. The Completion is Algebraically Closed

Theorem. *The universal field $\mathbf{C}_p$ is algebraically closed.*

PROOF. Let L ($\subset \Omega_p$) be a finite — hence algebraic — extension of $\mathbf{C}_p$. We can apply the general form of Krasner's lemma to the extension $\mathbf{C}_p \subset \Omega_p$, since we already know that

- the field $\mathbf{C}_p$ is complete,
- the field $\mathbf{Q}_p^a$ is algebraically closed,
- he field Ω_p has an absolute value extending the p-adic one.

Assume that $L = \mathbf{C}_p(a)$ is generated by an algebraic element a of degree $n \geq 1$ and let $f \in \mathbf{C}_p[X]$ be the monic irreducible polynomial of a. By density of the algebraic closure $\mathbf{Q}_p^a$ of $\mathbf{Q}_p$ in $\mathbf{C}_p$, we can choose (1.5) a polynomial $g \in \mathbf{Q}_p^a[X]$ sufficiently close to f in order to ensure that a root of g generates L over $\mathbf{C}_p$. But $\mathbf{Q}_p^a$ is algebraically closed, so that g has all its roots in $\mathbf{Q}_p^a$, and this proves that f has degree 1: $L = \mathbf{C}_p$. ∎

Comment. We have not used the possibility of extending the absolute value of $\mathbf{C}_p$ to finite extensions of this field, since we work in the field Ω_p constructed in (2.2). The general possibility of extending absolute values for finite (algebraic) extensions — where the base field is not locally compact — involves other algebraic techniques.

3.4. The Field $\mathbf{C}_p$ is not Spherically Complete

Proposition. *The universal field $\mathbf{C}_p$ is not spherically complete.*

PROOF. Here is an argument showing the existence of strictly decreasing sequences of closed balls of $\mathbf{C}_p$ having an empty intersection (without explicitly constructing one such sequence!).

Let $r_n \to r > 0$ be a strictly decreasing sequence of $\Gamma = p^{\mathbf{Q}} = |\mathbf{C}_p^{\times}|$

$$r_0 > r_1 > \cdots > r_n > \cdots > \lim r_n = r > 0.$$

In the ball $B = B_{\leq r_0}(0)$ we can choose two closed disjoint balls B_0 and B_1 with the same radius $r_1 < r_0$. In each of these we can select two closed disjoint balls of radii $r_2 < r_1$, say

$$B_{i0} \text{ and } B_{i1} \text{ closed and disjoint in } B_i.$$

Continuing these choices, we define sequences of closed balls having decreasing radii given by the sequence (r_n) and satisfying in particular

$$B_i \supset B_{ij} \supset \cdots \supset B_{ij\cdots k} \supset B_{ij\cdots kl} \supset \cdots$$

(with multi-indices equal to 0 or 1). By construction, two balls having distinct multi-indices of the same length are disjoint. If $(i) = (i_1, i_2, \ldots)$ is a binary sequence we can define

$$B_{(i)} = \bigcap_{n \geq 1} B_{i_1 \cdots i_n}.$$

Such an intersection is either empty or is a closed ball of radius $r = \lim r_n$ having for center any element in it, as always in the ultrametric case. In any case, all $B_{(i)}$ are *open* subsets of $\mathbf{C}_p$ (this is where $r = \lim r_n > 0$ is used). If two sequences (i) and (j) are distinct — say $i_n \neq j_n$ — then by construction $B_{i_1 \cdots i_n}$ and $B_{j_1 \cdots j_n}$ are disjoint, and a fortiori $B_{(i)} \subset B_{i_1 \cdots i_n}$, $B_{(j)} \subset B_{j_1 \cdots j_n}$ are disjoint. Since the metric space $\mathbf{C}_p$ is separable, the *uncountable family* of disjoint open sets $(B_{(i)})$ can only be a *countable set* of distinct open sets (any countable dense subset must meet all nonempty open sets). This forces most of the $B_{(i)}$ to be empty! ∎

A pictorial representation of the preceding proof is sketched in the exercises.

3.5. *The Field $\mathbf{C}_p$ is Isomorphic to the Complex Field $\mathbf{C}$*

The result of this section will not be used in this book. It gives the answer to a natural question, namely: What is the algebraic structure of the field $\mathbf{C}_p$?

Let us start by the determination of the cardinality of the field $\mathbf{C}_p$.

Lemma. *The field $\mathbf{C}_p$ has the power of the continuum.*

Proof. The unit ball of $\mathbf{Q}_p$ is $\mathbf{Z}_p \cong \prod_{n\geq 0}\{0, 1, \ldots, p-1\}$, hence has the power of the continuum c: numeration in base p gives a 1-1 correspondence with $[0, 1] \subset \mathbf{R}$ except for contably many overlaps, so these sets have the same cardinality. (In fact, each $\mathbf{Z}_p$ is homeomorphic to the Cantor set: Exercise 13 of Chapter I.) The field $\mathbf{Q}_p$ itself has the same power, since it is the countable union of balls $p^m\mathbf{Z}_p$ (each having cardinality c). All finite extensions of $\mathbf{Q}_p$ have the same power. The algebraic closure $\mathbf{Q}_p^a$ of $\mathbf{Q}_p$ still has the same power (the ring of polynomials in one variable over $\mathbf{Q}_p$ has also the power of the continuum). Finally, a countable product $(\mathbf{Q}_p^a)^{\mathbf{N}}$ cannot have bigger cardinality. Such a product contains all Cauchy sequences of $\mathbf{Q}_p^a$, and

$$\operatorname{Card}(\mathbf{C}_p) \leq \operatorname{Card}((\mathbf{Q}_p^a)^{\mathbf{N}}) = c.$$ ∎

Recall the terminology used for field extensions. A transcendence basis of a field extension K/k is a family $(X)_{i\in I}$ in K such that *the subfield $k(X_i)_{i\in I} \subset K$ is a purely transcendental extension of k, and $K/k(X_i)_{i\in I}$ is an algebraic extension.*

Here are some general results of Steinitz concerning field theory:

Two algebraic closures of a field k are k-isomorphic.
Every field extension K/k has a transcendence basis.
Two transcendence bases of K/k have the same cardinality.

For example, let $\mathbf{Q}^a$ be the algebraic closure of $\mathbf{Q}$ in $\mathbf{C}_p$ and $\mathbf{Q}^b$ the algebraic closure of $\mathbf{Q}$ in $\mathbf{C}$. Then there is an isomorphism

$$\mathbf{Q}^a \xrightarrow{\sim} \mathbf{Q}^b.$$

These fields are countable. But the fields $\mathbf{C}_p$ and $\mathbf{C}$ have the power of the continuum, hence the same transcendence degree (over the prime field $\mathbf{Q}$ or its algebraic closure).

Theorem. *The fields* $\mathbf{C}$ *and* $\mathbf{C}_p$ *are isomorphic.*

PROOF. Any extension of the rational field $\mathbf{Q}$ having the power of the continuum has a transcendence basis having this cardinality. By the above lemma the transcendence degrees of $\mathbf{C}$ and $\mathbf{C}_p$ over $\mathbf{Q}$ (or its algebraic closure) are the same, and we can select transcendence bases $(X)_{i\in I}$ in $\mathbf{C}$ and resp. $(Y)_{i\in I}$ in $\mathbf{C}_p$ (indexed by the same set). Now, $\mathbf{C}$ is an algebraic closure of $\mathbf{Q}(X_i)_{i\in I}$ and $\mathbf{C}_p$ is an algebraic closure of $\mathbf{Q}(Y_i)_{i\in I}$. Hence these two algebraic closures are isomorphic. ∎

As a consequence, we can view the field $\mathbf{C}_p$ as the complex field $\mathbf{C}$ endowed with an exotic topology. But the preceding considerations *do not lead to a canonical isomorphism* between these universal fields: The *axiom of choice* has to be used to show the existence of such an isomorphism.

Field $\supset B_{\le 1} \supset B_{<1}$	Residue field	Nonzero $\lvert\,.\,\rvert$	Properties
$\mathbf{Q}_p \supset \mathbf{Z}_p \supset p\mathbf{Z}_p$	$\mathbf{F}_p$	$p^{\mathbf{Z}}$	locally compact
$K \supset R \supset P = \pi R$	$\mathbf{F}_q\ (q = p^f)$	$\lvert\pi\rvert^{\mathbf{Z}} = p^{\frac{1}{e}\mathbf{Z}}$	$ef = \dim_{\mathbf{Q}_p} K < \infty$ locally compact
$\mathbf{Q}_p^a \supset A^a \supset M^a$	$k^a = \mathbf{F}_p^a = \mathbf{F}_{p^\infty}$	$p^{\mathbf{Q}}$	algebraically closed not locally compact
$\mathbf{C}_p \supset \mathbf{A}_p \supset \mathbf{M}_p$	$\mathbf{F}_p^a = \mathbf{F}_{p^\infty}$	$p^{\mathbf{Q}}$	algebraically closed complete
$\Omega_p \supset A_\Omega \supset M_\Omega$	k_Ω uncountable	$\mathbf{R}_{>0}$	algebraically closed spherically complete

4. Multiplicative Structure of $\mathbf{C}_p$

4.1. Choice of Representatives for the Absolute Value

Definition. *Let G be an abelian group written multiplicatively and $n \geq 2$ an integer. We say that*

1. G *is* n-divisible *if for each $g \in G$, there is $x \in G$ with $x^n = g$,*
2. G *is* uniquely n-divisible *if for each $g \in G$, there is a unique $x \in G$ with $x^n = g$,*
3. G *is* divisible *if it is n-divisible for all $n \geq 2$.*

A simple application of Zorn's lemma will show the possibility of extending all homomorphisms having a divisible group as target.

Theorem. *Let G be a divisible abelian group. For each abelian group H and homomorphism $\varphi : H_0 \to G$ of a subgroup $H_0 \subset H$, there is a homomorphism $\psi : H \to G$ extending φ.*

Proof. Consider all homomorphisms $H_0 \subset H' \xrightarrow{\varphi'} G$ (H' is a subgroup of H containing H_0) extending a given homomorphism $\varphi : H_0 \to G$. There will be a *maximal* one ψ for the order relation given by *continuation*: Every totally ordered set of extensions has an upper bound, defined in the obvious way on the union of the increasing chain of subgroups. I claim that the domain of such a maximal homomorphism is the whole group H. Indeed, if the domain of an extension φ' is a proper subgroup $H' \subset H$, let us show that it is not maximal. For this purpose, select any element $g \in H, g \notin H'$ and consider the subgroup H'' generated by H' and g, namely the image of the homomorphism

$$(\ell, x') \mapsto g^\ell x' : \mathbf{Z} \times H' \to H.$$

When the only power of the element g that lies in H' is the trivial one, the subgroup H'' is isomorphic to $\mathbf{Z} \times H'$, and an extension of φ' is given by

$$\varphi''(g^\ell x') := \varphi'(x').$$

If other powers of g lie in H', the inverse image of H' by the homomorphism $\ell \mapsto g^\ell : \mathbf{Z} \to H$ is a nontrivial subgroup $m\mathbf{Z} \subset \mathbf{Z}$ ($m > 0$) (in other words, g^m is the smallest positive power of g in H'). In this case, we choose an mth root $z \in G$ of $\varphi'(g^m) \in G$ such that $z^m = \varphi'(g^m)$. We can define the extension $\varphi'' : H'' \to G$ by

$$\varphi''(g^\ell x') := z^\ell \varphi'(x').$$

This is well-defined because if $g^{\ell_1} x'_1 = g^{\ell_2} x'_2$ ($x'_i \in H'$), we have

$$g^{\ell_1 - \ell_2} = x'_2 (x'_1)^{-1} \in H';$$

hence $\ell_1 - \ell_2 = km$ is a multiple of m and

$$\varphi'(g^{\ell_1-\ell_2}) = \varphi'(g^{km}) = \varphi'(g^m)^k = (z^m)^k = z^{mk},$$
$$\varphi'(x_2')\varphi'(x_1')^{-1} = \varphi'(x_2'(x_1')^{-1}) = \varphi'(g^{\ell_1-\ell_2}) = z^{\ell_1-\ell_2},$$

and finally

$$z^{\ell_1}\varphi'(x_1') = z^{\ell_2}\varphi'(x_2'). \qquad \blacksquare$$

Remarks. (1) For an additively written abelian group G, divisibility requires that all equations $nx = a$ ($x \in G, n$ positive integer) have (at least) one solution $x \in G$, hence the terminology. For example, the additive groups $\mathbf{Q}$ and $\mathbf{R}$ are divisible, but $\mathbf{Z}$ is not a divisible group.

(2) An abelian group G having the extension property mentioned in the statement of the theorem is called *injective group* or *injective* $\mathbf{Z}$*-module*.

Application. The universal field $\mathbf{C}_p$ is algebraically closed; hence the multiplicative group $\mathbf{C}_p^\times$ is divisible. The homomorphism $\varphi : \mathbf{Z} \to \mathbf{C}_p^\times$ defined by $\varphi(n) = p^n \in \mathbf{C}_p^\times$ has an extension $\psi : \mathbf{Q} \to \mathbf{C}_p^\times$. This extension is one-to-one, since its kernel is a subgroup of $\mathbf{Q}$ with $\ker\psi \cap \mathbf{Z} = \{0\}$. The image of ψ is a discrete subgroup $\Gamma \subset \mathbf{C}_p^\times$ isomorphic to the subgroup $p^{\mathbf{Q}} \subset \mathbf{R}_{>0}$. Instead of $\psi(r)$ we shall often write $p^r \in \mathbf{C}_p$ and $\psi(\mathbf{Q}) = p^{\mathbf{Q}}$. But — although the notation does not emphasize it — this subgroup $p^{\mathbf{Q}} \subset \mathbf{C}_p^\times$ depends on a sequence of choices of roots of p in $\mathbf{C}_p$ and is not *canonical*. When we consider $p^{\mathbf{Q}}$ as a subgroup of $\mathbf{C}_p^\times$, we have to remember that $|p^a| = 1/p^a > 0$. This subgroup is a complement to the kernel

$$\mathrm{U}(1) = \{x \in \mathbf{C}_p : |x| = 1\} \subset \mathbf{C}_p^\times$$

of the absolute value. In particular, we have a direct product decomposition

$$\mathbf{C}_p^\times = \Gamma \cdot \mathrm{U}(1) \cong p^{\mathbf{Q}} \times \mathrm{U}(1)$$

(analogous to polar coordinates in $\mathbf{C}^\times$) given by

$$x = r \cdot (x/r) \mapsto (|x|, x/r) \quad (r \in \Gamma,\ |x| = |r|,\ x/r \in \mathrm{U}(1)).$$

Since both $\mathbf{A}_p$ and $\mathbf{M}_p$ are clopen subsets of the metric space $\mathbf{C}_p$, the subgroup $\mathrm{U}(1) = \mathbf{A}_p - \mathbf{M}_p$ is clopen and the preceding product is a *topological isomorphism.*

4.2. *Roots of Unity*

A first analysis of the structure of the group of units

$$\mathrm{U}(1) = \mathbf{A}_p - \mathbf{M}_p \subset \mathbf{C}_p^\times$$

is made by looking at the reduction mod $\mathbf{M}_p$. The restriction of the (ring) homomorphism

$$\varepsilon : \mathbf{A}_p \to \mathbf{A}_p/\mathbf{M}_p = \mathbf{F}_{p^\infty}$$

(where $\mathbf{F}_{p^\infty}$ is an algebraic closure of $\mathbf{Z}_p/p\mathbf{Z}_p = \mathbf{F}_p$) to units gives a surjective (group) homomorphism (II.4.3)

$$\varepsilon : \mathrm{U}(1) \to \mathbf{F}_{p^\infty}^\times$$

with kernel $\overset{-1}{\varepsilon}(1) = 1 + \mathbf{M}_p \subset \mathrm{U}(1)$, whence a canonical isomorphism

$$\mathrm{U}(1)/(1+\mathbf{M}_p) \cong \mathbf{F}_{p^\infty}^\times.$$

In the algebraically closed field $\mathbf{C}_p$, we can find roots of unity of all orders, so that $\mu = \mu(\mathbf{C}_p)$ is isomorphic to the group of roots of unity in the complex field. There is a canonical product decomposition of this group,

$$\mu = \mu_{(p)} \cdot \mu_{p^\infty} \quad \text{(direct product)},$$

where $\mu_{(p)}$ is the subgroup consisting of the roots of unity of order prime to p, and μ_{p^∞} the subgroup consisting of the pth power roots of unity (in $\mathbf{C}_p$).

The restriction of the reduction homomorphism ε gives an isomorphism of this subgroup $\mu_{(p)}$ with $\mathbf{F}_{p^\infty}^\times$, and hence a direct product decomposition

$$\mathrm{U}(1) = \mu_{(p)} \cdot (1+\mathbf{M}_p) \subset \mathbf{C}_p^\times.$$

On the other hand,

$$\mu_{p^\infty} \subset (1+\mathbf{M}_p) \cap \mathbf{Q}_p^a.$$

Let us recall the more precise result established in (II.4.4).

Theorem. *Let $\zeta \in \mu_{p^\infty} \subset \mathbf{C}_p$ be a root of unity having order p^t $(t \geq 1)$. Then*

$$|\zeta - 1| = |p|^{1/\varphi(p^t)} < 1 \quad (\varphi(p^t) = p^{t-1}(p-1)). \qquad \blacksquare$$

For a subextension K of $\mathbf{C}_p$, the link with the notation used in (II.4.3) is

$\mu_{(p)}(K) = \mu_{(p)} \cap K$: roots of unity (in K) having order prime to p,

$\mu_{p^\infty}(K) = \mu_{p^\infty} \cap K$: pth power roots of unity (in K).

4.3. Fundamental Inequalities

In the preceding section (4.2) — based on II.4.4 — we recalled the estimates for absolute values of pth powers. Such estimates form a recurring theme of p-adic analysis, and we give a few more precise forms of these estimates for convenient reference. The first one is purely algebraic.

Fundamental Inequalities: First form. *Denote by $I = (p, T)$ the ideal of the ring $\mathbf{Z}[T]$ generated by the prime p and the indeterminate T. Then*

$$(1+T)^{p^n} - 1 \in T \cdot I^n \quad (n \geq 0).$$

Proof. For $n = 0$, the assertion is a tautology, and we proceed by induction on $n \geq 0$. Assume that $(1+T)^{p^n} = 1 + Tu$ for some $u \in I^n$. Hence

$$(1+T)^{p^{n+1}} = (1+Tu)^p = 1 + pTuv + T^p u^p$$

for some polynomial $v \in \mathbf{Z}[T]$. But

$$pTu \in T \cdot pI^n \subset T \cdot I^{n+1},$$
$$T^p u^p = T \cdot T^{p-1} u^p \in T \cdot I^{n+1}$$

(since $p \geq 2$), and the sum $pTu + T^p u^p$ belongs to $T \cdot I^{n+1}$ as expected. ■

Let us replace the indeterminate T by an element $t \in \mathbf{A}_p \subset \mathbf{C}_p$. Since each element in I^n is a sum of terms containing factors $p^i T^{n-i}$ for $0 \leq i \leq n$, the ultrametric inequality shows that all elements obtained have an absolute value smaller than or equal to the maximum of $|p^i t^{n-i}|$, and we see that we have obtained the following inequality.

Fundamental Inequalities: Second form. *Let $t \in \mathbf{C}_p$, $|t| \leq 1$. Then*

$$|(1+t)^{p^n} - 1| \leq |t| \cdot (\max(|t|, |p|))^n \quad (n \geq 0)$$

(cf. (V.4.3)). ■

Other forms are often used (they are not completely equivalent to the preceding ones, but also admit useful applications). We mention them briefly.

Third form. *Let K be a finite extension of $\mathbf{Q}_p$, $K \supset R \supset P$. Then*

$$(1+P)^{p^n} \subset 1 + P^{n+1} \quad (n \geq 0).$$

If $P = \pi R$ and $|\pi| = \theta < 1$ (generator of the discrete group $|K^\times| \subset \mathbf{R}_{>0}$), then in K the announced inclusion is equivalent to

$$|t| \leq \theta \Longrightarrow |(1+t)^{p^n} - 1| \leq \theta^{n+1}. \quad ■$$

This third form follows from the first one (replace T by π) but is less precise than the second form because

$$p \in P, \ |p| = \theta^e$$

and $\theta = |p|^{1/e} > |p|$ if $e > 1$.

Fourth form. *With the same assumptions as in the third form, we have*

$$(1+t)^n \equiv 1+nt \pmod{pntR}$$

if $t \in 2pR$ ($n \in \mathbf{N}$, $\mathbf{Z}$ *or even* $\mathbf{Z}_p$). ■

If we look at the first term only in the expansion

$$(1+t)^n - 1 - nt = n(n-1)t^2/2 + \cdots,$$

we find that for $t/2 \in pR$,

$$\frac{n(n-1)t^2}{2} = (n-1)\cdot nt \cdot \frac{t}{2} \in nt \cdot pR.$$

It only remains to check that the next terms are not competitive. Since we shall not need this form before Chapter VII, we refrain from giving a proof now. It will be obtained by a general method in (V.3.6). ■

4.4. *Splitting by Roots of Unity of Order Prime to p*

We have a direct product decomposition (4.2)

$$\mathbf{U}(1) \cong \mu_{(p)} \times (1+\mathbf{M}_p)$$

of the multiplicative subgroup defined by $|x|=1$ in $\mathbf{C}_p^\times$. The corresponding projection $\mathbf{U}(1) \to \mu_{(p)}$ is the *Teichmüller character*. It can be made explicit in several forms. Let $|x|=1$ and $K=\mathbf{Q}_p(x)$ have residue degree f. The residue field $k=R/P$ of K has order $q=p^f$, and the reduction homomorphism ε sends the given unit x to an element $\varepsilon(x) \in \mathbf{F}_q^\times$ of order dividing $q-1$ (II.4.3). Hence

$$\varepsilon(x)^{q-1} = 1, \; x^{q-1} \equiv 1 \pmod{P}.$$

The fundamental inequality (second form) shows that the pth powers of $x^{q-1} = 1+t$ ($t \in P \subset K$ or $t \in \mathbf{M}_p \subset \mathbf{C}_p$) tend to 1:

$$x^{(q-1)p^n} \to 1 \quad (n \to \infty).$$

A fortiori, taking $n = fm$,

$$\frac{x^{q^{m+1}}}{x^{q^m}} = x^{(q-1)q^m} \to 1 \quad (m \to \infty).$$

Say $x^{q^{m+1}} = x^{q^m}(1+\varepsilon_m)$ where $\varepsilon_m \to 0$. Hence $x^{q^{m+1}} - x^{q^m} = x^{q^m}\varepsilon_m \to 0$, and the Cauchy sequence $(x^{q^m})_{m\geq 0}$ has a limit ζ in the complete (locally compact) field $K \subset \mathbf{C}_p$. Obviously, $\zeta^q = \zeta$ and

$$\zeta = \lim_{m\to\infty} x^{q^m} = x + (x^q - x) + (x^{q^2} - x^q) + \cdots \equiv x \pmod{P}.$$

The map

$$x \mapsto \zeta = \omega(x) = \lim_{m\to\infty} x^{q^m}$$

defines a homomorphism $\mathbf{U}(1) \cap K^\times \to \mu_{q-1} \subset K^\times$ that corresponds to the projection on the first factor in the direct product decomposition (II.4.3)

$$\mathbf{U}(1) \cap K^\times \cong \mu_{q-1} \times (1+P).$$

It is possible to give a formula working independently from the residue degree of $x \in \mathbf{U}(1)$. Indeed, if q is given, the subsequence $(x^{p^{n!}})$ has an end tail in (x^{q^m}).

We have obtained the following result.

Theorem. *Let $x \in \mathbf{C}_p$ with $|x| = 1$. Then the sequence $(x^{p^{n!}})$ converges to the unique root of unity that is congruent to $x \pmod{\mathbf{M}_p}$ and the homomorphism*

$$\omega : x \mapsto \zeta = \omega(x) = \lim_{m\to\infty} x^{p^{n!}}$$

corresponds to the projection on the first factor in the direct product decomposition

$$\mathbf{U}(1) \cong \mu_{(p)} \times (1 + \mathbf{M}_p).$$ ■

4.5. *Divisibility of the Group of Units Congruent to 1*

In this section we investigate the divisibility properties of the multiplicative group $1 + \mathbf{M}_p$.

Proposition 1. *The group $1 + \mathbf{M}_p$ is divisible. For each $m \geq 2$ prime to p, it is uniquely m-divisible.*

Proof. It is enough to prove that the group $1 + \mathbf{M}_p$ is p-divisible and uniquely m-divisible for each m prime to p.

(1) Let $1+t \in 1+\mathbf{M}_p$ and select a root $x \in \mathbf{C}_p$ of $X^p - (1+t)$: this is possible, since this field is algebraically closed. Since $|x|^p = |x^p| = |1+t| = 1$, we have $|x| = 1$: $x \in \mathbf{U}(1)$. Now

$$(x \bmod \mathbf{M}_p)^p = x^p \bmod \mathbf{M}_p = 1 \in k$$

implies $x \bmod \mathbf{M}_p = 1$, since k has characteristic p. This proves $x = 1 + s \in 1 + \mathbf{M}_p$.

(2) Let $1 + t \in 1 + \mathbf{M}_p$ and select a positive integer m prime to p. We are looking for a root of the polynomial $f(X) = X^m - (1 + t)$. We already have an approximate root $y = 1$ for which the derivative mX^{m-1} does not vanish mod $\mathbf{M}_p$ (p does not divide m):

$$f(y) = 1 - (1+t) = -t,\ f'(y) = m,\ |f'(y)| = 1.$$

Thus we have $|f(y)/f'(y)^2| = |-t| < 1$, and Hensel's lemma (II.1.5) is applicable: There is a unique root of f in the open ball of center 1 and radius 1. ■

In fact, for each $\zeta \in \mu_m \subset \mu_{(p)} \subset \mathbf{F}_{p^\infty}^\times$, there is one root x of f with $x \equiv \zeta \pmod{\mathbf{M}_p}$. These m roots of f are all the roots of this polynomial, and for each given $\zeta \in \mu_m$ there can be only *one* root of f congruent to this root of unity ζ.

For later reference, let us formulate explicitly the following characterization of the topological torsion of $\mathbf{C}_p^\times$.

Proposition 2. *For $x \in \mathbf{C}_p$ we have*

$$x \in 1 + \mathbf{M}_p \iff x^{p^n} \to 1 \quad (n \to \infty).$$

PROOF. If $x = 1 + t \in 1 + \mathbf{M}_p$, the sequence

$$x^{p^n} - 1 = (1+t)^{p^n} - 1$$

tends to 0 by the fundamental inequality (4.3) (second form). Conversely, assume that $x^{p^n} \to 1$ (for some $x \in \mathbf{C}_p$) and take an integer n such that x^{p^n} belongs to the open neighborhood $1 + \mathbf{M}_p$ of 1 in $\mathbf{C}_p$. Since we have proved in (4.1) that there is a torsion-free subgroup Γ ($\cong p^{\mathbf{Q}}$) of $\mathbf{C}_p^\times$ and a direct-product decomposition

$$\mathbf{C}_p^\times = \Gamma \cdot \mu_{(p)} \cdot (1 + \mathbf{M}_p),$$

we see that $x \in \mu_{(p)} \cdot (1 + \mathbf{M}_p)$. The first component ζ of x is trivial simply because it has an order prime to p:

$$x^{p^n} \in 1 + \mathbf{M}_p \Longrightarrow \zeta^{p^n} = 1 \Longrightarrow \zeta = 1.$$ ■

Observe that the convergent sequence $(x^{p^n})_{n \geq 0}$ is *eventually constant* precisely when x is a pth power root of unity

$$x \in \mu_{p^\infty} \subset 1 + \mathbf{M}_p.$$

Appendix to Chapter 3: Filters and Ultrafilters

A.1. Definition and First Properties

Let X be a set. A family $\mathcal{F}$ of subsets of X is a *filter* when

0. $X \in \mathcal{F}, \emptyset \notin \mathcal{F}$,
1. $A \in \mathcal{F}, B \in \mathcal{F} \Longrightarrow A \cap B \in \mathcal{F}$,
2. $A \in \mathcal{F}, A' \supset A \Longrightarrow A' \in \mathcal{F}$.

If there is a filter on a set X, then this set is not empty by the condition 0. The condition 1 shows (by induction) that the intersection of a finite family of subsets $A_i \in \mathcal{F}$ is an element of the filter $\mathcal{F}$ and in particular is not empty. The intersection of all elements of $\mathcal{F}$ may be empty, in which case we say that this filter is *free*.

Example. Let X be a subset of a topological space Y, choose a point $y \in \overline{X} - X$, and define a filter $\mathcal{F}$ on X as follows:

$$\mathcal{F} = \{V \cap X : V \text{ is a neighborhood of } y \text{ in } Y\}.$$

This example is typical, since if $\mathcal{F}$ is any free filter on a set X, we can define a topology on the disjoint union $Y = X \sqcup \{\omega\}$ by specifying its open sets:

$$\text{subsets of } X \text{ and subsets } A \cup \{\omega\} \ (A \in \mathcal{F}).$$

The topology induced on X is the discrete one, but ω is in the closure of X in Y, and the filter on X attached to ω is precisely $\mathcal{F}$.

A family $\mathcal{B} \subset \mathcal{F}$ is a basis of this filter if any $A \in \mathcal{F}$ contains a $B \in \mathcal{B}$.

Lemma. *Let $\mathcal{B}$ be a family of nonempty subsets of a set X such that*

$$A \in \mathcal{B},\ B \in \mathcal{B} \Longrightarrow \text{there exists } C \in \mathcal{B} \text{ such that } C \subset A \cap B.$$

Then the family of subsets of X containing elements of $\mathcal{B}$ is a filter having $\mathcal{B}$ as a basis. ∎

The filter constructed in the previous lemma is called the *filter generated* by $\mathcal{B}$.

Lemma. *Let $\mathcal{F}$ be a filter on a set X and let $f : X \to Y$ be a map. Then the family $f(\mathcal{F}) = \{f(A) : A \in \mathcal{F}\}$ is a filter on $f(X)$ and a basis of a filter on Y.* ∎

Example. Let $\mathcal{F}$ be a free filter on $\mathbf{N}$. Choose for each $n \in \mathbf{N}$ an element $A_n \in \mathcal{F}$ such that $n \notin A_n$. Hence $A = \bigcap_{1 \le n < N} A_n \in \mathcal{F}$ and $[N, \infty) \supset A$, and hence $[N, \infty) \in \mathcal{F}$. Any free filter on $\mathbf{N}$ contains all subsets $[N, \infty)$ $(N \in \mathbf{N})$.

More generally, let X be an infinite set. Then any free filter on X contains all cofinite subsets (i.e. complements of finite subsets) as elements. The cofinite subsets form the *Fréchet filter* on X.

A.2. Ultrafilters

The inclusion relation for families $\mathcal{F} \subset \mathcal{P}(X)$ is an order relation, and if $\mathcal{F}' \supset \mathcal{F}$, we say that $\mathcal{F}'$ is *finer* than $\mathcal{F}$. For example, any free filter on X is finer than the Fréchet filter.

In an obvious sense, the subsets of a finer filter $\mathcal{F}'$ are smaller than those of $\mathcal{F}$.[1]

Definition. *Maximal filters are called ultrafilters.*

[1]Compare with coffee powder, where finer grinding also provides finer granules!

Any totally ordered sequence of filters on a set X has a majorant (the union in $\mathcal{P}(X)$), and by Zorn's lemma, any filter is contained in a maximal one. For example, the Fréchet filter on X is contained in an ultrafilter (necessarily free).

Theorem. *Let $\mathcal{F}$ be a filter on a set X. Then $\mathcal{F}$ is an ultrafilter precisely when the following criterion is satisfied:*

$$\textit{for each } Y \subset X \textit{ either } Y \in \mathcal{F} \textit{ or } Y^c = X - Y \in \mathcal{F}.$$

Proof. If the condition is satisfied, $\mathcal{F}$ is obviously maximal. Conversely, assume that there is a subset $Y \subset X$ with $Y \notin \mathcal{F}$ and $Y^c \notin \mathcal{F}$. Observe that all $A \in \mathcal{F}$ meet Y:

$$Y^c \notin \mathcal{F} \underset{A \in \mathcal{F}}{\Longrightarrow} Y^c \not\supset A \Longrightarrow Y \cap A \neq \emptyset.$$

Define $\mathcal{F}' \supset \mathcal{F}$ as follows:

$$\mathcal{F}' = \{A' \subset X : A' \supset A \cap Y \text{ for some } A \in \mathcal{F}\}.$$

Hence $\mathcal{F}'$ is a filter, and $Y \in \mathcal{F}'$. Since $Y \notin \mathcal{F}$, $\mathcal{F}'$ is strictly finer than $\mathcal{F}$, proving that this last filter was not maximal. ■

Corollary 1. *Let $\mathcal{U}$ be an ultrafilter on a set X. If $A_1, \ldots, A_n$ is a finite family of subsets of X such that $\bigcup_{1 \le i \le n} A_i \in \mathcal{U}$, then there exists at least one index i for which $A_i \in \mathcal{U}$.*

Proof. It is enough to prove the assertion for two subsets (by induction). If $A \notin \mathcal{U}$ and $B \notin \mathcal{U}$, we infer from the above criterion that $A^c \in \mathcal{U}$, $B^c \in \mathcal{U}$; hence $(A \cup B)^c = A^c \cap B^c \in \mathcal{U}$, and $A \cup B \notin \mathcal{U}$. ■

Corollary 2. *Let $f : X \to Y$ and let $\mathcal{U}$ be an ultrafilter on the set X. Then $f(\mathcal{U})$ is an ultrafilter on $f(X)$ and a basis of an ultrafilter on Y.*

Proof. It is enough to prove the assertion when f is surjective. For any $A \subset Y$, either $f^{-1}(A)$ or $f^{-1}(A)^c = f^{-1}(A^c)$ belongs to $\mathcal{U}$; hence

$$\text{either } A = f(f^{-1}(A)) \text{ or } A^c = f(f^{-1}(A^c)) \text{ belongs to } \mathcal{U}.$$

By the criterion, $f(\mathcal{U})$ is an ultrafilter on Y. ■

A.3. Convergence and Compactness

Definition. *Let X be a topological space. A filter $\mathcal{F}$ on X is said to converge to a point $x \in X$ if it is finer than the filter of neighborhoods of this point, namely, when each neighborhood of x in X contains a subset $A \in \mathcal{F}$.*

For example, the filter of neighborhoods of a point converges to this point. In a Hausdorff space, a convergent filter can converge to at most one point.

Let X be a compact space. Then for each filter $\mathcal{F}$ on X, the family $(\overline{A})_{A\in\mathcal{F}}$ has nonempty finite intersections; hence by compactness

$$\Omega := \bigcap_{A\in\mathcal{F}} \overline{A} \neq \emptyset.$$

If U is any open set containing Ω, then

$$U^c \cap \Omega = \emptyset \Longrightarrow U^c \cap \bigcap_{j\in J} \overline{A_j} = \emptyset \Longrightarrow U \supset \bigcap_{j\in J} \overline{A_j} \supset \bigcap_{j\in J} A_j \ (\in \mathcal{F})$$

for some finite family (A_j) of subsets $A_j \in \mathcal{F}$. This proves that U contains an element of $\mathcal{F}$ and this filter is finer than the filter of neighborhoods of Ω.

Theorem. *In a compact space, every ultrafilter converges.*

Proof. Let $\mathcal{U}$ be an ultrafilter on the compact space X and choose x in the nonempty intersection $\bigcap_{A\in\mathcal{U}} \overline{A}$. The nonempty subsets

$$U \cap V \quad (U \in \mathcal{U}, V \text{ neighborhood of } x)$$

generate a filter finer than $\mathcal{U}$, hence equal to $\mathcal{U}$. Hence this ultrafilter converges to x, and a posteriori

$$\bigcap_{A\in\mathcal{U}} \overline{A} = \{x\}.$$

■

Application. Let $\mathcal{U}$ be an ultrafilter on the set $\mathbf{N}$ of natural numbers and let $(a_n)_{n\geq 0}$ be a bounded sequence of real numbers. Then $\lim_{\mathcal{U}} a_n$ exists and

$$\inf_n a_n \leq \lim_{\mathcal{U}} a_n \leq \sup_n a_n.$$

Proof. Since the sequence $(a_n)_{n\geq 0}$ is bounded, then

$$-\infty < \alpha := \inf_n a_n \leq \beta := \sup_n a_n < \infty,$$

and this sequence defines a map

$$n \mapsto a_n : \mathbf{N} \to [\alpha, \beta] \subset \mathbf{R}$$

taking its values in a compact space. The image of the ultrafilter $\mathcal{U}$ is a basis of an ultrafilter in the compact space $[\alpha, \beta]$; hence it converges in this space. ■

A.4. *Circular Filters*

Let $K = \Omega_p$ be the spherically complete extension of $\mathbf{Q}_p$ constructed in (III.2). Recall that $|\Omega_p^\times| = \mathbf{R}_{>0}$ and the residue field k_Ω is infinite.

To each closed ball $B \subset K$ we associate a filter $\mathcal{F}_B$ on K defined as follows:

If the ball B is a single point $\{a\}$, we take for $\mathcal{F}_B$ the filter of neighborhoods of this point, generated by the $B_{<\varepsilon}(a)$ ($\varepsilon > 0$).

If $B = B_{\leq r}(a)$ has positive radius r, we take for $\mathcal{F}_B$ the filter generated by the subsets

$$A(\varepsilon, a_1, \dots, a_n) = B_{\leq r+\varepsilon}(a) - \bigcup_{1 \leq i \leq n} B_{<r-\varepsilon}(a_i) \quad (0 < \varepsilon < r,\ a_i \in B).$$

When ε decreases and/or the number of points n increases, these subsets decrease, and we see that these subsets make up a basis of a filter. The filter $\mathcal{F}_B$ generated by this basis is the *circular filter* associated to the closed ball B.

By definition, the subset $A(\varepsilon, a_1, \dots, a_n)$ contains

$$r < |x - a| < r + \varepsilon \quad (x \in K)$$

(observe that this set is independent of the choice of center $a \in B$). Also, for any $b \in B$ there is a $\delta > 0$ such that

$$\{x \in K : r - \delta < |x - b| < r\} \subset A(\varepsilon, a_1, \dots, a_n).$$

Lemma. *Let B be a closed ball of radius $r > 0$ and choose $a \in B$. Then a basis of the circular filter $\mathcal{F}_B$ is given by the following subsets*

$$A'(\varepsilon, a_1, \dots, a_n) = \{r - \varepsilon < |x - a| < r + \varepsilon\} - \bigcup_{\text{finite}} B_{<r-\varepsilon}(a_i),$$

where the a_i are chosen on the sphere $S_r(a) : |x - a| = r$ and $0 < \varepsilon < r$. ■

Here, replacing ε by a smaller one, we may even assume that the points a_i satisfy $|a_i - a_j| = r$ ($i \neq j$).

The preceding definitions can be relativized to a subset $X \subset K = \Omega_p$. Assume

$$X \cap A \neq \emptyset \text{ for all } A \in \mathcal{F}_B,$$

so that $\mathcal{F}_B$ induces a filter on X. Then this induced filter $\mathcal{F}_B(X) = \mathcal{F}_B \cap X$ is still called a *circular filter* on X.

For example, let $X = \mathbf{C}_p$. When the closed ball $B \subset K$ does not meet $\mathbf{C}_p$, we have $r := d(B, \mathbf{C}_p) > 0$, and if $\delta(B) = r$, the trace of $\mathcal{F}_B$ on $\mathbf{C}_p$ is a *circular filter without a center* in $\mathbf{C}_p$.

Exercises for Chapter 3

1. Prove that $\mathbf{Q}_p^a$ is not complete by considering the series $\sum_{(p,n)=1} p^n p^{1/n}$.
(*Hint.* Let x be the sum in a completion of $\mathbf{Q}_p^a$ and let K be the completion of $\mathbf{Q}_p(x)$.

Show by induction that all $p^{1/n} \in K$ for $(n, p) = 1$, and hence K is not algebraic over $\mathbf{Q}_p$.)

2. Let K be an algebraically closed valued field. Prove that its completion $\widehat{K}$ is also algebraically closed.
(*Hint.* Let $f(X) = X^n + a_{n-1}X^{n-1} + \cdots + a_1X + a_0 \in \widehat{K}[X]$ and select monic polynomials $f_j(X) = X^n + a_{n-1,j}X^{n-1} + \cdots + a_{1,j}X + a_{0,j} \in K[X]$ converging coefficientwise to f. Then $\delta_j := \|f_{j+1} - f_j\| = \max_i |a_{i,j+1} - a_{i,j}| \to 0\ (j \to \infty)$. Choose inductively a root x_j (in the algebraically closed field K) of f_j so that $(x_j)_j$ is a Cauchy sequence (cf. III.1.5) and hence converges in the completion $\widehat{K}$ to a root of f. This type of proof also appears in (VI.2.2).)

3. Let X be the real Banach space consisting of sequences $x = (x_n)_{n\geq 0}$ of real numbers converging to zero with the norm $\|x\| = \sup|x_n| = \max|x_n|$. Consider the sequence in X defined by

$$a_0 = (0), \quad a_n = (1+1, 1+\tfrac{1}{2}, \ldots, 1+\tfrac{1}{n}, 0, 0, \ldots) \quad (n \geq 1)$$

so that $\|a_n\| = 2\ (n \geq 1)$. Show that with the induced metric, the set $A = \{a_n : n \geq 0\}$ is an complete ultrametric space which is not spherically complete.
(*Hint.* The induced metric on A is given by

$$d(a_n, a_{n+k}) = \|a_n - a_{n+k}\| = 1 + \tfrac{1}{n+1} \quad (n \geq 0,\ k \geq 1).$$

What is the closed ball of center a_n and radius $1 + \frac{1}{n}$?)

4. (*a*) Let X be a complete metric space having the following property: Any decreasing sequence of possible values of the distance function converges to 0. Show that X is spherically complete.
(*b*) If a complete metric space is not spherically complete, show that we can replace its metric by a uniformly equivalent one δ for which it is spherically complete. (Hint: For given x and y, define $\delta(x, y) = 2^n$, where the integer $n \in \mathbf{Z}$ is chosen so that $d(x, y) \leq 2^n < 2d(x, y)$. Then use (*a*).)

5. Prove that the residue field of Ω_p is uncountable.
(*Hint.* Each sequence $\mathbf{N} \to \mu_{(p)} \subset \mathbf{Q}_p^a$ leads to a nonzero element of the residue field k_Ω of Ω_p. If $\mathbf{N} \to k_\Omega$ is any map, use Cantor's diagonal procedure as in (I.1.1) to define an element not contained in the image.)

6. There are many possible choices of copies of $p^{\mathbf{Q}}$ in $\mathbf{C}_p$. Let φ_n denote the homomorphism $x \mapsto x^n$: $\mathbf{C}_p^\times \to \mathbf{C}_p^\times\ (n \geq 1)$, then $\ker \varprojlim \varphi_n = \varprojlim \mu_{n!}$ gives a parametrization of choices. (Recall that a countable projective limit of surjective maps is surjective (4.3).)

7. Let K be an extension of $\mathbf{Q}_p$ with $|K^\times|$ dense in $\mathbf{R}_{>0}$. Recall (exercise of Chapter II) that the tree T_K is the ordered set of closed balls of K. This tree comes with a projection $\delta : T_K \to \mathbf{R}_{\geq 0}$. For $r > 0$, the fiber $\delta^{-1}(r) = K/B_{\leq r}$ is the uniformly discrete quotient group of closed balls of radius r.
   (*a*) Show that the maximal totally ordered subsets of T_K are isomorphic to either $[0, \infty)$ or to $(0, \infty)$: Let us call these subsets *maximal branches*, and in the first case, we say that the corresponding branch bears a fruit. The projection by δ of a maximal branch is either an isomorphism with the interval $[0, \infty)$ or an isomorphism with an interval (r, ∞) $(r \geq 0)$: The fruit of a branch can lie only above $r = 0$.

(*b*) Show that K is complete exactly when all maximal branches having a projection containing $(0, \infty)$ do bear a fruit, i.e., are isomorphic to $[0, \infty)$ by projection.

(*c*) Assume that the field K is separable, so that all fibers $\delta^{-1}(r)$ $(r > 0)$ are countable. Show that such a field cannot be spherically complete.

(*Hint.* The set of distinct branches having a nonempty intersection with any $\delta^{-1}[r', r'']$ for some fixed $r' < r''$ is uncountable.)

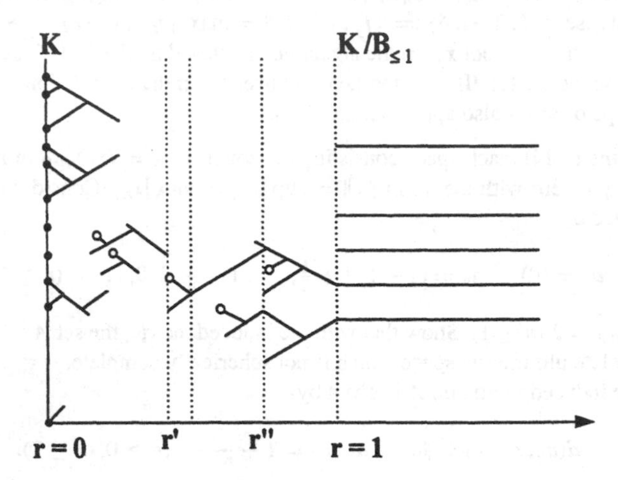

Tree of K: Fruits, branches, and holes

(*d*) Define an action of the 2×2 upper triangular matrix group $T_2^+(K) \subset \mathrm{Gl}_2(K)$ on T_K (cf. (VI.3.I)). When $K = \Omega_p$, show that this action is transitive on the subtree defined by $\delta > 0$.

8. Let $a \in \Omega_p - \mathbf{C}_p$ and $r := d(a, \mathbf{C}_p) > 0$. Show that the cases $S_r(a) \cap \mathbf{C}_p = \emptyset$ and $S_r(a) \cap \mathbf{C}_p \neq \emptyset$ both occur.
(*Hint.* Choose first $a \in \Omega_p$ with $|a| = 1$ and residue class $\tilde{a} \in k_\Omega$ not algebraic over the prime field: In this case $r = 1$ and the sphere $S_1(a)$ meets $\mathbf{C}_p$. On the other hand, select a decreasing sequence of closed balls $B_{\leq r_n}(a_n)$, $r_n \searrow 1$ having an empty intersection in $\mathbf{C}_p$ and choose a in the intersection of the same balls of Ω_p: The sphere $S_1(a)$ does not meet $\mathbf{C}_p$.)

9. Let K be an ultrametric field. Assume that both k (the residue field) and $|K^\times|$ are countable. Prove that for fixed $r > 0$, the set of dressed balls of radius r is also countable.
(*Hint.* Observe that the set of open balls of radius r is countable. Define a surjective map from the set of open balls to the set of closed balls of the same radius.)

10. Let K be an ultrametric field with $|K^\times|$ dense in $\mathbf{R}_{>0}$. For real $t \geq 1$ let $\mathcal{P}_t$ denote the partition of the closed unit ball $A = \{x \in K : |x| \leq 1\}$ into its cosets mod the additive subgroup $B_{\leq 1/t} = \{x \in A : |x| \leq 1/t\}$. The family $(\mathcal{P}_t)$ indexed by $t \in [0, \infty)$ has the property

$$\text{for } s > t \geq 1,\ \mathcal{P}_s \text{ is } \textit{strictly finer} \text{ than } \mathcal{P}_t.$$

(The "continuous family" $(\mathcal{A}_t)_{t \geq 1}$ of associated σ-algebras is a *filtration* of the space A in the sense used in the theory of stochastic processes.)

11. Let us denote by $B_{\leq r}$ $(r \geq 0)$ the additive subgroup $|x| \leq r$ (in $\mathbf{C}_p$ or in any ultrametric field having dense valuation).
    (*a*) For $0 < r < s$, show that the subgroup $B_{\leq r}$ of $B_{\leq s}$ has no supplement: $B_{\leq s}$ is not a direct product of $B_{\leq r}$ with another subgroup. In other words, the short exact sequence
    $$0 \to H = B_{\leq r} \to G = B_{\leq s} \to G/H \to 0$$
    does not split. (*Hint.* For all $x \in G = B_{\leq s}$, $p^n x \to 0$.)
    (*b*) For $0 < r < s < 1$ show that the multiplicative subgroup $1 + B_{\leq r}$ of $1 + B_{\leq s}$ has no supplement. (*Hint.* If $|x| < 1$, then $(1+x)^{p^n} \to 1$.)
    (*c*) For $0 < s^2 \leq r < s < 1$, prove that there is a canonical isomorphism
    $$(1 + B_{\leq s})/(1 + B_{\leq r}) \xrightarrow{\sim} B_{\leq s}/B_{\leq r}.$$
    (*Hint.* Consider the homomorphism $x \mapsto t = x - 1 \bmod B_{\leq r}$.)
    (*d*) We have $\mu_{p^\infty} \cap (1 + B_{<r_p}) = \{1\}$, but the direct product $\mu_{p^\infty} \cdot (1 + B_{<r_p})$ is a proper subgroup of $1 + \mathbf{M}_p$. Show that $1 + B_{<r_p}$ has no supplement in $1 + \mathbf{M}_p$ and more precisely, μ_{p^∞} is maximal among the subgroups $H \subset 1 + \mathbf{M}_p$ such that $H \cap (1 + B_{<r_p}) = \{1\}$. (*Hint.* The sequence $(1+t)^{p^n} \to 1$ is eventually stationary precisely when $1 + t \in \mu_{p^\infty}$.)

12. Prove the first form of the fundamental inequalities by induction, using $a = (1+T)^{p^n}$ and the factorization
    $$a^p - 1 = (a-1)(1 + a + \cdots + a^{p-1}),$$
    where each $a^k \in 1 + I$ $(k \geq 1)$ so that $1 + a + \cdots + a^{p-1} \in p + I = I$. (Observe that the case $n = 1$ of the statement is crucial, and the induction step is based solely on it!)

4

Continuous Functions on $\mathbf{Z}_p$

The goal of this chapter is the study of continuous functions on subsets of the p-adic field $\mathbf{Q}_p$ with values in an extension of $\mathbf{Q}_p$. Since $\mathbf{Q}_p$ admits a partition into clopen balls $x + \mathbf{Z}_p$ $(x \in \mathbf{Q}_p/\mathbf{Z}_p = \mathbf{Z}[1/p]/\mathbf{Z})$, it is enough to study continuous functions on $\mathbf{Z}_p$. Thus, we shall typically study continuous functions $\mathbf{Z}_p \to \mathbf{C}_p$. Since the natural numbers $\mathbf{N}$ form a dense subset of the ring $\mathbf{Z}_p$, we shall start by the study of functions on $\mathbf{N}$ or $\mathbf{Z}$ and with values in any abelian group.

In classical analysis, real- or complex-valued functions that are continuous on an interval can be uniformly approximated by polynomial functions (theorem of Weierstrass). But there is no canonical series representation for them. It is a specific feature of p-adic analysis that continuous functions $\mathbf{Z}_p \to \mathbf{C}_p$ have a canonical Mahler series representation. As has been noticed and proved by L. van Hamme, many systems of polynomials can also be used instead of the binomial system. This leads us into the umbral calculus, where suitable systems are found.

Due to the granular structure of $\mathbf{Z}_p$, the locally constant functions also constitute a dense subspace of $C(\mathbf{Z}_p; \mathbf{C}_p)$ (these functions correspond to the step functions on an interval in the classical theory). A *basis* of this space consisting of characteristic functions of suitable balls has been devised by M. van der Put.

1. Functions of an Integer Variable

1.1. *Integer-Valued Functions on the Natural Integers*

A polynomial $f(x) \in \mathbf{Q}[x]$ can take integral values on all natural integers even if its coefficients are not integers. For example $n^2 \equiv n \pmod 2$ shows that $\frac{1}{2}x^2 - \frac{1}{2}x$ is such a polynomial. More generally, $n^p \equiv n \pmod{p}$ shows that the polynomial $\frac{1}{p}x^p - \frac{1}{p}x$ is also such a polynomial.

The study of these polynomials is based on the following observation. Each *binomial polynomial*

$$\binom{x}{n} = \frac{x(x-1)\cdots(x-n+1)}{n!} \in \mathbf{Q}[x] \quad (n \geq 0)$$

defines an integer-valued function $\mathbf{N} \to \mathbf{N}$. This (and the theorem below) explains their central role in this chapter.

The first binomial polynomials are

$$\binom{x}{0} = 1, \quad \binom{x}{1} = x, \quad \binom{x}{2} = \frac{x^2}{2} - \frac{x}{2}.$$

One can read the sequence of values given by $\binom{\cdot}{n}$ in Pascal's triangle: The first values are 0 (outside of the triangle)

$$\binom{0}{n} = 0, \ \binom{1}{n} = 0, \ldots, \binom{n-1}{n} = 0, \ \binom{n}{n} = 1, \ \binom{n+1}{n} = n, \quad \text{etc.}$$

In the figure below, we exhibit the values of the binomial polynomials in vertical columns, with special attention to $\binom{n}{4}$.

$$\begin{array}{ccccccccc}
 & & & & \binom{n}{4} & & & & \\
 & & & & \downarrow & & & & \\
1 & & & & 0 & & & & \\
1 & 1 & & & 0 & & & & \\
1 & 2 & 1 & & 0 & & & & \\
1 & 3 & 3 & 1 & 0 & & & & \\
1 & 4 & 6 & 4 & 1 & & & & \\
1 & 5 & 10 & 10 & 5 & 1 & & & \\
1 & 6 & 15 & 20 & 15 & 6 & 1 & & \\
1 & 7 & 21 & 35 & 35 & 21 & 7 & 1 & \\
1 & 8 & 28 & 56 & 70 & 56 & 28 & 8 & 1 \\
 & & & & \vdots & & & &
\end{array}$$

Values of the binomial polynomials as vertical columns

On the other hand, introduce the *finite-difference operator* ∇ defined by

$$(\nabla f)(x) = f(x+1) - f(x).$$

(This is a discrete analogue of the gradient operator, whence the notation; we keep Δ for a discrete analogue of the Laplace operator.) This forward-difference operator acts on any function f on $\mathbf{N}$ taking values in an abelian group. An abelian group can always be considered as a $\mathbf{Z}$-module, and conversely, any $\mathbf{Z}$-module is an abelian group. Thus we shall now consider functions $f : \mathbf{N} \to M$ where M is a **Z**-module. The action of the finite-difference operator on the binomial functions is easily determined: An elementary computation shows that

$$\nabla\binom{x}{0} = 0, \quad \nabla\binom{x}{i} = \binom{x}{i-1} \quad (i \geq 1).$$

The binomial polynomials behave with respect to the difference operator as the polynomials $x^i/i!$ do with respect to the derivation operator:

$$D(x^0) = 0, \quad D\left(\frac{x^i}{i!}\right) = \frac{ix^{i-1}}{i!} = \frac{x^{i-1}}{(i-1)!} \quad (i \geq 1).$$

This analogy will be exploited and generalized.

Theorem. *Let M be any abelian group and let $f : \mathbf{N} \to M$ be an arbitrary map. Then there is a unique sequence $(m_i)_{i\geq 0}$ of M such that*

$$f(x) = \sum_{i\geq 0} m_i \binom{x}{i} = \sum_{0\leq i\leq x} m_i \binom{x}{i} \quad (x \in \mathbf{N}).$$

For $x \in \mathbf{N}$ only finitely many terms of the sum are nonzero, and $m_i = (\nabla^i f)(0)$.

Proof. Since $\binom{x}{i} = 0$ for $x = 0$ and $i \geq 1$, we see that $m_0 = f(0)$ is uniquely defined. The finite-difference operator can be used repeatedly to bring any coefficient into the *constant term* position:

$$\nabla f(x) = \sum_{i\geq 1} m_i \nabla\binom{x}{i} = \sum_{i\geq 1} m_i \binom{x}{i-1},$$

$$\nabla^k f(x) = \sum_{i\geq k} m_i \binom{x}{i-k}.$$

Hence $m_k = \nabla^k f(0)$. These computations already prove the *uniqueness* of the coefficients m_k and show how they have to be computed. Conversely, if the function f is given, let us compute the iterated differences $\nabla^k f(0) \in M$ and define $g(x) = \sum_{0\leq i\leq x} \nabla^i f(0)\binom{x}{i}$, $\varphi = f - g$. The iterated differences of φ vanish at the origin by construction: $\varphi(0) = 0$, $\varphi(1) - \varphi(0) = 0$, whence $\varphi(1) = 0, \ldots,$ from which it is apparent that φ vanishes at the points $0, 1, 2, \ldots$. More formally, one can establish by induction the general formula

$$\nabla^k \varphi(0) = \sum_{i\leq k} (-1)^i \binom{k}{i} \varphi(k-i) = \varphi(k) + \cdots.$$

The induction hypothesis $\varphi(j) = 0$ for all $j < k$ and $\nabla^k\varphi(0) = 0$ implies

$$\varphi(k) = -\sum_{1\leq i\leq k} (-1)^i \binom{k}{i} \varphi(k-i) = 0 \quad (k \geq 0).$$

Hence $\varphi \equiv 0$, as expected. This proves $f = g$ and the *existence* of an expansion of the desired form. ∎

Comments. (1) The preceding proof shows that the expansion of f is simply $f = \sum_{i\geq 0} m_i \binom{\cdot}{i}$. This series converges pointwise: Although infinitely many coefficients m_i will be nonzero in general, for each fixed $x \in \mathbf{N}$ the sum $\Sigma_{i\geq 0}\, m_i f_i(x)$ is a finite sum. Let us introduce the *Pochhammer symbol*

$$(x)_0 = 1,\ (x)_i = x(x-1)\cdots(x-i+1) \quad (i \geq 1),$$

so that

$$\nabla(x)_i = i(x)_{i-1} \text{ and } \binom{x}{i} = \frac{(x)_i}{i!}.$$

The preceding series expansion of f takes the form

$$f = \sum_{i\geq 0} \frac{\nabla^i f(0)}{i!} \cdot (x)_i,$$

which is strikingly similar to the *Taylor-MacLaurin power series* of an analytic function (of a real or complex variable).

(2) The formulas

$$\nabla^k f(0) = \sum_{i\leq k} (-1)^{k-i} \binom{k}{i} f(i)$$

correspond to the formal power series identity

$$\sum_{k\geq 0} \nabla^k f(0) \frac{x^k}{k!} = e^{-x} \cdot \sum_{n\geq 0} f(n) \frac{x^n}{n!}$$

between these two generating functions.

1.2. Integer-Valued Polynomial Functions

We shall denote by $L = L(\mathbf{Z}) \subset \mathbf{Q}[x]$ the $\mathbf{Z}$-module consisting of polynomial functions taking integer values on the natural integers:

$$L = \{f \in \mathbf{Q}[x] : f(\mathbf{N}) \subset \mathbf{Z}\}.$$

We have seen in (1.1) that $\mathbf{Z}[x] \subset L$ is a proper inclusion: All binomial functions belong to L.

Theorem. *The* $\mathbf{Z}$*-module* L *consisting of polynomial functions* $f \in \mathbf{Q}[x]$ *integer-valued on* $\mathbf{N}$ *is free, with a basis given by the binomial polynomials* $\binom{\cdot}{i}$.

Proof. Let f be an integer-valued polynomial. Obviously, all the iterated differences of f have the same property, and in particular the coefficients m_i of the series expansion of f are rational integers. On the other hand, the iterated differences $\nabla^i f$ will vanish identically if the exponent i is greater than the degree of f. Hence the series $\sum m_i \binom{\cdot}{i}$ is a finite sum, and the uniqueness of the representation has been proved in (1.1). ∎

Corollary 1. *If a polynomial $f \in \mathbf{Q}[x]$ takes integral values on* $\mathbf{N}$, *it also takes integral values on* $\mathbf{Z}$.

Proof. It is enough to check this property for the basis of L consisting of the binomial polynomials. If $x = -m$ is a negative integer, then

$$\binom{-m}{i} = -m(-m-1)\cdots(-m-i+1)/i! = (-1)^i \binom{m+i-1}{i} \in \mathbf{Z}.$$

Hence the $\binom{\cdot}{i}$ and all $f \in L$ define functions $\mathbf{Z} \to \mathbf{Z}$. ∎

Corollary 2. *If a polynomial of degree $d \geq 0$ (with rational coefficients) takes integral values on $d+1$ consecutive integers, then it takes integral values on all integers.*

Proof. Let f take integral values on the integers $a, a+1, \ldots, a+d$ and consider its translate $g(x) = f(x-a)$ which takes integral values on the first integers $0, 1, \ldots, d$. Hence the first iterated differences of g at the origin are also integers, and if f is a polynomial of degree d, so is g. The expansion $g = \sum_{i \leq d} \nabla^i g(0) \binom{\cdot}{i}$ shows that $g \in L$. ∎

Definition. *Let M be a* $\mathbf{Z}$*-module. The M-valued* polynomial functions *are those that have a finite expansion in the basis consisting of binomial polynomials.*

Since the polynomial functions with values in M are the finite sums $\sum m_i \binom{\cdot}{i}$, the mapping

$$\mathrm{Maps}(\mathbf{N}; M) \to M^{\mathbf{N}} : f \mapsto (\nabla^i f(0))_{i \geq 0}$$

induces a bijection between the polynomial functions and $M^{(\mathbf{N})}$: The subspace of the product consisting of families with only finitely many nonzero entries.

1.3. Periodic Functions Taking Values in a Field of Characteristic p

We shall have to consider the case where the $\mathbf{Z}$-module M is a vector space over the finite field $\mathbf{F}_p$. To start with, let us take $M = \mathbf{F}_p$.

Proposition. *For $i < p^t$, the functions $\widetilde{\binom{\cdot}{i}} : \mathbf{Z} \to \mathbf{F}_p,\ x \mapsto \binom{x}{i} \bmod p$ are periodic of period $T = p^t$. They make up a basis of this space of T-periodic maps.*

PROOF. The binomial coefficients are best described by their generating function

$$(1+u)^x = \sum_{i\geq 0} \binom{x}{i} u^i.$$

The identity $(1+u)^{x+T} = (1+u)^x \cdot (1+u)^T$ combined with the congruence

$$(1+u)^{p^t} \equiv 1 + u^{p^t} \pmod{p}$$

leads to

$$(1+u)^{x+p^t} \equiv (1+u)^x \cdot (1+u^{p^t}) \pmod{p}.$$

For $i < p^t$, the coefficients of u^i in $(1+u)^{x+p^t}$ and in $(1+u)^x$ are the same mod p; hence

$$\binom{x+p^t}{i} \equiv \binom{x}{i} \pmod{p} \quad (i < p^t).$$

To prove the second part of the statement, consider the linear map

$$j : \mathrm{Maps}_{T\text{-periodic}}(\mathbf{Z}; \mathbf{F}_p) \to \mathbf{F}_p^T,\ f \mapsto (\nabla^i f(0))_{0\leq i<T}.$$

If $\nabla^i f(0) = 0$ for $0 \leq i < T$, then f vanishes at the points $0, 1, \ldots, T-1$ (1.1), hence vanishes identically by T-periodicity. This proves that the linear map j is *injective*. Since both spaces $\mathrm{Maps}_{T\text{-periodic}}(\mathbf{Z}; \mathbf{F}_p)$ and $\mathbf{F}_p^T$ have the same dimension over $\mathbf{F}_p$ (even the same number of elements, since this field is finite), it is *bijective*. It will be enough to check that the image of the set of binomial polynomials $\binom{\cdot}{i}$ is the canonical basis of the target

$$\widetilde{\tbinom{\cdot}{i}} \text{ is a polynomial of degree } i \implies \nabla^k \widetilde{\tbinom{\cdot}{i}} = 0 \quad \text{for } k > i,$$
$$\nabla^k \widetilde{\tbinom{\cdot}{i}} = \widetilde{\tbinom{\cdot}{i-k}} \text{ vanishes at 0} \quad \text{for } k < i, \quad \nabla^k \widetilde{\tbinom{\cdot}{k}}(0) = 1. \qquad \blacksquare$$

Remark. It is not difficult to prove periodicity of the binomial coefficients relative to nonprime moduli. For example,

$$x \mapsto \binom{x}{i} \bmod m$$

is periodic of period m^i.

Theorem. *Let M be a vector space over $\mathbf{F}_p$ and $f : \mathbf{Z} \to M$ a function that is periodic of period $T = p^t$ (for some $t \geq 0$). Then f can be uniquely written in the form*

$$f = \sum_{0\leq i<T} \binom{\cdot}{i} m_i \quad (m_i \in M).$$

In other words,

$$\text{Maps}_{T\text{-periodic}}(\mathbf{Z}; M) = \bigoplus_i M_i$$

is the direct sum of the subspaces

$$M_i = \binom{\cdot}{i} M \subset \text{Maps}(\mathbf{Z}; M) \quad (0 \le i < T).$$

PROOF. A T-periodic map on $\mathbf{Z}$ is a map on the finite quotient $\mathbf{Z}/T\mathbf{Z}$, and hence takes only finitely many values. This reduces the proof to the finite-dimensional case. The map

$$j : \text{Maps}_{T-\text{periodic}} \to M^T, \; f \mapsto (\nabla^i f(0))_{0 \le i < T}$$

is linear and injective. Since both spaces $\text{Maps}_{T\text{-periodic}}$ and M^T have the same dimension over $\mathbf{F}_p$ (even the same number of elements, since this field is finite), it is bijective. ∎

1.4. Convolution of Functions of an Integer Variable

Let A be a commutative ring and $f, g : \mathbf{N} \to A$ two functions. We define their *shifted convolution product* by[1]

$$f \underline{*} g(n) = \sum_{i+j=n-1} f(i)g(j) = \sum_{0 \le i \le n-1} f(i)g(n-1-i) \quad (n \ge 1)$$

and $f \underline{*} g(0) = 0$. This is a commutative, associative, and distributive product on $\text{Maps}(\mathbf{N}, A)$.

Proposition. *The iterated differences of a shifted convolution product are given by*

$$\nabla^n(f \underline{*} g) = \textstyle\sum_{i+j=n-1} \nabla^i f \cdot \nabla^j g(0) + f \underline{*} \nabla^n g \quad (n \ge 1).$$

PROOF. It will be practical to use the notation f_1 for a unit translate of a function f:

$$f_1(n) = f(n+1) \quad (n \ge 0)$$

(the value $f(0)$ is lost). With this notation the difference operator is expressed by

$$\nabla f = f_1 - f.$$

[1] The usual convolution product is defined by $f * g(n) = \sum_{i+j=n} f(i)g(j)$ $(n \ge 0)$.

Let us evaluate the translate of a shifted convolution product:

$$\begin{aligned}(f \underline{*} g)_1(n) &= f \underline{*} g(n+1) = \sum_{i+j=n} f(i)g(j) \\ &= f(n)g(0) + \sum_{i+j=n-1} f(i)g(j+1) \\ &= f(n)g(0) + (f \underline{*} g_1)(n),\end{aligned}$$

whence

$$\nabla(f \underline{*} g) = (f \underline{*} g)_1 - f \underline{*} g = f \cdot g(0) + f \underline{*} \nabla g.$$

Iterating the preceding formula, we obtain

$$\begin{aligned}\nabla^2(f \underline{*} g) &= \nabla(f \cdot g(0) + f \underline{*} \nabla g) \\ &= \nabla f \cdot g(0) + \nabla(f \underline{*} \nabla g) \\ &= \nabla f \cdot g(0) + f \cdot \nabla g(0) + f \underline{*} \nabla^2 g.\end{aligned}$$

By induction, we obtain

$$\nabla^n(f \underline{*} g) = \sum_{i+j=n-1} \nabla^i f \cdot \nabla^j g(0) + f \underline{*} \nabla^n g,$$

which expresses $\nabla^n(f \underline{*} g)$ as a sum of $f \underline{*} \nabla^n g$ and a linear combination of the finite differences $\nabla^i f$ $(i < n)$ of f. ■

1.5. Indefinite Sum of Functions of an Integer Variable

If the finite-difference operator ∇ is to be compared to the derivation operator — pursuing the analogy — we should construct an inverse of it, corresponding to integration. It is clear that for any function $f : \mathbf{N} \to A$, there is a unique *primitive* $F : \mathbf{N} \to A$ satisfying

$$\nabla F = f \text{ and } F(0) = 0.$$

These conditions indeed imply

$$f(0) = \nabla F(0) = F(1) - F(0) = F(1)$$

and then

$$f(n) = F(n+1) - F(n), \; F(n+1) = F(n) + f(n).$$

By induction, $F(n+1) = \sum_{0 \leq i \leq n} f(i)$.

Definition. *The indefinite sum operator S is defined by*

$$Sf(0) = 0 \text{ and } Sf(n) = \sum_{0 \leq i < n} f(i) \quad (n \geq 1).$$

If we use the shifted convolution product introduced in the preceding section, we see that $Sf = 1 \underline{*} f$, where 1 represents the constant function 1 on $\mathbf{N}$. In fact,

$$\nabla(1 \underline{*} f)(n) = \sum_{i \le n} f(i) - \sum_{i<n} f(i) = f(n), \quad \nabla(1 \underline{*} f) = f.$$

Examples. (1) Let $f = 1$ be the constant unit function. Then $S1(n) = n$.

(2) Let f $(= S1)$ be the identity function $\mathbf{N} \to \mathbf{N}$. Then

$$Sf(n) = \sum_{i<n} i = \binom{n}{2} = \frac{n(n-1)}{2}.$$

(3) More generally, let $f = \binom{\cdot}{k}$ be the kth binomial polynomial $\mathbf{N} \to \mathbf{N}$. We have seen in (1.1) that $\nabla\binom{\cdot}{k} = \binom{\cdot}{k-1}$ and $\binom{0}{k} = 0$ $(k \ge 1)$. Hence

$$S\binom{\cdot}{k-1} = \binom{\cdot}{k} \qquad (k \ge 1).$$

This property can be read in Pascal's triangle. Consider, for example, the two consecutive sequences

$$f_2 : 0, 0, 1, 3, 6, 10, \ldots,$$
$$f_3 : 0, 0, 0, 1, 4, 10, \ldots.$$

The differences of the second one indeed give the first one.

(4) Consider now $f(n) = n^2$, the square function. In this case

$$Sf(n) = \sum_{i<n} i^2 = 1^2 + 2^2 + \cdots + (n-1)^2.$$

Since

$$2\binom{n}{2} = n^2 - n = n^2 - \binom{n}{1},$$

we have

$$f(n) = n^2 = 2\binom{n}{2} + \binom{n}{1}$$

and

$$Sf(n) = 2\binom{n}{3} + \binom{n}{2},$$

$$\begin{aligned} 6Sf(n) &= 2n(n-1)(n-2) + 3n(n-1) \\ &= n(n-1)[2n-4+3] = n(n-1)(2n-1). \end{aligned}$$

We have obtained the well-known formula

$$Sf(n) = \sum_{i<n} i^2 = \tfrac{1}{6} n(n-1)(2n-1).$$

This way of proceeding is similar to the general procedure that consists in writing the binomial series expansion of $F = Sf$,

$$F = F(0) + \nabla F(0)\binom{\cdot}{1} + \nabla^2 F(0)\binom{\cdot}{2} + \nabla^3 F(0)\binom{\cdot}{3},$$

using $F(0) = 0$ and $\nabla F = f$; hence

$$\nabla F(0) = f(0) = 0, \quad \nabla^2 F(0) = \nabla f(0) = 1, \quad \nabla^3 F(0) = \nabla^2 f(0) = 2.$$

The preceding examples lead to the following result.

Proposition 1. *If the function f of an integer variable is given by*

$$f(n) = \sum_{i \geq 0} c_i \binom{n}{i},$$

then

$$F(n) = Sf(n) = \sum_{i \geq 0} c_i \binom{n}{i+1}. \qquad \blacksquare$$

The preceding examples also show that

$$1 \underline{*} 1 = \mathrm{id}, \quad 1 \underline{*} 1 \underline{*} 1(n) = \binom{n}{2}, \ldots, \quad \underbrace{1 \underline{*} 1 \underline{*} \cdots \underline{*} 1}_{k+1 \text{ factors}}(n) = \binom{n}{k}.$$

A few more formulas may be useful. By definition $f = \nabla(Sf) = \nabla(1 \underline{*} f)$. Let us compute $S(\nabla f)$:

$$\begin{aligned} S(\nabla f)(n) = 1 \underline{*} \nabla f(n) &= \sum_{0 \leq k < n} \nabla f(k) \\ &= \sum_{0 \leq k < n} [f(k+1) - f(k)] = f(n) - f(0). \end{aligned}$$

If we denote by P_0 the projection on constant functions defined by

$$P_0 : \mathrm{Maps}\,(\mathbf{N}; A) \to A, \quad f \mapsto f(0) \cdot 1$$

we have obtained $S \circ \nabla = \mathrm{id} - P_0$. Hence the following proposition.

Proposition 2. *The indefinite-sum and finite-difference operators are linked by the formulas*

$$\nabla \circ S = \mathrm{id}, \quad S \circ \nabla = \mathrm{id} - P_0, \quad \nabla \circ S - S \circ \nabla = P_0. \qquad \blacksquare$$

The identity $S(\nabla f) = f - f(0) \cdot 1$ gives a first-order limited expansion of f if we only rewrite it $f = f(0) \cdot 1 + S(\nabla f)$. This point of view has been generalized by van Hamme.

Theorem. *For every function f of an integer variable and every integer $n \geq 0$, we have*

$$f = f(0) \cdot 1 + \nabla f(0) \cdot \binom{\cdot}{1} + \nabla^2 f(0) \cdot \binom{\cdot}{2} + \cdots + \nabla^n f(0) \cdot \binom{\cdot}{n} + R_{n+1} f$$

with the van Hamme form for the remainder $R_{n+1} f = \nabla^{n+1} f \mathbin{\underline{*}} \binom{\cdot}{n}$.

PROOF. The case $n = 0$ has already been obtained: $R_1 f = S\nabla f = 1 \mathbin{\underline{*}} \nabla f$. For $n \geq 1$ we can use the identity

$$\nabla^n (f \mathbin{\underline{*}} g) = \sum_{i+j=n-1} \nabla^i f \cdot \nabla^j g(0) + f \mathbin{\underline{*}} \nabla^n g$$

proved in (1.4). Let us apply

$$\nabla^{n+1} (g \mathbin{\underline{*}} f) = \sum_{i+j=n} \nabla^i g \cdot \nabla^j f(0) + g \mathbin{\underline{*}} \nabla^{n+1} f$$

to the function $g(x) = \binom{x}{n}$, for which $\nabla^i g(x) = \binom{x}{n-i} = \binom{x}{j}$. We find that

$$\nabla^{n+1} \left(\binom{x}{n} \mathbin{\underline{*}} f \right) = \sum_{0 \leq j \leq n} \binom{x}{j} \cdot \nabla^j f(0) + \binom{x}{n} \mathbin{\underline{*}} \nabla^{n+1} f.$$

But the left-hand side is

$$\nabla^{n+1} \left(\binom{x}{n} \mathbin{\underline{*}} f \right) = \nabla^{n+1} (1 \mathbin{\underline{*}} 1 \mathbin{\underline{*}} \cdots \mathbin{\underline{*}} 1 \mathbin{\underline{*}} f) = \nabla^{n+1} (S^{n+1} f) = f,$$

whence the result, since $\nabla \circ S = \mathrm{id}$. ∎

2. Continuous Functions on $\mathbf{Z}_p$

2.1. *Review of Some Classical Results*

Let us recall the basic property of uniform convergence.

Theorem. *Let X be a topological space, M a* complete *metric space, and $(f_n)_{n \geq 0}$ a sequence of continuous maps $X \to M$. If*

$$d(f_m, f_n) := \sup_{x \in X} d_M(f_m(x), f_n(x)) \to 0 \quad (m, n \to \infty),$$

then the sequence $(f_n)_{n \geq 0}$ has a limit that is a continuous function $f : X \to M$.

PROOF. Fix momentarily $x \in X$. Then $(f_n(x))_{n \geq 0}$ is a Cauchy sequence in the complete space M; hence it converges. Let $f(x) = \lim_{n \to \infty} f_n(x)$ denote its limit.

This defines a function $f : X \to M$. We have to prove that this function is continuous. But for each positive $\varepsilon > 0$ there is a rank $N = N_\varepsilon$ such that

$$d_M(f_m(y), f_n(y)) \le \sup_X d_M(f_n(x), f(x)) = d(f_m, f_n) \le \varepsilon$$

$(m, n \ge N,\ y \in X)$. Letting $m \to \infty$ we infer

$$d_M(f(y), f_n(y)) \le \varepsilon \quad (n \ge N,\ y \in X),$$

and hence

$$d(f, f_n) = \sup_{y \in X} d_M(f(y), f_n(y)) \le \varepsilon \quad (n \ge N).$$

This proves that the sequence $(f_n)_{n \ge 0}$ converges *uniformly* to f and implies the expected continuity: let us recall this point. For $a, y \in X$, we write

$$d_M(f(y), f(a)) \le d_M(f(y), f_n(y)) + d_M(f_n(y), f_n(a)) + d_M(f_n(a), f(a));$$

whence

$$d_M(f(y), f(a)) \le \varepsilon + d_M(f_n(y), f_n(a)) + \varepsilon \quad (n \ge N).$$

Let us choose and fix an integer $n \ge N$. If $a \in X$, the continuity of the function f_n assures us that there is a neighborhood V of a in X such that

$$y \in V \Longrightarrow d_M(f_n(y), f_n(a)) \le \varepsilon.$$

The preceding inequality shows that

$$d_M(f(y), f(a)) \le 3\varepsilon \quad (y \in V),$$

and hence f is continuous at the point a (for any $a \in X$). ■

Another classical result for continuous functions $f : \mathbf{Z}_p \to \mathbf{R}$ is the following. If we fix a continuous *injective* function $\varphi : \mathbf{Z}_p \to \mathbf{R}$ (for example, a linear model of $\mathbf{Z}_p$ (I.2.3) corresponds to such a function), then f can be uniformly approximated by polynomial expressions in φ. Indeed, the algebra of polynomials in φ is a subalgebra of the algebra of continuous functions over the compact space $\mathbf{Z}_p$, which *separates points.* The *Stone-Weierstrass theorem* implies that this subalgebra is dense for uniform convergence.

Finally, let $f : \mathbf{Z}_p \to \mathbf{C}_p$ be a continuous function. Then $|f| : \mathbf{Z}_p \to \mathbf{R}$ is continuous, and since $\mathbf{Z}_p$ is a compact space, $\sup |f| = \max |f|$ is attained at some point $x \in \mathbf{Z}_p$. More precisely, $f(\mathbf{Z}_p)$ is a compact subset of $\mathbf{C}_p$ and the proposition in (II.1.1) shows that

$$\{|f(x)| \ne 0 : x \in \mathbf{Z}_p\} \text{ is discrete in } \mathbf{R}_{>0}.$$

In particular, for every $\varepsilon > 0$ there are only finitely many possible real values of $|f(x)|$ satisfying $|f(x)| \ge \varepsilon$.

2.2. *Examples of p-adic Continuous Functions on* $\mathbf{Z}_p$

The definition of a topological ring A shows that any polynomial $f \in A[X]$ gives rise to a continuous polynomial function $A \to A$. In particular, if $f \in \mathbf{C}_p[X]$ is a polynomial with coefficients in $\mathbf{C}_p$, it gives rise to a continuous function $\mathbf{Z}_p \to \mathbf{C}_p$ by restriction. Since $x \in \mathbf{Z}_p$ implies $|x| \le 1$, any power series $\sum_{i\ge 0} a_i x^i$ with $a_i \in \mathbf{C}_p$ and $|a_i| \to 0$ converges uniformly, and hence defines a continuous function $\mathbf{Z}_p \to \mathbf{C}_p$. For any continuous function $f : \mathbf{Z}_p \to \mathbf{C}_p$, we define its sup norm by

$$\|f\| = \sup_{x\in\mathbf{Z}_p} |f(x)| = \max_{x\in\mathbf{Z}_p} |f(x)| \quad (<\infty).$$

Finally, let us give examples of continuous functions $\mathbf{Z}_p \to \mathbf{C}_p$ of an apparently different type. If $x = \sum_{i\ge 0} a_i p^i \in \mathbf{Z}_p$, we define $f(x) = \sum_{i\ge 0} a_i p^{2i}$. This defines a continuous function $\mathbf{Z}_p \to \mathbf{Z}_p$ with

$$|f(x) - f(y)| = |x - y|^2.$$

This estimate shows that f is even differentiable at every point with $f' \equiv 0$, but f is not locally constant. We shall come back to this example later on. If we put $f(x) = \sum a_i p^{mi}$, we have similarly

$$|f(x) - f(y)| = |x - y|^m,$$

and with $f(x) = \sum a_i p^{i!}$, then for any $m \ge 1$,

$$|f(x) - f(y)| \le |x - y|^m$$

if $|x - y|$ is small enough.

2.3. *Mahler Series*

The binomial polynomials define continuous functions

$$\binom{\cdot}{k} : \mathbf{Z}_p \to \mathbf{Z}_p, \quad x \mapsto \binom{x}{k}.$$

Since $\mathbf{N}$ is dense in $\mathbf{Z}_p$, we have $\left\|\binom{\cdot}{k}\right\| = \sup_{\mathbf{N}} \left|\binom{n}{k}\right| \le 1$. In fact, $\binom{k}{k} = 1$ proves that

$$\left\|\binom{\cdot}{k}\right\| = 1 \quad (k \ge 0).$$

As noted in the previous section, for any sequence $(a_i)_{i\ge 0}$ in $\mathbf{C}_p$ with $|a_i| \to 0$, the series $\sum_{k\ge 0} a_k \binom{\cdot}{k}$ defines a continuous function $f : \mathbf{Z}_p \to \mathbf{C}_p$. It is quite remarkable that *conversely, every continuous function* $\mathbf{Z}_p \to \mathbf{C}_p$ *can be so represented.*

This result has been obtained by Mahler and will be established below. For example, it is applicable to all locally constant functions.

Definition. *A Mahler series is a series* $\sum_{k\geq 0} a_k\binom{\cdot}{k}$ *with coefficients* $|a_k| \to 0$ *in* $\mathbf{C}_p$ *(or* Ω_p*).*

Comment. If a series $\sum_{k\geq 0} a_k\binom{\cdot}{k}$ *converges simply* at each $x \in \mathbf{Z}_p$, it *converges uniformly* on $\mathbf{Z}_p$ and is a Mahler series. In fact, assume that it *converges at the single point* -1. This implies $a_k\binom{-1}{k} \to 0$, and since $\binom{-1}{k} = (-1)^k$, we see that $|a_k| = \left|a_k\binom{-1}{k}\right| \to 0$: the series converges uniformly.

Example. Let $t \in \mathbf{M}_p$ — namely, $t \in \mathbf{C}_p$, $|t| < 1$ — and consider the sequence $a_k = t^k$, which tends to 0. The Mahler series $\sum_{k\geq 0} t^k\binom{x}{k}$ converges uniformly to a continuous function $f : \mathbf{Z}_p \to \mathbf{C}_p$. Since $(1+t)^n = \sum_{0\leq k\leq n}\binom{n}{k}t^k$ for integers $n \geq 1$, the preceding continuous function extends $n \mapsto (1+t)^n$, and it is still denoted by

$$(1+t)^x = \sum_{k\geq 0} t^k\binom{x}{k} \qquad (x \in \mathbf{Z}_p).$$

2.4. *The Mahler Theorem*

Keeping the preceding notation concerning the binomial polynomials $f_k(x) = \binom{x}{k}$ and the sup norm $\|f\| = \sup_{\mathbf{Z}_p} |f(x)|$ for continuous functions on $\mathbf{Z}_p$, we intend to prove the following general result.

Theorem 1. *Let* $f : \mathbf{Z}_p \to \mathbf{C}_p$ *be a continuous function and put* $a_k = \nabla^k f(0)$. *Then* $|a_k| \to 0$, *and the series* $\sum_{k\geq 0} a_k\binom{\cdot}{k}$ *converges uniformly to* f. *Moreover,* $\|f\| = \sup_{k\geq 0} |a_k|$.

Proof. Since the function f is continuous, $f(\mathbf{Z}_p)$ is a compact subset of $\mathbf{C}_p$ and $|f(\mathbf{Z}_p)|$ has at most 0 as an accumulation point in $\mathbf{R}_{\geq 0}$. Without loss of generality we may assume $f \neq 0$ and replace f by $f/f(x_0)$, where $x_0 \in \mathbf{Z}_p$ is chosen with $|f(x_0)|$ maximal. Hence we shall assume $\|f\| = 1$ from now on: The image of f is contained in the unit ball $\mathbf{A}_p$ of $\mathbf{C}_p$. Let us consider the quotient $E = \mathbf{A}_p/p\mathbf{A}_p$ ($p\mathbf{A}_p = B_{\leq|p|}(\mathbf{C}_p)$) as a vector space over the prime field $\mathbf{F}_p$. Then the composite $\varphi = (f \bmod p) : \mathbf{Z}_p \to \mathbf{A}_p \to E$ is continuous (takes only finitely many values, is locally constant) and is not identically zero. Since $\mathbf{Z}_p$ is compact, it is uniformly continuous and uniformly locally constant. This means that for t suitably large, φ is constant on cosets mod $p^t\mathbf{Z}_p$. Hence φ is T-periodic on $\mathbf{Z}$, where $T = p^t$ with values in the vector space E. By (1.3) we can write

$$\varphi = \sum_{k<T} \alpha_k \widetilde{\binom{\cdot}{k}} \qquad (\alpha_k \in E).$$

Taking representatives $a_k^0 \in \mathbf{A}_p$ for the α_k, the difference

$$f - \sum_{k<T} a_k^0 f_k$$

has values in $p\mathbf{A}_p$. By the competition principle, at least one $|a_k^0| = 1$, and

$$|a_k^0| \leq 1, \quad \max |a_k^0| = 1.$$

By construction $\|f - \sum_{k<T} a_k^0 \binom{\cdot}{k}\| = r \leq |p|$. If $f - \sum_{k<T} a_k^0 \binom{\cdot}{k}$ is not 0, we can iterate the procedure on this difference and find $S > T$ and coefficients a_k^1 $(k < S)$ with

$$|a_k^1| \leq r, \quad \max |a_k^1| = r,$$

and

$$\left\| \left(f - \sum_{k<T} a_k^0 \binom{\cdot}{k} \right) - \sum_{k<S} a_k^1 \binom{\cdot}{k} \right\| = r' \leq |p^2|.$$

We can even write

$$\left\| f - \sum_{k<S} (a_k^0 + a_k^1) \binom{\cdot}{k} \right\| = r' \leq |p^2|$$

if we agree to define $a_k^0 = 0$ for $k \geq T$. It is obvious that this procedure leads to convergent series

$$a_k = a_k^0 + a_k^1 + \cdots \in \mathbf{C}_p, \quad |a_k^n| \leq |p^n| \to 0,$$
$$|a_k| \leq 1 \quad (k < T), \quad |a_k| \leq r \quad (T \leq k < S), \text{ etc.},$$

and also $\sup_{k>0} |a_k| = \sup_{k<T} |a_k| = 1 = \|f\|$. The proof of the theorem is therefore complete, since $\|f - \sum_{k\geq 0} a_k \binom{\cdot}{k}\| < |p|^m$ for all positive integers m. ∎

Corollary. *For any continuous function $f : \mathbf{Z}_p \to \mathbf{C}_p$, there is a sequence of polynomials $f_n \in \mathbf{C}_p[x]$ that converges uniformly to f.* ∎

Theorem 2. *Let $f : \mathbf{N} \to \mathbf{C}_p$ be any map and define $a_k = \nabla^k f(0)$. Then the following properties are equivalent:*

(i) $|a_k| \to 0$ when $k \to \infty$.
(ii) The Mahler series $\sum_{k\geq 0} a_k \binom{\cdot}{k}$ converges uniformly.
(iii) f admits a continuous extension to $\mathbf{Z}_p \to \mathbf{C}_p$.
(iv) f is uniformly continuous (for the p-adic topology on $\mathbf{N}$).
(v) $\|\nabla^k f\| \to 0$ when $k \to \infty$.

Proof. Here is a complete scheme of implications.

(*i*) $\Rightarrow$ (*ii*) We have

$$\left|a_k\binom{x}{k}\right| \leq |a_k| \left\|\binom{\cdot}{k}\right\| = |a_k| \quad (x \in \mathbf{Z}_p),$$

hence the uniform convergence if $|a_k| \to 0$.

(*ii*) $\Rightarrow$ (*iii*) This is the basic property of uniform convergence reviewed in (3.1).

(*iii*) $\Leftrightarrow$ (*iv*) On a compact metric space, any continuous function is uniformly continuous.

(*iii*) $\Rightarrow$ (*v*) Apply the Mahler theorem to the continuous extension of f to $\mathbf{Z}_p$ (still denoted by f):

$$f = \sum_{k\geq 0} a_k \binom{\cdot}{k} \qquad (a_k = \nabla^k f(0)).$$

Since $\nabla\binom{\cdot}{k} = \binom{\cdot}{k-1}$, we have

$$\nabla f = \sum_{k\geq 1} a_k \binom{\cdot}{k-1}$$

and by induction

$$\nabla^j f = \sum_{k\geq j} a_k \binom{\cdot}{k-j}.$$

By the same theorem

$$\|\nabla^j f\| = \sup_{k\geq j} |a_k| \to 0.$$

In particular, $|a_j| = |\nabla^j f(0)| \leq \|\nabla^j f\| \to 0$; hence (*v*) $\Rightarrow$ (*i*). ∎

2.5. Convolution of Continuous Functions on $\mathbf{Z}_p$

As an application of the Mahler theorem, we show that the (shifted) convolution product defined in (1.4) for functions of an integer variable $\mathbf{N} \to \mathbf{C}_p$ extends to $\mathbf{Z}_p \to \mathbf{C}_p$. In turn, this result allows us to give an explicit estimate for the remainder in a finite Mahler expansion. By definition,

$$f \underline{*} g(n) = \sum_{i+j=n-1} f(i)g(j),$$

$$|f \underline{*} g(n)| \leq \max_{i+j=n-1} |f(i)g(j)| \leq \|f\|\|g\|,$$

and

$$\|f \underline{*} g\| \leq \|f\|\|g\|.$$

Proposition. *Let f and g be two continuous maps $\mathbf{Z}_p \to \mathbf{C}_p$. Then the shifted convolution product $f \underline{*} g$ has a continuous extension $\mathbf{Z}_p \to \mathbf{C}_p$.*

PROOF. By (3.4) (Theorem 2, (*i*) $\Rightarrow$ (*iii*)), the existence of a continuous extension of $f \underline{*} g$ (initially only defined on $\mathbf{N}$) will follow from $\nabla^k(f \underline{*} g)(0) \to 0$. To prove this convergence, let us come back to the formula (proved in 1.4)

$$\nabla^{2n+1}(f \underline{*} g) = \sum_{i+j=2n} \nabla^i f \cdot \nabla^j g(0) + f \underline{*} \nabla^{2n+1} g,$$

$$\nabla^{2n+1}(f \underline{*} g)(0) = \sum_{i+j=2n} \nabla^i f(0) \cdot \nabla^j g(0) + (f \underline{*} \nabla^{2n+1} g)(0).$$

For any bounded function h, the ultrametric property gives $\|\nabla h\| \le \|h\|$, and we can estimate

$$\nabla^{2n+1}(f \underline{*} g)(0) = T_1 + T_2 + T_3,$$

where

$$T_1 = \sum_{n \le i \le 2n} \nabla^i f(0) \cdot \nabla^{2n-i} g(0),$$

$$T_2 = \sum_{n < j \le 2n} \nabla^{2n-j} f(0) \cdot \nabla^j g(0),$$

$$T_3 = (f \underline{*} \nabla^{2n+1} g)(0),$$

as follows:

$$|T_1| \le \|\nabla^n f\| \cdot \|g\|,$$

$$|T_2| \le \|f\| \cdot \|\nabla^n g\|,$$

$$|T_3| \le \|f\| \cdot \|\nabla^{2n+1} g\| \le \|f\| \cdot \|\nabla^n g\|.$$

Altogether, this shows that

$$\begin{aligned} |\nabla^{2n+1}(f \underline{*} g)(0)| &\le \max(|T_1|, |T_2|, |T_3|) \\ &\le \max(\|\nabla^n f\| \cdot \|g\| \,,\ \|f\| \cdot \|\nabla^n g\|) \to 0. \end{aligned}$$

Similar estimates can be made for $|\nabla^{2n}(f \underline{*} g)(0)|$, and we prove thereby the required convergence: $\nabla^k(f \underline{*} g)(0) \to 0$. ■

Corollary 1. *Any continuous $f : \mathbf{Z}_p \to \mathbf{C}_p$ has limited Mahler expansions*

$$f = f(0) + \nabla f(0) \cdot \binom{\cdot}{1} + \nabla^2 f(0) \cdot \binom{\cdot}{2} + \cdots$$

$$+ \nabla^n f(0) \cdot \binom{\cdot}{n} + R_{n+1} f \quad (n \ge 1)$$

with the van Hamme form of the remainder

$$R_{n+1}f = \nabla^{n+1}f \,\underline{*}\binom{\cdot}{n}, \quad \|R_{n+1}f\| \le \|\nabla^{n+1}f\| \to 0 \quad (n \to \infty).$$

PROOF. The announced formulas hold on $\mathbf{N}$ by the preceding section. Taking $g = \binom{\cdot}{n}$ in $\|f \,\underline{*}\, g\| \le \|f\|\,\|g\|$, we see that they extend continuously to $\mathbf{Z}_p$ by the proposition. ∎

Another application of the Mahler theorem (or of the possibility of extending the convolution product to continuous functions over $\mathbf{Z}_p$) is given by the following corollary.

Corollary 2. *For any continuous function $f : \mathbf{Z}_p \to \mathbf{C}_p$, the indefinite sum $Sf = f \,\underline{*}\, 1$ of f extends continuously to $\mathbf{Z}_p$. More precisely, if $f = \sum_{k\ge 0} a_k \binom{\cdot}{k}$ is the Mahler expansion of f, then*

$$Sf = 1 \,\underline{*}\, f = \sum_{k\ge 0} a_k \binom{\cdot}{k+1}, \quad \|Sf\| = \|f\|.$$

PROOF. We have noticed that

$$S\binom{\cdot}{k} = 1 \,\underline{*}\binom{\cdot}{k} = \binom{\cdot}{k+1},$$

whence the result. ∎

Corollary 3. *The only linear form $\varphi : C(\mathbf{Z}_p; K) \to K$ that is invariant under translation is the trivial one $\varphi = 0$.*

PROOF. In fact, we prove that if $\varphi(F) = \varphi(F_1)$ for all $F \in C(\mathbf{Z}_p; K)$, where $F_1(x) = F(x+1)$, then $\varphi = 0$. Indeed, take any $f \in C(\mathbf{Z}_p; K)$. There exists an $F \in C(\mathbf{Z}_p; K)$ with $f = \nabla F = F_1 - F$ (take $F = Sf$), and thus

$$\varphi(f) = \varphi(F_1 - F) = \varphi(F_1) - \varphi(F) = 0. \qquad ∎$$

Corollary 4. *Let $\sigma : \mathbf{Z}_p \to \mathbf{Z}_p$, $x \mapsto -1-x$, be the canonical involution (I.1.2). Then $S(f \circ \sigma)(x) = -Sf(-x)$.*

PROOF. For integers $n, m \ge 1$ we have

$$Sf(n+m) - Sf(n) = f(n) + \cdots + f(n+m-1).$$

By density of the integers $n \ge 1$ in $\mathbf{Z}_p$ and continuity of both sides, we get more generally

$$Sf(x+m) - Sf(x) = f(x) + \cdots + f(x+m-1) \quad (x \in \mathbf{Z}_p).$$

Take now $x = -m$ in this equality:

$$\begin{aligned} Sf(0) - Sf(-m) &= f(-m) + \cdots + f(-1) \\ &= f(\sigma(m-1)) + \cdots + f(\sigma(0)) \\ &= S(f \circ \sigma)(m). \end{aligned}$$

Since $Sf(0) = 0$, the result follows. ■

Example. Let $a = 1 + t \in 1 + M \subset \mathbf{C}_p$ and take

$$f(x) = a^x = (1+t)^x = \sum_{k \geq 0} t^k \binom{x}{k}.$$

Then we have

$$Sf(x) = \sum_{k \geq 0} t^k \binom{x}{k+1} = \frac{(1+t)^x - 1}{t} \quad (t \neq 0),$$

$$Sf(x) = \frac{a^x - 1}{a - 1} \quad (a \neq 1).$$

3. Locally Constant Functions on $\mathbf{Z}_p$

3.1. *Review of General Properties*

When X is a topological space and E any set, a map $f : X \to E$ is *locally constant* if for each $x \in X$

$$V_x = \{y \in X : f(y) = f(x)\}$$

is a neighborhood of x. Equivalently, one can require $f^{-1}(e)$ open in X for each $e \in E$, or even $f^{-1}(A)$ open in X for each subset $A \subset E$. In other words, locally constant functions $f : X \to E$ are continuous functions when E is endowed with the discrete topology. On a *connected* space, a locally constant function is *constant* (take $x \in X$, put $e = f(x) \in E$, $A = E - \{e\}$, and consider the partition of the connected space X into two disjoint open sets $f^{-1}(e)$ and $f^{-1}(A)$: Since $f^{-1}(e) \neq \emptyset$, $f^{-1}(A)$ must be empty and $f \equiv e$ is constant). A locally constant function f on a compact space X can take only a finite number of values ($f(X)$ must be compact and discrete).

Lemma. *If X is a compact metric space, a locally constant function f on X is uniformly locally constant when there exists $\delta > 0$ such that*

$$d(x, y) < \delta \Longrightarrow f(x) = f(y).$$

PROOF. Give E the discrete metric. Since X is compact, $f : X \to E$ is uniformly continuous. Hence there is a $\delta > 0$ such that

$$d(x,y) < \delta \Longrightarrow d(f(x), f(y)) < 1 \Longrightarrow f(x) = f(y),$$

and the conclusion follows. ∎

The set of functions $X \to E$ is denoted by $\mathcal{F}(X;E)$, and when $E = K$ is a field, $\mathcal{F}(X;K) = \mathcal{F}(X)$ is a vector space over K (omitted from the notation if this field is implicit from the context). When X is a compact ultrametric space, the locally constant functions $X \to K$ form a K-vector subspace $\mathcal{F}^{\mathrm{lc}}(X)$ of $\mathcal{F}(X)$. The characteristic functions of clopen balls of X form a system of generators of $\mathcal{F}^{\mathrm{lc}}(X)$.

3.2. *Characteristic Functions of Balls of* $\mathbf{Z}_p$

We are interested in locally constant functions on $X = \mathbf{Z}_p$ taking values in any abelian group M (this abelian group will typically be an extension K of $\mathbf{Q}_p$). Let us start by the study of the (uniformly) locally constant functions $f \in \mathcal{F}^{\mathrm{lc}}(\mathbf{Z}_p; M)$ satisfying

$$|x - y| \le |p^j| = \frac{1}{p^j} \Longrightarrow f(x) = f(y)$$

for some fixed integer $j \ge 0$. These are the functions that are constant on all closed balls of radius $r_j = 1/p^j$. Since the balls in question are the cosets of $p^j\mathbf{Z}_p$ in $\mathbf{Z}_p$, these functions are the elements of the vector space

$$F_j = \mathcal{F}(\mathbf{Z}/p^j\mathbf{Z}) = \mathcal{F}(\mathbf{Z}_p/p^j\mathbf{Z}_p) \subset \mathcal{F}^{\mathrm{lc}}(\mathbf{Z}_p; K).$$

In fact, we have a partition

$$\mathbf{Z}_p = \coprod_{0 \le i < p^j} (i + p^j\mathbf{Z}_p)$$

into balls of radius r_j, and

$$i + p^j\mathbf{Z}_p = B_{\le p^{-j}}(i) \quad (0 \le i < p^j)$$

is an enumeration of these balls in $\mathbf{Z}_p$. For fixed j the characteristic functions

$$\varphi_{i,j} = \text{characteristic function of the ball } B_{\le 1/p^j}(i) \quad (0 \le i < p^j)$$

make up a basis of the finite-dimensional space F_j. When we let j increase, the subspaces F_j also increase, and

$$\mathcal{F}^{\mathrm{lc}}(\mathbf{Z}_p; K) = \bigcup_{j \ge 0} F_j.$$

Unfortunately, the previously given basis of F_j has no element in common with the basis constructed similarly in F_{j-1}. A clever way of constructing coherent bases of the spaces F_j — where the basis of F_j extends the basis of F_{j-1} — has been devised by M. van der Put. Let

$$\psi_i = \begin{cases} \varphi_{0,0} = 1 & \text{characteristic function of } \mathbf{Z}_p \quad (i = 0) \\ \varphi_{i,1} & \text{characteristic function of } i + p\mathbf{Z}_p \quad (1 \le i < p) \\ \varphi_{i,2} & \text{characteristic function of } i + p^2\mathbf{Z}_p \quad (p \le i < p^2), \text{ etc.} \end{cases}$$

Generally,

$$\psi_i = \varphi_{i,j} \text{ characteristic function of } i + p^j\mathbf{Z}_p \text{ if } p^{j-1} \le i < p^j.$$

Since absolute values of elements of $\mathbf{Z}_p$ can only be powers of p, we have

$$|x| < \frac{1}{i} \iff |x| \le \frac{1}{p^j} \quad (p^{j-1} \le i < p^j),$$

and $\psi_i = \varphi_{i,j}$ is also the characteristic function of the ball

$$B_i = \{x \in \mathbf{Z}_p : |x - i| < 1/i\}$$

(with the convention $B_0 = \mathbf{Z}_p$ for $i = 0$).

On the other hand, the indices i in the range $p^{j-1} \le i < p^j$ are precisely those that admit an expansion of *length* j in base p, namely an expansion of the form

$$i = i_0 + i_1 p + \cdots + i_{j-1}p^{j-1} \quad (0 \le i_\ell \le p - 1,\ i_{j-1} \ne 0).$$

Definition. *The* length *of an integer $i \ge 1$ is the integer $\nu = \nu(i) \ge 1$ such that the expansion of i in base p has digits $i_\ell = 0$ for $\ell \ge \nu$, while $i_{\nu-1} \ne 0$.*

With this definition, the *van der Put sequence* is defined by

$$\psi_i = \varphi_{i,\nu(i)} : \text{characteristic function of } i + p^{\nu(i)}\mathbf{Z}_p.$$

Here are the first few functions:

$$\begin{array}{ccccccccc} 1 \\ \varphi_{0,1} & \cdots & \varphi_{p-1,1} \\ \varphi_{0,2} & \cdots & \varphi_{p-1,2} & \varphi_{p,2} & \cdots & \varphi_{p^2-1,2} \\ \varphi_{0,3} & \cdots & \varphi_{p-1,3} & \cdots & \cdots & \cdots & \varphi_{p^2,3} & \cdots & \varphi_{p^3-1,3} \\ \vdots & & \vdots & & \vdots & \vdots & \vdots & & \vdots \\ \varphi_{0,j} & \cdots & \cdots & \cdots & \cdots & \cdots & \cdots & \cdots & \cdots \end{array}$$

The sequence $(\psi_i)_{i<p^j}$ appears at the top of this triangular table of characteristic functions.

Proposition. *The sequence $(\psi_i)_{0 \le i < p^j}$ is a basis of F_j $(j \ge 0)$.*

PROOF. For fixed $j \geq 0$, the components of any $f \in F_j$ in the known basis $\varphi_{i,j}$ $(i < p^j)$ of F_j are the (constant) values of f on the balls $i + p^j\mathbf{Z}_p$:

$$f = \sum_{i<p^j} f(i)\varphi_{i,j}.$$

In particular, for $f = \psi_\ell$, the characteristic function of $B_\ell = \ell + p^\nu\mathbf{Z}_p$, we have a sum of the form

$$\psi_\ell = \sum_{\cdots} \varphi_{i,j},$$

where the indices that occur are the same as those occurring in the partition

$$B_\ell = \coprod_{\cdots} (i + p^j\mathbf{Z}_p).$$

They are the indices i such that $0 \leq i < p^j$ and $i \equiv \ell \pmod{p}^\nu$ (in order to have $i \in B_\ell$). These indices can be listed:

$$i = \ell,\ \ell + p^\nu,\ \ell + 2p^\nu,\ \ldots.$$

The first one is ℓ itself, and they are all greater than or equal to ℓ. The matrix of the components of the ψ_ℓ in the basis $\varphi_{i,j}$ is lower triangular with 1's on its diagonal (all its entries are 0's and 1's). This matrix U has determinant 1 and hence is invertible: The ψ_ℓ $(0 \leq \ell < p^j)$ form a basis of F_j. If we write $U = I + N$, the matrix N is lower triangular with 0's on its diagonal and hence is *nilpotent*: A power of N vanishes. This proves that

$$U^{-1} = I - N + N^2 - \cdots + (-1)^m N^m \text{ if } N^{m+1} = 0.$$

In particular, the inverse U^{-1} of U has integral entries: The components of the $\varphi_{i,j}$ in the basis (ψ_ℓ) are also integers. ∎

Here is an even more precise result.

Proposition. *If $f = \sum a_i\psi_i \in F_j$, the coefficients are given by*

$$a_0 = f(0) \text{ and } a_n = f(n) - f(n_-) \quad (n \geq 1),$$

where $n_- = n - n_{\nu-1}p^{\nu-1}$ denotes the integer of length strictly smaller than n obtained by deleting its top digit in base p.

PROOF. We have already observed that $f(0) = a_0$. Fix a positive integer n and consider the sum $f(n) = \sum_{i<p^j} a_i\psi_i(n)$ in which $\psi_i(n) = 0$ or 1. More precisely,

$$\begin{aligned}\psi_i(n) = 1 &\iff n \in B_i\\ &\iff n \equiv i \bmod p^{\nu(i)}\\ &\iff \text{the digits of } n \text{ and } i\\ &\qquad \text{are the same up to } \nu(i)\\ &\iff i \text{ is an initial partial sum of } n.\end{aligned}$$

This shows that

$$f(n) = a_0 + (*) + a_n,$$

whereas

$$f(n_-) = a_0 + (*).$$

Hence $f(n) - f(n_-) = a_n$ as claimed in the proposition. ■

Corollary. *When $f = \sum a_i \psi_i \in F_j$ takes its values in an ultrametric field, we have*

$$\|f\| = \max |a_i|.$$

PROOF. For each $x \in \mathbf{Z}_p$ we have $\psi_i(x) = 0$ or 1: $|\psi_i(x)| \leq 1$ and

$$|f(x)| = |\sum a_i \psi_i(x)| \leq \max |a_i|.$$

This proves

$$\|f\| = \sup |f(x)| \leq \max |a_i|.$$

Conversely, $a_0 = f(0) \implies |a_0| \leq \|f\|$, and for $n \geq 1$,

$$|a_n| = |f(n) - f(n_-)| \leq \max(|f(n)|, |f(n_-)|) \leq \|f\|,$$

hence $\max |a_n| \leq \|f\|$. ■

Since $(\psi_i)_{i\geq 0}$ is a basis of $\mathcal{F}^{\mathrm{lc}}(\mathbf{Z}_p; K) = \bigcup_{j\geq 0} F_j$, it is easy to generalize the preceding results to all locally constant functions (taking their values in an extension K of $\mathbf{Q}_p$).

Theorem. *Let $f : \mathbf{Z}_p \to K$ be a locally constant function. Define*

$$a_0 = f(0), \quad a_n = f(n) - f(n_-) \quad (n \geq 1).$$

Then $f = \sum a_i \psi_i$ is a finite sum and $\|f\| = \sup_i |a_i|$. ■

3.3. *The van der Put Theorem*

We are now able to give the main result, namely the representation of any continuous $f : \mathbf{Z}_p \to K$ where K is a complete extension of $\mathbf{Q}_p$.

Theorem. *Let $f : \mathbf{Z}_p \to K$ be a continuous function. Define*

$$a_0 = f(0), \quad a_n = f(n) - f(n_-) \quad (n \geq 1).$$

Then $|a_n| \to 0$, *and* $\sum a_i \psi_i$ *converges uniformly to* f. *Moreover,*

$$\|f\| = \sup_i |a_i| = \max_i |a_i|.$$

PROOF. Since $|n - n_-| \to 0$ $(n \to \infty)$ and f is uniformly continuous, we have $|a_n| = |f(n) - f(n_-)| \to 0$ $(n \to \infty)$, and the series converges uniformly. The sum of this series is a continuous function,

$$g = \sum a_i \psi_i.$$

We still have to prove $f = g$. Since these functions are continuous, it is enough to show that their restrictions to the dense subset $\mathbf{N}$ are the same. The obvious equality $f(0) = a_0 = g(0)$ can be used as the first step in an induction on n. Let $g_j = \sum_{i<p^j} a_i \psi_i$. For $n < p^j$ we have

$$\begin{aligned} f(n) - f(n_-) &= a_n && \text{(by definition)} \\ &= \text{coefficient of } g_j && (\text{since } n < p^j) \\ &= g_j(n) - g_j(n_-), \\ f(n) - g_j(n) &= f(n_-) - g_j(n_-). \end{aligned}$$

This shows that if f and g_j agree on $\{0, 1, 2, \ldots, n-1\}$, they will also agree at the point n (provided that $n < p^j$). As a consequence, for all integers $n \in \mathbf{N}$, $f(n) = \lim_j g_j(n) = g(n)$ (with a stationary convergence). As mentioned, this proves $f = g$. The equality $\|f\| = \sup_i |a_i|$ is obtained exactly as in the case f locally constant. ∎

4. Ultrametric Banach Spaces

In this section K *will always denote a* complete *ultrametric extension of* $\mathbf{Q}_p$.

We have already given in (II.3.1) the formal properties of ultrametric norms on $\mathbf{Q}_p$-vector spaces, and we have studied finite-dimensional such spaces over K. Here we turn to infinite-dimensional ones.

We shall simply say *normed space* for ultrametric normed space over K, and *Banach space* for *complete* normed space.

4.1. *Direct Sums of Banach Spaces*

The *direct sum* of a family $(E_i)_{i\in I}$ of normed spaces is the algebraic direct sum of this family,

$$\bigoplus_{i\in I} E_i = \{(x_i) : \text{only finitely many } x_i \neq 0\} \subset \prod_{i\in I} E_i$$

equipped with the sup norm on the components,

$$\|x\| = \sup_i \|x_i\| = \max_i \|x_i\| \quad \text{if } x = (x_i).$$

When $(E_i)_{i\in I}$ is a family of Banach spaces, it is convenient to consider a completion of the preceding direct sum. Here is the construction. The *support* of a family $x = (x_i)_{i\in I} \in \prod_{i\in I} E_i$, is $I_x = \{i \in I : x_i \neq 0\} \subset I$, and $\|x_i\| \to 0$ means

$$\text{for all } \varepsilon > 0, \text{ the set } I_x(\varepsilon) := \{i \in I : \|x_i\| > \varepsilon\} \text{ is finite.}$$

If $\|x_i\| \to 0$, then the *support* I_x of the family x is *at most countable*, since it is the countable union of the finite sets $I_x(1/n)$ $(n \geq 1)$.

Definition. *The* Banach direct sum *of the family* $(E_i)_{i\in I}$ *of Banach spaces* E_i *is the normed space*

$$\widehat{\bigoplus}_{i\in I} E_i \subset \prod_{i\in I} E_i$$

consisting of the families $x = (x_i)$ *such that* $\|x_i\| \to 0$, *equipped with the sup norm*

$$\|x\| = \|(x_i)\| := \sup_i \|x_i\| = \max_i \|x_i\|.$$

This terminology is justified by the following result.

Theorem. *The Banach direct sum of a family* $(E_i)_{i\in I}$ *of Banach spaces is a completion of the normed direct sum of the family.*

Proof. The set of families x such that $\|x_i\| \to 0$ is a vector subspace of the product $\prod_{i\in I} E_i$, and the algebraic direct sum is dense in it. Let us show that the Banach direct sum is *complete*,

$$\bigoplus_{i\in I} E_i \subset \widehat{\bigoplus}_{i\in I} E_i \subset \prod_{i\in I} E_i.$$

Let $n \mapsto x^{(n)} = (x_i^{(n)})_{i\in I}$ be a Cauchy sequence in the direct sum. For each $i \in I$, $n \mapsto x_i^{(n)}$ is a Cauchy sequence in E_i. Let x_i be its limit. For given $\varepsilon > 0$, there is an integer N_ε such that

$$\|x^{(n)} - x^{(m)}\| \leq \varepsilon \quad (n, m \geq N_\varepsilon).$$

A fortiori, for all $i \in I$,

$$\|x_i^{(n)} - x_i^{(m)}\| \leq \varepsilon \quad (n, m \geq N_\varepsilon).$$

Letting $m \to \infty$, we obtain

$$\|x_i^{(n)} - x_i\| \leq \varepsilon \quad (n \geq N_\varepsilon,\ i \in I). \qquad (*)$$

Since $\|x_i^{(n)}\| \leq \varepsilon$ outside a finite set J (depending on ε and n), we also have

$$\|x_i\| \leq \max(\|x_i^{(n)}\|, \|x_i^{(n)} - x_i\|) \leq \varepsilon \quad (i \notin J).$$

This proves that the family $x := (x_i)$ is in $\widehat{\bigoplus}_{i\in I} E_i$. Coming back to the inequality $(*)$, we see that

$$\|x^{(n)} - x\| = \sup_i \|x_i^{(n)} - x_i\| \le \varepsilon \quad (n \ge N_\varepsilon).$$

This proves $x^{(n)} \to x$ in $\widehat{\bigoplus}_{i\in I} E_i$. ∎

Example. When all Banach spaces $E_i = E$ are equal, the algebraic direct sum is also denoted by

$$\bigoplus_{i\in I} E = E^{(I)} \subset E^I = \prod_{i\in I} E.$$

Its completion is the space of sequences in E converging to 0: we denote this space by $c_0(I; E)$. (The notation $c_0(K)$ is similar to the classical (c_0) introduced by S. Banach when $K = \mathbf{C}$.) When $E = K$ is the base field, or when $I = \mathbf{N}$, we drop them from the notation if there is no risk of confusion:

$$c_0(I) = c_0(I; K),\ c_0(E) = c_0(\mathbf{N}; E),\ c_0 = c_0(\mathbf{N}) = c_0(\mathbf{N}; K).$$

We can now formulate a few consequences of the theorem.

Corollary 1. *Let E be a Banach space. Then $c_0(I; E)$ is a completion of $E^{(I)} \subset E^I$ for the sup norm.* ∎

Corollary 2. *Let E be a Banach space. Then the sum map $E^{(I)} \to E$ has a unique continuous extension $\Sigma : c_0(I; E) \to E$.*

PROOF. The sum $x = (x_i) \mapsto \sum_{i\in I} x_i : E^{(I)} \to E$ is a contracting linear map

$$\left\|\sum_{i\in I} x_i\right\| \le \sup_i \|x_i\| = \|x\| \quad (x \in E^{(I)}).$$

It has a unique continuous extension Σ. This extension is also a contracting linear map by density and continuity. Hence we have more generally

$$\left\|\sum_{i\in I} x_i\right\| \le \sup_i \|x_i\| = \|x\| \quad (x \in c_0(I; E)).$$ ∎

This sum Σ can be computed using any ordering of the index set I and any grouping $I = \coprod_j I_j$: The equality for families with finite support extends by continuity to the completion $c_0(I; E)$ (cf. (II.1.2)).

Corollary 3 (Universal Property of Direct Sums). *Let ε_j denote the canonical injection of a factor into the direct sum $E_j \to \bigoplus_{i\in I} E_i \subset \widehat{\bigoplus}_{i\in I} E_i$. Then for each Banach space E and family (f_j) consisting of linear contractions*

$f_j : E_j \to E$, there is a unique linear contraction f such that the following diagram is commutative:

$$\begin{array}{ccccc} E_j & \overset{\varepsilon_j}{\to} & \bigoplus_{i\in I} E_i & \subset & \widehat{\bigoplus}_{i\in I} E_i \\ & f_j \searrow & \downarrow \oplus f_i & \swarrow f & \\ & & E & & \end{array}$$

PROOF. Under the assumptions made,

$$(x_i) \mapsto (f_i x_i) : \widehat{\bigoplus}_{i\in I} E_i \to c_0(I; E)$$

is a linear contracting map, and composition with the sum Σ yields the unique solution to the factorization problem

$$f = \Sigma \circ (f_i), \quad fx = \sum_i f_i x_i. \qquad \blacksquare$$

4.2. Normal Bases

When E and F are (ultrametric) normed spaces over K, we denote by $L(E; F)$ the normed vector space of *continuous linear maps* $T : E \to F$. Recall that a linear map is continuous precisely when it is *continuous at the origin*, or, equivalently, when it is *bounded*:

$$\|T\| := \sup_{x\neq 0} \frac{\|Tx\|}{\|x\|} < \infty.$$

By definition, we have

$$\|Tx\| \le \|T\| \|x\| \quad (x \in E).$$

This shows that T is a *contraction* precisely when $\|T\| \le 1$.

Comment. The inequality $\|Tx\| \le \|T\|\|x\|$ $(x \in E)$ shows that $\|Tx\| \le \|T\|$ when $\|x\| \le 1$, and hence $\sup_{\|x\|\le 1} \|Tx\| \le \|T\|$. But contrary to classical functional analysis, this inequality can be a strict inequality: When $1 \notin \|E\|$, the unit sphere $\|x\| = 1$ is empty, closed and open unit balls coincide, and $\sup_{\|x\|\le 1} = \sup_{\|x\|<1}$. For the operator $T = \mathrm{id}$ (and $\|E - \{0\}\|$ discrete in $\mathbf{R}_{>0}$) we have

$$\sup_{\|x\|\le 1} \|x\| = \sup_{\|x\|<1} \|x\| < 1 \neq \|\mathrm{id}\| = 1.$$

Proposition 1. *If F is complete, then $L(E; F)$ is also complete.*

PROOF. Let (T_n) be a Cauchy sequence in $L(E; F)$. For each $x \in E$, $(T_n(x))$ is a Cauchy sequence in the complete space F, and hence has a limit Tx which obviously depends linearly on $x \in E$. This defines a linear map $T : E \to F$. Let $\varepsilon > 0$ be given. There exists an integer N_ε such that $\|T_n - T_m\| \le \varepsilon$ for all

$n, m \geq N_\varepsilon$. Letting $m \to \infty$ we deduce $\|Tx - T_m x\| \leq \varepsilon \|x\|$ for all $n, m \geq N_\varepsilon$. This proves that the operator $T - T_m$ is continuous (bounded); hence $T = T_m + (T - T_m)$ is continuous. Moreover, $\|T - T_m\| \leq \varepsilon$ when $m \geq N_\varepsilon$. This shows that $\|T - T_m\| \to 0$, $T_m \to T$ $(m \to \infty)$, and everything is proved. ■

Corollary. *For any normed space E, the topological dual $E' = L(E; K)$ is a Banach space.* ■

Example. Let I be any index set and E a Banach space. The vector space of bounded sequences $\mathbf{a} = (a_i)_{i \in I}$ in E with the norm $\|\mathbf{a}\| = \sup_i \|a_i\|$ is a normed space, denoted by $l^\infty(I; E)$ (it is complete: cf. exercise).

The universal property of a direct sum consisting of factors E_i all equal to the same Banach space E and for linear forms $\varphi_i : E \to K$ leads to the following statement.

Proposition 2. *The topological dual of the space $c_0(I; E)$ is canonically isomorphic as a normed space to $l^\infty(I; E')$.*

Proof. If φ is a continuous linear form on $c_0(I; E)$, we let $\varphi_i = \varphi \circ \varepsilon_i$ denote the restriction of φ to the ith factor E in $c_0(I; E)$ (families having a zero component for all indices except i). Since $\|\varphi_i\| \leq \|\varphi\|$, we get a bounded family $(\varphi_i) \in l^\infty(I; E')$. Conversely, if $(\varphi_i) \in l^\infty(I; E')$, we can define a linear form $\varphi = \Sigma \varphi_i$ on $c_0(I; E)$ by the formula $(a_i) \mapsto \sum_i \varphi_i(a_i)$ (a summable series, since the sequence φ_i is bounded and $\|a_i\| \to 0$). Both maps

$$\varphi \mapsto (\varphi \circ \varepsilon_i), \quad (\varphi_i) \mapsto \Sigma \varphi_i$$

are linear and decrease norms. Hence they are inverse isometries. ■

In other words, the bilinear map

$$((a_i), (\varphi_i)) \mapsto \sum_i \varphi_i(a_i), \quad c_0(I; E) \times l^\infty(I; E') \to K$$

is a *duality pairing* that proves the proposition.

Corollary. *The space $l^\infty(I) = l^\infty(I; K)$ is a Banach space.* ■

In the space $c_0 = c_0(I)$, the family of elements $\mathbf{e}_i = (\delta_{ij})_{j \geq 0}$ (Kronecker symbol) has the following basic property. Each sequence $\mathbf{a} = (a_n)_{n \geq 0} \in c_0$ is the sum of a unique convergent series

$$\mathbf{a} = \sum_{n \geq 0} a_n \mathbf{e}_n,$$

and $\|\mathbf{a}\| = \sup_{n\geq 0} |a_n| = \max_{n\geq 0} |a_n|$. We say that this family of elements $\mathbf{e}_i = (\delta_{ij})$ constitutes the *canonical basis* of this space (in spite of the fact that it is not a vector space basis: In linear algebra, linear combinations are always assumed to be *finite* linear combinations).

This leads to the following definition.

Definition. *A* normal basis *in an ultrametric Banach space E is a family $(\mathbf{e}_i)_{i\in I}$ of elements of E such that*

- *each $\mathbf{x} \in E$ can be represented by a convergent series $\mathbf{x} = \sum_I x_i\mathbf{e}_i$ where the sequence of components $|x_i| \to 0$,*
- *in any representation $\mathbf{x} = \sum_I x_i\mathbf{e}_i$ we have $\|\mathbf{x}\| = \sup_{i\in I} |x_i|$.*

A normal basis is sometimes called an *orthonormal basis*. In particular, for each convergent series $\sum_I x_i\mathbf{e}_i$, the set of nonzero components is at most countable, as observed earlier. If $(\mathbf{e}_i)_I$ is a normal basis, we have $\|\mathbf{e}_i\| = 1$ for each $i \in I$. On the other hand,

$$\mathbf{x} = \sum_i x_i\mathbf{e}_i = \sum_i y_i\mathbf{e}_i \Longrightarrow \sum_i (x_i - y_i)\mathbf{e}_i = \mathbf{0},$$

and by the second postulated property of a normal basis,

$$\sup_{i\in I} |x_i - y_i| = \|\mathbf{0}\| = 0, \Longrightarrow x_i = y_i \quad (i \in I),$$

whence the *uniqueness of representations in normal bases*. All properties of normal bases are summarized in the following obvious result.

Proposition 3. *Let E be an ultrametric Banach space having a normal basis $(\mathbf{e}_i)_{i\in I}$. Then the mapping $(x_i) \mapsto \sum_{i\in I} x_i\mathbf{e}_i$ defines a linear bijective isometry $c_0(I;K) \xrightarrow{\sim} E$. Conversely, any linear bijective isometry $c_0(I;K) \xrightarrow{\sim} E$ defines a normal basis in E, namely the image of the canonical basis of $c_0(I;K)$.* ∎

Example 1. The Banach spaces $c_0(I;K)$ supply examples of ultrametric spaces with normal bases. In particular, when the index set I is finite, we get the (finite) product spaces K^n with the sup norm (cf. exercises).

Example 2. Let $E = C(\mathbf{Z}_p;K)$ be the space of continuous functions $\mathbf{Z}_p \to K$ (where K is a complete extension of $\mathbf{Q}_p$) equipped with the sup norm. The Mahler theorem (2.4) asserts that the binomial polynomials constitute a normal basis of E: The map

$$c_0(K) \to E : (a_k) \mapsto \sum_{k\geq 0} a_k f_k = \sum_{k\geq 0} a_k \binom{\cdot}{k}$$

is a *bijective linear isometry.* The van der Put theorem (3.3) asserts that the sequence $(\psi_j)_{j\geq 0}$ constitutes another normal basis of E. These two normal bases are quite different in nature.

4.3. Reduction of a Banach Space

Let K be a complete extension of $\mathbf{Q}_p$ and E an ultrametric Banach space over K. We keep the general notation for

- $A = B_{\leq 1}(K)$: maximal subring of K,
- $M = B_{<1}(K)$: maximal ideal of A,
- $k = A/M$: residue field of K.

Moreover, we consider the closed unit ball in E, $E_1 = \{v \in E : \|v\| \leq 1\}$, as an A-module and $ME_1 = \{\lambda x : \lambda \in M,\ x \in E_1\}$ as an A-submodule (ME_1 is obviously an additive subgroup of E_1). As a consequence, $\widetilde{E} = E_1/ME_1$ is a k-vector space.

Remark. We have quite generally $ME_1 \subset B_{<1}(E)$. This inclusion is in general a *strict inclusion.* For example consider any finite, ramified extension K of $\mathbf{Q}_p$ as a Banach space over $\mathbf{Q}_p$. Its open unit ball is strictly larger than $pB_{\leq 1}(K)$: The open unit ball contains an element of norm $|p|^{1/e}$, while all elements of $pB_{\leq 1}(K)$ have norms $\leq |p| < |p|^{1/e}$.

Lemma. *If either* $\|E\| = |K|$, *or* $|K^\times|$ *is dense in* $\mathbf{R}_{>0}$, *then* $ME_1 = B_{<1}(E)$.

PROOF. In the first case, if $\lambda = \|x\| < 1$, we can write $x = \lambda \cdot (x/\lambda) \in ME_1$. In the second case, if $\|x\| < 1$ we can choose a scalar $\lambda \in K$ with $\|x\| \leq |\lambda| < 1$ and still write $x = \lambda \cdot (x/\lambda) \in ME_1$. ∎

Proposition. *If* $(\mathbf{e}_i)_{i\in I}$ *is a normal basis of* E, *then* $(\mathbf{e}_i \bmod ME_1)_{i\in I}$ *is a basis of the* k-*vector space* $\widetilde{E}$.

PROOF. Define $\varepsilon_i = (\mathbf{e}_i \bmod ME_1) \in \widetilde{E}$. These elements *generate* $\widetilde{E}$: If $\widetilde{x} = (x \bmod ME_1) \in \widetilde{E}$, we can write $x = \sum x_i \mathbf{e}_i$ with all $|x_i| \leq 1$ and only finitely many $|x_i| = 1$, giving rise to a *finite* linear combination $\widetilde{x} = \sum_i (x_i \bmod M)\varepsilon_i$. On the other hand, take a linear combination $\sum_i \alpha_i \varepsilon_i = 0 \in \widetilde{E} = E_1/ME_1$ ($\alpha_i \in k$, only finitely many nonzero such coefficients). We can choose scalars $a_i \in A \subset K$ with $\alpha_i = (a_i \bmod M)$ and $a_i = 0$ if $\alpha_i = 0$. By assumption

$$\sum_i a_i e_i \in ME_1 \subset B_{<1}(E),$$

namely $\|\sum_i a_i e_i\| < 1$. By definition of a normal basis,

$$\sup_i |a_i| = \left\|\sum_i a_i e_i\right\| < 1$$

and $|a_i| < 1$ for all i. This proves that all $\alpha_i = (a_i \bmod M) = 0 \in k$ and the linear combination is trivial. The family $(\varepsilon_i)_I$ is *free* and is a basis of the reduced vector space $\widetilde{E}$. ■

4.4. A Representation Theorem

With the same notation as before, observe that the closed balls $B_{\leq r}(E)$ $(r \geq 0)$ and the open balls $B_{<r}(E)$ $(r > 0)$ are A-modules and for any ideal I of A,

$$B_{\leq r}(E)/I B_{\leq r}(E) \text{ is an } A/I\text{-module.}$$

In particular, if the ideal I is principal, say $I = (\xi)$ with $|\xi| < 1$, then

$$B_{\leq r}(E)/\xi B_{\leq r}(E) \text{ is an } A/(\xi)\text{-module.}$$

Let us generalize the expansion theorem (II.1.4) to the vector case.

Theorem. *Let E be an ultrametric Banach space, $\xi \in K$, $|\xi| < 1$, and choose a set of representatives $S \subset B_{\leq r}(E)$ for the classes* mod $\xi B_{\leq r}(E)$. *Assume that $0 \in S$. Then every element $x \in B_{\leq r}(E)$ can be represented uniquely as the sum of a convergent series*

$$x = \sum_{i \geq 0} a_i \xi^i \quad (a_i \in S).$$

Proof. Take for a_0 the (unique) representative in S with $x - a_0 \in \xi B_{\leq r}(E)$. Hence $x - a_0 = r_1 = \xi x_1$ for some $x_1 \in B_{\leq r}(E)$. One can proceed similarly for x_1 and find elements $a_1 \in S$, $x_2 \in B_{\leq r}(E)$ with $x_1 - a_1 = \xi x_2$, namely

$$x = a_0 + a_1 \xi + \xi^2 x_2.$$

Iterating the construction, we obtain a series $\sum_{i \geq 0} a_i \xi^i$, which converges to x. For this part of the proof, the completeness of E is not needed, since the element x, the limit of partial sums, is known a priori. But when E is complete, *every series* $\sum_{i \geq 0} a_i \xi^i$ *with coefficients $a_i \in S$ is convergent, since* $|a_i \xi^i| \leq r|\xi|^i \to 0$. The *uniqueness statement* is immediately verified. Indeed, if $\sum_{i \geq 0} a_i \xi^i = 0$, we have

$$a_0 = -\sum_{i \geq 1} a_i \xi^i \in \xi B_{\leq r}(E),$$

hence $a_0 = 0$, since this representative is in S. By induction, all $a_i = 0$. ■

4.5. The Monna-Fleischer Theorem

In a Banach space $c_0(I; K)$, we have

$$\|\mathbf{x}\| = \sup_I |x_i| = \max_I |x_i| \in |K|.$$

Hence if an ultrametric Banach space admits a normal basis, we have

$$\|E\| = |K|, \quad \|E - \{0\}\| = |K^\times|.$$

Theorem. *Let K be a complete ultrametric field with $|K^\times|$ discrete in $\mathbf{R}_{>0}$ and E an ultrametric Banach space over K. Then E admits a normal basis precisely when $\|E\| = |K|$.*

Proof. The preliminary comment proves the *necessity* of the condition. Conversely, let us show why it is *sufficient.* Since K has a discrete valuation, its maximal subring $A = R$ is principal, with maximal ideal $M = P = \pi R$. Then $PE_1 = \pi E_1$. Let us choose and fix a system of representatives $S \subset R$ for the classes mod P, with $0 \in S$. Also choose and fix a basis $(\varepsilon_i)_I$ of the k-vector space $\widetilde{E} = E_1/\pi E_1$ with liftings $\mathbf{e}_i \in E_1$, so $\varepsilon_i = \mathbf{e}_i \bmod \pi E_1$. I claim that $(\mathbf{e}_i)_I$ is a normal basis of E. Consider first the case of a vector $\mathbf{x} \in E_1$: $\|\mathbf{x}\| \le 1$. The vector $\widetilde{\mathbf{x}} = (\mathbf{x} \bmod \pi E_1)$ can be expanded in the k-basis $(\varepsilon_i)_I$, say $\widetilde{\mathbf{x}} = \sum \alpha_i \varepsilon_i$ (only finitely many $\alpha_i \neq 0$). Consider the representatives

$$a_i^{(0)} \in S, \; a_i^{(0)} \equiv \alpha_i \pmod{P};$$

hence $a_i^{(0)} = 0$ except for finitely many indices. If $\|\mathbf{x}\| = 1$, at least one $|a_i^{(0)}| = 1$ and all $|a_i^{(0)}| \le 1$. We have

$$\mathbf{r}_1 = \mathbf{x} - \sum a_i^{(0)} \mathbf{e}_i \in B_{<1}(E).$$

By the lemma in (4.3), we have $B_{<1}(E) = PE_1 = \pi E_1$, and the same construction with the vector $\pi^{-1}(\mathbf{x} - \sum a_i^{(0)} \mathbf{e}_i) \in E_1$ gives a family

$$a_i^{(1)} \in S, \quad a_i^{(1)} \neq 0 \text{ for finitely many indices only,}$$

such that

$$\mathbf{x} = \sum a_i^{(0)} \mathbf{e}_i + \pi \sum a_i^{(1)} \mathbf{e}_i + \mathbf{r}_2 \quad (\mathbf{r}_2 \in \pi^2 E_1).$$

By iteration, we obtain a sequence $\mathbf{r}_n \in \pi^n E_1$ and convergent series

$$a_i = a_i^{(0)} + \pi a_i^{(1)} + \cdots \in R \subset K$$

giving a representation $x = \sum_I a_i \mathbf{e}_i$ with $a_i \to 0$ (for fixed j, only finitely many $a_i^{(j)} \neq 0$). At each step of the iteration we have to choose a scalar λ_n such that

$$\|\lambda_n(\pi^{-n}\mathbf{r}_n)\| = 1.$$

This is possible by the assumption $\|E\| = |K|$. ∎

4.6. *Spaces of Linear Maps*

Let K be a complete ultrametric field and E, F two ultrametric Banach spaces over K. Assume that E admits a normal basis and fix an isomorphism $c_0(J) \cong E$. Then any linear map $T : E \to F$ furnishes a family of $\mathbf{f}_j = T\mathbf{e}_j \in F$, namely the image of the normal basis of E — canonical basis of $c_0(J)$. When the linear map T is continuous, this family is bounded

$$\|\mathbf{f}_j\| \leq \|T\|\|\mathbf{e}_j\| = \|T\|.$$

We thus obtain a linear map

$$L(E;F) \to l^\infty(J;F) \ : \ T \mapsto (\mathbf{f}_j)_J.$$

Proposition 1. *Assume that E admits a normal basis and fix an isomorphism $c_0(J) \cong E$. Then the map*

$$L(E;F) \to l^\infty(J;F)$$

defined above is an isometric isomorphism.

PROOF. We have already seen that $\|(\mathbf{f}_j)_J\| \leq \|T\|$. Conversely,

$$\mathbf{x} = \sum_j x_j \mathbf{e}_j \Longrightarrow T(\mathbf{x}) = \sum_j x_j \mathbf{f}_j \ \text{(this sum converges!)},$$

$$\|T(\mathbf{x})\| \leq \sup_j \|x_j \mathbf{f}_j\| \leq \sup_j |x_j| \sup_j \|\mathbf{f}_j\| = \|x\| \sup_j \|\mathbf{f}_j\|$$

whence $\|T\| \leq \sup_j \|\mathbf{f}_j\| = \|(\mathbf{f}_j)_J\|$. Observe that for any choice of bounded family $\mathbf{f}_j \in F$, there is a $T \in L(E;F)$ with $T\mathbf{e}_j = \mathbf{f}_j$ ($j \in J$), so that the map $L(E;F) \to l^\infty(J;F)$ is surjective. ∎

In particular, for $F = K$, we get the following result (cf. Proposition 1 in (4.2)).

Corollary. *There is a canonical isometric isomorphism*

$$(c_0(J))' \cong l^\infty(J).$$

∎

Assume now symmetrically that F has a normal basis and fix an isomorphism $c_0(I) \cong F$. The linear maps $T : E \to F \cong c_0(I)$, $\mathbf{x} \mapsto T(\mathbf{x}) = (\varphi_i(\mathbf{x}))$ give a family $(\varphi_i)_I$ of linear forms $\varphi_i : E \to K$. If T is continuous, so are the linear forms φ_i and $\|T(\mathbf{x})\| = \sup_i |\varphi_i(\mathbf{x})|$,

$$\begin{aligned}
\|T\| &= \sup_{x \neq 0} \|T(\mathbf{x})\|/\|\mathbf{x}\| \\
&= \sup_{x \neq 0} \sup_i |\varphi_i(\mathbf{x})|/\|\mathbf{x}\| \\
&= \sup_i \sup_{x \neq 0} |\varphi_i(\mathbf{x})|/\|\mathbf{x}\| = \sup_i \|\varphi_i\|.
\end{aligned}$$

This proves the following proposition.

Proposition 2. *The linear map*

$$L(E; c_0(I)) \to l^\infty(I; E') : T \mapsto (\varphi_i)$$

is isometric, but not always surjective. ∎

When both E and F have normal bases, we get the following statement.

Proposition 3. *When* $E = c_0(J)$ *and* $F = c_0(I)$*, we can make canonical identifications*

$$L(E; F) = l^\infty(J; c_0(I)) \subset l^\infty(I; E') = l^\infty(I; l^\infty(J)) \cong l^\infty(I \times J).$$ ∎

In other words, when normal bases are chosen, continuous linear maps $E \to F$ are represented by bounded matrices with columns in $c_0(I) \cong F$.

More particularly, if T is continuous and of rank less than or equal to 1, we can write

$$T(\mathbf{x}) = \varphi(\mathbf{x})\mathbf{a} = (\varphi(\mathbf{x})a_i)_I$$

for some $\varphi \in E'$. In this case, $\varphi_i(\mathbf{x}) = \varphi(\mathbf{x})a_i$, $\|\varphi_i\| = |a_i|\|\varphi\| \to 0$. This proves that the image of T belongs to the closed subspace $c_0(I; E')$. By linearity, the same property will hold for any continuous linear map T *of finite rank*:

$$L_{\mathrm{fr}}(E; c_0(I)) \to c_0(I; E') : T \mapsto (\varphi_i).$$

Definition. *A* completely continuous *linear map* $T : E \to F$ *is a linear map that can be approximated (uniformly on the unit ball) by finite-rank continuous linear maps.*

If we denote by $L_{\mathrm{cc}}(E, c_0(I))$ the space of completely continuous maps $E \to F$, then $T \mapsto (\varphi_i)$ defines an isometric map $L_{\mathrm{cc}}(E, c_0(I)) \to c_0(I, E')$. It is *surjective*: It is enough to check that the image of the finite-rank operators is dense in the target space. But if (φ_i) is an arbitrary sequence of continuous linear forms on E with $\|\varphi_i\| \to 0$, and $\varepsilon > 0$ is given, there is a finite subset $J \subset I$ such that $\|\varphi_i\| \leq \varepsilon$ for $i \notin J$. Define $\psi_i = \varphi_i$ for $i \in J$ and $\psi_i = 0$ for $i \notin J$. Then (ψ_i) is the image of a continuous finite-rank operator and $\|(\varphi_i) - (\psi_i)\| \leq \varepsilon$.

Comment. One can show that when K is a locally compact field, the completely continuous maps $T : E \to F$ are precisely the linear maps that transform bounded sets of E into relatively compact sets in F. These transformations are classically called *compact linear maps.* In the general case, the distinction between compact and completely continuous operators has been studied in detail and has led to the definition of *compactoids*.

4.7. The p-adic Hahn-Banach Theorem

Let E be a normed space, $V \subset E$ a vector subspace, and $\varphi : V \to K$ a continuous linear form

$$\|\varphi\| = \sup_{0 \neq x \in V} \frac{|\varphi(x)|}{\|x\|} < \infty.$$

Is there a continuous linear form $\overline{\varphi} : E \to K$ extending φ? If the answer is positive, can we find a linear form $\overline{\varphi}$ with the same norm ?

Theorem (Ingleton). *Let V be a subspace of a normed space E. When the base field K is* spherically complete, *the restriction map*

$$\psi \mapsto \varphi = \psi|_V, \; E' \to V'$$

is surjective. Moreover, for each $\varphi \in V'$, it is possible to find an extension $\psi = \overline{\varphi}$ with $\|\psi\| = \|\varphi\|$.

PROOF. (*a*) Let us show first that a continuous linear form on a subspace $V \neq E$ can be extended to $V + Ka$ (for any $a \in E - V$) without increasing its norm. The definition of $\psi = \overline{\varphi}$ has to satisfy

$$\|\psi(x + \lambda a)\| \leq \|\varphi\| \cdot \|x + \lambda a\| \quad (x \in V, \lambda \in K).$$

For $\lambda = 0$ this is satisfied, since $\psi|_V = \varphi$. When $\lambda \neq 0$, we may divide by $-\lambda$ and see that it is sufficient to find a linear form ψ with

$$\|\psi(x - a)\| \leq \|\varphi\| \cdot \|x - a\| \quad (x \in V),$$
$$\|\varphi(x) - \psi(a)\| \leq \|\varphi\| \cdot \|x - a\| := r_x \quad (x \in V).$$

In other words, we have to choose $\alpha = \psi(a)$ in the intersection of the balls $B_x = B_{\leq r_x}(\varphi(x)) \subset K$. For any pair of points $x, y \in V$, $\varphi(x) \in B_x$ and $\varphi(y) \in B_y$ are at distance

$$|\varphi(x) - \varphi(y)| \leq \|\varphi\| \cdot \|x - y\| \leq \|\varphi\| \max(\|x - a\|, \|y - a\|) = \max(r_x, r_y).$$

This proves that the smallest among the balls B_x and B_y is contained in the largest: $B_x \cap B_y \neq \emptyset$. Since we are assuming that the field K is spherically complete, the intersection $\bigcap_{x \in V} B_x$ is not empty and any α in this intersection is a possible choice for $\alpha = \psi(a)$.

(*b*) Consider now the set of pairs (V', φ') consisting of a vector subspace $V' \supset V$ and an extension φ' of φ to V' with the same norm as φ. This set of pairs is ordered by the relation

$$(V'', \varphi'') \succ (V', \varphi') \Longleftrightarrow V'' \supset V' \text{ and } \varphi''|_{V'} = \varphi'.$$

Any linearly ordered set of such pairs has an upper bound. By Zorn's lemma, there is a maximal pair. By the first part, this maximal pair is defined on the whole space E. ∎

5. Umbral Calculus

The Mahler theorem (3.4) has been generalized by L. van Hamme. To be able to give this generalization, we have to briefly review *umbral calculus*: This term has its origin in the nineteenth century, when formal computations were used with little justification. Today, it refers to an algebraic treatment of polynomials, power series, and identities between them.

5.1. Delta Operators

Let K be a field of characteristic 0 and $K[X]$ the vector space of polynomials (in one variable) with coefficients in K. The translations τ_a $(a \in K)$ are the linear operators in $K[X]$ defined by

$$(\tau_a f)(X) = f(X + a).$$

We shall often identify the indeterminate X with a variable x (in K or in an extension of K: Since K is infinite, there is no danger in identifying formal polynomials and polynomial functions on K). Since the degree of the zero polynomial is not defined, let us adopt the ad hoc convention $\deg(0) = -1$: This allows us to speak of the *subspace* of polynomials having degree less than or equal to n for any $n \geq 0$. The unit translation will also be denoted by $\tau_1 = E$.

Definition. *A* delta operator *is a linear endomorphism δ of $K[X]$ such that*

(1) *δ commutes with all translations τ_a $(a \in K)$,*
(2) *$\delta(X) = c \in K^\times$ is a nonzero constant.*

Proposition. *Let δ be a delta operator in $K[X]$. Then*

(1) *$\delta(a) = 0$ for all constants $a \in K$,*
(2) *if f is a nonconstant polynomial, then $\deg(\delta f) = \deg f - 1$.*

PROOF. By hypothesis, $\delta(X) = c \neq 0$, and by translation,

$$c = \tau_a c = \tau_a \delta X = \delta \tau_a X = \delta(X + a) = \delta X + \delta a = c + \delta a.$$

Hence $\delta a = 0$ for all constants $a \in K$. To prove the second point, it will suffice to show that

$$\deg(\delta X^n) = n - 1 \quad (n \geq 1).$$

Fix an integer $n \geq 1$ and put $\delta X^n = f(X)$. Then

$$f(X+a) = \tau_a f(X) = \tau_a \delta X^n = \delta\tau_a(X^n) = \delta(X+a)^n$$
$$= \delta \sum \binom{n}{k} a^k \cdot X^{n-k} = \sum \binom{n}{k} a^k \cdot \delta(X^{n-k}),$$

and for $X = 0$,

$$f(a) = \sum \binom{n}{k} a^k \delta(X^{n-k})(0),$$

or

$$f(X) = \sum \binom{n}{k} \delta(X^{n-k})(0) \cdot X^k.$$

We see that f is a polynomial of degree less than or equal to n with a coefficient of X^n given by $\delta(1)(0) = \delta(1) = 0$ (using the first part, already proved). The coefficient of X^{n-1} is $n\delta X(0) = nc \neq 0$ (the field K has characteristic 0). Hence $f(X) = \delta X^n$ is a polynomial of degree $n-1$. ■

Corollary. *The image by a delta operator of the subspace of polynomials of degrees less than or equal to n ($n \geq 1$) is the subspace of polynomials of degrees less than or equal to $n-1$.*

PROOF. The dimension of the image of a linear operator is equal to the dimension of the source minus the dimension of the kernel. The assertion follows from the proposition. ■

Examples. (1) The differentiation operator D is itself a delta operator. More generally, if $a \in K$, the operator $\tau_a D = D\tau_a$ is a delta operator.

(2) The finite difference operators (recall $E = \tau_1$: Unit translation)

$$\nabla = \nabla_+ = \tau_1 - \mathrm{id} = E - \mathrm{id},$$
$$\nabla_- = \mathrm{id} - \tau_{-1} = \tau_{-1}\nabla,$$

and $\tau_a\nabla_\pm$ are delta operators. When $a \neq b$, the operators $\tau_a - \tau_b$ are also delta operators.

(3) Any formal power series in D of order 1, namely

$$\delta = \sum_{i\geq 1} c_i D^i = c_1 D + \cdots \in K[[D]] \quad (c_1 \neq 0),$$

defines a delta operator. For example

$$\log(1+D) = D - \tfrac{1}{2}D^2 + \tfrac{1}{3}D^3 - \cdots,$$
$$e^D - 1 = D + \tfrac{1}{2!}D^2 + \tfrac{1}{3!}D^3 + \cdots,$$
$$D^2/(e^D - 1) = D - \tfrac{1}{2}D^2 + \cdots$$

are delta operators.

5.2. *The Basic System of Polynomials of a Delta Operator*

Let δ be a delta operator. If f is a nonzero polynomial, there is a polynomial g such that $\delta(g) = f$ (and necessarily $\deg g = \deg f + 1$ by the preceding section). This polynomial g is determined up to an additive constant, since the kernel of δ consists of the constants. Replacing g by $g - g(0)$, we see that there exists a unique polynomial g such that

$$\delta(g) = f, \ g(0) = 0 \ \text{(normalization)},$$

and the degree of this polynomial is one more than the degree of f.

Definition. *The* basic system $(p_n)_{n\geq 0}$ *corresponding to a delta operator* δ *is the system of polynomials such that*

1. $\deg p_n = n \quad (n \geq 0)$,
2. $\delta p_n = np_{n-1} \quad (n \geq 1)$,
3. $p_0 = 1,\ p_n(0) = 0 \quad (n \geq 1)$.

Starting with $p_0 = 1$ there is a unique polynomial p_1 (of degree 1) such that $\delta(p_1) = 1$ and $p_1(0) = 0$. Proceeding inductively, there is a unique polynomial p_n (of degree n) such that $\delta(p_n) = np_{n-1}$ and $p_n(0) = 0$. Hence the definition characterizes a unique system of polynomials for any delta operator. Explicit formulas for computing these polynomials will be given in (5.5) and (6.2). Any basic system constitutes a K-basis of the vector space $K[X]$.

For example, the basic system of the delta operator D (derivation) is the system $(x^n)_{n\geq 0}$ of powers of x. For the operator $\delta = \nabla$ the basic system is

$$(x)_n = x(x-1)\cdots(x-n+1) \quad \text{(Pochhammer symbol)}$$

with the convention $(x)_0 = 1$. We indeed have (1.1)

$$\nabla(x)_n = n(x)_{n-1} \quad (n \geq 1),$$

and $(x)_n$ vanishes at $x = 0$ if $n \geq 1$. For $\delta = \nabla_-$ the basic sequence consists of the polynomials $p_n(x) = x(x+1)\cdots(x+n-1)$. For every basic sequence, we have

$$\delta^k p_n = n(n-1)\cdots(n-k+1)\cdot p_{n-k} \quad (k \leq n).$$

In particular,

$$\delta^n p_n = n!\cdot p_0 = n!$$

Generalized Taylor Expansion. *Let* δ *be a delta operator and* $(p_n)_{n\geq 0}$ *its basic system in* $K[X]$. *Then we have general expansions*

$$f(x+y) = \sum_{k\geq 0} \frac{\delta^k f(x)}{k!}\cdot p_k(y) \quad (f \in K[X]).$$

We can indeed write the tautology

$$p_n = \sum_{k \leq n} \frac{\delta^k(p_n)(0)}{k!} \cdot p_k,$$

where all coefficients of the p_k are 0 except for p_n, which is 1. For any linear combination f of the p_n we obtain by linearity

$$f = \sum_{k \geq 0} \frac{\delta^k(f)(0)}{k!} \cdot p_k.$$

Replacing f by one translate $\tau_x f$ in the preceding equality, we obtain

$$\tau_x f = \sum_{k \geq 0} \frac{\delta^k(\tau_x f)(0)}{k!} \cdot p_k = \sum_{k \geq 0} \frac{\tau_x(\delta^k f)(0)}{k!} \cdot p_k$$

$$= \sum_{k \geq 0} \frac{\delta^k f(x)}{k!} \cdot p_k,$$

which is the announced generalized Taylor expansion. ■

In particular, if we take for f the polynomial p_n, we obtain the following equalities.

Binomial Identities. *Any basic sequence of a delta operator satisfies the following identities:*

$$p_n(x + y) = \sum_{0 \leq k \leq n} \binom{n}{k} \cdot p_k(x) p_{n-k}(y). \quad ■$$

The binomial identity can be written in the mnemonic way

$$p_n(x + y) = \text{“}(p(x) + p(y))^n\text{”}$$

where one must remember to replace powers by indices in the binomial expansion of the right-hand side.

5.3. *Composition Operators*

Definition. *A* composition operator *is an endomorphism T of $K[X]$ that commutes with translations.*

We shall determine all composition operators. More precisely, we shall establish the following result.

Theorem. *The following properties of an endomorphism $T\colon K[X] \to K[X]$ are equivalent: They characterize composition operators.*

(*i*) *T commutes with the unit translation.*
(*ii*) *T commutes with all translations.*

(*iii*) *For all delta operators* δ, T *can be written as a formal power series in* δ: $T = \varphi(\delta) \in K[[\delta]]$.

(*iv*) $T = \varphi(D) \in K[[D]]$ *is a formal power series in the derivation* D.

(*v*) T *commutes with the derivation* $TD = DT$.

(*vi*) T *commutes with any delta operator.*

PROOF. (*i*) $\Rightarrow$ (*ii*) Write $\tau_1 = E$ and use the Taylor series expansion around n:

$$E^n f(x) = f(n+x) = \sum_{k \geq 0} f^{(k)}(n) \cdot \frac{x^k}{k!}.$$

By the commutation hypothesis $TE = ET$ we infer

$$E^n Tf = TE^n f = \sum_{k \geq 0} f^{(k)}(n) \cdot \frac{T(x^k)}{k!}.$$

Put $g = Tf$ and consider the polynomial in two variables

$$F(x, y) = g(x+y) - \sum_{k \geq 0} f^{(k)}(y) \cdot \frac{T(x^k)}{k!}.$$

We have seen that $F(x, n) = 0$ for all positive integers n and all x. If x is fixed, the corresponding polynomial in y has infinitely many roots and is consequently identically zero. This proves $F = 0$. Hence $\tau_y g - T\tau_y f = 0$, or $\tau_y Tf - T\tau_y f = 0$. Since this is valid for all polynomials f, we see that the operator T commutes with translations.

(*ii*) $\Rightarrow$ (*iii*) Let T be a composition operator and δ a delta operator. Write the generalized Taylor formula using the basic sequence (p_k) corresponding to δ:

$$\tau_y f(x) = f(x+y) = \sum \delta^k f(y) \cdot \frac{p_k(x)}{k!}$$

(first for fixed y and variable x). We have

$$\tau_y f = \sum \frac{p_k}{k!} \cdot \delta^k f(y).$$

Let us apply the composition operator T to this polynomial,

$$\tau_y Tf = T\tau_y f = \sum \frac{Tp_k}{k!} \cdot \delta^k f(y)$$

and evaluate it at the origin (we now fix $x = 0$ and consider a variable y):

$$(Tf)(y) = (\tau_y Tf)(0) = (T\tau_y f)(0) = \sum \frac{(Tp_k)(0)}{k!} \cdot \delta^k f(y).$$

Hence

$$Tf = \sum \frac{(Tp_k)(0)}{k!} \cdot \delta^k f,$$

and finally,

$$T = \sum \frac{(Tp_k)(0)}{k!} \cdot \delta^k \in K[[\delta]].$$

The coefficients of the expansion of a composition operator T as a power series in the delta operator δ — say $T = \sum a_k \delta_k$ — are given by $a_k = (Tp_k)(0)/k!$. In particular, let us remember that for the case $\delta = D$ these coefficients are

$$a_k = \frac{(T(x^k))(0)}{k!}.$$

Since obviously (*ii*) $\Rightarrow$ (*i*) and (*iii*) $\Rightarrow$ (*iv*) $\Rightarrow$ (*v*) $\Rightarrow$ (*vi*) it only remains to prove (*vi*) $\Rightarrow$ (*ii*) to accomplish the full cycle of equivalences. Let δ be a delta operator and write the generalized Taylor formula for an arbitrary polynomial:

$$\begin{aligned} f(x+y) &= \sum \delta^k f(x) \cdot p_k(y)/k!, \\ \tau_y f(x) &= \sum \delta^k f(x) \cdot p_k(y)/k!, \\ \tau_y f &= \sum p_k(y)/k! \cdot \delta^k f, \\ \tau_y &= \sum p_k(y)/k! \cdot \delta^k. \end{aligned}$$

This shows that all translations can be expressed as formal power series in any delta operator. As a consequence, if an operator commutes with a delta operator, it commutes with all translations. ■

For convenience, let us use the following notation for the *commutant* of a subset A of the endomorphism ring of $K[X]$:

$$A' = \{T \in \text{End}\, K[X] : TS = ST \ \text{ for all } S \in A\}.$$

The commutant of A' is the *bicommutant* — or *double commutant* — of A: It is denoted by $A'' = (A')'$.

Corollary. *In the endomorphism ring of $K[X]$, the commutant of a delta operator δ can be identified with the ring $K[[D]]$. In particular, this commutant is commutative, and the bicommutant of any delta operator can be identified with the ring $K[[D]]$.*

PROOF. The equivalences (*v*) $\Leftrightarrow$ (*vi*) $\Leftrightarrow$ (*iv*) of the theorem show that the commutant of the derivation, or of any delta operator, coincides with the ring of power series in the derivation D. This ring is independent of the delta operator in question and is commutative; hence by (*ii*) $\Leftrightarrow$ (*v*) $\Leftrightarrow$ (*vi*) we have

$$\{D\}' = \{\tau_y : y \in K\}' = \{\delta\}' \subset \text{End}\, K[X].$$

Since $\{D\}'$ is commutative, $\{D\}' \subset \{D\}''$. On the other hand, the operation that consists in taking the commutant obviously reverses inclusions, and

$$D \in \{D\}' \Longrightarrow \{D\}' \supset \{D\}''. \qquad \blacksquare$$

Let T be a nonzero composition operator. We can write $T = \sum_{j \geq \nu} a_j D^j$ with a first nonzero coefficient $a_\nu \neq 0$ $(\nu \geq 0)$. In this case, we say that the composition operator T has *order* ν and we write $T = D^\nu S = SD^\nu$ with a composition operator S of order 0, namely, S is *invertible*. Since the kernel of the operator D^ν consists of the polynomials of degree less than ν, we infer

$$\nu = \dim \ker D^\nu = \dim \ker T.$$

This equality shows that the order of a composition operator is independent of the delta operator used to represent it as a power series. On the other hand, the delta operators are the composition operators of order 1.

If $T = \sum a_j D^j$ and $T' = \sum b_j D^j$ are two composition operators, then $T \circ T'$ is also a composition operator, and its formal power series is obtained by *multiplication of the formal power series giving T and T'*.

5.4. The van Hamme Theorem

Let T be a continuous endomorphism of the Banach space $C(\mathbf{Z}_p)$ of continuous functions on $\mathbf{Z}_p$ (and values in a fixed complete extension K of $\mathbf{Q}_p$). Let us recall the definitions of the norms

$$\|f\| = \sup |f(x)| = \max |f(x)|,$$

sup = max taken on the compact space $\mathbf{Z}_p$,

$$\|T\| = \sup_{f \neq 0} \|Tf\|/\|f\| = \sup_{\|f\|=1} \|Tf\|.$$

When this continuous endomorphism commutes with the unit translation operator $E = \tau_1$, it also commutes with the (forward) difference operator $\nabla = \tau_1 - \mathrm{id}$ and its powers. Hence T leaves the subspaces $\ker \nabla^n \subset C(\mathbf{Z}_p)$ invariant.

Lemma. *The subspace* $\ker \nabla^n$ *of* $C(\mathbf{Z}_p)$ *consists of all polynomials of degree strictly smaller than n.*

Proof. The statement is obvious for $n = 0$ and 1. In fact, if $\nabla^n f = 0$, the finite difference theory applied to the restriction of f on $\mathbf{N}$ shows that this restriction is a polynomial p of degree smaller than n. Hence we have $f = p$ on $\mathbf{N}$ and also on $\mathbf{Z}_p$ by continuity and density of $\mathbf{N}$ in $\mathbf{Z}_p$. $\blacksquare$

As a consequence, any continuous endomorphism T of $C(\mathbf{Z}_p)$ commuting with E (or equivalently, with ∇) leaves the polynomial subspace

$$\Pi = K[X] = \bigcup_{n\geq 0} \ker \nabla^n \subset C(\mathbf{Z}_p)$$

invariant and induces a composition operator in this space. Let us expand this composition operator as a formal power series in the delta operator ∇

$$T|_\Pi = \sum_{n\geq \nu} \alpha_n \nabla^n \in K[[\nabla]].$$

If $T \neq 0$, the order ν of T is the index of the smallest nonzero coefficient. Since $\|\nabla\| = 1$, the ultrametric inequality shows that

$$\|T\| \leq \sup |\alpha_n|.$$

On the other hand, the basic polynomial sequence of the delta operator ∇ is the sequence $(x)_n = x(x-1)\cdots(x-n+1)$, and the coefficients α_n are given by the formula

$$\alpha_n = \frac{T((x)_n)}{n!}.$$

In particular for $n = 1$, $|\alpha_1| = \|T(x)\| \leq \|T\|\|x\| = \|T\|$. If we assume $\|T\| = |\alpha_1|$, we see that $|\alpha_n| \leq |\alpha_1|$ for all $n \geq 1$. The main step of van Hamme's generalization of the Mahler theorem can now be given.

Proposition. *Let T be a continuous endomorphism of $C(\mathbf{Z}_p)$ that commutes with ∇. Assume $T(1) = 0$ and $\|T\| = |\alpha_1| = 1$, so that T induces a delta operator on $K[X]$ with basic polynomial sequence $(p_n)_{n\geq 0}$:*

$$p_0 = 1,\ \deg p_n = n,\ T(p_n) = np_{n-1} \text{ and } p_n(0) = 0 \quad (n \geq 1).$$

Then $\|p_n/n!\| = 1$.

Proof. Let us use the renormalized $q_n = p_n/n!$, so that by definition

$$q_0 = 1,\ \deg q_n = n,\ T(q_n) = q_{n-1},\ \text{and } q_n(0) = 0 \quad (n \geq 1).$$

We have to prove $\|q_n\| = 1$ $(n \geq 1)$. Replacing T by T/α_1, we may assume $\alpha_1 = 1$,

$$1 = \|q_0\| = \|Tq_1\| \leq \|q_1\| = \|Tq_2\| \leq \|q_2\| \leq \cdots.$$

Now by assumption, $T = \nabla + \alpha_2\nabla^2 + \cdots = \nabla(I + \alpha_2\nabla + \cdots) = \nabla U$ with an invertible composition operator $U = (I + \alpha_2\nabla + \cdots)$, $\|U\| = 1$. Define $V = U^{-1} = I - \alpha_2\nabla + \cdots$ also with $\|V\| = 1$. We claim that there is a suitable continuous invertible composition operator S, with $\|S\| = 1$, such that

$$q_n = SV^n(f_n), \tag{$*$}$$

where $f_n = \binom{\cdot}{n}$ denote momentarily the binomial polynomials: $\nabla f_n = f_{n-1}$. First, for any composition operator S of order 0, the preceding definition leads to polynomials with $\deg q_n = n$. Moreover,

$$Tq_n = T \circ SV^n(f_n) = \nabla U \circ SV^n(f_n),$$

and since $U = V^{-1}$ and all the operators in question commute,

$$Tq_n = SV^{n-1} \circ \nabla(f_n) = SV^{n-1}(f_{n-1}) = q_{n-1}.$$

There only remains to construct a suitable invertible composition operator S with $\|S\| = 1$ so that the formula $(*)$ furnishes polynomials with $q_n(0) = 0$ $(n \geq 1)$. Let us take

$$S = I - \nabla\frac{V'}{V} = I - \nabla V'U,$$

where V' is given by the formally derived power series in ∇ giving V. Namely,

$$V = I + \sum_{n\geq1} \beta_n \nabla^n \Longrightarrow V' = \sum_{n\geq1} n\beta_n \nabla^{n-1}.$$

Now we have

$$\begin{aligned} SV^n(f_n) &= \left(I - \nabla\frac{V'}{V}\right) \circ V^n(f_n) \\ &= (V^n - \nabla V^{n-1}V')(f_n). \end{aligned}$$

Recall that all the operators are formal power series in ∇, and $\nabla^k f_n = f_{n-k}$ vanishes at the origin for $k < n$. The only interesting term is thus the monomial containing $\nabla^n f_n$. But if $\varphi = \varphi(t)$ is a formal power series, the formal power series

$$\varphi^n - t\varphi^{n-1}\varphi' = \varphi^n - (t/n)(\varphi^n)' = \psi - (t/n)\psi'$$

has a coefficient of t^n equal to 0. Since this is the constant term in $SV^n(f_n)$, this proves that $q_n(0) = 0$. All operators used in the definition of S have norm less than or equal to 1; hence $\|S\| \leq 1$, $\|q_n\| \leq \|S\|\|V^n\|\|f_n\| = 1$. ∎

Theorem. *Let T be a continuous endomorphism of $C(\mathbf{Z}_p)$ that commutes with ∇. Assume $T(1) = 0$ and $\|T\| = |T(x)| = 1$. Define the polynomial sequence*

$$q_0 = 1, \ \deg q_n = n, \ T(q_n) = q_{n-1}, \ q_n(0) = 0 \quad (n \geq 1).$$

Then each continuous function $f \in C(\mathbf{Z}_p)$ can be expanded in a generalized Mahler series

$$f(x) = \sum_{n\geq0} c_n q_n$$

with $c_n = (T^n f)(0) \to 0$ and $\|f\| = \sup_{n\geq0} |c_n|$.

PROOF. Using the notation of the preceding proposition, we have $T = \nabla U$ with U invertible and $\|U\| = 1$. Hence

$$|T^n f(0)| \le \|T^n f\| = \|U^n \nabla^n f\| \le \|\nabla^n f\| \to 0$$

(by the Mahler theorem). It will be enough to establish all statements for the polynomial functions f, since the general case will result from this by density and continuity. The generalized Taylor expansion of a polynomial f takes the form

$$f = \sum_{n\ge0} (T^n f)(0) \cdot \frac{p_n}{n!} = \sum_{n\ge0} (T^n f)(0) \cdot q_n.$$

From $\|q_n\| = 1$ follows quite generally

$$\|f\| \le \sup |c_n|,$$

and from the asserted formula for the coefficients,

$$|c_n| = |(T^n f)(0)| \le \|T^n f\| \le \|T^n\| \|f\| \le \|T\|^n \|f\| \le \|f\|,$$

whence conversely $\sup |c_n| \le \|f\|$ and finally $\sup |c_n| = \|f\|$. ∎

Comment. The generalized Mahler expansion is not valid for the delta operator D (differentiation): This operator does not extend continuously to all of $C(\mathbf{Z}_p)$. On the other hand, its renormalized basic sequence is $q_n(x) = x^n/n!$, and even if a series expansion $f(x) = \sum_{n\ge0} c_n x^n/n!$ converges uniformly, $\|f\| = \sup |f(x)| = \max |f(x)|$ is not usually equal to $\sup |c_n|$. The delta operator D on $K[X]$) has a power series expansion in ∇ with coefficients

$$\begin{aligned}\alpha_n &= D\left(\frac{(x)_n}{n!}\right)(0) = \text{coefficient of } x \text{ in } \binom{x}{n} \\ &= \text{constant coefficient of } (x-1)\cdots(x-n+1)/n! = (-1)^{n-1}/n.\end{aligned}$$

In particular, $\alpha_1 = 1$, but $|\alpha_n| > 1$ when n is a multiple of p, so that the theorem is not applicable.

5.5. *The Translation Principle*

To illustrate an important principle we begin with a particular case.

Example. We know that the basic sequence for the delta operator D is the sequence of powers. The basic sequence corresponding to a translate $\tau_a D$ of D is

$$p_n(x) = x(x - na)^{n-1} \quad (n \ge 1).$$

Indeed, we have

$$\begin{aligned} Dp_n &= (x-na)^{n-1} + (n-1)x(x-na)^{n-2} \\ &= (x-na)^{n-2}[x-na+nx-x] = n(x-na)^{n-2}[x-a], \end{aligned}$$

whence

$$\tau_a Dp_n = n(x+a-na)^{n-2}[x+a-a] = np_{n-1}.$$

To be able to prove the *general translation principle* we need a couple of easy results.

Lemma. *Let $T = \varphi(D) = \sum_{n\geq 0} a_n D^n$ be a composition operator and let M_x be the multiplication by x operator $f \mapsto xf$. Then*

$$TM_x - M_xT = \varphi'(D).$$

Proof. By definition,

$$(DM_x - M_xD)f = (xf)' - xf' = f,$$

whence $DM_x - M_xD = I$ (identity operator). Similarly,

$$(D^nM_x - M_xD^n)f = (xf)^{(n)} - xf^{(n)} = nf^{(n-1)},$$

whence $D^nM_x - M_xD^n = nD^{(n-1)}$. This is the particular case $T = D^n$ of the expected formula. The general case results by additivity, since for any polynomial f, $(TM_x - M_xT)(f)$ is a finite sum

$$\sum a_n(D^nM_x - M_xD^n)f = \sum a_n nD^{(n-1)}f$$

(only terms with $n \leq \deg(f) + 1$ really occur). ∎

Comment. One can define the *Pincherle derivative* $T' = TM_x - M_xT$ of any composition operator. For $T = \varphi(D)$ the lemma shows that $T' = \varphi'(D)$. A similar result has been used for a long time in quantum theory: If M_f denotes the multiplication operator by a polynomial f, then

$$DM_f - M_fD = M_{f'}: \text{multiplication operator by the derivative } f'.$$

Observe that in (5.4) we have used a different derivative, namely a derivative with respect to a series expansion in the operator ∇. For this reason, it is always necessary to specify with respect to which delta operator the derivative of a composition operator is taken.

Proposition. *Let δ be a delta operator and write $\delta = D\varphi(D)$ with an invertible power series φ. Then the basic sequence of polynomials of δ is given by*

$$p_n = x\varphi(D)^{-n}(x^{n-1}) \quad (n \geq 1).$$

PROOF. Since $\varphi(D)$ as well as $\varphi(D)^{-n}$ are invertible operators, $\varphi(D)^{-n}(x^{n-1})$ is a polynomial of degree $n-1$ and the polynomial $p_n = x\varphi(D)^{-n}(x^{n-1})$ has degree n. Since x divides p_n, we obviously have $p_n(0) = 0$. It only remains to check that $\delta p_n = np_{n-1}$. By definition, $p_n = M_x\varphi(D)^{-n}(x^{n-1})$, so that $\delta p_n = D\varphi(D)M_x\varphi(D)^{-n}(x^{n-1})$. Using the lemma, we can write

$$\begin{aligned} M_x\varphi(D)^{-n}(x^{n-1}) &= \varphi(D)^{-n}M_x(x^{n-1}) - [\varphi(D)^{-n}]'(x^{n-1}) \\ &= \varphi(D)^{-n}(x^n) + n[\varphi(D)^{-n-1}](x^{n-1}). \end{aligned}$$

Hence

$$\begin{aligned} \delta p_n &= D\varphi(D)M_x\varphi(D)^{-n}(x^{n-1}) \\ &= D\varphi(D)[\varphi(D)^{-n}(x^n) + n[\varphi(D)^{-n-1}](x^{n-1})] \\ &= \varphi(D)^{-n+1}(Dx^n) + n\varphi(D)^{-n}(Dx^{n-1}) \\ &= \varphi(D)^{-n+1}(nx^{n-1}) + n(n-1)\varphi(D)^{-n}(x^{n-2}) \\ &= [n\varphi(D)^{-n+1}M_x + n(n-1)\varphi(D)^{-n}](x^{n-2}). \end{aligned}$$

Using again the lemma to bring the operator M_x into the first position, we obtain

$$\begin{aligned} \delta p_n &= [M_x n\varphi(D)^{-n+1} + (n\varphi(D)^{-n+1})' + n(n-1)\varphi(D)^{-n}](x^{n-2}) \\ &= nM_x\varphi(D)^{-(n-1)}(x^{n-2}) + [-n(n-1)\varphi(D)^{-n} + n(n-1)\varphi(D)^{-n}](x^{n-2}) \\ &= nM_x\varphi(D)^{-(n-1)}(x^{n-2}) = np_{n-1}. \end{aligned}$$

■

The Translation Principle. *Let δ be a delta operator and $(p_n)_{n\geq 0}$ its basic sequence. Then the basic sequence of the translate delta operator $\tau_a\delta$ is given by $p_0 = 1$ and*

$$\widetilde{p}_n = \frac{x}{x-na}p_n(x-na) \quad (n \geq 1).$$

PROOF. By the explicit formula of the proposition with $\delta = D\varphi(D)$, we have

$$\begin{aligned} \widetilde{p}_n &= x[\tau_a\varphi(D)]^{-n}(x^{n-1}) \\ &= x\tau_{-na}\varphi(D)^{-n}(x^{n-1}) = x\tau_{-na}[(1/x)p_n], \end{aligned}$$

from which the translation principle follows. ■

Observe that since the polynomial p_n is divisible by x ($n \geq 1$), $p_n(x-na)$ is divisible by $x - na$ and the polynomial $\widetilde{p}_n$ is divisible by x: It vanishes at the origin as required. Several cases of this translation principle are interesting. For example, the case $a = -1$ leads to the backward difference operator $\tau_{-1}\nabla = \nabla_-$, while $a = -\frac{1}{2}$ leads to a *centered* difference operator.

Umbral calculus

Delta operator (IV.5)	Basic sequence of polynomials (IV.5.2)	Related sequences (IV.6.1)
$D = d/dx$	$(x^n)_{n\geq 0}$	Appell sequences $Dp_n = np_{n-1}$
	$T \downarrow$ umbral operator	
δ	$(p_n)_{n\geq 0}$	Sheffer sequences $\delta s_n = ns_{n-1}$
$\tau_{-y}\delta$ (IV.5.5)	$\left(\frac{x}{x+ny}p_n(x+ny)\right)_{n\geq 0}$ translation principle	

Binomial identity: $p_n(x+y) =$ "$(p(x)+p(y))^n$,"

Appell sequences: $p_n(x+y) =$ "$(p(x)+y)^n$,"

Sheffer sequences: $s_n(x+y) =$ "$(s(x)+p(y))^n$,"

cf. (V.5.4), (V.5.5) for the example of the Bernoulli numbers and polynomials.

6. Generating Functions

6.1. Sheffer Sequences

In this section δ will be a fixed delta operator, and $(p_k)_{k\geq 0}$ will denote its basic sequence. Recall the generalized Taylor series (5.2)

$$f(x) = \sum_{k\geq 0} \frac{(\delta^k f)(0)}{k!} \cdot p_k(x),$$

valid for any polynomial $f \in K[X]$.

Definition. *A* Sheffer sequence *(relative to δ) is any sequence of polynomials $(s_n)_{n\geq 0}$ such that*

1. $\deg s_n = n$ *for all* $n \geq 0$,
2. $\delta s_n = n \cdot s_{n-1}$ *for all* $n \geq 1$.

The constant s_0 is nonzero. If $(s_n)_{n\geq 0}$ is a Sheffer sequence, we have

$$\delta^k s_n = n(n-1)\cdots(n-k+1) \cdot s_{n-k} = (n)_k \cdot s_{n-k} \quad (k \leq n).$$

The generalized Taylor expansion

$$f(x+y) = \sum \delta^k f(y)/k! \cdot p_k(x)$$

gives for $f = s_n$

$$s_n(x+y) = \sum \binom{n}{k} p_k(x) \cdot s_{n-k}(y). \tag{S}$$

This formula generalizes the binomial identity for the basic sequence (p_k).

Definition. *An* Appell sequence *is a Sheffer sequence corresponding to the derivation operator D.*

The Appell sequences (p_n) are characterized by the relations

1. $\deg p_n = n$ *for all* $n \geq 0$,
2. $p_n' = n \cdot p_{n-1}$ *for all* $n \geq 1$.

The Appell sequences satisfy (S), which is in this case

$$p_n(x+y) = \sum_{0 \leq k \leq n} \binom{n}{k} x^k \cdot p_{n-k}(y).$$

This identity may be symbolically written

$$p_n(x+y) = \text{“}(x + p(y))^n\text{,”}$$

where we interpret exponents of the binomial expansion of the right-hand side as indices.

Proposition. *Let S be an invertible composition operator. Then the polynomial sequence $s_n = S(p_n)$ is a Sheffer sequence. Conversely, if (s_n) is a Sheffer sequence, the endomorphism S of $K[X]$ that sends the basis (p_n) onto the basis (s_n) is an invertible composition operator.*

PROOF. To check the first statement, we compute δs_n using the fact that δ and S commute:

$$\delta s_n = \delta S p_n = S \delta p_n = S(np_{n-1}) = nS(p_{n-1}) = ns_{n-1}.$$

Conversely, for $n \geq 0$ we have

$$S\delta p_n = Snp_{n-1} = ns_{n-1} = \delta s_n = \delta S p_n.$$

Since the polynomials p_n make up a basis of $K[X]$, this proves that S and δ commute. Hence S is a composition operator (5.3). ∎

6.2. *Generating Functions*

Let us still consider a fixed delta operator δ with basic system of polynomials $(p_n)_{n\geq 0}$. Let S be an invertible composition operator. The polynomial system $S^{-1}p_n = s_n$ is a Sheffer sequence, and we are going to determine more explicitly the *exponential generating function*

$$\sum_{n\geq 0} s_n(x)\frac{z^n}{n!} = F_S(x, z)$$

(where z is a new indeterminate, a variable in $\mathbf{C}$ or $\mathbf{C}_p, \ldots$). We know that

$$\delta = \varphi(D) \in K[[D]],\ \varphi(0) = 0 \text{ and } \varphi'(0) \neq 0,$$
$$S = \psi(D) \in K[[D]],\ \psi(0) \neq 0.$$

By (5.3) the formal power series corresponding to the composition operator $\tau_x S^{-1}$ is

$$\tau_x S^{-1} = \sum \tau_x S^{-1}(p_n)(0)\frac{\delta^n}{n!} = \sum S^{-1}(p_n)(x)\frac{\delta^n}{n!}$$
$$= \sum s_n(x)\frac{\delta^n}{n!} = F_S(x, \delta).$$

On the other hand, the formal power series (in D resp. δ) corresponding to δ can be computed as follows. Firstly, we have seen that

$$\tau_x = \sum p_n(x)\frac{\delta^n}{n!} = \sum x^n \frac{D^n}{n!} = \exp(xD),$$

and secondly,

$$\tau_x = \tau_x S^{-1} \circ S = F_S(x, \delta) \circ \psi(D).$$

Since $\delta = \varphi(D)$, or equivalently $D = \varphi^{-1}(\delta)$ (a systematic characterization of invertibility of formal power series is given in (VI.1.3) Theorem 1), a comparison of the two expressions for τ_x furnishes

$$F_S(x, \varphi(D)) \cdot \psi(D) = \exp(xD).$$

With the formal power series $\delta = \varphi(D)$, we can express $D = \varphi^{-1}(\delta)$ and come back to the above expression:

$$F_S(x, z) \cdot \psi(\varphi^{-1}(z)) = \exp(x\varphi^{-1}(z)),$$
$$F_S(x, z) = \sum s_n(x)\frac{z^n}{n!} = \frac{1}{\psi(\varphi^{-1}(z))} \cdot \exp(x\varphi^{-1}(z)).$$

We can deduce several useful identities from this one. For example, derivation with respect to x leads to

$$\partial_x F_S(x,z) = \sum_{n\geq 0} s_n'(x)\frac{z^n}{n!}$$

$$= \frac{\varphi^{-1}(z)}{\psi(\varphi^{-1}(z))}\cdot \exp(x\varphi^{-1}(z)),$$

$$\sum_{n\geq 1} s_n'(0)\frac{z^n}{n!} = \frac{\varphi^{-1}(z)}{\psi(\varphi^{-1}(z))}.$$

In particular, for the basic sequence $s_n = p_n$ which corresponds to the identity composition operator $S = \mathrm{id}$, hence to the formal power series $\psi \equiv 1$,

$$\sum_{n\geq 0} p_n(x)\frac{z^n}{n!} = \exp(x\varphi^{-1}(z)),$$

$$\sum_{n\geq 1} p_n'(0)\frac{z^n}{n!} = \varphi^{-1}(z).$$

The first identity gives an algorithm for the computation of the basic sequence $(p_n)_{n\geq 0}$. Here are a few examples.

Example 1. Let us consider the delta operator

$$\nabla = \nabla_+ = \tau - 1 = e^D - 1 = \varphi(D),$$

for which

$$z = \varphi(u) = e^u - 1, \quad e^u = z + 1, \quad u = \log(1+z) = \varphi^{-1}(z).$$

We have

$$\exp(x\varphi^{-1}(z)) = \exp(x\log(1+z)) = (1+z)^x$$

$$= \sum_{n\geq 0}\binom{x}{n} z^n = \sum_{n\geq 0} (x)_n \frac{z^n}{n!}.$$

The basic polynomials for this delta operator ∇ are simply the Pochhammer polynomials

$$p_0 = 1, \quad p_n(x) = (x)_n = x(x-1)\cdots(x-n+1) \quad (n \geq 1).$$

Example 2. As with the delta operator $\nabla_- = 1-\tau_- = 1-e^{-D}$, which corresponds to $z = \varphi(u) = 1 - e^{-u}$, $u = \log 1/(1-z) = \varphi^{-1}(z)$, we have

$$\exp(x\varphi^{-1}(z)) = (1-z)^{-x} = \sum \binom{-x}{n}(-z)^n.$$

The basic polynomials are now

$$p_0 = 1, \; p_n(x) = x(x+1)\cdots(x+n-1) = (-1)^n(-x)_n \quad (n \geq 1).$$

Another example of the general formulas follows (cf. the exercises for the Fibonacci numbers and the Gould polynomials).

6.3. The Bell Polynomials

The *Bell polynomials* $B_n(x)$ can be defined by their generating function

$$\sum_{n\geq 0} B_n(x)\frac{z^n}{n!} = \exp[x(e^z - 1)].$$

This generating function has the required form for a basic sequence of polynomials of a delta operator. We can indeed take

$$u = \varphi^{-1}(z) = e^z - 1 = (\tau_{-1} \circ \exp)(z)$$

and hence

$$z = \varphi(u) = (\log \circ \tau_1)(u) = \log(1+u).$$

This shows that the delta operator δ that leads to this generating function is

$$\delta = \varphi(D) = \log(1+D) = D - \frac{1}{2}D^2 + \frac{1}{3}D^3 - \cdots.$$

The following formulas result from the general theory:

$$B_n(x+y) = \sum_{0\leq k\leq n} \binom{n}{k} B_k(x)B_{n-k}(y),$$

$$\sum_{n\geq 1} B_n'(0)\frac{z^n}{n!} = \varphi^{-1}(z) = e^z - 1,$$

whence $B_n'(0) = 1$ $(n \geq 1)$. The polynomials B_n are monic polynomials having zero constant term if $n \geq 1$. The first ones are

$$B_0 = 1, \quad B_1(x) = x, \quad B_2(x) = x + x^2, \quad B_3(x) = x + 3x^2 + x^3,$$
$$B_4(x) = x + 7x^2 + 6x^3 + x^4.$$

If we take the derivative of the generating function (with respect to z), we obtain the relation

$$B_{n+1}(x) = x \sum_{0\leq k\leq n} \binom{n}{k} B_k(x),$$

from which these polynomials are easily computed inductively.

Comment. The special values $B_n = B_n(1)$ of the Bell polynomials are the *Bell numbers.* They represent the numbers of distinct partitions of the set $\{1, 2, \ldots, n\}$ into nonempty subsets. The first ones are

$$B_0 = 1, \quad B_1 = 1, \quad B_2 = 2, \quad B_3 = 5, \quad B_4 = 15,$$

where, for example, the five partitions of $\{1, 2, 3\}$ are

$$\begin{gathered} \{1\}, \{2\}, \{3\} \\ \{1, 2\}, \{3\} \\ \{2, 3\}, \{1\} \\ \{1, 3\}, \{2\} \\ \{1, 2, 3\}. \end{gathered}$$

Exercises for Chapter 4

1. (*a*) Suppose that we define the notion of Banach space E over an ultrametric field K simply as a *complete normed K-vector space.* Try to prove that if E is a Banach space of positive dimension over K, then K is *complete.*
   (*b*) If you cannot prove (*a*), think of the following examples: K is a noncomplete ultrametric field and $E = \widehat{K}$ is its completion with the norm given by the extension of the absolute value. This is a Banach space over K. For example, take $K = \mathbf{Q}$ with the p-adic absolute value and $E = \mathbf{Q}_p$ as Banach space over $\mathbf{Q}$, or $K = \mathbf{Q}_p^a$ (algebraic closure of $\mathbf{Q}_p$) and $E = \mathbf{C}_p$ as Banach space over $\mathbf{Q}_p^a$. What is happening?

2. Let $(E_i)_{i \in I}$ be a family of Banach spaces. Define the Banach product of this family as the normed vector subspace

$$\breve{\prod_{i \in I}} E_i \subset \prod_{i \in I} E_i$$

consisting of the bounded families $x = (x_i)$, equipped with the sup norm

$$\|x\| = \|(x_i)\| := \sup_i \|x_i\|.$$

In particular, $\widehat{\bigoplus}_{i \in I} E_i$ is a normed vector subspace of $\breve{\prod}_{i \in I} E_i$.
   (*a*) Show that this Banach product is *complete* and hence is a Banach space. Observe that $l^\infty(I; E) = \breve{\prod}_{i \in I} E$ and conclude that $l^\infty(I; E)$ is complete for any Banach space E.
   (*b*) Show that the dual of $\widehat{\bigoplus}_{i \in I} E_i$ is canonically isomorphic to $\breve{\prod}_{i \in I} E_i'$.
   (*c*) Formulate the universal properties of the direct sum and Banach product as canonical isometries

$$L(E; \breve{\prod} E_i) \cong \breve{\prod} L(E; E_i),$$

$$L(\widehat{\bigoplus} E_i; E) \cong \breve{\prod} L(E_i; E).$$

   [The second isomorphism for $E = K$ gives (*b*).]

3. Let $E = K^n$ (for some $n \geq 1$) with the sup norm. If $(\varepsilon_j)_{1 \leq j \leq n}$ is a normal basis of E, show that the matrix having the ε_j for columns is in $\mathrm{Gl}_n(A)$: It has entries in A and determinant in $A^\times$. Conversely, any normal basis is obtained in this form.

4. When $|K^\times| = |\pi|^{\mathbf{Z}}$ is discrete but the condition $\|E\| = |K|$ is not satisfied, show that the norm of E can be replaced by an *equivalent one* that satisfies it. Take either
$$\|\mathbf{x}\|' = \sup\{|\lambda| : \lambda \in K,\ |\lambda| \leq \|\mathbf{x}\|\}$$
or
$$\|\mathbf{x}\|'' = \inf\{|\lambda| : \lambda \in K,\ |\lambda| \geq \|\mathbf{x}\|\},$$
for which
$$|\pi|\|\mathbf{x}\| \leq \|\mathbf{x}\|' \leq \|\mathbf{x}\| \leq \|\mathbf{x}\|'' \leq |\pi|^{-1}\|\mathbf{x}\|.$$
Since sup = max and inf = min, these new norms take their values in $|K|$.

5. Let $(u_i)_{i \in I}$ be a family of continuous operators in an unltrametric Banach space E such that for each $x \in E$, $\varphi(x) := \sup_{i \in I} \|u_i(x)\| < \infty$. Show that $\sup_{i \in I} \|u_i\| < \infty$. (*Hint.* Consider the subsets $E_n \subset E$ defined by $\varphi \leq n$ and use the Baire property for the union $\bigcup_{n \geq 1} E_n = E$, or copy the proof of the Banach-Steinhaus theorem from any book on functional analysis !)

6. (*a*) Assume $f \in \mathbf{Q}[X]$, $f(0) \in \mathbf{Z}$, and that ∇f takes integral values on all natural numbers. Prove that f also takes integral values on $\mathbf{N}$.
   (*b*) Let the polynomial $f \in \mathbf{Q}[X]$ take integral values on all natural numbers: $f(\mathbf{N}) \subset \mathbf{Z}$. Prove that f also takes integral values on all integers: $f(\mathbf{Z}) \subset \mathbf{Z}$.
   (*Hint.* Show that $f(\mathbf{Z}) \subset \mathbf{Q} \cap \mathbf{Z}_p$ for all primes p.)

7. The maximal number of electrons on atomic layers is given by the following sequence
$$K : 2,\ L : 8,\ M : 18,\ N : 32, \ldots.$$
What is the next one? Find a polynomial formula $f(n)$ giving these values.
(*Hint.* Compute the finite differences to determine the simplest polynomial f taking these prescribed values.)

8. The maximal number of regions in the plane $\mathbf{R}^2$ determined by n lines is given by (make pictures!)

n	0	1	2	3	4	...
$f(n)$	1	2	4	7	11	...

Find a polynomial formula for $f(n)$.

9. Let $f(n)$ denote the maximal number of regions determined in the unit disk $|z| \leq 1$ (of the complex plane $\mathbf{C}$) by connecting n distinct points on its boundary $|z| = 1$ by lines. Show that this sequence starts as follows:

n	1	2	3	4	5	6	...
$f(n)$	1	2	4	8	16	31	...

Find a polynomial formula for $f(n)$.

10. Consider the Fibonacci sequence as a function of an integer variable $n \to f_n$:
$$f_0 = 0, \quad f_1 = 1, \quad f_{n+1} = f_n + f_{n-1} \quad (n \geq 1).$$
Does this function extend continuously to any $\mathbf{Z}_p$ (p prime)?

11. What are the Mahler series expansions of the following polynomial functions:

$$f(x) = 2x^2 - 1, \quad g(x) = 4x^3 - 3x, \quad f(g(x)).$$

12. Let $x^n = \sum_{k \geq 0} a_{n,k} \binom{x}{k}$ be the Mahler series of the continuous function $x^n \colon \mathbf{Z}_p \to \mathbf{Q}_p$. Hence $a_{n,k} = 0$ for $k > n$. Show that

$$a_{n,0} = \delta_n \ (=1 \text{ for } n = 0 \text{ and} = 0 \text{ for } n > 0),$$

$$a_{n,k} = \sum_{0 \leq j \leq k} (-1)^{k-j} \binom{k}{j} j^n,$$

$$a_{n,k} = k(a_{n-1,k} + a_{n-1,k-1}) \quad (k \geq 1, n \geq 1).$$

Show also that when p is an odd prime,

$$a_{pk} \equiv 0 \pmod p \quad (2 \leq k \leq p-1).$$

The $a_{n,k}/k!$ are the Stirling numbers of the second kind; cf. (VI.4.7).

13. Let $f : \mathbf{Z}_p \to \mathbf{Q}_p$ be continuous, given by a Mahler series

$$f(x) = \sum_{n \geq 0} a_n \binom{x}{n}.$$

What is the Mahler series of the function xf?

14. Prove the following formula:

$$\binom{x}{n}' = \binom{x}{n-1} - \frac{1}{2}\binom{x}{n-2} + \frac{1}{3}\binom{x}{n-3} \mp \cdots + \frac{(-1)^{n-1}}{n}$$

for $n \geq 1$.

15. Show that the series

$$\sum_{n \geq 0} \frac{1}{p^n} \binom{x}{p^{2n} - 1}.$$

converges for all $x \in \mathbf{Z}_p, x \neq 1$. The sum $f(x)$ defines a continuous unbounded function $\mathbf{Z}_p - \{-1\} \to \mathbf{Q}_p$.

16. Let $a \in 1 + \mathbf{M}_p$ and m a positive integer prime to p. Show that there is a unique mth root of a in $1 + \mathbf{M}_p$.
(*Hint.* Consider the series expansion $(1+t)^{1/m} = \sum_{k \geq 0} \binom{1/m}{k} t^k$.)

17. Let $f : \mathbf{Z}_p \to \mathbf{C}_p$ be a continuous function and $F = Sf$ its indefinite sum (IV.1.5).
(*a*) Show that there are uniform estimates

$$|F(n + p^\nu) - F(n)| \leq \varepsilon_\nu \quad (n \in \mathbf{N}),$$

where $\varepsilon_\nu \to 0$ ($\nu \to \infty$). (*Hint.* In a sum $F(n + p^\nu) - F(n) = \sum_{n \leq i < p^\nu} f(i)$ group the indices i in question into cosets mod $p^s\mathbf{Z}$. Let C be one such coset and pick one $c \in C$,

$$\sum_{x \in C} f(x) = \sum_{x \in C} (f(x) - f(c)) + p^{\nu - s} f(c).$$

Hence $|\sum_{x \in C} f(x)| \le \max_{x \in C} (|f(x) - f(c)|, |p^{\nu-s} f(c)|)$.)

(*b*) Show that for every given $\varepsilon > 0$, there is an integer ν such that

$$|F(n + kp^\nu) - F(n)| \le \varepsilon \quad (n, k \in \mathbf{N}).$$

(*c*) Prove directly (i.e., without the Mahler theorem) that for any continuous function $f : \mathbf{Z}_p \to \mathbf{C}_p$, the indefinite sum F of f extends continuously to $\mathbf{Z}_p$. (Corollary 2 in 3.5).

18. Show that the finite sums

$$\sum c_i \tau_{a_i} \quad \left(\sum c_i = 0, \ \sum a_i \neq 0\right)$$

are delta operators (notations of 5.1).

19. Let us define the Bell-Carlitz polynomials B_n^c by their generating function

$$\exp\left(xz + (e^z - 1)\right) = \sum_{n \ge 0} B_n^c(x) \frac{z^n}{n!}.$$

Hence $B_n^c(0) = B_n$ ($=B_n(1)$ cf. (IV.6.3)) are the usual Bell numbers.

(*a*) Prove that

$$B_n^c(x) = \sum_{0 \le j \le n} \binom{n}{j} B_{n-j} x^j, \quad B_n^c(1) = B_{n+1}.$$

Hence these polynomials interpolate consecutive values in the sequence $(B_n)_{n \ge 0}$.

(*b*) Prove that the sequence $(B_n^c)_{n \ge 0}$ is an Appell sequence (IV.6.1). (*Hint.* Differentiate the expression found under (*a*).)

20. Consider the power series expansion $(1-t-t^2)^{-1} = \sum_{n \ge 0} a_n t^n$. Show that $a_n = f_{n+1}$, where $(f_n)_{n \ge 0}$ is the Fibonacci sequence

$$f_0 = 0, \quad f_1 = 1, \quad f_{n+1} = f_n + f_{n-1} \quad (n \ge 1).$$

Define a sequence (p_n) of polynomials by the identity

$$\exp\left(x \log \frac{1}{1 - z - z^2}\right) = \sum p_n(x) \frac{z^n}{n!},$$

so that $a_n = p_n(1)/n!$. Show that this generating function corresponds to the choice

$$u = \varphi^{-1}(z) = \log 1/(1 - z - z^2), \quad e^{-u} - 1 = -z - z^2, \quad \delta = \varphi(D).$$

Show that

$$\delta = \varphi(D) = \sum_{k \ge 1} \frac{1}{2} \binom{2k}{k} \cdot (-\nabla_-)^k$$

(the operator $-\nabla_-$ is simply given by $f \mapsto f(x-1) - f(x)$).

21. Show that the basic sequence of polynomials corresponding to the translate delta operator $\tau_{-\varepsilon}\nabla$ is

$$p_n(x) = x \cdot (x + n\varepsilon - 1)_{n-1} \quad (n \geq 1).$$

The renormalized polynomials $q_n(x) = p_n(x)/n!$ are the *Gould polynomials*

$$q_n(x) = \frac{x}{x+n\varepsilon} \cdot \frac{(x+n\varepsilon)_n}{n!} = \frac{x}{x+n\varepsilon}\binom{x+n\varepsilon}{n} = \frac{x}{n}\binom{x+n\varepsilon-1}{n-1}.$$

(*Hint.* Check by induction that

$$\nabla p_n = n(x+\varepsilon)(x+\varepsilon+(n-1)\varepsilon-1)_{n-2} = n\tau_\varepsilon(p_{n-1})$$

and hence $\tau_{-\varepsilon}\nabla p_n = np_{n-1}$, then use the translation principle (5.5).) Write explicitly the binomial identity for the Gould polynomials.
Show that the delta operators $\tau_{-\varepsilon}\nabla$ satisfy the condition $\|T\| = |\alpha_1| = 1$ of the van Hamme theorem (5.4), and hence they give rise to uniformly convergent expansions of all continuous functions $\mathbf{Z}_p \to K$ (complete extension of $\mathbf{Q}_p$).

5

Differentiation

Calculus in the p-adic domain is rather straightforward. Let us emphasize, however, that:

- A function with a continuous derivative is not necessarily strictly differentiable.
- The mean value theorem is valid provided the increment is small enough: $|h| \leq r_p$.
- The radius of convergence of the exponential series is $r_p < \infty$.

In this chapter the field K will denote a complete extension of $\mathbf{Q}_p$, e.g., $K = \mathbf{C}_p$ or Ω_p.

1. Differentiability

1.1. Strict Differentiability

Let $X \subset K$ be a subset *with no isolated point.*

> **Definition.** *A function $f : X \to K$ is said to be* differentiable *at a point $a \in X$ if the difference quotients $(f(x) - f(a))/(x - a)$ have a limit $\ell = f'(a)$ as $x \to a$ $(x \neq a)$ in X.*

Equivalently, one can require the existence of a *limited expansion of the first order*

$$f(x) = f(a) + (x - a)f'(a) + (x - a)\phi(x) \text{ where } \phi(x) \to 0 \quad (x \to a).$$

Example 1. Let $(B_n)_{n\geq 1}$ be the sequence of open balls

$$B_n = \{x \in \mathbf{Z}_p : |x - p^n| < |p^{2n}|\} \subset \{x \in \mathbf{Z}_p : |x| = |p^n|\}$$

and f the function on $\mathbf{Z}_p$ vanishing outside $\bigcup B_n$ (a disjoint union) with values

$$f(x) = p^{2n} \quad (x \in B_n).$$

Then f is constant on each open ball B_n and hence is locally constant outside the origin. Consequently f is differentiable at each $x \neq 0$ with $f'(x) = 0$. At the origin $\lim_{x\to 0}(f(x) - f(0))/x = \lim_{x\to 0} f(x)/x$ exists and is zero, so that f is also differentiable at this point with $f'(0) = 0$. In this example, $f' = 0$ (identically), f' is continuous, a situation classically denoted by $f \in C^1$, but the difference quotients

$$\frac{f(y) - f(x)}{y - x} = \frac{f(x) - f(y)}{x - y}$$

take the value 1 on the pairs $x = x_n = p^n$, $y = y_n = p^n - p^{2n}$, which are arbitrarily close to the origin.

Example 2. Let $f : \mathbf{Z}_p \to \mathbf{Z}_p$ be the continuous function defined by

$$x = \sum_{n\geq 0} a_n p^n \mapsto f(x) = \sum_{n\geq 0} a_n p^{2n}.$$

Then f is differentiable at all points $x \in \mathbf{Z}_p$ with $f'(x) = 0$. Again $f' = 0 \in C^1$, but f is injective, and hence far from being locally constant.

The preceding examples show that the notion of differentiability at each point of a set X is not very useful, even if we require these derivatives to vary continuously, and we shall introduce a stronger condition.

Definition. *We say that f is* strictly differentiable *at a point $a \in X$ — and denote this property by $f \in S^1(a)$ — if the difference quotients*

$$\Phi f(x, y) = \frac{f(x) - f(y)}{x - y}$$

have a limit $\ell = f'(a)$ as $(x, y) \to (a, a)$ (x and y remaining distinct).

Classically, i.e., for a function $f : I \to \mathbf{R}$ (where $I \subset \mathbf{R}$ is an open interval), if $f'(a)$ exists at each $a \in I$ *and $f' : a \mapsto f'(a)$ is continuous*, then f is *strictly differentiable* at all points $a \in I$. The examples preceding the definition show that in ultrametric analysis, the situation is different and we have to assume strict differentiability to get interesting results.

Proposition 1. *Let $f : X \to K$ be strictly differentiable at a point $a \in X$ with $f'(a) \neq 0$. Then there is a neighborhood V of a in X such that the restriction of $f/f'(a)$ to V is isometric.*

PROOF. Since $f \in S^1(a)$, for each $\varepsilon > 0$ there is a neighborhood V_ε of a for which

$$|\Phi f(x, y) - f'(a)| < \varepsilon \text{ if } x \in V_\varepsilon \text{ and } y \in V_\varepsilon.$$

Let us take $\varepsilon = |f'(a)|$ ($\neq 0$ by assumption) and V the corresponding neighborhood. Then

$$|\Phi f(x, y) - f'(a)| < |f'(a)| \neq 0, \text{ if } (x, y) \in V \times V$$

and there is a competition between the terms $\Phi f(x, y)$ and $f'(a)$

$$|\Phi f(x, y)| = |f'(a)| \text{ for } (x, y) \in V \times V.$$

Hence $|f(x) - f(y)| = |f'(a)| \cdot |x - y|$ for $(x, y) \in V \times V$. ∎

Corollary. *If $f \in S^1(a)$ and $f'(a) \neq 0$, then there is a neighborhood V of $a \in X$ in which f is injective.* ∎

Theorem. *Assume that the function f is defined in a neighborhood of $a \in K$ and strictly differentiable at this point with $f'(a) \neq 0$. Choose an open ball B containing a such that*

$$\sigma = \sup_{x \neq y \in B} \left| \frac{f(x) - f(y)}{x - y} - f'(a) \right| < |f'(a)|.$$

Then f maps each open ball contained in B onto an open ball, namely

$$B_{<\varepsilon}(b) \subset B \Longrightarrow f(B_{<\varepsilon}(b)) = B_{<\varepsilon'}(f(b)) \quad (\varepsilon' = |f'(a)|\varepsilon).$$

PROOF. Put $s = f'(a) \neq 0$. As in the preceding proposition, we have

$$\left| \frac{f(x) - f(y)}{x - y} \right| = |s| \quad (x \neq y \in B),$$

and f/s is an isometry in the ball B. This already proves

$$f(B_{<\varepsilon}(b)) \subset B_{<|s|\varepsilon}(f(b)).$$

To prove that this inclusion is an equality, we select any $c \in B_{<|s|\varepsilon}(f(b))$, namely $|f(b) - c| < |s|\varepsilon$, and show that the equation $f(x) = c$ has a solution x with $|x - b| < \varepsilon$. Equivalently, we show that the map $\varphi(x) = x - (f(x) - c)/s$ has a fixed point x with $|x - b| < \varepsilon$. Observe first that $\varphi(B_{<\varepsilon}(b)) \subset B_{<\varepsilon}(b)$:

$$\begin{aligned} \varphi(x) - b &= x - b - (f(x) - c)/s \\ &= x - b - (f(x) - f(b))/s - (f(b) - c)/s, \\ |\varphi(x) - b| &\leq \max(|x - b|, |f(x) - f(b)|/|s|, |f(b) - c|/|s|) < \varepsilon. \end{aligned}$$

Now we prove that φ is a *contracting map* with contraction ratio $\sigma/|s| < 1$:

$$\varphi(x) - \varphi(y) = x - y - \frac{f(x) - f(y)}{s}$$
$$= \frac{x-y}{s}\left(s - \frac{f(x) - f(y)}{x-y}\right),$$
$$|\varphi(x) - \varphi(y)| \leq \frac{|x-y|}{|s|}\sigma = \frac{\sigma}{|s|}\cdot|x-y|.$$

Since the ball $B_{<\varepsilon}(b)$ is closed in the complete space K, the mapping φ has a unique fixed point in this ball and the theorem is completely proved. ■

Observe that this theorem is a generalization of Hensel's lemma (II.1.5) (here f is not a polynomial): The function $f - c$ has a zero $x \in B$, or $f(x) = c$, as soon as $|f(b) - c|$ is small enough for some $b \in B$.

Let us turn to strict differentiability on a subset X *having no isolated point.* Since X is a metric space, it is Hausdorff and the diagonal of X is closed in $X \times X$. The open subset $X \times X - \Delta_X$ is *dense* in the product $X \times X$.

Proposition 2. *For $f : X \to K$, the following properties are equivalent:*

(i) *$f \in S^1(a)$ for all $a \in X$.*
(ii) *The function Φf, initially defined only on $X \times X - \Delta_X$, admits a continuous extension $\widetilde{\Phi}$ to $X \times X$.*
(iii) *f is differentiable at each point $a \in X$ and there is a continuous function α on $X \times X$ vanishing on Δ_X with*
$f(y) = f(x) + (y-x)f'(x) + (y-x)\alpha(x, y) \quad (x, y \in X)$.

Proof. The implication $(i) \Longrightarrow (ii)$ is given by the *double limit theorem,* which we recall: *Let X_0 be a dense subset of a topological space X, Y a metric space, and f a continuous map $X_0 \to Y$ such that for each $x \in X$*

$$z \in X_0 \text{ and } z \to x \text{ implies } f(z) \text{ has a limit } g(x) \in Y.$$

Then the extension $g : X \to Y$ is continuous. (More generally, the conclusion is valid when the target space Y is a *regular space*, i.e., a topological space in which every point has a fundamental system of neighborhoods consisting of closed sets.)

The implication $(ii) \Longrightarrow (i)$ is obvious.

Finally, if Φf has a continuous extension $\widetilde{\Phi}$, it has a unique one by the density of $X \times X - \Delta_X$ in $X \times X$. Since we can write

$$f(y) = f(x) + (y-x)\Phi f(x, y)$$
$$= f(x) + (y-x)f'(x) + (y-x)\underbrace{[\Phi f(x,y) - f'(x)]}_{\alpha(x,y)},$$

it is obvious that $(ii) \Leftrightarrow (iii)$. ■

Definition. *We shall say that* f *is* strictly differentiable *on* X *— notation* $f \in S^1(X)$ *or even* $f \in S^1$ *— when the conditions of Proposition* 2 *are satisfied.*

When $f \in S^1$, $f'(x) = \widetilde{\Phi}(x, x)$ is continuous and $f \in C^1$, but strict differentiability is a stronger condition, justifying a specific notation. Strict differentiability furnishes *coherent* limited expansions, and if

$$M = \sup_{x \neq y} |\Phi f(x, y)| = \sup_{x,y} |\widetilde{\Phi}(x, y)| < \infty,$$

we have

$$|f(x) - f(y)| \leq M|x - y|.$$

1.2. Granulations

The theorem of the preceding section is particularly interesting when the field K is locally compact, namely when it is a *finite* extension of $\mathbf{Q}_p$. Let us come back to the usual notation for this case:

$$K \supset R \supset P = \pi R, \quad k = R/P = \mathbf{F}_q.$$

If $r \in |K^\times|$, every ball $B_{\leq r}(a)$ is a disjoint union of q open balls $B_i = B_{<r}(a_i) = B_{\leq \theta r}(a_i)$ (with $\theta = |\pi| < 1$) and any set containing q distinct points $x_i \in B_{\leq r}(a)$ with

$$|x_i - x_j| \geq r \quad (i \neq j)$$

contains at most one point in each B_i, hence exactly one point in each B_i.

Proposition. *Let* K *be a finite extension of* $\mathbf{Q}_p$ *and* $f : \Omega \to K$ *be an isometry where* Ω *is some compact open subset of* K. *Then* f *maps the balls contained in* Ω *onto balls of* K.

PROOF. If $B_{\leq r}(a)$ is a ball contained in Ω, it is clear that

$$f(B_{\leq r}(a)) \subset B_{\leq r}(b) \quad (b = f(a)).$$

There remains to prove the surjectivity $f(B_{\leq r}(a)) = B_{\leq r}(b)$. But if we take a partition of $B_{\leq r}(a)$ consisting of smaller disjoint balls, say $B'_i = B_{\leq \varepsilon}(a_i)$ with $\varepsilon = |\pi|^\nu r$, the images $x_i = f(a_i)$ of chosen points $a_i \in B'_i$ form a system of q^ν points in $B_{\leq r}(f(a))$ with

$$|x_i - x_j| = |a_i - a_j| > \varepsilon \quad (i \neq j).$$

Hence the image $f(B_{\leq r}(a))$ contains a point in each smaller ball of the partition of $f(B_{\leq r}(a))$ into q^ν balls of radius $\varepsilon = |\pi|^\nu r \leq r$ $(j \geq 0)$. This shows that the image of $B_{\leq r}(a)$ by f meets all closed balls of positive radius. Hence this image $f(B_{\leq r}(a))$ is dense in $B_{\leq r}(f(a))$. Since it is compact, it is closed, and the proposition is established. ∎

Definition. *A* granulation *of an open compact set* $\Omega \subset K$ *is a finite partition of* Ω *into balls* $B_{\leq r}(a_i)$ *of the same radius* $r > 0$.

Since two balls B_1, B_2 having a common point satisfy either $B_1 \subset B_2$ or $B_2 \subset B_1$, two granulations are always comparable: One is finer than the other. Every ball of the coarser one is a disjoint union of some power q^ν of balls of the finer one. Now observe that $q^\nu \equiv 1 \pmod{p-1}$, so that the numbers of balls in the two granulations differ only by a multiple of $p - 1$. This number of balls is well-defined modulo $p - 1$.

Definition. *For any open compact set* Ω *in a finite extension* K *of* $\mathbf{Q}_p$, *we define the* type $\tau(\Omega) \in \mathbf{Z}/(p-1)\mathbf{Z}$ *of* Ω *to be the class* $\text{mod}(p-1)$ *of the number of balls in any granulation of* Ω.

For example, the type of $\mathbf{Z}_p$ is $p \equiv 1$ and the type of $\mathbf{Z}_p^\times$ is $p - 1 \equiv 0$.
It is obvious that the type is *additive* for disjoint unions:

$$\tau(\Omega \amalg \Omega') = \tau(\Omega) + \tau(\Omega') \in \mathbf{Z}/(p-1)\mathbf{Z}.$$

Consequently, to compute the type of any open compact set Ω, it is enough to know the cardinality of any partition of Ω into balls (allowing unequal radii). The following theorem summarizes the preceding comments.

Theorem. *Let* Ω *be an open compact subset of a finite extension* K *of* $\mathbf{Q}_p$ *and* f *an injective strictly differentiable map* $\Omega \to K$. *If* f' *vanishes nowhere, then* Ω *and* $f(\Omega)$ *have the same type.*

PROOF. From $f \in S^1(a)$ and $f'(a) \neq 0$ we infer that there is a neighborhood V (for example an open ball) of a in Ω such that any ball in V is transformed by f into a ball of $f(V)$. ∎

Corollary. *Let* $p > 2$, *and* $f : \mathbf{Z}_p \to \mathbf{Z}_p^\times$ *be an injective strictly differentiable map with nowhere vanishing* f'. *Then* f *is not surjective.* ∎

1.3. Second-Order Differentiability

With the same notation as in (1.1), we define

$$\Phi_2 f(x, y, z) = \frac{\Phi f(x, z) - \Phi f(y, z)}{x - y}$$

when x, y, and z are distinct. Since we can also write

$$\Phi_2 f(x, y, z) = \frac{f(x)}{(x-y)(x-z)} + \frac{f(y)}{(y-x)(y-z)} + \frac{f(z)}{(z-x)(z-y)},$$

this function $\Phi_2 f$ is symmetric in x, y, and z.

Definition. *We say that a function f is* twice strictly differentiable *at a point $a \in X$ — and denote this property by $f \in S^2(a)$ — if $\Phi_2 f(x, y, z)$ tends to a limit as $(x, y, z) \to (a, a, a)$, x, y, and z remaining distinct.*

Proposition 1. *If $f \in S^2(a)$, then $f \in S^1(a)$.*

PROOF. Let us take two pairs (x, y) and $(z, t) \in X \times X - \Delta_X$ in the vicinity of (a, a) and estimate the difference

$$\begin{aligned}\Phi f(x, y) - \Phi f(z, t) &= \Phi f(x, y) - \Phi f(z, y) + \Phi f(z, y) - \Phi f(z, t)\\ &= (x - z)\Phi_2 f(x, z, y) + (y - t)\Phi_2 f(y, t, z).\end{aligned}$$

If we assume $f \in S^2(a)$, then $\Phi_2 f$ will remain bounded in a neighborhood of (a, a, a), say $|\Phi_2 f| \le M$, when the three variables of $\Phi_2 f$ are close enough to a. In particular if x, y, z, and t are near enough to a, we have

$$|\Phi f(x, y) - \Phi f(z, t)| \le M \max(|x - z|, |y - t|),$$

a quantity that tends to zero when (x, y) and (z, t) tend to (a, a). Since the target of Φf is a complete space, the Cauchy criterion is valid and shows that this function Φf has a limit as $(x, y) \to (a, a)$. ■

As in (1.1) (Proposition 2), the double limit theorem shows that the following two properties are equivalent:

(i) *$f \in S^2(a)$ for all $a \in X$.*
(ii) *The function $\Phi_2 f$, initially defined only on triples with distinct entries, admits a continuous extension to $X \times X \times X$.*

We shall say that the function f is *twice strictly differentiable* — notation $f \in S^2(X)$ or even $f \in S^2$ — when these conditions are satisfied.

Proposition 2. *If $f \in S^2$, then $f' \in S^1$.*

PROOF. We have to prove that the difference quotients

$$\Phi(f')(x, y) = \frac{f'(x) - f'(y)}{x - y}$$

have a continuous extension across the diagonal of $X \times X$. By assumption, there is a continuous function $\widetilde{\Phi}_2$ that extends $\Phi_2 f$ to $X \times X \times X$, and we have

$$\Phi f(x, z) - \Phi f(y, z) = (x - y) \cdot \widetilde{\Phi}_2(x, y, z).$$

In this expression we let $z \to x$. We know that $\Phi f(x, z)$ tends to $f'(x)$ and

$$f'(x) - \Phi f(y, x) = (x - y) \cdot \widetilde{\Phi}_2(x, y, x).$$

Since the order of the variables in Φf, $\Phi_2 f$, and $\widetilde{\Phi}_2$ is irrelevant, we can write

$$f'(x) = \Phi f(x, y) + (x - y) \cdot \widetilde{\Phi}_2(x, x, y),$$

and interchanging x and y,

$$f'(y) = \Phi f(y, x) + (y - x) \cdot \widetilde{\Phi}_2(y, y, x).$$

Subtracting these expressions, we obtain

$$f'(x) - f'(y) = (x - y)[\widetilde{\Phi}_2(x, x, y) + \widetilde{\Phi}_2(x, y, y)],$$
$$\Phi f'(x, y) = \widetilde{\Phi}_2(x, x, y) + \widetilde{\Phi}_2(x, y, y).$$

This shows that $\Phi f'$ admits a continuous extension to $X \times X$: $f' \in S^1$. Moreover,

$$f''(a) = (f')'(a) = \Phi f'(a, a) = 2\widetilde{\Phi}_2(a, a, a). \qquad \blacksquare$$

1.4. *Limited Expansions of the Second Order*

It is also possible to characterize the second-order differentiability by means of limited expansions (this will not be used later and may be skipped).

Proposition. *In order for a function f to be in the class S^2, it is necessary and sufficient that it admits a limited expansion*

$$f(x) = f(y) + (x - y) \cdot \alpha(y) + (x - y)^2 \beta(x, y),$$

where α and β are two continuous functions.

Proof. (*a*) Suppose first that $f \in S^2 \subset S^1$. In the formula

$$\Phi_2 f(x, y, z) = \frac{\Phi f(x, z) - \Phi f(y, z)}{x - y} \quad (x, y, z \text{ distinct}),$$

we can let $z \to y$. In the limit, we get

$$\widetilde{\Phi}_2 f(x, y, y) = \frac{\Phi f(x, y) - \widetilde{\Phi} f(y, y)}{x - y} = \frac{\Phi f(x, y) - f'(y)}{x - y} \quad (x \neq y),$$

namely

$$\Phi f(x, y) = f'(y) + (x - y)\widetilde{\Phi}_2 f(x, y, y).$$

Coming back to the definition of Φf, we have

$$f(x) - f(y) = (x - y) f'(y) + (x - y)^2 \widetilde{\Phi}_2 f(x, y, y).$$

This gives an expansion of the desired form with

$$\alpha = f' \text{ and } \beta(x, y) = \widetilde{\Phi}_2 f(x, y, y).$$

(*b*) Conversely, let us postulate the existence of a limited expansion as in the statement of the proposition; hence

$$\Phi f(x, y) = \alpha(y) + (x - y)\beta(x, y) \quad (x \neq y).$$

If x, y, and z are distinct, we have

$$\Phi f(x, z) = \alpha(z) + (x - z)\beta(x, z),$$
$$\Phi f(y, z) = \alpha(z) + (y - z)\beta(y, z),$$

whence by subtraction (and division by $x - y$),

$$\Phi_2 f(x, y, z) = \frac{x - z}{x - y}\beta(x, z) + \frac{z - y}{x - y}\beta(y, z)$$
$$= \lambda\beta(x, z) + \mu\beta(y, z)$$

(where $\lambda + \mu = 1$). Let us choose a point $a \neq x$ and subtract the same quantity $\beta(x, a) = (\lambda + \mu)\beta(x, a)$ from both members:

$$\Phi_2 f(x, y, z) - \beta(x, a) = \lambda[\beta(x, z) - \beta(x, a)] + \mu[\beta(y, z) - \beta(x, a)].$$

It is clear that

$$y \to a \text{ and } z \to a \Longrightarrow \mu \to 0 \text{ and } \Phi_2 f(x, y, z) \to \beta(x, a)$$

(observe that $|\lambda| = 1$ as soon as $\max(|z - a|, |y - a|) < |x - a|$). When x, y, and $z \to a$ (while remaining distinct), we even see that $\Phi_2 f(x, y, z) \to \beta(a, a)$: In the region U: $\max(|x - z|, |y - z|) \leq |x - y|$, in which $|\mu|$ and $|\lambda|$ are less than or equal to 1 we have

$$|\Phi_2 f(x, y, z) - \beta(a, a)| \leq$$
$$\max(|\beta(x, z) - \beta(x, a)|, |\beta(y, z) - \beta(x, a)|, |\beta(x, a) - \beta(a, a)|).$$

In this region $\Phi_2 f(x, y, z) \to \beta(a, a)$ (x, y, and z distinct $\to a$). Since $\Phi_2 f$ is symmetric in its three variables, we can estimate the difference

$$|\Phi_2 f(x, y, z) - \beta(a, a)|$$

by first permuting the variables in order to bring them into the region in which the preceding estimates have been made. ∎

Caution. A function f on $\mathbf{Z}_p$ can have a derivative $f' \in S^1$ without being twice strictly differentiable, namely with $f \notin S^2$. One can think of a function f with vanishing derivative at each point, hence with $f' \equiv 0 \in S^2$, but that is not strictly differentiable at a point: We have given an example of such a function, locally constant outside the origin, in (1.1).

1.5. Differentiability of Mahler Series

Let f be a continuous function on $\mathbf{Z}_p$ and choose $y \in \mathbf{Z}_p$. We can write the Mahler expansion of the continuous function $x \mapsto f(x+y)$ as

$$f(x+y) = \sum_{k\geq 0} c_k(y)\binom{x}{k} \quad \text{with } c_k(y) = (\nabla^k f)(y) \to 0.$$

Theorem. *Let f be a continuous function on $\mathbf{Z}_p$. Then f is differentiable at y precisely when*

$$|(\nabla^k f)(y)/k| \to 0 \quad (k \to \infty).$$

In this case $f'(y) = \sum_{k\geq 1}(-1)^{k-1}(\nabla^k f)(y)/k$.

Proof. Replacing f by its translate $x \mapsto f(x+y)$ we see that it is enough to prove the theorem when $y = 0$. Now, since $c_0 = f(0)$, we have

$$\frac{f(x)-f(0)}{x} = \sum_{k\geq 1}\frac{c_k}{x}\binom{x}{k} = \sum_{k\geq 1}\frac{c_k}{k}\binom{x-1}{k-1}.$$

If $|c_k/k| \to 0$ (when $k \to \infty$), the Mahler series

$$\sum_{k\geq 1}(c_k/k)\binom{y}{k-1} = g(y)$$

represents a continuous function of $y \in \mathbf{Z}_p$. In particular, $f'(0)$ exists and

$$f'(0) = g(-1) = \sum_{k\geq 1}\binom{-1}{k-1}\frac{c_k}{k} = \sum_{k\geq 1}(-1)^{k-1}\frac{c_k}{k}.$$

Conversely, if f is differentiable at the origin, the function g defined by $g(0) = f'(0)$ and $g(x) = (f(x)-f(0))/x$ for $x \neq 0$ is continuous on $\mathbf{Z}_p$ and possesses a Mahler expansion

$$g(x) = \sum_{k\geq 0}\gamma_k\binom{x}{k} \quad (\text{where } \gamma_k = \nabla^k g(0) \to 0).$$

We deduce

$$f(x) - f(0) = xg(x) = \sum_{k\geq 0}\gamma_k x\binom{x}{k}.$$

But

$$x\binom{x}{k} = (x-k)\binom{x}{k} + k\binom{x}{k} = (k+1)\binom{x}{k+1} + k\binom{x}{k}.$$

Hence we can write

$$\begin{aligned} f(x) &= f(0) + xg(x) \\ &= c_0 + \sum_{k\geq 0} \gamma_k \left((k+1)\binom{x}{k+1} + k\binom{x}{k} \right) \\ &= c_0 + \sum_{k\geq 1} k(\gamma_{k-1} + \gamma_k)\binom{x}{k}. \end{aligned}$$

By uniqueness of the Mahler coefficients of f, we deduce $c_k = k(\gamma_k + \gamma_{k-1})$ $(k \geq 1)$ and in particular $c_k/k = \gamma_k + \gamma_{k-1} \to 0$ $(k \to \infty)$. ∎

Comment. For any integer $k \geq 1$ we have $|k| = p^{-v(k)} \geq 1/k$, or equivalently, $1/|k| \leq k$. Hence $|c_k/k| \leq k|c_k|$, and the condition $k|c_k| \to 0$ implies $|c_k/k| \to 0$. This stronger condition will imply *strict differentiability* of the Mahler series.

Let us first give a statement concerning Mahler series of *Lipschitz functions.*

Definition. *A function $f : X \to K$ (as in (1.1)) is* Lipschitz *when there exists a constant M with*

$$|f(x) - f(y)| \leq M|x - y| \quad (x, y \in X).$$

Since the smallest bound M is

$$\|\Phi f\| := \sup_{x\neq y} |\Phi f(x, y)|,$$

Lipschitz functions are also characterized by $|\Phi f|$ *bounded.* We shall denote by $\mathrm{Lip}\,(\mathbf{Z}_p)$ the subspace of $C(\mathbf{Z}_p)$ consisting of Lipschitz functions. By definition,

$$S^1(\mathbf{Z}_p) \subset \mathrm{Lip}\,(\mathbf{Z}_p) \subset C(\mathbf{Z}_p).$$

Proposition. *A function $f = \sum_{k\geq 0} c_k \binom{\cdot}{k} \in C(\mathbf{Z}_p)$ is Lipschitz precisely when $\{k|c_k|\}_{k\geq 0}$ is bounded, namely*

$$|\Phi f| \text{ bounded} \iff \text{the sequence } k|c_k| \text{ is bounded in } \mathbf{R}_{\geq 0}.$$

The proof of this proposition is based on the following lemma.

Lemma. *For $k \geq 1$ and $p^j \leq k < p^{j+1}$, we have*

$$\left|\binom{x}{k} - \binom{y}{k}\right| \leq p^j|x - y| \quad (x, y \in \mathbf{Z}_p).$$

Comment. More precisely, when k is in the quoted interval, its expansion in base p has the form

$$k = k_0 + k_1 p + \cdots + k_j p^j \quad (0 \leq k_i < p,\ k_j \neq 0)$$

($j+1$ digits), and we call k_- the integer $k_0+k_1p+\cdots+k_{j-1}p^{j-1} < p^j$ (at most j digits). Then

$$k - k_- = k_j p^j, \quad |k - k_-| = p^{-j},$$

and the statement of the lemma can be written uniformly for all integers k:

$$\left|\binom{x}{k} - \binom{y}{k}\right| \leq |k - k_-|^{-1} \cdot |x - y|.$$

This lemma shows that $|x - y| < |p^j| = p^{-j}$ implies $\left|\binom{x}{k} - \binom{y}{k}\right| < 1$. For example, if $y = x + p^h$ for some $h > j$,

$$\left|\binom{x}{k} - \binom{x+p^h}{k}\right| < 1 \quad (\text{for } k < p^h).$$

This is the p^h-periodicity of the binomial functions (IV.1.3)

$$x \mapsto \binom{x}{k} \bmod p \quad (k < p^h)$$

already exploited in the proof of the Mahler theorem.

Proof of the Lemma. The formal identity $(1+T)^{x+y} = (1+T)^x(1+T)^y$ leads to the well-known relation

$$\binom{x+y}{k} = \sum_{i+j=k} \binom{x}{i}\binom{y}{j}$$

(first for positive integers x and y but also by density and continuity for p-adic integers x and y). Write then

$$\binom{x}{k} = \binom{x-y+y}{k} = \sum_{i+j=k} \binom{x-y}{i}\binom{y}{j} = \binom{y}{k} + \sum_{1\leq i\leq k} \binom{x-y}{i}\binom{y}{k-i}.$$

Thus

$$\binom{x}{k} - \binom{y}{k} = \sum_{1\leq i\leq k} \binom{x-y}{i}\binom{y}{k-i} = \sum_{1\leq i\leq k} \frac{x-y}{i}\binom{x-y-1}{i-1}\binom{y}{k-i},$$

and it only remains to estimate

$$\left|\sum_{1\leq i\leq k} \frac{1}{i}\binom{x-y-1}{i-1}\binom{y}{k-i}\right| \leq \max_{1\leq i\leq k} \frac{1}{|i|} = \frac{1}{\min_{1\leq i\leq k} |i|}.$$

It is clear that for $p^j \le k < p^{j+1}$, the minimum in question is attained for $i = p^j$ with $|i| = |p^j| = p^{-j}$. The lemma follows. ∎

Remark. A slightly less precise inequality, namely

$$\left|\binom{x}{k} - \binom{y}{k}\right| \le k \cdot |x - y|,$$

would be sufficient for our study of S^1 functions.

Proof of the Proposition. Let us write the difference quotients

$$\begin{aligned}\Phi f(x+h, x) = \frac{f(x+h) - f(x)}{h} &= \frac{1}{h}\sum_{k\ge 1} c_k \left(\binom{x+h}{k} - \binom{x}{k}\right)\\ &= \frac{1}{h}\sum_{k\ge 1} c_k \left(\sum_{i+j=k}\binom{x}{i}\binom{h}{j} - \binom{x}{k}\right)\\ &= \sum_{k\ge 1}\frac{c_k}{h}\sum_{0\le i<k}\binom{x}{i}\binom{h}{k-i}\end{aligned}$$

for $h \ne 0$. We observe that

$$\left|\frac{c_k}{h}\binom{x}{i}\binom{h}{k-i}\right| \le \frac{|c_k|}{|h|} \to 0 \quad (k \to \infty)$$

uniformly in i (and fixed $h \ne 0$). The double family

$$\frac{c_k}{h}\binom{x}{i}\binom{h}{k-i} = \frac{c_k}{k-i}\binom{x}{i}\binom{h-1}{k-i-1}$$

is thus summable in any order, and in particular, it is equal to a double series over the indices $i \ge 0$ and $j = k - i - 1 \ge 0$. Replacing h by $y + 1$, we obtain

$$\Phi f(x+y+1, x) = \sum_{i,j\ge 0}\frac{c_{i+j+1}}{j+1}\binom{x}{i}\binom{y}{j} \quad (y \ne -1).$$

Firstly, the ultrametric inequality gives

$$|\Phi f(x+y+1, x)| \le \sup_{i,j\ge 0}\left|\frac{c_{i+j+1}}{j+1}\right|,$$

and hence

$$\|\Phi f\| = \sup_{x,y\ne -1}|\Phi f(x+y+1, x)| \le \sup_{i,j\ge 0}\left|\frac{c_{i+j+1}}{j+1}\right|.$$

In fact, the preceding expansion is a *false* Mahler series in y because it is not valid for all values of $y \in \mathbf{Z}$. Nevertheless, it holds for all $(x, y) \in \mathbf{N} \times \mathbf{N}$, and this proves that the coefficients are given by the finite differences on the integers. Hence secondly,

$$\left|\frac{c_{i+j+1}}{j+1}\right| \leq \sup_{x,y\neq -1} |\Phi f| = \|\Phi f\| \ (\leq \infty)$$

and

$$\sup_{i,j\geq 0} \left|\frac{c_{i+j+1}}{j+1}\right| \leq \|\Phi f\|.$$

Altogether, we have

$$\sup_{i,j\geq 0} \left|\frac{c_{i+j+1}}{j+1}\right| = \|\Phi f\| \ (\leq \infty).$$

In particular, considering only the subset of indices (i, j) for which $i + j + 1 = n$,

$$\sup\left(|c_n|, |c_n/2|, \ldots, |c_n/n|\right) \leq \|\Phi f\|.$$

On the left we have

$$|c_n| \sup_{i\leq n} 1/|i| = |c_n| \cdot p^s \quad (p^s \leq n < p^{s+1}).$$

Call κ_n the highest power of p that is less than or equal to n. The preceding considerations prove that

$$\|\Phi f\| = \sup_{i,j\geq 0} \left|\frac{c_{i+j+1}}{j+1}\right| = \sup_{n\geq 1} \kappa_n |c_n|.$$

Since $\kappa_n \leq n < p\kappa_n$, the proposition follows. ∎

Corollary 1. *Let* $f \in \mathrm{Lip}(\mathbf{Z}_p)$ *and* $f = \sum c_n \binom{\cdot}{n}$ *its Mahler expansion. Then*

$$\|\Phi f\| = \sup_{n\geq 1} \kappa_n |c_n| < \infty.$$

∎

The number $\|\Phi f\|$ does not define a norm on the vector space $\mathrm{Lip}(\mathbf{Z}_p)$ because Φf vanishes for constant functions: It is only a *seminorm*. In order to have a norm, we take

$$\|f\|_1 = \sup\left(|f(0)|, \|\Phi f\|\right).$$

Since $f(0) = c_0$, we define in an ad hoc way the value $\kappa_0 = 1$ in order to have

$$\|f\|_1 = \sup_{n\geq 0} \kappa_n |c_n|.$$

Corollary 2. *Let $f \in \mathrm{Lip}(\mathbf{Z}_p)$ and Sf its indefinite sum. Then $Sf \in \mathrm{Lip}(\mathbf{Z}_p)$ and*

$$\|f\|_1 \leq \|Sf\|_1 \leq p\|f\|_1.$$

Proof. We have

$$\|f\|_1 = \sup_{n\geq 0} \kappa_n |a_n|,$$
$$\|Sf\|_1 = \sup_{n\geq 1} \kappa_n |a_{n-1}|$$

by Corollary 2 in (IV.3.5). Now observe that

$$\kappa_{n-1} \leq \kappa_n \leq p\kappa_{n-1},$$

whence the assertion. ■

Corollary 3. *The map*

$$f = \sum_{n\geq 0} c_n \binom{\cdot}{n} \mapsto \left(\frac{c_n}{\kappa_n}\right)_{n\geq 0}$$

is an isomorphism between the normed spaces $(\mathrm{Lip}(\mathbf{Z}_p),\ \|\,.\,\|_1)$ *and* ℓ^∞. *The functions*

$$1 \text{ and } \kappa_n \binom{x}{n} \quad (n \geq 1)$$

correspond to the "canonical basis" of ℓ^∞.

Here, κ_n (highest power of p that is less than or equal to n) is considered as an element of $\mathbf{Z}_p$: Its absolute value is $|\kappa_n| = 1/\kappa_n \in \mathbf{R}_{>0}$.

Proof. Any $f \in \mathcal{C}(\mathbf{Z}_p)$ is given by a Mahler series

$$f = \sum_{n\geq 0} c_n \binom{\cdot}{n} \qquad (|c_n| \to 0).$$

When f is Lipschitz, we write this series

$$f = \sum_{n\geq 0} \frac{c_n}{\kappa_n} \kappa_n \binom{\cdot}{n} \in \mathrm{Lip}(\mathbf{Z}_p)$$

with

$$\|f\|_1 = \sup(|f(0)|, \|\Phi f\|) \overset{\text{Cor. 1}}{=} \sup_{n\geq 0} \kappa_n |c_n| = \sup_{n\geq 0} |c_n/\kappa_n|;$$

hence the result. ■

1.6. *Strict Differentiability of Mahler Series*

Theorem. *For a continuous function* $f = \sum_{k\geq 0} c_k \binom{\cdot}{k} \in C(\mathbf{Z}_p)$, *we have*

$$k|c_k| \to 0 \quad (k \to \infty) \Longrightarrow f \in S^1(\mathbf{Z}_p).$$

Proof. We have

$$\Phi f(x, y) = \sum_{k\geq 1} c_k \left(\binom{x}{k} - \binom{y}{k} \right) \Big/ (x - y),$$

and thanks to the lemma (in its weak form),

$$\sup_{x,y} \left| c_k \left(\binom{x}{k} - \binom{y}{k} \right) \Big/ (x - y) \right| \leq k|c_k|.$$

If $k|c_k| \to 0$, Φf is a continuous function as a sum of a uniformly convergent series with continuous terms: The polynomial $\binom{x}{k} - \binom{y}{k}$ in x and y vanishes identically on the diagonal $x = y$ and is divisible by $x - y$, whence $\left(\binom{x}{k} - \binom{y}{k}\right)/(x - y)$ is also a polynomial function. ∎

It is possible to prove conversely

$$f \in S^1(\mathbf{Z}_p) \Longrightarrow k|c_k| \to 0 \quad (k \to \infty).$$

Corollary. *Let* $f \in S^1(\mathbf{Z}_p)$ *and* Sf *its indefinite sum. Then* $Sf \in S^1(\mathbf{Z}_p)$. ∎

With the preceding results, it is easy to construct examples of continuous functions on $\mathbf{Z}_p$ exhibiting various behaviors (as far as differentiability is concerned).

Example 1. Let the Mahler coefficients c_k of a continuous function f be

$$c_k = \begin{cases} p^j & \text{if } k = p^j, \\ 0 & \text{if } k \text{ is not a power of } p, \end{cases}$$

so that

$$|c_k/k| \text{ takes alternatively values 0 and 1.}$$

Hence $|c_k/k|$ does not tend to 0, thereby proving that f is not differentiable at the origin. But Φf is bounded, since $k|c_k|$ (taking values 0 and 1 only) is bounded.

Example 2. As in the preceding example, but with

$$c_k = \begin{cases} p^{2j} & \text{if } k = p^j, \\ 0 & \text{if } k \text{ is not a power of } p. \end{cases}$$

Then

$$|c_k/k| \text{ takes alternatively values 0 and } |p^j| \to 0,$$

so that f is differentiable at the origin. Here $|c_k/k| = k|c_k|$ and $f \in S^1$.

2. Restricted Formal Power Series

2.1. A Completion of the Polynomial Algebra

Recall that in this chapter K denotes a complete extension of $\mathbf{Q}_p$. A formal power series with coefficients in a subring R of the field K is a sequence $(a_n)_{n\geq 0}$ of elements of R. However, when we use the product

$$(a_n)_{n\geq 0} \cdot (b_n)_{n\geq 0} = (c_n)_{n\geq 0} \text{ with } c_n = \sum_{i+j=n} a_i b_j \quad (n \geq 0)$$

we prefer the series representation $f(X) = \sum_{n\geq 0} a_n X^n$ instead of $(a_n)_{n\geq 0}$. The set of formal power series is a ring and an R-algebra denoted by $R[[X]]$. Recall that the formal power series ring with integral coefficients has already been considered in (I.4.8); we shall come back to a more systematic study of formal power series rings in (VI.1). The particular formal power series having coefficients $a_n \to 0$ are called *restricted formal power series*, or more simply *restricted (power) series*. The restricted formal power series with coefficients in K form a vector subspace of $K[[X]]$ denoted by $K\{X\}$ and isomorphic to the Banach space $c_0(K)$ (IV.4.1). This subspace is a completion of the polynomial space for the Gauss norm — sup norm on the coefficients —

$$K[X] \subset K\{X\} \subset K[[X]].$$

We still call *Gauss norm* the extension

$$\|f(X)\| := \sup_{n\geq 0} |a_n| = \max_{n\geq 0} |a_n| \quad (f(X) = \sum_{n\geq 0} a_n X^n \in K\{X\}).$$

Lemma. *For two restricted power series f and g, we have*

$$\|fg\| \leq \|f\|\|g\|.$$

PROOF. Let $f(X) = \sum_{n\geq 0} a_n X^n$, $g(X) = \sum_{n\geq 0} b_n X^n$ be two polynomials. Their product $h = fg$ is the polynomial $h(X) = \sum_{n\geq 0} c_n X^n$ having coefficients $c_n = \sum_{i+j=n} a_i b_j$. Since $|c_n| \leq \max_{i+j=n} |a_i||b_j| \leq \|f\|\|g\|$,

$$\|fg\| = \max_{n\geq 0} |c_n| \leq \|f\|\|g\| \quad (f, g \in K[X]).$$

Hence multiplication is (uniformly) continuous in $K[X]$ and extends continuously to the completion $K\{X\}$ with the same inequality. ∎

Hence $K\{X\}$ is a ring and a *Banach algebra*. This is the *Tate algebra in one variable over K*.

As usual, we denote by A the maximal subring of K, M the maximal ideal of A, and $k = A/M$ the residue field of K. The unit ball $A\{X\} \subset K\{X\}$ is a subring, since

$$\|f\| \leq 1, \ \|g\| \leq 1 \Longrightarrow \|fg\| \leq 1.$$

From this it follows that the reduction (IV.4.3) of the Banach space $K\{X\}$, the quotient of its closed unit ball by its open unit ball, is the polynomial ring over the residue field

$$A\{X\}/M\{X\} = k[X].$$

Let $|x| < 1$ (x in K or K': complete extension of K) and $f = \sum_{n\geq 0} a_n X^n$ a restricted formal power series. Then $|a_n x^n| \to 0 \, (n \to \infty)$, so that $f(x) = \sum_{n\geq 0} a_n x^n$ converges and f defines a function on the unit ball of K (or K')

$$f : A \to K : x \mapsto \sum_{n\geq 0} a_n x^n.$$

The sup norm of this function satisfies

$$\|f\|_{C_b(A,K)} = \sup_A |f(x)| \leq \sup_{n\geq 0} |a_n| = \|f\|_{K\{X\}}.$$

In particular, the series $\sum_{n\geq 0} a_n x^n$ converges uniformly on the unit ball A, and f defines a *continuous function* on this ball. The linear map

$$K\{X\} \to C_b(A; K) : \sum_{n\geq 0} a_n X^n \mapsto f$$

is a contracting map of Banach spaces.

Example. Let $K = \mathbf{Q}_p$ and consider the polynomial (restricted formal power series) $X - X^p$ of norm 1 in $\mathbf{Q}_p\{X\}$. Since $x^p \equiv x \pmod{p}$ for all $x \in \mathbf{Z}_p$, we have $|x^p - x| \leq |p| = 1/p$ when $x \in \mathbf{Z}_p$ and the norm of the continuous function $x \mapsto x^p - x$ on $\mathbf{Z}_p$ is $1/p < 1$.

Theorem. *If the residue field k of K is infinite, the canonical embedding $K\{X\} \to C_b(A; K)$ is isometric:*

$$\sup_{x\in A} |f(x)| = \|f\|_{K\{X\}}.$$

PROOF. If $f = 0$, then $\|f\| = 0$ and there is nothing to prove. Otherwise, we can replace f by f/a_n where $|a_n| = \|f\|$. Thus we may assume that $\|f\| = 1$. In this case the image of $f \in A\{X\}$ in the quotient is a nonzero polynomial

$$\widetilde{f} \in A\{X\}/M\{X\} = k[X],$$

and since the residue field k is infinite, we can find $\alpha \in k^\times$ with $\tilde{f}(\alpha) \neq 0$. Taking any $a \in A^\times \subset K^\times$ with residue class α in $k^\times$, we have

$$|a| = 1 \text{ and } |f(a)| = 1,$$

whence $\sup_{|x|\leq 1} |f(x)| = \max_{|x|\leq 1} |f(x)| = \max_{|x|=1} |f(x)| = 1 = \|f\|$. ■

The preceding proof shows more: For $f \in K\{X\}$, we have $\sup_A |f| = \max_A |f|$ in spite of the fact that the unit ball A is generally not compact. Moreover, the maximum of $|f(x)|$ ($|x| \leq 1$) is attained at a point x with $|x| = 1$.

Recall that we denote by $\mathbf{A}_p$ the maximal subring of $\mathbf{C}_p$ (closed unit ball) and by $\mathbf{M}_p$ the maximal ideal of $\mathbf{A}_p$ (open unit ball).

Corollary. *We have*

$$\sup_{n\geq 0} |a_n| = \sup_{x\in\mathbf{A}_p} \left|\sum a_n x^n\right| = \max_{x\in\mathbf{A}_p} \left|\sum a_n x^n\right|,$$

and this maximum is attained on $\mathbf{A}_p^\times = \mathbf{A}_p - \mathbf{M}_p$, which is the unit sphere $|x| = 1$ in $\mathbf{C}_p$. The canonical embedding

$$K\{X\} \to C_b(\mathbf{A}_p; \mathbf{C}_p)$$

is isometric. ■

2.2. Numerical Evaluation of Products

Let $f(X) = \sum_{n\geq 0} a_n X^n$ and $g(X) = \sum_{n\geq 0} b_n X^n$ be two restricted power series. Their formal product is the power series

$$h(X) = \sum_{n\geq 0} c_n X^n,$$

where

$$c_n = \sum_{0\leq i\leq n} a_i b_{n-i} \quad (n \geq 0).$$

As we have seen in the previous section, it is again a restricted power series.

Theorem. *Let $f(X), g(X) \in K\{X\}$ be two restricted power series and let $h(X)$ be their formal product. Then $h(X) \in K\{X\}$, and the evaluation of this formal product can be made according to the usual product*

$$h(x) = f(x)g(x) \quad (|x| \leq 1).$$

PROOF. Replacing $a_n x^n$ by a_n and similarly $b_n x^n$ by b_n, we see that it is sufficient to prove the statement for $x = 1$. With $c_n = \sum_{0\leq i\leq n} a_i b_{n-i}$ we have to prove

if $\sum_{n\geq 0} a_n$ *and* $\sum_{n\geq 0} b_n$ *converge, then* $\sum_{n\geq 0} c_n$ *converges and*

$$\sum_{n\geq 0} c_n = \sum_{n\geq 0} a_n \cdot \sum_{n\geq 0} b_n.$$

Because multiplication is continuous,

$$\sum_{i\geq 0} a_i \cdot \sum_{j\geq 0} b_j - \sum_{i\leq N} a_i \cdot \sum_{j\leq N} b_j \to 0 \quad (N \to \infty).$$

Let us show now that

$$\sum_{i\leq N} a_i \cdot \sum_{j\leq N} b_j - \sum_{k\leq N} c_k \to 0.$$

Choose N_ε large enough to ensure

$$|a_i| \leq \varepsilon,\ |b_i| \leq \varepsilon \quad (i > N_\varepsilon).$$

Now the difference

$$\sum_{i\leq N} a_i \cdot \sum_{j\leq N} b_j - \sum_{k\leq N} c_k$$

is the sum of the terms $a_i b_j$ corresponding to pairs (i, j) in the square $0 \leq i, j \leq N$ above the diagonal, namely with $i + j > N$. The contribution of these terms is less than or equal to εC if C is an upper bound for the coefficients and $N \geq 2N_\varepsilon$ because at least one index i or j will be greater than or equal to N_ε. This proves the theorem. ∎

Observe that the classical result concerning *absolutely convergent* series cannot be applied here, since we only assume $|a_n| \to 0$ and $|b_n| \to 0$ (but $\sum |a_n|$ and/or $\sum |b_n|$ may diverge). On the other hand, due to the ultrametric inequality, it is now easier to estimate tails of sums!

Corollary. *The canonical map* $K\{X\} \to C_b(A; K)$ *is a norm-decreasing homomorphism of K-algebras.* ∎

This isomorphism is isometric when the residue field k of K is infinite (and also when $|K^\times|$ is dense in $\mathbf{R}_{\geq 0}$, as we shall see later (VI.1.4)). The identification of a restricted formal power series $f(X)$ with the function f that it defines on the unit ball A will often be made.

2.3. *Equicontinuity of Restricted Formal Power Series*

Let us still identify $K\{X\}$ with a normed subspace of $C_b(\mathbf{A}_p; \mathbf{C}_p)$. With $A = B_{\leq 1}(K)$ as usual,

$$\sup_{x\in A} |f(x)| \leq \|f\|_{K\{X\}} := \sup_{n\geq 0} |a_n| = \sup_{x\in \mathbf{A}_p} |f(x)| \quad (f = \sum a_n X^n \in K\{X\}).$$

Proposition 1. *The unit ball in $K\{X\}$ is uniformly equicontinuous. More precisely, $K\{X\} \subset \mathrm{Lip}\,(A)$ and $\|\Phi f\| \le \|f\|_1 \le \|f\|$ for $f \in K\{X\}$. In particular,*

$$|f(x+h) - f(x)| \le |h|\,\|f\| \text{ if } |x| \le 1 \text{ and } |h| \le 1.$$

PROOF. Write $f = \sum a_n X^n$, so that

$$f(x) - f(y) = \sum_{n\ge 0} a_n(x^n - y^n) = (x-y)\sum_{n\ge 1} a_n(x^{n-1} + \cdots + y^{n-1}).$$

If $|x| \le 1$ and $|y| \le 1$, the ultrametric inequality gives $|x^{n-1} + \cdots + y^{n-1}| \le 1$, and the result follows. ∎

In a similar vein, let us derive the following inequalities.

Proposition 2. *If $f \in K\{X\}$, then*

$$|f(x+y) - f(x) - f(y) + f(0)| \le \|f\| \cdot |xy| \quad (|x| \le 1,\ |y| \le 1).$$

If moreover f is odd, then

$$|f(x+y) - f(x) - f(y)| \le \|f\| \cdot |xy(x+y)| \quad (|x| \le 1,\ |y| \le 1).$$

PROOF. With the same notation as before,

$$f(x+y) - f(x) - f(y) + f(0) = \sum_{n\ge 2} a_n((x+y)^n - x^n - y^n)),$$

whence the result, since each term $(x+y)^n - x^n - y^n$ is divisible by xy. When f is odd,

$$f(x+y) - f(x) - f(y) = \sum_{n \text{ odd } \ge 0} a_n((x+y)^n - x^n - y^n)).$$

Only the terms with n odd and greater than or equal to 3 remain in the sum, and for these

$$(x+y)^n - x^n - y^n = xy(x+y)p_n(x)$$

for some integral polynomials $p_n \in \mathbf{Z}[X]$. Hence $\|p_n\| \le 1$. ∎

Remark. Although it is uniformly equicontinuous, the unit ball of $\mathbf{C}_p\{X\}$ is not precompact in $\mathcal{C}_b(\mathbf{A}_p; \mathbf{C}_p)$: The Ascoli theorem is not applicable, since $\mathbf{A}_p$ is not locally compact. For example, the infinite sequence $(X^n)_{n\ge 0}$ satisfies

$$\|X^n - X^m\| = 1 \quad (n \ne m)$$

and hence contains no convergent subsequence. However, if we consider the restriction of these continuous functions to the (compact) unit ball R of a finite extension K of $\mathbf{Q}_p$, the preceding sequence admits a convergent subsequence, namely $(X^{q^m})_{m\geq 0}$, where q is the cardinality of the residue field of K. In fact, the subsequence $(X^{p^{n!}})$ converges uniformly on every unit ball $B_{\leq 1}(0; K)$, provided that K has finite residue degree over $\mathbf{Q}_p$ (III.4.4).

2.4. *Differentiability of Power Series*

The formal derivation operator $f \mapsto f'$ is continuous and contracting on $K\{X\}$ simply since $|na_n| \leq |a_n| \to 0$ and

$$\|f'\| = \sup |na_n| \leq \sup |a_n| = \|f\|.$$

We are interested in strict differentiability; hence we look at the differential quotients

$$\Phi f(x, y) = \frac{f(x) - f(y)}{x - y} \quad (x \neq y).$$

When $f \in K\{X\}$, Proposition 1 of (2.3) shows that

$$\Phi f(X, Y) = \sum_{n\geq 1} a_n \sum_{0\leq i\leq n-1} X^{n-1-i} Y^i$$

is a formal power series in two variables and coefficients tending to zero. Thus Φf has a continuous extension to $A \times A$ that is a sum of a uniformly convergent power series (in two variables). The value on the diagonal is

$$\Phi f(X, X) = \sum_{n\geq 1} na_n X^{n-1}.$$

This proves the following result.

Theorem 1. *Let $f \in K\{X\}$. Then f defines a strictly differentiable function on the unit ball A of K: $f \in S^1(A)$. The derivative of f is given by the restricted formal power series*

$$f' = \Phi f|_\Delta = \sum_{n\geq 1} na_n X^{n-1} \in K\{X\}. \qquad \blacksquare$$

It is easy to give more precise estimates for the convergence:

$$\begin{aligned}\Phi f(x, y) - \sum_{n\geq 1} na_n \xi^{n-1} &= \sum_{n\geq 1} a_n (x^n - y^n)/(x - y) - \sum_{n\geq 1} na_n \xi^{n-1} \\ &= \sum_{n\geq 1} a_n (x^{n-1} + \cdots + y^{n-1} - n\xi^{n-1}) \to 0\end{aligned}$$

when $(x, y) \to (\xi, \xi)$. In fact,

$$(x^{n-1} + \cdots + y^{n-1} - n\xi^{n-1}) = \sum_{i+j=n-1} (x^i y^j - \xi^{n-1}),$$

and by (2.3),

$$\begin{aligned}|x^i y^j - \xi^{n-1}| &= |x^i y^j - \xi^i \xi^j| \\ &\le \max(|x^i||y^j - \xi^j|, |x^i - \xi^i||\xi^j|) \\ &\le \max(|y - \xi|, |x - \xi|) \quad (x, y, \xi \in A).\end{aligned}$$

We have obtained

$$|\Phi f(x, y) - f'(\xi)| \le \|f\| \cdot \max(|y - \xi|, |x - \xi|).$$

Theorem 2. *A restricted formal power series $f = \sum a_n X^n$ defines a twice strictly differentiable function on the unit ball A of K: $f \in S^2(A)$.*

PROOF. As we have seen in the proof of the preceding theorem,

$$\Phi f(x, y) = \sum_{n\ge 1} a_n \sum_{i+j=n-1} x^i y^j = a_1 + a_2(x + y) + \cdots,$$

and hence, for distinct x, y, z,

$$\begin{aligned}\Phi_2 f(x, y, z) &= \frac{\Phi f(x, z) - \Phi f(y, z)}{x - y} \\ &= \sum_{n\ge 2} a_n \sum_{i+j=n-1} \frac{x^i - y^i}{x - y} z^j \\ &= \sum_{n\ge 2} a_n \sum_{k+\ell+m=n-2} x^k y^\ell z^m.\end{aligned}$$

Since $|a_n| \to 0$, this series converges uniformly on $A \times A \times A$ and represents a continuous extension of $\Phi_2 f$. ∎

Generalization. Let us just indicate here that differentiability of restricted power series is not limited to order two. In fact, one can define *higher-order difference quotients* inductively by

$$\Phi_k f(x_0, x_1, \ldots, x_k) := \frac{\Phi_{k-1} f(x_0, x_2, \ldots, x_k) - \Phi_{k-1} f(x_1, x_2, \ldots, x_k)}{x_0 - x_1}$$

($x_0 \ne x_1$). The expressions $\Phi_k f$ are symmetric in their $k + 1$ variables, and an easy computation shows that

$$\Phi_k f(x_0, x_1, \ldots, x_k) = \sum_i \left(\frac{f(x_i)}{\prod_{j \ne i}(x_j - x_i)} \right).$$

Taking $f(x) = x^N$ we obtain

$$\Phi_k f(x_0, x_1, \ldots, x_k) = \sum_{i_0+i_1+\cdots+i_k=N-k} x_0^{i_0} x_1^{i_1} \cdots x_k^{i_k},$$

a sum of all homogeneous monomials of total degree $N-k$. In this case $\Phi_k f$ has an obvious extension to the diagonal (all $x_i = x$):

$$\Phi_k f(x, x, \ldots, x) = x^{N-k} \cdot \# \left\{ \begin{array}{c} \text{monomials of degree } N-k \\ \text{in } k+1 \text{ variables} \end{array} \right\}$$

$$= \binom{N}{k} x^{N-k} = \frac{(x^N)^{(k)}}{k!}.$$

The equality $\Phi_k f(x, x, \ldots, x) = f^{(k)}(x)/k!$ remains true by linearity when f is a polynomial or even a restricted series. These functions are of class S^k on A.

2.5. *Vector-Valued Restricted Series*

Let E be an ultrametric Banach space over K. A *restricted vector series* (with coefficients in E) is a formal power series

$$f(X) = \sum_{n\geq 0} a_n X^n,$$

where $a_n \in E$, $\|a_n\| \to 0$. We can still define the Gauss norm of such a restricted series f by

$$\|f\| := \sup_{n\geq 0} \|a_n\|.$$

Hence the normed space of restricted vector series (with coefficients in E) is a Banach space isometric to $c_0(E)$. If the indeterminate X is replaced by a variable $x \in A \subset K$, the restricted series $\sum_{n\geq 0} a_n X^n$ gives rise to a continuous vector-valued function $f : A \to E$, which we can write as $f(x) = \sum_{n\geq 0} x^n a_n$ (not that it matters, but we may prefer to write scalar multiplications on the left), for which

$$\sup_{|x|\leq 1} \|f(x)\| \leq \sup_{n\geq 0} \|a_n\| = \|f\|.$$

That is, the linear map $c_0(E) \to C_b(A, E)$ is continuous and contracting.

Proposition. *When the residue field k of K is infinite, the canonical map* $c_0(E) \to C_b(A, E)$ *is an isometry.*

PROOF. Assume $\|f\| = c > 0$ and look at the A-module $B_{\leq c}(E)$: Its quotient $\widetilde{E} = B_{\leq c}(E)/B_{<c}(E)$ is a vector space over $k = A/M$. The restricted series f with coefficients in $B_{\leq c}(E)$ has a polynomial image $\widetilde{f} = \sum \widetilde{a_n} X^n$ having at least one nonzero coefficient, since $\|f\| = c$. We can choose a k-linear form φ on $\widetilde{E}$ such that $\varphi(\widetilde{a_n}) \neq 0$ for such a coefficient. The scalar polynomial $\varphi \circ \widetilde{f} = \sum \alpha_n X^n = \sum \varphi(\widetilde{a_n}) X^n$ is not identically zero and there is an element $\alpha \in k^\times$ such that $\varphi \circ \widetilde{f}(\alpha) \neq 0$. A fortiori $\widetilde{f}(\alpha) \neq 0$ and $\|f(a)\| = c$ for every $a \in A$, a in the coset α (mod M). ∎

3. The Mean Value Theorem

3.1. The p-adic Valuation of a Factorial

Since the formula for the p-adic order of $n!$ will play an essential role, we review it.

Lemma. *Let $n \geq 1$ be an integer and let $S_p(n)$ be the sum of the digits of n in base p. Then the p-adic order of $n!$ is given by*

$$\text{ord}_p(n!) = \frac{n - S_p(n)}{p-1}.$$

PROOF. We have to compute

$$\text{ord}_p(n!) = \sum_{1 \leq k \leq n} \text{ord}_p(k).$$

Let us fix an integer $k \leq n$ say with order $\text{ord}_p(k) = \nu$ and write its expansion in base p:

$$k = k_\nu p^\nu + \cdots + k_\ell p^\ell \quad (\nu \leq \ell,\ k_\nu \neq 0).$$

Then

$$k - 1 = (p-1) + \cdots + (p-1)p^{\nu-1} + (k_\nu - 1)p^\nu + \cdots + k_\ell p^\ell,$$

and hence

$$S_p(k-1) = \nu(p-1) + S_p(k) - 1.$$

Equivalently,

$$\nu = \text{ord}_p(k) = \frac{1}{p-1}(1 + S_p(k-1) - S_p(k)).$$

Summing over all values of $k \leq n$ we obtain a telescoping sum

$$\text{ord}_p(n!) = \frac{1}{p-1} \sum_{1 \leq k \leq n} (1 + S_p(k-1) - S_p(k)) = \frac{1}{p-1}(n - S_p(n)). \quad \blacksquare$$

ALTERNATIVE PROOF. A more traditional way of obtaining the same formula goes as follows. The number of integers k with fixed $\nu = \text{ord}_p(k)$ that appear in the product $n!$ is equal to the number of multiples of p^ν that are not multiples of $p^{\nu+1}$ (and are less than or equal to n), namely

$$\left[\frac{n}{p^\nu}\right] - \left[\frac{n}{p^{\nu+1}}\right],$$

where $[x]$ denotes the integral part of the real number x. Hence

$$\begin{aligned}\mathrm{ord}_p(n!) &= \sum_{\nu\geq 1} \nu\left(\left[\frac{n}{p^\nu}\right]-\left[\frac{n}{p^{\nu+1}}\right]\right)\\ &= \left[\frac{n}{p}\right]+\left[\frac{n}{p^2}\right]+\left[\frac{n}{p^3}\right]+\cdots = \sum_{j\geq 1}\left[\frac{n}{p^j}\right].\end{aligned}$$

Let us write n in base p as $n = n_0 + n_1 p + n_2 p^2 + \cdots$ (a finite sum). Then

$$\begin{aligned}\left[\frac{n}{p}\right] &= n_1 + n_2 p + n_3 p^2 + \cdots,\\ \left[\frac{n}{p^2}\right] &= n_2 + n_3 p + n_4 p^2 + \cdots,\\ \cdots &\quad \cdots\end{aligned}$$

Hence

$$\begin{aligned}n &= n_0 + p\left[\frac{n}{p}\right],\\ \left[\frac{n}{p}\right] &= n_1 + p\left[\frac{n}{p^2}\right],\\ \cdots &\quad \cdots\\ \left[\frac{n}{p^j}\right] &= n_j + p\left[\frac{n}{p^{j+1}}\right],\\ \cdots &\quad \cdots\end{aligned}$$

Summing all these, we obtain

$$\begin{aligned}n + \mathrm{ord}_p(n!) &= S_p(n) + p\,\mathrm{ord}_p(n!),\\ n - S_p(n) &= (p-1)\,\mathrm{ord}_p(n!)\end{aligned}$$

and the result follows. ■

3.2. *First Form of the Theorem*

As already recalled in (2.1) the field K is assumed to be a complete extension of $\mathbf{Q}_p$ (e.g., $\mathbf{C}_p$ or Ω_p). Even for polynomial functions f, the following form of the mean value theorem,

$$|f(h) - f(0)| \leq |h| \cdot \|f'\|$$

does not hold without restriction. Recall that for polynomials f (or more generally for restricted power series), we use the sup norm on the coefficients (Gauss norm). If the residue field of K is infinite, this norm coincides with the sup norm of $|f|$ on the unit ball of K (2.1).

For example, if $f(t) = t^p$, we have $f'(t) = pt^{p-1}$; hence $\|f'\| = |p| = 1/p < 1$. And with $h = 1$,

$$1 = |f(1) - f(0)| > |h| \cdot \|f'\| = 1/p.$$

However, we show below that there is a universal bound (depending on the prime p but not on the restricted series f) such that *the mean value theorem holds for* $|h| \leq r_p$. The preceding example can be used to discover the limitation in size of the increment h. In order to have

$$|f(h) - f(0)| \leq |h| \cdot \|f'\|$$

in this particular case, we must have

$$|h|^p \leq |h| \|f'\| = |h||p|,$$

whence the restriction $|h| \leq |p|^{1/(p-1)}$.

Let us recall that we have introduced a special notation (II.4.4) for this absolute value:

$$r_p := |p|^{1/(p-1)}, \quad |p| \leq r_p < 1.$$

It will play an important part from now on. Observe that

$$r_2 = \tfrac{1}{2}, \quad r_p > \tfrac{1}{p} \quad (p \text{ odd prime}),$$

and also $r_p \nearrow 1$ when the prime p increases (whereas $|p| = 1/p \searrow 0$).

Theorem. *Let $f(X) \in K\{X\}$ be a restricted power series and also denote by f the corresponding function $t \mapsto f(t) = \sum_{n \geq 0} a_n t^n$ on the unit ball A of K. Then*

$$|f(t + h) - f(t)| \leq |h| \cdot \|f'\|$$

for all $t, h \in K$ with $|t| \leq 1$ and $|h| \leq r_p = |p|^{1/(p-1)}$.

First Proof. (1) Let us establish first the result for a polynomial f. The Taylor formula permits us to compute the difference $f(t + h) - f(t)$ as

$$f(t + h) - f(t) = \sum_{k \geq 1} h^k \cdot \frac{D^k f(t)}{k!} = h \sum_{k \geq 1} \frac{h^{k-1}}{k!} \cdot D^{k-1} f'(t),$$

so that

$$|t| \leq 1 \Longrightarrow |f(t + h) - f(t)| \leq |h| \sup_{k \geq 1} \left| \frac{h^{k-1}}{k!} \right| \|D^{k-1} f'\|.$$

Since

$$\|Df\| = \|f'\| = \sup_{k \geq 1} |ka_k| \leq \sup_{k \geq 0} |a_k| = \|f\|,$$

we see that $\|D\| \le 1$, $\|D^k\| \le 1$ $(k \ge 1)$. In particular, $\|D^{k-1}f'\| \le \|f'\|$ and

$$\|D^{k-1}f'\| \le \sup_{n\ge k} |a_n| \overset{k\to\infty}{\longrightarrow} 0 \quad (= 0 \text{ for } k > \deg f).$$

The result will be proved if we can show that $|h^{k-1}/k!| \le 1$ (for all $k \ge 1$). This condition for $k = p$ requires $|h|^{p-1} \le |p|$, i.e., $|h| \le r_p$. When it is satisfied, we have

$$|h|^{k-1} \le |p|^{(k-1)/(p-1)} \le |k!|,$$

simply since

$$\mathrm{ord}_p(k!) = \frac{k - S_p(k)}{p-1} \le \frac{k-1}{p-1}.$$

(2) Consider now the general case of a restricted series $f(t) = \sum_{k\ge 0} a_k t^k$. Without loss of generality we may assume $|f(t+h) - f(t)| \ne 0$, hence f not constant. Consider the polynomials $f_n(t) = \sum_{k\le n} a_k t^k$. We have

$$\|f - f_n\| = \sup_{k>n} |a_k| \to 0$$

as well as

$$\|f_n\| = \sup_{k\le n} |a_k| = \|f\|,$$

$$\|f_n'\| = \sup_{k\le n} \|k a_k\| = \|f'\|$$

for all large n. Take t and h as in the first part. The convergence

$$f_n(t+h) - f_n(t) \to f(t+h) - f(t) \ne 0$$

implies $|f_n(t+h) - f_n(t)| = |f(t+h) - f(t)|$ for all large n. Hence, using such a large value of the integer n, and using the result for polynomials, we have

$$|f(t+h) - f(t)| = |f_n(t+h) - f_n(t)| \le |h| \cdot \|f_n'\| = |h| \cdot \|f'\|. \qquad \blacksquare$$

SECOND PROOF. Let us observe that

$$\frac{D^k(x^m)}{k!} = \binom{m}{k} \cdot x^{m-k},$$

so that the operator $D^k/k!$ transforms polynomials with integer coefficients into polynomials with integer coefficients: $\|D^k/k!\| \le 1$, $\|D^k\| \le |k!| \to 0$. Better still: If g is any restricted series, it is obvious that $\|D^k(g)/k!\| \to 0$, since the first k coefficients of g are destroyed by the operator D^k (while the other coefficients of g are multiplied by integers under the effect of the operator $D^k/k!$). Coming back to the expression

$$f(t+h)-f(t)=h\sum_{k\geq 1}\frac{h^{k-1}}{k}\cdot\frac{D^{k-1}f'(t)}{(k-1)!}$$

we see that the mean value theorem will be proved if we show $|h^{k-1}/k|\leq 1$. For $k=p$ this condition requires $|h|\leq|p|^{1/(p-1)}=r_p$ as before. When it is satisfied, take an integer k, put $\nu=\mathrm{ord}_p(k)$, and write $k=p^\nu m\geq p^\nu$. We have

$$\left|\frac{h^{k-1}}{k}\right|\leq\frac{r_p^{k-1}}{|p^\nu|}\leq r_p^{p^\nu-1}\cdot|p|^{-\nu}=|p|^\mu.$$

The exponent is

$$\mu=\frac{p^\nu-1}{p-1}-\nu=(1+p+\cdots+p^{\nu-1})-\nu\geq 0,$$

and the proof is completed. ∎

Remark. Let E be an ultrametric Banach space over K and

$$f(X)=\sum_{k\geq 0}a_kX^k$$

a restricted power series *with coefficients in E*. Then we can view f as the vector-valued function

$$t\mapsto\sum_{k\geq 0}t^ka_k,\ A\to E$$

on the unit ball A of K (2.5). Then the mean value theorem immediately furnishes

$$\|f(t+h)-f(t)\|\leq|h|\cdot\|f'\|$$

for all $t,h\in K$ such that $|t|\leq 1$ and $|h|\leq r_p=|p|^{1/(p-1)}$. In fact, simply replace in the above proof $|a_k|$ by $\|a_k\|$, $|f(t)|$ by $\|f(t)\|$, etc. whenever necessary.

3.3. *Application to Classical Estimates*

Let us apply the mean value theorem (3.2) to the polynomials $f(t)=(1+t)^{p^n}$ ($n\geq 1$). Since $f'(t)=p^n(1+t)^{p^n-1}$, we have $\|f'\|=|p^n|$ and hence

$$|(1+t)^{p^n}-1|\leq|t|\cdot|p^n|\text{ for }|t|\leq r_p.$$

Recall that the fundamental inequality (III.4.3) in its second form gives

$$|(1+t)^{p^n}-1|\leq|t|\cdot(\max(|t|,|p|))^n,$$

hence the same inequality only when $|t| \le |p|$. Since

$$1/p = |p| \le r_p < 1$$

with strict inequalities for all $p \ge 3$, the mean value estimate is sharper for odd primes (in the indicated region).

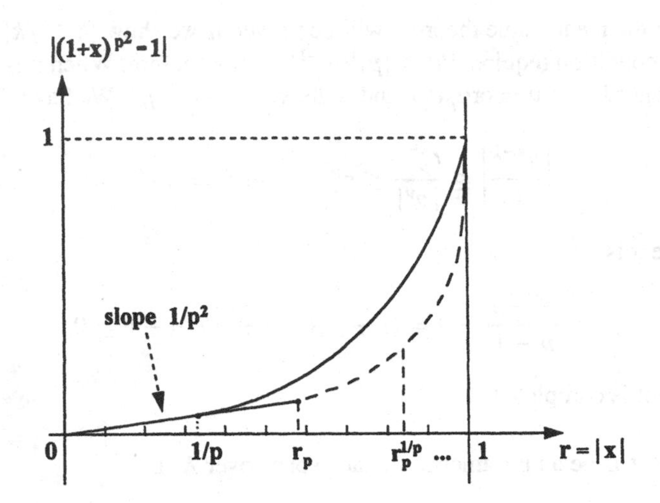

An application of the mean value theorem

With the Newton polygon method (VI.1.6) we shall be able to compute more explicitly these absolute values $|(1+t)^{p^n} - 1|$. Let us simply observe now that this absolute value vanishes when $(1+t)^{p^n} = 1$, namely when $1 + t \in \mu_{p^n}$. The smallest $|t|$ for which this occurs is $1 + t \in \mu_p$, and as we have seen in (II.4.4), this implies $|t| = |p|^{1/(p-1)} = r_p$.

Let us give another application to binomial coefficients. Define successively two polynomials g and f (having integral coefficients) by

$$(1+X)^p = 1 + p \cdot g(X) + X^p,$$
$$f(t) = (1 + tg(X) + X^p)^n,$$

so that

$$f(0) = (1+X^p)^n,$$
$$f(p) = (1+X)^{pn}.$$

Here we consider $f : \mathbf{Q}_p \to E$, where $E \subset \mathbf{Q}_p[X]$ denotes the finite-dimensional subspace consisting of polynomials of degree less than or equal to np. The mean value theorem (vector form) leads to an estimate of the norm of

$$f(p) - f(0) = (1+X)^{pn} - (1+X^p)^n = \sum \binom{np}{j} X^j - \sum \binom{n}{k} X^{pk}.$$

Since $\|f'\| \leq |n|$, we have $\|f(p) - f(0)\| \leq |np|$ and in particular, looking at the coefficient of X^{pk}

$$\binom{pn}{pk} \equiv \binom{n}{k} \pmod{np\mathbf{Z}_p}.$$

On the other hand, when j is not divisible by p, the coefficient of X^j satisfies

$$\binom{np}{j} \equiv 0 \pmod{np\mathbf{Z}_p}.$$

This last congruence was obvious a priori, since

$$\binom{np}{j} = \frac{np}{j}\binom{np-1}{j-1} \in np\mathbf{Z}_p.$$

3.4. *Second Form of the Theorem*

Let us give a closely related form of the mean value theorem for series converging in $M \subset A \subset K$. Assume that $f(t) = \sum_{k \geq 0} a_k t^k$ converges whenever $|t| < 1$. More precisely, let us assume that the coefficients $a_k \in K$ satisfy

$$|a_k| r^k \to 0 \text{ for all positive real numbers } r < 1.$$

The variable t itself can be taken in the field K or any extension of this field, e.g., in $\mathbf{C}_p$. When $\tau \in \mathbf{M}_p$ we can consider the restricted series $f_\tau \in \mathbf{C}_p\{X\}$ defined by

$$f_\tau(X) = \sum_{k \geq 0} a_k \tau^k X^k.$$

If $|t| < 1$ we have $f(t) = f_\tau(t/\tau)$ as soon as the element $\tau \in \mathbf{M}_p$ is chosen such that $|t| \leq |\tau| < 1$. Obviously;

$$f'(t) = \frac{f'_\tau(t/\tau)}{\tau} \quad \text{and} \quad \sup_{|t| \leq |\tau|} |f'(t)| = \frac{\|f'_\tau\|}{|\tau|}.$$

The mean value theorem for the restricted series f_τ now gives

$$\left| f_\tau\left(\frac{t}{\tau} + \frac{h}{\tau}\right) - f_\tau\left(\frac{t}{\tau}\right) \right| \leq \|f'_\tau\| \cdot \left|\frac{h}{\tau}\right|.$$

Equivalently,

$$|f(t+h) - f(t)| \leq |h| \cdot \sup_{|t| \leq |\tau|} |f'(t)|.$$

All this is valid whenever $|h/\tau| \leq r_p$ for some $\tau \in \mathbf{M}_p$. We can find such a $\tau \in \mathbf{M}_p$ exactly when $|h| < r_p$. Let us summarize this result in the case $K = \mathbf{C}_p$, using the notation

$$\|g\|_{<1} := \sup_{|t|<1} |g(t)| \quad (\leq \infty).$$

Theorem. *Let $f \in \mathbf{C}_p[[X]]$ be a formal power series that converges in the open unit ball $\mathbf{M}_p$. Assume that $\|f'\|_{<1} < \infty$. Then we have*

$$|f(t+h) - f(t)| \leq |h| \cdot \|f'\|_{<1}$$

for all $t \in \mathbf{M}_p$ and $|h| < r_p$. ∎

In this case, the mean value theorem holds in $\mathbf{M}_p$ for increments $h \in \mathbf{C}_p$ satisfying $|h| < r_p$ (notice the strict inequality). Examples of this situation are given by power series $f \in \mathbf{Z}_p[[X]]$ (or more generally in $A[[X]]$): $\|f'\|_{<1} \leq 1$. Take for instance $f = \sum_{k \geq 0} X^k = 1/(1-X)$. For all $|t| < 1$ we have $|f(t+h) - f(t)| = |h|/|(1-t)(1-t-h)| = |h|$ $(t,\ h \in \mathbf{M}_p)$, simply since $|1-t| = |1-t-h| = 1$ *the strongest wins!* Here $|f'(t)| = 1/|1-t|^2 = 1$ and $\|f'\|_{<1} = 1$.

3.5. A Fixed-Point Theorem

Theorem. *Let K be a finite extension of $\mathbf{Q}_p$, $R = B_{\leq 1}(K)$ its closed unit ball, and $f \in K\{T\}$ a restricted formal power series with $\|f\| \leq 1$. Assume*

$$\|f'\| < 1 \text{ and } \inf_{x \in R} |f(x) - x| \leq r_p = |p|^{1/(p-1)}.$$

Then f has a fixed point in R.

Proof. The function f defines a continuous map from the unit ball R of K into itself. Since $|K^\times|$ is discrete in $\mathbf{R}_{>0}$ (we are assuming that the field K is a *finite* extension of $\mathbf{Q}_p$), there is a point $x_0 \in R$ with $|f(x_0) - x_0| \leq r_p$. Define inductively $x_{n+1} = f(x_n)$ for $n \geq 0$. By the mean value theorem,

$$\begin{aligned} |x_{n+1} - x_n| &= |f(x_n) - f(x_{n-1})| \\ &\leq |x_n - x_{n-1}| \cdot \|f'\| \leq |x_n - x_{n-1}|, \end{aligned}$$

and we see by induction that $|x_{n+1} - x_n| \leq r_p$ $(n \geq 0)$. If $x_n = x_{n-1}$ for one positive n, we are done. Otherwise, $|x_n - x_{n-1}| \neq 0$ for all positive integers, and as before,

$$\begin{aligned} |x_{n+1} - x_n| &= |f(x_n) - f(x_{n-1})| \\ &\leq |x_n - x_{n-1}| \cdot \|f'\| < |x_n - x_{n-1}| \leq r_p. \end{aligned}$$

The strictly decreasing sequence $|x_{n+1} - x_n|$ in the discrete subgroup $|K^\times| \subset \mathbf{R}_{>0}$ has to tend to 0: (x_n) is a Cauchy sequence. The limit of this sequence is a fixed point of f in R. ∎

Comment. To show that the hypotheses are necessary, let us consider the function $f(T) = T^p + 1 \in \mathbf{Q}_p[T] \subset \mathbf{Q}_p\{T\}$. We have

$$f'(T) = pT^{p-1}, \quad \|f'\| = \tfrac{1}{p} \leq r_p < 1.$$

This function f has no fixed point in the unit ball $\mathbf{Z}_p$ of $\mathbf{Q}_p$. In fact,

$$f(x) - x = x^p - x + 1 \equiv 1 \pmod{p} \quad (x \in \mathbf{Z}_p)$$

so that $\inf_{x \in \mathbf{Z}_p} |f(x) - x| = 1$, and the second assumption of the theorem is not satisfied. (However, $x^p - x + 1 = 0$ certainly has a root in a suitable finite extension of $\mathbf{Q}_p$, and f has a fixed point in the unit ball of such an extension.)

3.6. *Second-Order Estimates*

Let us keep the notation of the general mean value theorem (3.2).

Theorem. *We have*

$$|f(t+h) - f(t) - f'(t)h| \leq |h^2/2| \cdot \|f''\|$$

whenever $t, h \in K$ satisfy $|t| \leq 1$ and

$$|h| \leq |\sqrt{2}| \text{ if } p = 2, \quad |h| \leq |p|^{1/(p-2)} \text{ if } p \text{ is an odd prime.}$$

PROOF. As in (3.2), it is enough to prove this theorem for polynomials. Let us write the Taylor series of f at the point t:

$$\begin{aligned} f(t+h) - f(t) - f'(t)h &= \sum_{k \geq 2} h^k \cdot D^k f(t)/k! \\ &= h^2 \sum_{k \geq 2} \frac{h^{k-2}}{k!} \cdot D^{k-2} f''(t) \\ &= h^2 \sum_{k \geq 2} \frac{h^{k-2}}{k(k-1)} \cdot \frac{D^{k-2} f''}{(k-2)!}(t). \end{aligned}$$

As in the proof of the mean value theorem (3.2) we have

$$\left\| \frac{D^{k-2} f''}{(k-2)!} \right\| \leq \|f''\| \quad (k \geq 2), \quad \text{and} \quad \left\| \frac{D^{k-2} f''}{(k-2)!} \right\| \to 0 \quad (k \to \infty).$$

(1) For $p \neq 2$ it only remains to prove

$$\left| \frac{h^{k-2}}{k(k-1)} \right| \leq 1 \quad (k \geq 2).$$

For $k = p$ this requires $|h^{p-2}| \leq |p|$, which is the condition given in the theorem. When it is satisfied, if $\nu = \mathrm{ord}_p(k) \geq 1$, we have $k \geq p^\nu$ and $|k-1| = 1$, whence

$$\left| \frac{h^{k-2}}{k(k-1)} \right| \leq \frac{|h|^{p^\nu - 2}}{|p|^\nu} \leq |p|^e$$

with an exponent

$$e = \frac{p^\nu - 2}{p - 2} - \nu \geq \frac{p^\nu - 1}{p - 1} - \nu \geq 0$$

(the linear fractional transformation $x \mapsto \frac{p^\nu - x}{p - x}$ increases when $x \leq p$ — vertical asymptote — since for $x = 0$ it takes a value $p^{\nu-1} \geq 1$ above the horizontal asymptote). In the case $\nu = \mathrm{ord}_p(k-1) \geq 1$, we have $k \geq p^\nu + 1$, and the preceding estimates are satisfied. Finally, when $\mathrm{ord}_p(k-1) = \mathrm{ord}_p(k) = 0$, we have $|k(k-1)| = 1$, and the proof is complete.

(2) In the case $p = 2$, we take a factor $h^2/2$ in front of the above Taylor expansion, and it only remains to prove

$$\left| \frac{h^{k-2}}{k(k-1)/2} \right| \leq 1 \quad (k \geq 2)$$

for $|h| \leq |\sqrt{2}|$. For $k = 4$ this already requires $|h^2/2| \leq 1$, which furnishes the restriction $|h| \leq |\sqrt{2}|$. Conversely, assume that this condition is satisfied. For $\nu = \mathrm{ord}_2(k) \geq 1$,

$$k \geq 2^\nu \text{ and } |k(k-1)/2| = |2|^{\nu-1},$$

whence

$$\left| \frac{h^{k-2}}{k(k-1)/2} \right| \leq \frac{|h|^{2^\nu - 2}}{|2|^{\nu-1}} \leq |2|^{2^{\nu-1}-1-(\nu-1)} = |2|^e.$$

Since we are assuming $\nu \geq 1$, the exponent e is equal to $2^{\nu-1} - \nu \geq 0$. One can treat the case $\nu = \mathrm{ord}_2(k-1) \geq 1$ in a similar way. ∎

Comment. The condition on the absolute value of the increment $|h|$ is less restrictive for $p = 2$, but the inequality is also weaker in this case, since the denominator 2 in $|h^2/2|$ is important (it is irrelevant for odd primes p).

Corollary. *Let K be a finite extension of $\mathbf{Q}_p$, R its ring of integers with maximal ideal P. For $n \in \mathbf{N}$ (or even $n \in \mathbf{Z}_p$), we have*

$$(1 + x)^n \equiv 1 + nx \pmod{pnxR}$$

as soon as $x \in 2pR$.

Proof. We take $f(T) = (1 + T)^n$, so that

$$f''(T) = n(n-1)(1+T)^{n-2} \quad (n \geq 2),$$

$$\|f''\| = |n(n-1)| \quad (n \geq 0).$$

For $|x| \le |p|^{1/(p-2)}$ (resp. $\le |\sqrt{2}|$ if $p = 2$) we have

$$|(1+x)^n - 1 - nx| \le \left|\frac{x^2}{2}\right| \cdot \|f''\| \le |nx| \cdot \left|\frac{x}{2}\right|.$$

Since $|x/2| \le |p|$ when $x \in 2pR$, the preceding inequality furnishes the expected statement. ∎

This corollary gives the fourth form of the fundamental inequality, mentioned in (III.4.3).

4. The Exponential and Logarithm

4.1. Convergence of the Defining Series

Theorem. *The series $\sum_{k\ge1}(-1)^{k-1}x^k/k$ converges precisely when $|x| < 1$. The series $\sum_{k\ge0} x^k/k!$ converges precisely when $|x| < r_p = |p|^{1/(p-1)}$.*

PROOF. Since $|k| = 1$ for all integers k prime to p, in order to have convergence of the first series, the condition $|x^k/k| \to 0$ implies $|x| < 1$. Conversely, when $|x| < 1$,

$$\left|\frac{x^k}{k}\right| \le k|x|^k \to 0 \quad (k \to \infty).$$

For the second series

$$\left|\frac{x^k}{k!}\right| = |x|^k|p|^{-\mathrm{ord}_p(k!)} = |p|^{k\cdot\mathrm{ord}_p x - \mathrm{ord}_p(k!)},$$

we use (3.1) for the p-adic valuation of factorials. The exponent is

$$k\left(\mathrm{ord}_p(x) - \frac{1}{p-1}\right) + \frac{S_p(k)}{p-1}.$$

Since $S_p(k) = 1$ when $k = p^j$ $(j \ge 0)$ is a power of p, we have

$$\left|\frac{x^k}{k!}\right| \to 0 \quad \Longleftrightarrow \quad k\left(\mathrm{ord}_p(x) - \frac{1}{p-1}\right) \to \infty,$$

and this happens precisely when $\mathrm{ord}_p(x) - \frac{1}{p-1} > 0$, namely when

$$\mathrm{ord}_p(x) > \frac{1}{p-1}, \qquad |x| < |p|^{1/(p-1)}$$

as asserted. ∎

By analogy with the classical case, we shall write

$$\log(1+x) = \sum_{k\geq 1} (-1)^{k-1} \frac{x^k}{k},$$

$$e^x = \exp(x) = \sum_{k\geq 0} \frac{x^k}{k!}$$

for the sums of these series whenever they converge. Strictly speaking, we should mention the dependence on the prime p and, for example, write $\log_p(1+x)$ and $\exp_p(x)$.

Comments. (1) In the p-adic domain, the exponential function is not an entire function: The convergence of the exponential series is limited by the radius r_p:

$$r_2 = \tfrac{1}{2} \quad \text{and} \quad \tfrac{1}{p} < r_p < 1 \quad (p \text{ odd prime}),$$

which we have already encountered as a limitation for the increments in the mean value theorem (3.2) and (3.4). A heuristic explanation for this apparent coincidence is furnished by the Taylor series, when expressed in terms of the differential operator $D = d/dx$. Quite formally, we have

$$f(x+h) = \sum_{k\geq 0} \frac{h^k D^k f(x)}{k!} = \exp(hD)(f)(x).$$

On polynomial functions, or more generally on restricted power series, we have seen that $\|D\| = 1$, so that the series for $\exp(hD)$ converges for $|h| < r_p$ (as we have just seen). However, observe that the first form (3.2) of the mean value theorem holds even up to $|h| \leq r_p$. In the classical case, the exponential is an entire function, and there is no limitation for the size of the increments in the mean value theorem.

(2) Since $|x| < r_p < 1$ is required for convergence of the exponential, there is no number $e = \exp(1)$ defined in $\mathbf{Q}_p$. For $p \geq 3$, however, $r_p > 1/p = |p|$, and $\exp(p)$ is well-defined by the series: One could select a definition of a number $e = e_p$ as a pth root of $\exp(p)$. Similarly, when $p = 2$ one could define e as fourth root of $\exp(4)$. However, there is no canonical choice for these roots.

(3) The series defining the functions log and exp have rational coefficients. Hence for each complete extension L of $\mathbf{Q}_p$,

$$x \in B_{<1}(L) \Longrightarrow \log(1+x) \in L,$$

$$x \in B_{<r_p}(L) \Longrightarrow e^x = \exp(x) \in L.$$

4.2. *Properties of the Exponential and Logarithm*

Proposition 1. *For $|x| < r_p$ we have*

$$|\log(1+x)| = |x|, \quad |\exp(x)| = 1, \quad |1 - \exp(x)| = |x|.$$

PROOF. For $k \geq 1$ we have $S_p(k) \geq 1$ and hence $\text{ord}_p(k!) \leq (k-1)/(p-1)$. We infer

$$|k| \geq |k!| \geq |p|^{\frac{k-1}{p-1}} = r_p^{k-1},$$

$$|x^k/k| \leq |x^k/k!| \leq (|x|/r_p)^{k-1} \cdot |x| < |x| < 1$$

for $k \geq 2$ and $0 < |x| < r_p$. Hence the absolute values of the terms in the series

$$1 + x + \sum_{k \geq 2} \frac{x^k}{k!} = \exp(x),$$

$$x + \sum_{k \geq 2} (-1)^{k-1} \frac{x^k}{k} = \log(1+x)$$

are strictly smaller than the first ones. By the ultrametric character of the absolute value, the strongest (we underline it!) wins:

$$\exp(x) = \underline{1} + x + \sum_{k \geq 2} \cdots \Longrightarrow |\exp(x)| = 1,$$

$$\exp(x) - 1 = \underline{x} + \sum_{k \geq 2} \cdots \Longrightarrow |\exp(x) - 1| = |x|,$$

$$\log(1+x) = \underline{x} + \sum_{k \geq 2} \cdots \Longrightarrow |\log(1+x)| = |x|$$

if $|x| < r_p$. ■

Corollary. *The only zero of* $\log(1+x)$ *in the ball* $|x| < r_p$ *is* $x = 0$. ■

In fact, we shall prove a stronger result:

$x \mapsto \log(1+x)$ *is injective in the ball* $B_{<r_p}$.

Proposition 2. *For two indeterminates X and Y, we have the following formal identities:*

$$\exp(X+Y) = \exp(X) \cdot \exp(Y),$$
$$\log \exp(X) = X,$$
$$\exp \log(1+X) = 1 + X.$$

PROOF. The first identity is easily obtained if we observe that the product of two monomials $X^i/i!$ and $Y^j/j!$ is

$$\frac{X^i Y^j}{i!j!} = \binom{i+j}{i} \frac{X^i Y^j}{(i+j)!}.$$

Grouping the terms with $i + j = n$ leads to a sum $(X+Y)^n/n!$. Let us turn to the second identity. In the series $\log(1+X) = \sum_{n\geq 1} a_n X^n$ we would like to substitute $X = e^Y - 1 = b_1 Y + b_2 Y^2 + \cdots$ ($b_1 = 1$). We have to expand the following expression and group the powers of Y:

$$\sum_{n\geq 1} a_n \left(b_1 Y + b_2 Y^2 + \cdots\right)^n = \sum_{n\geq 1} c_n Y^n.$$

Here are the first coefficients:

$$c_1 = a_1 b_1, \quad c_2 = a_1 b_2 + a_2 b_1^2, \quad c_3 = a_1 b_3 + a_2 \cdot 2 b_1 b_2 + a_3 b_1^3.$$

More generally, we see that

$$c_n = a_1 b_n + a_2(\cdots) + \cdots + a_{n-1}(\cdots) + a_n b_1^n.$$

For $2 \leq j \leq n-1$ the coefficient of a_j is a polynomial in $b_1, \ldots, b_{n-1}$ with integral coefficients (of total degree j). The problem is to evaluate the polynomials c_n at the rational values

$$a_n = \frac{(-1)^{n-1}}{n}, \quad b_n = \frac{1}{n!} \quad (n \geq 1).$$

The result of this computation is known: Identical computations are classically made for the substitution of the real-valued power series $x = e^y - 1$ in the real-valued $\log(1+x)$ (convergent for $|x| < 1$). But it is established in any calculus course that the result is $\log(e^y) = y$. Hence all evaluations of the polynomials c_n vanish for $n \geq 2$, and the expected formula is proved. The third identity is treated similarly. ∎

Remark. The preceding proof is surprising: It relies on real analysis for a purely formal result that is applied to p-adic series. It was our purpose to deal with the exponential and logarithm function in an elementary way — before treating power series systematically — and thus we had to give an ad hoc proof for this inversion. But a more systematic treatment of formal power series will give us an opportunity to present an independent proof of this property with no reference to real analysis (VI.1).

Proposition 3. *For $|x| < r_p$ and $|y| < r_p$ we have*

$$\exp(x+y) = \exp(x) \cdot \exp(y),$$
$$\log \exp(x) = x,$$
$$\exp \log(1+x) = 1 + x.$$

Proof. Observe first that if a_n and $b_n \to 0$, then the family $(a_n b_m)_{n,m\geq 0}$ is summable. In particular, its sum is independent of the way terms are grouped before summing. Hence the first identity holds as soon as the variables x and y are in the domain of convergence of the p-adic exponential

$$\exp(x) \cdot \exp(y) = \exp(x+y).$$

Let us check the second identity: We have to show that it is legitimate to substitute a value $x \in \mathbf{C}_p$, $|x| < r_p$ in the formal identity

$$X = \log e^X = \log(1 + e(X)),$$

where

$$e(X) = \sum_{n \geq 1} \frac{X^n}{n!} = e^X - 1.$$

The substitution in the sum can be made by addition of two contributions:

$$x = \left[\sum_{n \leq N} \frac{(-1)^{n-1}}{n} e(X)^n\right]_{X=x} + \left[\sum_{m>N} \frac{(-1)^{m-1}}{m} e(X)^m\right]_{X=x}.$$

In the first *finite* sum, the substitution can obviously be made in each term according to

$$e(X)^n|_{X=x} = \left(x + \frac{x^2}{2!} + \cdots\right)^n = e(x)^n \quad (|x| < r_p).$$

Since $|e(x)| = |x| < r_p < 1$, we have

$$\sum_{n \leq N} \frac{(-1)^{n-1}}{n} e(x)^n \to \log(1 + e(x)) = \log e^x \quad (N \to \infty).$$

The proof of the second identity will be completed if we show that the second contribution is arbitrarily small (for large N). But when $|x| < r_p$, each monomial appearing in the computation of $e(x)$ satisfies $|x^i/i!| < r_p$ (because $i \geq 1$), and each monomial appearing in the computation of $e(x)^m$ has an absolute value less than r_p^m. All individual monomials appearing in the evaluation of the second contribution $\sum_{m>N} \cdots$ have an absolute value smaller than

$$\sup_{m>N} \left|\frac{(-1)^{m-1}}{m}\right| r_p^m.$$

Since the power series for the logarithm converges, it is possible to choose N large enough to ensure that all $|1/m| r_p^m$ $(m > N)$ are arbitrarily small and that the same holds for their sum (independently of the groupings made to compute it). Again, the verification of the third identity is similar. ∎

Corollary. (*a*) *The exponential map defines an isometric homomorphism*

$$\exp : B_{<r_p} \xrightarrow{\sim} B_{<r_p}(1) = 1 + B_{<r_p} \subset \mathbf{C}_p^\times.$$

(*b*) *The homomorphism* $\log : 1 + \mathbf{M}_p \to \mathbf{C}_p$ *is surjective.*

Proof. (*a*) The fact that the exponential map is injective in its domain of definition results from the equality $\log e^x = x$. Better still, the exponential is an isometry:

$$|e^x - e^y| = |e^y||e^{x-y} - 1| = |e^{x-y} - 1| = |x - y|.$$

The inverse of the isometry

$$B_{<r_p} \xrightarrow{\sim} 1 + B_{<r_p}$$

is the restriction of the logarithm to the ball $B_{<r_p}(1) \subset 1 + \mathbf{M}_p$. In particular, we have

$$\log(1+x)(1+y) = \log(1+x) + \log(1+y) \quad (x, y \in B_{<r_p}).$$

But the power series

$$f(x, y) = \log(1+x)(1+y) - \log(1+x) - \log(1+y) = \sum_{n,m \geq 0} a_{mn} x^m y^n$$

converges for $|x| < 1$ and $|y| < 1$. Since

$$m!n!a_{mn} = \left. \frac{\partial^{m+n} f}{\partial^m x \partial^n y} \right|_{(x,y)=(0,0)} = 0,$$

we conclude that the logarithm is a homomorphism in its ball of convergence:

$$\log(1+x)(1+y) = \log(1+x) + \log(1+y) \quad (x, y \in B_{<1}).$$

(*b*) If $x \in \mathbf{C}_p$, choose a sufficiently large integer n in order to ensure that $|p^n x| < r_p$. Hence

$$p^n x = \log \exp p^n x, \quad x = \log \xi$$

for a p^nth root $\xi \in 1 + \mathbf{M}_p$ of $\exp p^n x \in 1 + \mathbf{M}_p$ (III.4.5). ■

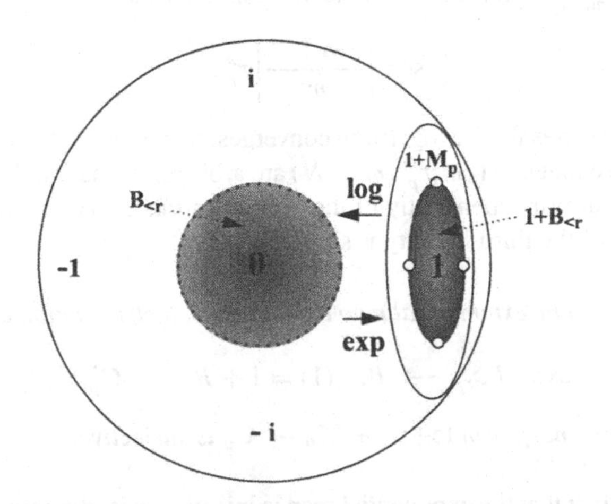

The unit ball in $\mathbf{C}_5$

Theorem. *The logarithm defines a homomorphism* $\log : 1+\mathbf{M}_p \to \mathbf{C}_p$. *Its kernel is the subgroup* μ_{p^∞}. *Its restriction to* $1 + B_{<r_p}$ *is an isometry (hence injective).*

PROOF. There only remains to establish the statement concerning the kernel of the logarithm. Let $x = 1+t \in 1+\mathbf{M}_p$ be in this kernel. We know that $x^{p^n} \to 1$ (when $n \to \infty$) (III.4.5: Proposition 2). Take n large enough so that $|x^{p^n} - 1| < r_p$. Since x^{p^n} is still in the kernel, we now have $x^{p^n} = 1$ by the corollary of the first proposition. ∎

4.3. Derivative of the Exponential and Logarithm

The exponential and logarithm are strictly differentiable functions in their disk of convergence (2.4), and

$$[\exp x]' = \sum_{k\geq 1} \frac{kx^{k-1}}{k!} = \sum_{k\geq 1} \frac{x^{k-1}}{(k-1)!} = \exp x,$$

$$[\log(1+x)]' = \sum_{k\geq 1} (-1)^{k-1} \frac{kx^{k-1}}{k} = \sum_{k\geq 1} (-1)^{k-1} x^{k-1} = \frac{1}{1+x}.$$

Proposition. *Let* $t \in \mathbf{M}_p$. *Then, the derivative of*

$$x \mapsto (1+t)^x : \mathbf{Z}_p \to \mathbf{C}_p$$

at the origin is $\log(1+t)$.

PROOF. By definition, we have a Mahler expansion

$$(1+t)^x = \sum_{k\geq 0} t^k \binom{x}{k} \quad (x \in \mathbf{Z}_p),$$

since $t \in \mathbf{M}_p \subset \mathbf{C}_p$. We deduce

$$(1+t)^x - 1 = \sum_{k\geq 1} t^k \binom{x}{k} = \sum_{k\geq 1} \frac{x}{k} \binom{x-1}{k-1} \cdot t^k,$$

$$\frac{(1+t)^x - 1}{x} = \sum_{k\geq 1} \binom{x-1}{k-1} \frac{t^k}{k},$$

$$\frac{(1+t)^x - 1}{x} - \log(1+t) = \sum_{k\geq 2} \left(\binom{x-1}{k-1} - (-1)^{k-1} \right) \frac{t^k}{k}.$$

When $|t| < 1$ we know that $t^k/k \to 0$, whereas

$$\left|\binom{x-1}{k-1} - (-1)^{k-1}\right| \le 1, \quad \binom{x-1}{k-1} \to (-1)^{k-1} \text{ when } x \to 0 \quad (k \ge 1).$$

This proves that

$$\frac{(1+t)^x - 1}{x} \to \log(1+t) \quad (x \to 0 \text{ in } \mathbf{Z}_p)$$

uniformly in t on any disk $B_{<r} \subset \mathbf{C}_p$ of radius $r < 1$. ∎

Comment. We can write

$$(d/dx)_{x=0}(1+t)^x = \log(1+t),$$

where the derivative is the limit of difference quotients taken with respect to increments in $\mathbf{Z}_p$. When $\log(1+t) \ne 0$ and $|h|$ is small enough, we have

$$\left|\frac{(1+t)^h - 1}{h}\right| = |\log(1+t)|.$$

This provides an improvement of the second form of the fundamental inequalities (III.4.3), in the region $r_p < |t| < 1$. But $|\log(1+t)|$ is arbitrarily large in this region, since $\log : 1 + \mathbf{M}_p \to \mathbf{C}_p$ is surjective (4.2).

4.4. *Continuation of the Exponential*

It is natural to try to construct a homomorphism

$$f : \mathbf{C}_p \to \mathbf{C}_p^\times$$

extending the exponential defined above by a series expansion. If such an extension exists, when $x \in \mathbf{C}_p$ we can choose a high power p^n of p so that $p^n x \in B_{<r_p}$ (the exponent n depends on x) and then

$$f(x)^{p^n} = f(p^n x) = \exp(p^n x).$$

In other words, $f(x)$ has to be a p^nth root of $\exp p^n x$ in the algebraically closed field $\mathbf{C}_p$. This can be done in a coherent way, thus furnishing a continuation of the exponential homomorphism.

Proposition. *There is a continuous homomorphism* $\mathrm{Exp} : \mathbf{C}_p \to 1 + \mathbf{M}_p$ *extending the exponential mapping, originally defined only on the ball* $B_{<r_p} \subset \mathbf{C}_p$.

PROOF. Recall that $1 + \mathbf{M}_p$ is a divisible group (III.4.5), and divisible groups are *injective* **Z**-*modules* (III.4.1), and hence enjoy an extension property for homomorphisms defined over subgroups. We can use this for the homomorphism

$$\exp : B_{<r_p} \longrightarrow 1 + B_{<r_p} \subset 1 + \mathbf{M}_p,$$

since its target is the divisible group $1 + \mathbf{M}_p$. ∎

For this corollary, only the p-divisibility of $1 + \mathbf{M}_p$ is used.

The usefulness of the extensions Exp is limited by the fact that none are canonical. However, since the logarithm is defined on the image of any extension, the composite $\log \circ \mathrm{Exp} : \mathbf{C}_p \to \mathbf{C}_p$ has a meaning. If $x \in \mathbf{C}_p$, let us choose an integer n such that $p^n x \in B_{<r_p}$, and consider the following equalities:

$$p^n \log \circ \mathrm{Exp}(x) = \log(\mathrm{Exp}\ x)^{p^n} = \log(\mathrm{Exp}(p^n x))$$
$$= \log(\exp(p^n x)) = p^n x.$$

Consequently, $\log \circ \mathrm{Exp}(x) = x$. We have obtained the following result.

Corollary. *The* $\log : 1 + \mathbf{M}_p \to \mathbf{C}_p$ *is inverse to all extensions*

$$\mathrm{Exp} : \mathbf{C}_p \to 1 + \mathbf{M}_p,$$

and any such extension is injective. ∎

Let us summarize the construction in a diagram of homomorphisms (of abelian groups):

$$\begin{array}{ccccc} \mathbf{C}_p & \stackrel{\mathrm{Exp}}{\longrightarrow} & 1+\mathbf{M}_p & \stackrel{\log}{\longrightarrow} & \mathbf{C}_p \\ \cup & & \cup & & \cup \\ B_{<r_p} & \stackrel{\exp}{\longrightarrow} & 1 + B_{<r_p} & \stackrel{\log}{\longrightarrow} & B_{<r_p}. \end{array}$$

Both composite arrows are identities.

4.5. *Continuation of the Logarithm*

Proposition. *There is a unique function $f : \mathbf{C}_p^\times \to \mathbf{C}_p$ having the properties*

(1) *f is a homomorphism: $f(xy) = f(x) + f(y)$,*
(2) *The restriction of f to $1 + \mathbf{M}_p = B_{<1}(1)$ coincides with the logarithm defined by its series expansion,*
(3) *$f(p) = 0$ (normalization).*

Proof. Let us start with the *uniqueness* statement. By (III.4.2), the subgroups $p^{\mathbf{Q}}\mu_{(p)}$ and $1+\mathbf{M}_p$ generate $\mathbf{C}_p^\times$. Hence it is enough to see that the given conditions imply that f vanishes on $p^{\mathbf{Q}}\mu$. This is obvious on the subgroup μ, since the field $\mathbf{C}_p$ (of characteristic 0) has no additive torsion. On the other hand, if $x \in p^{\mathbf{Q}}$, there will be an equation $x^a = p^b$ (with some integers a and b). Hence $af(x) = bf(p) = 0$

by the normalization condition, and $f(x) = 0$ as expected. For the *existence* part, it is enough to define f trivially by 0 on $p^{\mathbf{Q}}\mu$ observing that this definition is coherent with the logarithm on the intersection $p^{\mathbf{Q}}\mu \cap (1 + \mathbf{M}_p) = \mu_{p^\infty}$. But the subgroup μ_{p^∞} is precisely the kernel of the logarithm series (Theorem in (4.2)). ∎

The preceding continuation of the log function is called the *Iwasawa logarithm* and is denoted by Log.

Theorem. *The Iwasawa logarithm has the following properties:*

(1) *It is locally analytic: In the neighborhood of any $a \neq 0$*

$$\operatorname{Log} x = \operatorname{Log} a + \sum_{k \geq 1} \frac{(-1)^{k-1}}{k} \left(\frac{x-a}{a} \right)^k \qquad (|x - a| < |a|).$$

(2) *For $x \in \mathbf{Z}_p^\times$,*

$$\operatorname{Log} x = \frac{1}{1-p} \sum_{k \geq 1} \frac{(1 - x^{p-1})^k}{k}.$$

(3) *For any complete subfield K of $\mathbf{C}_p$, $\operatorname{Log}(K^\times) \subset K$.*
(4) *For every continuous automorphism σ of $\mathbf{C}_p$,*

$$\operatorname{Log}(x^\sigma) = (\operatorname{Log} x)^\sigma.$$

Proof. (1) When $a \neq 0$ let us simply write $x = a(1 + (x-a)/a)$ and

$$\operatorname{Log} x = \operatorname{Log} a + \log\left(1 + \frac{x-a}{a}\right),$$

so that the asserted expansion follows.

(2) If $x \in \mathbf{Z}_p^\times$, we have $x^{p-1} \equiv 1 \pmod{p}$ and

$$\operatorname{Log} x = \frac{1}{p-1} \operatorname{Log}(x^{p-1}) = \frac{1}{p-1} \log\left(1 + (x^{p-1} - 1)\right).$$

The series expansion is applicable to the last term and furnishes the announced expression.

(3) If $K \supset A \supset M$ and $x \in K^\times$, let us write

$$x = p^r \cdot \zeta \cdot u \quad (r \in \mathbf{Q},\ \zeta \in \mu_{(p)},\ u \in 1 + M),$$

so that $\operatorname{Log} x = \log u$. Hence we can find integers n and m with $x^n = p^m v$, where $v \in 1 + M$, and hence $n \operatorname{Log} x = \operatorname{Log} v$. Since $x^n \in K$, we have $v \in K$. Now, the coefficients of the series defining the logarithm are rational and K is complete: $\log v \in K$, and finally $\operatorname{Log} x = (\log v)/n \in K$.

(4) Under the stated conditions, we can consider the homomorphism

$$f : \mathbf{C}_p^\times \to \mathbf{C}_p, \quad x \mapsto \sigma^{-1}\operatorname{Log}(x^\sigma).$$

We know that $|x^\sigma| = |x|$ by uniqueness of extensions of absolute values first on $\mathbf{Q}_p^a$ (II.3.3) and then also on $\mathbf{C}_p$ by continuity. Since the coefficients of $\log(1+x)$ are rational numbers, the second condition of the proposition is verified by f. Hence f satisfies the three characteristic properties of the Iwasawa logarithm: It must coincide with it. ∎

Comments. (1) The product of the subgroups μ_{p^∞} and $1+B_{<r_p}$ is a direct product, since the intersection of these subgroups is trivial. But this product is not equal to $1+\mathbf{M}_p$. Indeed, the logarithm of an element in the product is in the ball $B_{<r_p}$,

$$\log\left(\mu_{p^\infty}\cdot(1+B_{<r_p})\right) = B_{<r_p},$$

whereas $\log(1+\mathbf{M}_p) = \mathbf{C}_p$. A different way of seeing this consists in observing that

$$1 \neq \zeta = 1+\xi \in \mu_{p^\infty} \Longrightarrow |\xi| = |p|^{1/\varphi(p^f)} \geq r_p$$

(ζ of order p^f) (II.4.4). Taking $x = 1+t \in 1+B_{<r_p}$, hence $|t| < |\xi|$, we have

$$\zeta x = 1+\xi+t+\xi t \quad \text{with } |\xi t| < \min(|\xi|, |t|).$$

Now

$$|\zeta x - 1| = |\xi| = |p|^{1/\varphi(p^f)}$$

has a very particular form. It is clear that we are not obtaining all elements of $1+\mathbf{M}_p$ in this form (in $\mathbf{M}_p$ the p-adic order is an arbitrary positive rational number).

(2) The rationality property of the logarithm shows that for every *finite* extension K of $\mathbf{Q}_p$ the logarithm furnishes an isometric isomorphism

$$\log : 1+B_{<r_p}(K) \xrightarrow{\sim} B_{<r_p}(K).$$

In particular,

$$\log : 1+p\mathbf{Z}_p \xrightarrow{\sim} p\mathbf{Z}_p \quad (\cong \mathbf{Z}_p) \quad (p \text{ odd prime}),$$

$$\log : 1+4\mathbf{Z}_2 \xrightarrow{\sim} 4\mathbf{Z}_2 \quad (\cong \mathbf{Z}_2).$$

In general, with the conventional notation

$$K \supset R \supset P = \pi R,$$

the multiplicative subgroup $1+P$ still contains its torsion subgroup (p-torsion) as a direct factor. Let m be the largest integer such that K has a p^mth root of unity: This torsion subgroup is μ_{p^m}; it is cyclic. One can show that there is an isomorphism

$$1+P \xrightarrow{\sim} \mathbf{Z}/p^m\mathbf{Z} \times R.$$

This results from the structure theorem for finitely generated $\mathbf{Z}_p$-modules: The ranks of the multiplicative $\mathbf{Z}_p$-modules

$$1+P, \quad 1+B_{<r_p}(K) \subset 1+P$$

are the same, since the quotient is finite, cyclic of order p^m (cf. A. Weil: *Basic Number Theory*). These results do not extend to *infinite* extensions of $\mathbf{Q}_p$ contained in $\mathbf{C}_p$.

(3) Let us still consider a *finite* extension K of $\mathbf{Q}_p$ in $\mathbf{C}_p$. When $P = B_{\leq r_p}(K)$, both cases

$$\mu_p \subset 1+P \text{ and } \mu_p \not\subset 1+P$$

can occur. For example, for $p=3$, consider as in (II.4.7) the two quadratic extensions $K_1 = \mathbf{Q}_3(\sqrt{-3})$ and $K_2 = \mathbf{Q}_3(\sqrt{3})$. Since the field $\mathbf{Q}(\sqrt{-3})$ contains a 6th root of unity (the ring of integers of this field is a hexagonal lattice in the complex field) and since $\sqrt{-1} \in \mathbf{Q}_3(\sqrt{3}, \sqrt{-3}) = K_1 \cdot K_2$, we see that $\mu_4 \subset K_1 \cdot K_2$ and necessarily

$$\mu_{12} = \mu_4 \cdot \mu_3 \subset K_1 \cdot K_2.$$

On the other hand, the order of $\mu_{(3)} \cap (K_1 \cdot K_2)$ is $\#(k^\times) = 3^f - 1$, hence of the form 2, 8, 26, ..., and the presence of a fourth root of unity implies that this order is greater than or equal to 8. In particular, $K_1 \neq K_2$. Since $\mathbf{Q}(\sqrt{-3}) = \mathbf{Q}(\zeta_3)$ we see that $\zeta_3 \in K_1$ but $\zeta_3 \notin K_2$. Nevertheless, quite generally,

$$\mu_{p^\infty}(K) = \mu_{p^\infty} \cap K \quad (\subset (1+\mathbf{M}_p) \cap K = 1+P).$$

If $P = \pi R$, then $\mu_{p^\infty}(K) \subset 1+B_{\leq|\pi|}(\mathbf{C}_p)$, so that the order of $\mu_{p^\infty}(K)$ is a divisor of the order of

$$\mu_{p^\infty} \cap (1+B_{\leq|\pi|}(\mathbf{C}_p)),$$

which is known, since the absolute values $|\zeta - 1|$ for $\zeta \in \mu_{p^\infty}$ are given by (III.4.2).

(4) Here is a diagram summing up the general situation:

$$\begin{array}{ccccc}
(1) & \hookrightarrow & 1+B_{<r_p} & \underset{\text{exp}}{\overset{\text{log}}{\rightleftarrows}} & B_{<r_p} \\
\cap & & \cap & & \cap \\
\mu_p & \hookrightarrow & 1+B_{\le r_p} & \xrightarrow{\text{log}} & B_{\le r_p} \\
\cap & & \cap & & \cap \\
\mu_{p^\infty} & \hookrightarrow & 1+\mathbf{M}_p & \underset{\text{Exp}}{\overset{\text{log}}{\rightleftarrows}} & \mathbf{C}_p \\
\cap & & \cap & & \cap \\
p^{\mathbf{Q}}\cdot\mu & \hookrightarrow & \mathbf{C}_p^\times & \xrightarrow{\text{Log}} & \mathbf{C}_p.
\end{array}$$

The lines consist of short exact sequences, split by the choice of a section Exp of log. Observe that the subgroup $p^{\mathbf{Q}}\cdot\mu$ is well-defined, independently from the choice of a copy of $p^{\mathbf{Q}} \subset \mathbf{C}_p^\times$.

Note. The possibility of extending the exponential to the whole of $\mathbf{C}_p$ had already been shown by M.-C. Sarmant(-Durix) in her doctoral thesis. We have followed the method of the book by W. Schikhof.

5. The Volkenborn Integral

5.1. Definition via Riemann Sums

Let K be a complete extension of $\mathbf{Q}_p$. We are going to define $\int f\,dx$ for certain functions $f : \mathbf{Z}_p \to K$. Unfortunately, Corollary 3 in (IV.3.5) shows that one cannot define nontrivial translation-invariant linear forms on $C(\mathbf{Z}_p)$. Let us recall this result (notation $f_1(x) = f(x+1)$).

Lemma. *If $\varphi : C(\mathbf{Z}_p) \to K$ is linear and translation-invariant, i.e.,*

$$\varphi(f_1) = \varphi(f) \textit{ for all } f \in C(\mathbf{Z}_p),$$

then $\varphi = 0$. ∎

Observe that we can define translation-invariant linear forms on $\mathcal{F}^{\text{lc}}(\mathbf{Z}_p)$, the space of locally constant functions on $\mathbf{Z}_p$ (IV.2.1). Indeed, we can construct such a linear form with $\varphi(1) = 1$. Translation invariance imposes the same value $1/p$ for the characteristic functions of the cosets of $p\mathbf{Z}_p$. More generally, this translation-invariant linear form should take the same value $1/p^n$ for the characteristic functions of the cosets $p^n\mathbf{Z}_p$. These functions have sup norm 1 but

$$|\varphi(f)| = |\tfrac{1}{p^n}| = p^n \text{ is arbitrarily large.}$$

This shows that this linear form is not continuous. Equivalently, we can define the p-adic "volume" of the balls $B_{\le|p^n|}(j)$ in $\mathbf{Z}_p$ to be

$$m(B_{\le|p^n|}(j)) = \tfrac{1}{p^n} \in \mathbf{Q}_p.$$

The next construction works for *differentiable* functions — at least $(Sf)'(0)$ should exist — and is not translation-invariant. It is convenient to deal with strictly differentiable functions $f \in S^1(\mathbf{Z}_p)$.

Let us start with the expressions

$$\frac{1}{p^n}\sum_{0\le j<p^n} f(j) = \sum_{0\le j<p^n} f(j)\,m(j+p^n\mathbf{Z}_p)$$

representing *Riemann sums* for f. The integral of f over $\mathbf{Z}_p$ will be defined as the limit ($n \to \infty$) of these sums, when it exists. The indefinite sum F of a function f has been defined in (IV.1.5) in order to have $\nabla F = f$ ($F(0) = 0$):

$$F = Sf = 1\underline{*}f: \quad F(k) = \sum_{0\le j<k} f(j).$$

Hence we have

$$\frac{1}{p^n}\sum_{0\le j<p^n} f(j) = \frac{F(p^n) - F(0)}{p^n}$$

(since $F(0) = 0$), and we see that the limit exists if *F is differentiable at the origin.*

When the function f has Mahler coefficients c_n, we know that the coefficients of Sf are simply shifted, and the differentiability of Sf at the origin is equivalent to the requirement

$$|c_{n-1}/n| \to 0 \quad (n \to \infty)$$

(Theorem 1 in (1.5)). This is the case if $f \in S^1(\mathbf{Z}_p)$.

Definition. *The* Volkenborn integral *of a function $f \in S^1(\mathbf{Z}_p)$ is by definition*

$$\int_{\mathbf{Z}_p} f(x)\,dx = \lim_{n\to\infty}\frac{1}{p^n}\sum_{0\le j<p^n} f(j) = (Sf)'(0).$$

If $f = c$ is a constant, then $\int_{\mathbf{Z}_p} f(x)\,dx = c$. Here is a main property of this integral.

Proposition 1. *(a) For $f \in S^1(\mathbf{Z}_p)$ we have*

$$\left|\int_{\mathbf{Z}_p} f(x)\,dx\right| \le p\,\|f\|_1.$$

(b) If $f_n \to f$ in S^1, namely $\|f_n - f\|_1 \to 0$, then

$$\int_{\mathbf{Z}_p} f_n(x)\,dx \to \int_{\mathbf{Z}_p} f(x)\,dx.$$

PROOF. By definition,

$$\left|\int_{\mathbf{Z}_p} f(x)\,dx\right| = |(Sf)'(0)| \leq \|Sf\|_1 = \sup(\|\Phi Sf\|, |Sf(0)|),$$

so that (a) follows from Corollary 2 in (1.5):

$$\|Sf\|_1 \leq p\|f\|_1.$$

(b) is a consequence of (a). ∎

Recall that we use the notation ∇f for a *discrete gradient* of a function f:

$$\nabla f(x) = f(x+1) - f(x).$$

Proposition 2. *For $f \in S^1(\mathbf{Z}_p)$ we have*

$$\int_{\mathbf{Z}_p} \nabla f(x)\,dx = f'(0).$$

PROOF. By definition,

$$\int_{\mathbf{Z}_p} \nabla f(x)\,dx = (S\nabla f)'(0) = (f - f(0))'(0) = f'(0),$$

since $S\nabla f = f - f(0)$ (Proposition 2 of (IV.1.5)). ∎

5.2. Computation via Mahler Series

The indefinite sum of a binomial function $\binom{\cdot}{n}$ is the next one $S\binom{\cdot}{n} = \binom{\cdot}{n+1}$. This observation makes it easy to compute the Volkenborn integral of a function $f \in S^1$ of which the Mahler expansion is known.

Proposition. *Let $\sum_{k\geq 0} c_k \binom{\cdot}{k}$ be the Mahler series of a strictly differentiable function $f \in S^1$. Then*

$$\int_{\mathbf{Z}_p} f(x)\,dx = \sum_{k\geq 0} (-1)^k c_k/(k+1).$$

PROOF. Since $f_n = \sum_{k\le n} c_k\binom{\cdot}{k}$ tends to f in $\|.\|_1$, we can simply integrate term by term:

$$\int_{\mathbf{Z}_p} f(x)\,dx = \sum_{k\ge 0} c_k \int_{\mathbf{Z}_p} \binom{x}{k} dx.$$

Now,

$$\int_{\mathbf{Z}_p} \binom{x}{k} dx = \binom{x}{k+1}'(0)$$

and

$$\binom{x}{k+1} = \frac{x}{k+1}\binom{x-1}{k}$$

implies

$$\binom{x}{k+1}'(0) = \lim_{x\to 0} \frac{1}{k+1}\binom{x-1}{k} = \frac{1}{k+1}\binom{-1}{k} = \frac{(-1)^k}{k+1}$$

(one can also apply directly Theorem 1 in (1.5) to the function Sf). ∎

Example. Let us fix $t \in \mathbf{M}_p \subset \mathbf{C}_p$, namely $|t| < 1$, and consider the function $f = f_t$ defined by

$$f(x) = (1+t)^x = \sum_{k\ge 0} t^k \binom{x}{k}.$$

Then

$$\int_{\mathbf{Z}_p} (1+t)^x\,dx = \sum_{k\ge 0} \frac{(-1)^k t^k}{k+1} = \frac{1}{t}\log(1+t) \quad (= 1 \text{ for } t = 0).$$

5.3. *Integrals and Shift*

A few more formulas for the Volkenborn integral will be useful. Recall that the translation operators τ_x have been defined in (IV.5.1) by

$$\tau_x f(t) = f(x+t).$$

In particular, for $\tau = \tau_1 = E$ (unit translation), $\tau f = f_1$. Let us also denote by D the differentiation operator, ∇ the finite difference operator, and S the indefinite sum. Obviously, D commutes with translations and consequently also with $\nabla = \tau - \mathrm{id}$.

Proposition 1. *Let $P_0 : f \mapsto f(0)\cdot 1$ be the projection of $S^1(\mathbf{Z}_p)$ onto constants. Then the following relations hold:*

(*a*) $S\tau = \tau S - P_0$.
(*b*) *DS commutes with all translations τ_x.*
(*c*) $SD = DS - P_0 DS$.

Proof. By definition, for integers $n \geq 1$,

$$S(\tau f)(n) = \sum_{0 \leq j < n} \tau f(j) = \sum_{0 \leq j < n} f(j+1)$$
$$= \sum_{0 < i \leq n} f(i) = Sf(n+1) - f(0) = \tau Sf(n) - f(0),$$

which proves $S\tau = \tau S - P_0$ (by density of the integers $n \geq 1$ in $\mathbf{Z}_p$ and continuity of the functions in question). On the other hand, differentiation of the function $S\tau f = \tau Sf - f(0)$ leads to $DS\tau f = D\tau Sf = \tau DSf$. Moreover, recall that $\nabla Sf = f$ but $S\nabla f = f - f(0)$ (IV.1.5). In other words,

$$\nabla S = \text{id}, \quad S\nabla = \text{id} - P_0.$$

We infer

$$SD = SD\nabla S = S\nabla DS = DS - P_0 DS.$$

The proposition is proved. ∎

Proposition 2. *Let $f \in S^1(\mathbf{Z}_p)$. Then*

(*a*) $\displaystyle\int_{\mathbf{Z}_p} \tau_x f(t)\, dt = (Sf)'(x)$,

(*b*) $\displaystyle S(f')(x) = \int_{\mathbf{Z}_p} f(x+t)\, dt - \int_{\mathbf{Z}_p} f(t)\, dt.$

Proof. Start with the definition $\int_{\mathbf{Z}_p} f(t)\, dt = (Sf)'(0)$. Hence

$$\int_{\mathbf{Z}_p} f(t+1)\, dt = \int_{\mathbf{Z}_p} \tau f(t)\, dt = (S\tau f)'(0) = DS\tau f(0)$$
$$= \tau DSf(0) = DSf(1) = (Sf)'(1).$$

The first formula (*a*) for a positive integer $x = n$ follows by iteration, and for any $x \in \mathbf{Z}_p$ by continuity and density (alternatively, one can do the same calculations with τ_x in place of τ). Recall now Proposition 2 of (5.1),

$$\int_{\mathbf{Z}_p} \nabla f(t)\, dt = f'(0),$$

and use a translation

$$\int_{\mathbf{Z}_p} \nabla f(t+x)\, dt = f'(x).$$

From this, it is obvious that $S(f')(n) = f'(0) + \cdots + f'(n-1)$ can be expressed as a telescoping sum

$$S(f')(n) = \int_{\mathbf{Z}_p} f(t+n)\,dt - \int_{\mathbf{Z}_p} f(t)\,dt \quad (1 \le n \in \mathbf{N}),$$

and by continuity and density of the positive integers n in $\mathbf{Z}_p$,

$$S(f')(x) = \int_{\mathbf{Z}_p} f(t+x)\,dt - \int_{\mathbf{Z}_p} f(t)\,dt \quad (x \in \mathbf{Z}_p). \qquad \blacksquare$$

Writing $f' = g$, we can choose any *primitive* $G \in S^1(\mathbf{Z}_p)$ of g and write

$$S(g)(x) = \int_{\mathbf{Z}_p} G(x+t)\,dt - \int_{\mathbf{Z}_p} G(t)\,dt.$$

Of course, two different primitives of a function g may differ by any function h having $h' \equiv 0$.

Proposition 3. *Let $f \in S^2(\mathbf{Z}_p) \subset S^1(\mathbf{Z}_p)$ and define $F(x) = \int_{\mathbf{Z}_p} f(x+t)\,dt$. Then $F \in S^1(\mathbf{Z}_p)$ and*

$$F'(x) = \int_{\mathbf{Z}_p} f'(x+t)\,dt.$$

PROOF. By Proposition 2 of (1.3),

$$f \in S^2(\mathbf{Z}_p) \quad \Longrightarrow \quad f' \in S^1(\mathbf{Z}_p),$$

so that

$$G(x) = \int_{\mathbf{Z}_p} f'(x+t)\,dt$$

defines a function $G \in S^1(\mathbf{Z}_p)$. Moreover, by Proposition 2 (*a*),

$$\int_{\mathbf{Z}_p} f'(x+t)\,dt = (Sf')'(x) = (DSDf)(x),$$

which proves $G = DSDf$. Now by the first proposition $SD = DS - P_0DS$ and

$$G = D(DS - P_0DS)f = DDSf = (Sf)'' = F',$$

because $F = (Sf)'$. $\blacksquare$

Proposition 4. *Let σ denote the involution* (I.1.2) $x \mapsto -1 - x$ *of* $\mathbf{Z}_p$. *Then*

$$\int_{\mathbf{Z}_p} (f \circ \sigma)\,dx = \int_{\mathbf{Z}_p} f\,dx.$$

PROOF. We have seen that

$$\int_{\mathbf{Z}_p} f\,dx = (Sf)'(0) = \lim_{h\to 0} \frac{Sf(h)}{h}.$$

Let us take $h = -p^n$ $(n \to \infty)$. Hence

$$(Sf)'(0) = \lim_{n\to\infty} \frac{Sf(-p^n)}{-p^n} = \lim_{n\to\infty} \frac{-Sf(-p^n)}{p^n}.$$

But by the Corollary 4 in (IV.3.5),

$$-Sf(-p^n) = S(f \circ \sigma)(p^n),$$

whence the result $(Sf)'(0) = (S(f \circ \sigma))'(0)$. ∎

Corollary *If f is an odd function, then*

$$\int_{\mathbf{Z}_p} f\,dx = -\frac{f'(0)}{2}.$$

PROOF. Quite generally, using Proposition 2 in (5.1) and Proposition 4, we have

$$f'(0) = \int_{\mathbf{Z}_p} (f(x+1) - f(x))\,dx = \int_{\mathbf{Z}_p} (f(-x) - f(x))\,dx.$$

Now, if f is odd, we obtain the announced result

$$f'(0) = \int_{\mathbf{Z}_p} -2f(x)\,dx.$$

∎

5.4. *Relation to Bernoulli Numbers*

In (5.2), we have proved

$$\int_{\mathbf{Z}_p} (1+t)^x\,dx = \frac{1}{t}\log(1+t)$$

for $|t| < 1$, $t \in \mathbf{C}_p$. Let us now choose $|t| < r_p$ and define $s = \log(1+t)$, so that

$$t = e^s - 1, \quad |s| = |t| < r_p.$$

The preceding formula now reads

$$\int_{\mathbf{Z}_p} e^{sx}\,dx = \frac{s}{e^s - 1}.$$

Classical Definition. *The* Bernoulli numbers *are the rational numbers b_k defined by the following generating function:*

$$\frac{t}{e^t - 1} = \sum_{k \geq 0} b_k \frac{t^k}{k!}.$$

Here are the first few values:

$$b_0 = 1,\ b_1 = -\tfrac{1}{2},\ b_2 = \tfrac{1}{6},\ b_3 = 0,\ b_4 = -\tfrac{1}{30},\ b_5 = 0,$$

and,

$$b_6 = \tfrac{1}{42},\ b_8 = -\tfrac{1}{30},\ b_{10} = \tfrac{5}{66},\ b_{12} = -\tfrac{691}{2730}.$$

Since we can also write $e^{tx} = \sum_{k \geq 0} t^k x^k / k!$ with a convergence in $S^1(\mathbf{Z}_p)$, we can integrate term by term (Proposition 1 in (5.1))

$$\sum \left(\int_{\mathbf{Z}_p} x^k dx \right) \frac{t^k}{k!} = \sum b_k \frac{t^k}{k!}$$

and identify the coefficients

$$b_k = \int_{\mathbf{Z}_p} x^k \, dx.$$

Observe that by definition $b_k \in \mathbf{Q}$, and these integrals are independent of the prime p used to compute them! Also, $|b_k| \leq p \|x^k\|_1 = p$ (still by Proposition 1 in (5.1)), namely $|pb_k| \leq 1$, i.e., $pb_k \in \mathbf{Z}_p \cap \mathbf{Q}$. In (5.5) we shall give a more precise result.

Proposition. *The Volkenborn integral of a restricted series $f = \sum_{n \geq 0} a_n x^n$ exists and can be computed term by term:*

$$\int_{\mathbf{Z}_p} f(x) \, dx = \sum_{n \geq 0} a_n b_n. \qquad \blacksquare$$

Here, the b_n are the Bernoulli numbers, and using $\int_{\mathbf{Z}_p} f(x) \, dx = -f'(0)/2$ for the odd functions $f(x) = x^{2k+1}$ (Corollary at the end of (5.3)), we obtain

$$b_1 = -\tfrac{1}{2}, \quad b_{2k+1} = 0 \quad (k \geq 1).$$

Classical Definition. *The* Bernoulli polynomials B_k *are defined by the following generating function:*

$$\sum_{k \geq 0} B_k(x) \frac{t^k}{k!} = \frac{te^{xt}}{e^t - 1}.$$

I hope that no confusion will arise between the Bernoulli and the Bell polynomials (also denoted by B_n in (IV.6.3)): The context should always explicitly specify which ones are under consideration!

Obviously, $b_k = B_k(0)$. Conversely, the definition

$$\sum_{k\geq 0} B_k(x)\frac{t^k}{k!} = e^{xt}\sum_{k\geq 0} b_k\frac{t^k}{k!}$$

$$= \sum_{j\geq 0, k\geq 0} \frac{x^j t^j}{j!} b_k \frac{t^k}{k!}$$

leads to an explicit expression of the Bernoulli polynomials (with Bernoulli numbers as coefficients):

$$B_n(x) = n! \sum_{j+k=n} b_k \frac{x^j}{j!k!}$$

$$= \sum_{0\leq j\leq n} \binom{n}{j} b_{n-j} x^j.$$

Thus B_n is a monic polynomial of degree n (equal to its index). This expansion is symbolically written "$B_n(x) = (b+x)^n$:" the binomial formula leads to the correct expression, provided that we interpret b^k as the kth Bernoulli number b_k.

Here are a few values:

$$B_0(x) = 1,\ B_1(x) = x - \tfrac{1}{2},\ B_2(x) = x^2 - x + \tfrac{1}{6}.$$

Returning to the Volkenborn integral, we have

$$\sum_{k\geq 0} B_k(x)\frac{t^k}{k!} = e^{tx}\int_{\mathbf{Z}_p} e^{ty}\,dy = \int_{\mathbf{Z}_p} e^{t(y+x)}\,dy.$$

Identification of the coefficients leads to the p-adic expression of the Bernoulli polynomials

$$B_k(x) = \int_{\mathbf{Z}_p} (x+y)^k\,dy.$$

The formula (5.3)

$$\int_{\mathbf{Z}_p} \nabla f(x+y)\,dy = f'(x)$$

for $f = x^k$ leads to

$$\int_{\mathbf{Z}_p} (x+y+1)^k\,dy - \int_{\mathbf{Z}_p} (x+y)^k\,dy = kx^{k-1},$$

namely

$$B_k(x+1) - B_k(x) = kx^{k-1}.$$

In particular, $B_k(1) = B_k(0)$ for $k \geq 2$, and these polynomials may be extended by 1-periodicity on $\mathbf{R}$. We obtain the continuous periodic functions $x \mapsto B_k(x - [x])$ $(k \geq 2)$ on the real line (as usual, $[x]$ denotes the integral part of a real number x so that $0 \leq x - [x] < 1$).

On the other hand, we can expand $(y + x + 1)^n = \sum \binom{n}{k}(y + x)^k$ and hence rewrite

$$(n+1)x^n = (x^{n+1})' = \int_{\mathbf{Z}_p} \sum_{k \leq n} \binom{n+1}{k} (x+y)^k \, dy$$

$$= (n+1)B_n(x) + \sum_{k \leq n-1} \binom{n+1}{k} B_k(x).$$

This gives a recurrence relation for the computation of these monic polynomials:

$$B_n(x) = x^n - \frac{1}{n+1} \sum_{k \leq n-1} \binom{n+1}{k} B_k(x).$$

In particular, for $x = 0$ and $n \geq 1$,

$$b_n = B_n(0) = -\frac{1}{n+1} \sum_{k \leq n-1} \binom{n+1}{k} b_k = -\sum_{k \leq n-1} \binom{n}{k-1} \frac{b_k}{k}.$$

Another relation for the Bernoulli polynomials is easily obtained from the fact that the integrals of f and of $f \circ \sigma$ are the same (Proposition 4 in (5.3)):

$$B_k(1-x) = \int_{\mathbf{Z}_p} (y + 1 - x)^k \, dy$$

$$= \int_{\mathbf{Z}_p} (-1 - y + 1 - x)^k \, dy$$

$$= (-1)^k \int_{\mathbf{Z}_p} (y + x)^k \, dy = (-1)^k B_k(x).$$

5.5. *Sums of Powers*

The above formula $\nabla B_k(x) = kx^{k-1}$, with $S\nabla = \mathrm{id} - P_0$, leads to $B_k(x) - b_k = kS(x^{k-1})$. Replacing k by $k + 1$, we obtain

$$S(x^k) = \frac{B_{k+1}(x) - b_{k+1}}{k+1} \qquad (k \geq 0).$$

This gives an explicit formula for the sums of powers:

$$S(x^k) = \frac{1}{k+1} \sum_{1 \le j \le k} \binom{k+1}{j} b_{k+1-j} x^j + \frac{x^{k+1}}{k+1}$$

$$= b_k x + \sum_{2 \le j \le k} \binom{k}{j-1} b_{k+1-j} \frac{x^j}{j} + \frac{x^{k+1}}{k+1}.$$

Here are a few explicit expressions for these sums of powers:

$$S_k(n) = \sum_{1 \le i < n} i^k = S(x^k)|_{x=n} \quad (k \ge 1),$$

$$\begin{aligned} S_1(n) &= \tfrac{1}{2} n(n-1), \\ S_2(n) &= \tfrac{1}{6} n(n-1)(2n-1), \\ S_3(n) &= \tfrac{1}{4} n^2(n-1)^2, \\ S_4(n) &= \tfrac{1}{5} n^5 - \tfrac{1}{2} n^4 + \tfrac{1}{3} n^3 - \tfrac{1}{30} n, \\ S_5(n) &= \tfrac{1}{6} n^6 - \tfrac{1}{2} n^5 + \tfrac{5}{12} n^4 - \tfrac{1}{12} n^2. \end{aligned}$$

(Observe that for $k = 0$, $S(x^0) = B_1(x) - b_1 = x$ gives a sum of powers $\sum_{i<n} i^0 = n$, which is correct if the summation is extended over the indices $0 \le i < n$ and x^0 is the constant 1, including $0^0 = 1$.) In the Archimedean theory, the main term is $x^{k+1}/(k+1)$: It gives the primitive of x^k, namely the area below the graph of $t \mapsto t^k$ between the values 0 and x. Here the main term will turn out to be $b_k x$.

Proposition. *When p is an odd prime, the sums of kth powers satisfy*

$$S_k(p) \equiv p b_k \pmod{pk\mathbf{Z}_p} \quad (k \ge 1),$$

while

$$S_k(2) = 1 \equiv 2b_k \pmod{k\mathbf{Z}_2} \quad (k \ge 1).$$

Proof. We have $pb_0 = p$, $pb_1 = -p/2$, which are both in $\mathbf{Z}_p$ (even if $p = 2$). We already know (5.4) that $pb_k \in \mathbf{Z}_p$ $(k \ge 0)$ (this also follows by induction, as the next argument shows). Since

$$S_k(p) = pb_k + \sum_{2 \le j \le k} \binom{k}{j-1} b_{k+1-j} \frac{p^j}{j} + \frac{p^{k+1}}{k+1}$$

$$= pb_k + pk \sum_{2 \le j \le k} \binom{k-1}{j-2} pb_{k+1-j} \frac{p^{j-2}}{j(j-1)} + pk \frac{p^k}{k(k+1)},$$

we have to show that

$$\sum_{2\le j\le k} \binom{k-1}{j-2} pb_{k+1-j} \frac{p^{j-2}}{j(j-1)} + \frac{p^k}{k(k+1)} \in \mathbf{Z}_p.$$

But the pb_{k+1-j} are in $\mathbf{Z}_p$ for $j \ge 2$, and for $p \ge 3$

$$\mathrm{ord}_p\, j(j-1) \le \mathrm{ord}_p\, j! = \frac{j - S_p(j)}{p-1} \le \frac{j-1}{p-1} < j-1$$

implies $\frac{p^{j-2}}{j(j-1)} \in \mathbf{Z}_p$. For $p = 2$,

$$\mathrm{ord}_2\, j(j-1) = \max(\mathrm{ord}_2 j, \mathrm{ord}_2(j-1)) \le j-1$$

with equality for $j = 2$. This explains the loss of one power of 2. ∎

Corollary. *For any prime p,* $b_{2n} \in \mathbf{Q} \cap p^{-1}\mathbf{Z}_p \quad (n \ge 1)$. ∎

Remarks. (1) For $p = 2$ and *odd* $k \ge 3$, the corresponding Bernoulli number is zero: The congruence $S_k(2) = 1 \equiv 0 = 2b_k$ holds mod $k\mathbf{Z}_2$, not mod $2k\mathbf{Z}_2$. For *even* $k = 2n \ge 2$, the same congruence forces b_{2n} to have an even denominator.

(2) For $p = 3$ and even $k = 2n \ge 2$, the congruence $S_{2n}(3) \equiv 3b_{2n} \pmod{3\mathbf{Z}_3}$ leads to $3b_{2n} \equiv 1 + 2^{2n} = 1 + 4^n \pmod{3\mathbf{Z}_3}$, and 3 appears in the denominator of b_{2n}. By the preceding remark, the factor 6 appears in the denominator of all b_{2n} $(n \ge 1)$.

(3) The property $pb_{2n} \in \mathbf{Z}_p$ (all primes p) means that the denominator of b_{2n} is a product of distinct primes (each prime occurring at most once). We have a more precise result.

Theorem (Clausen-von Staudt). *The denominator of the Bernoulli number b_{2n} is the product of the primes ℓ such that $\ell - 1$ divides $2n$. More precisely,*

$$b_{2n} = -\sum_{\ell \text{ prime: } \ell-1|2n} \frac{1}{\ell} + m_{2n} \quad (m_{2n} \in \mathbf{Z}).$$

PROOF. Let us start with the congruence $pb_{2n} \equiv S_{2n}(p) \pmod{p\mathbf{Z}_p}$. Now, it is easy to compute a sum $\sum_{0\le j<p} j^k$ mod p: This is a sum over the field $\mathbf{F}_p$. Put

$$s_k = (S_k(p) \bmod p) \in \mathbf{F}_p.$$

For each $0 \ne u \in \mathbf{F}_p$ we have

$$s_k = \sum_{x\in\mathbf{F}_p} x^k = \sum_{x\in\mathbf{F}_p} (xu)^k = u^k \sum_{x\in\mathbf{F}_p} x^k = u^k s_k,$$

whence

$$(1 - u^k)s_k = 0.$$

If k is not a multiple of $p-1$, we can choose $u \in \mathbf{F}_p^\times$ such that $u^k \neq 1$ (because $\mathbf{F}_p^\times$ is *cyclic*), and in this case we see that $s_k = 0$, namely $S_k(p) \equiv 0 \pmod p$. On the other hand, if $p - 1 \mid k$, then

$$s_k = \sum_{x \in \mathbf{F}_p} x^k = \sum_{0 \neq x \in \mathbf{F}_p} 1 = p - 1 = -1 \in \mathbf{F}_p \quad (k \geq 1).$$

This information can be gathered together in the following form:

$$p\left(b_k + \sum_{\ell \text{ prime: } \ell-1 | k=2n} \frac{1}{\ell}\right) \in \mathbf{Q} \cap p\mathbf{Z}_p.$$

Letting the prime p *vary* (an exception!), we obtain

$$b_k + \sum_{\ell \text{ prime: } \ell-1|k} \frac{1}{\ell} \in \mathbf{Q} \cap \bigcap_{p \text{ prime}} \mathbf{Z}_p = \mathbf{Z}.$$

■

5.6. *Bernoulli Polynomials as an Appell System*

In (5.3) we have proved

$$f'(x) = \int_{\mathbf{Z}_p} \nabla f(y+x)\, dy.$$

In the case of Bernoulli polynomials, this gives

$$B_k'(x) = \int_{\mathbf{Z}_p} \nabla B_k(y+x)\, dy = \int_{\mathbf{Z}_p} k(y+x)^{k-1}\, dy = kB_{k-1}(x).$$

Hence $(B_k)_{k \geq 0}$ is an Appell system of polynomials. In particular, it satisfies the modified binomial identity for Sheffer systems (IV.6.1). We can derive it immediately in our context:

$$\begin{aligned} B_n(x+y) &= \int_{\mathbf{Z}_p} (t+x+y)^n\, dt \\ &= \int_{\mathbf{Z}_p} \sum_{0 \leq k \leq n} \binom{n}{k} (t+x)^k y^{n-k}\, dt \\ &= \sum_{0 \leq k \leq n} \binom{n}{k} B_k(x) y^{n-k}. \end{aligned}$$

In umbral notation, we can write symbolically

$$B_n(x+y) = \text{“}(B(x)+y)^n,\text{”}$$

which generalizes (5.4) $B_n(y) =$ “$(b+y)^n$,” since $B_n(0) = b_n$. Let us give the relation between this system and composition operators. Let U be the operator on

$K[X]$ defined by

$$U(p)(x) = \int_{\mathbf{Z}_p} p(y+x)\,dy.$$

It is obvious that this operator U commutes with the unit translation $E = \tau_1 = \tau$, hence (IV.5.3) with all translations:

$$\tau U(p)(x) = U(p)(x+1) = \int_{\mathbf{Z}_p} p(y+x+1)\,dy = U(\tau p)(x).$$

By definition $U(x^k) = B_k$, and the system of Bernoulli polynomials is a Sheffer sequence (IV.6.1). Moreover, as we have seen in (IV.6.2), $\nabla = e^D - 1$. We deduce

$$\nabla(Up)(x) = \int_{\mathbf{Z}_p} (p(y+x+1) - p(y+x))\,dy \overset{(5.3)}{=} p'(x),$$

$$(e^D - 1)Up = Dp, \quad Up = \frac{D}{e^D - 1}p.$$

This is the expression of the composition operator U as a formal power series in the derivation D (IV.5.3).

Exercises for Chapter 5

A. Classical reminder. Let $f,\ g,\ h : \mathbf{R} \to \mathbf{R}$ denote the functions defined by

$$f(x) = \begin{cases} x^2 \sin(1/x) & \text{if } x \neq 0, \\ 0 & \text{if } x = 0, \end{cases}$$

and $g = f + x/2$, $h = f - x^2$.

(*a*) Prove that f is differentiable at every point with $f'(0) = 0$, but f is not strictly differentiable at the origin $f \notin S^1(0)$ and f' is not continuous.

(*b*) Prove that g is differentiable at every point with $g'(0) = \frac{1}{2}$, but there is no neighborhood of the origin in which g is increasing.

(*c*) Prove that h is differentiable at every point with $h'(0) = 0$ and there are infinitely many points in every neighborhhood of the origin at which h has a relative maximum.

B. Classical reminder (continued). Let f be a real-valued function defined in the neighborhood of a point $a \in \mathbf{R}$. Assume that $f \in S^1(a)$ and $f'(a) > 0$. Show that there is a neighborhood V of a such that the restriction of f to V is an increasing function and in particular is injective.

1. Discuss the continuity and differentiability at the origin of the following functions on $\mathbf{Z}_p$:

$$\sum_{n \geq 0} n! \binom{x}{n}, \quad \sum_{n \geq 0} p^r \binom{x}{p^r}.$$

2. Prove that the function S_p introduced in (V.3.1) — sum of digits in base p — satisfies

$$S_p(m+n) = S_p(m) + S_p(n) - (p-1)\,\mathrm{ord}_p\binom{m+n}{n},$$

$$S_p(m-n) = S_p(m) - S_p(n) + (p-1)\,\mathrm{ord}_p\binom{m}{n}.$$

3. Let $f : \mathbf{Z}_p \to \mathbf{Q}_p$ be defined by

$$f(x) = \begin{cases} p^n \text{ if } |x| = |p^n| = \frac{1}{p^n}, \\ 0 \quad \text{if } x = 0. \end{cases}$$

Then f is locally constant outside the origin, and $\lim_{x\to 0} |f(x)/x| = 1$. By refining the preceding definition, construct a function g that is locally constant outside the origin, also differentiable at the origin with $g'(0) = 1$.

4. Check that $|\log(1+x)| \le r_p$ when $|x| = r_p$. But show that $|\log(1+x)|$ is variable on the sphere $|x| = r_p$.

(*a*) For which values of $x \in \mathbf{C}_p$ do the following series converge?

$$\sin x = x - \frac{x^3}{3!} + \cdots = \sum_{n\ge 0} (-1)^n \frac{x^{2n+1}}{(2n+1)!},$$

$$\cos x = 1 - \frac{x^2}{2!} + \cdots = \sum_{n\ge 0} (-1)^n \frac{x^{2n}}{(2n)!}.$$

(*b*) In the disk of convergence, prove that

$$\sin^2 x + \cos^2 x = 1,$$

$$\sin x \cos y + \cos x \sin y = \sin(x+y),$$

$$\cos x \cos y - \sin x \sin y = \cos(x+y).$$

Compute the derivative of the functions sin and cos.

(*c*) Choose a square root i of -1 in $\mathbf{C}_p$ and prove that

$$\cos x + i \sin x = e^{ix} \quad (i \in \mathbf{C}_p,\ i^2 = -1).$$

(*d*) Check the estimates (give their domain of validity)

$$|\sin x| = |x|, \quad |\cos x| = 1, \quad |\cos x - 1| = ?$$

6. Prove that when $t \in \mathbf{M}_p$, $x \mapsto (1+t)^x$ is differentiable at the origin of $\mathbf{C}_p$: To compute the limit of differential quotients for $x \to 0$ (in $\mathbf{C}_p$ not only in $\mathbf{Z}_p$), use the expression $(1+t)^x = \exp(x \log(1+t))$ valid for small $|x|$.

7. The Chebyshev polynomials (of the first kind) can be defined by the classical formulas $T_n(\cos\theta) = \cos n\theta$ $(n \ge 0)$. Observe that $T_{mn}(x) = T_m(T_n(x))$. When p is an odd prime, prove that

$$T_p(x) \equiv x^p \pmod{p}$$

and with the mean value theorem, show that

$$T_{np}(x) \equiv T_n(x^p) \pmod{pn\mathbf{Z}_p(x)}$$

(what can you say about the case $p = 2$?).

8. Let us say that a polynomial $f(x) \in \mathbf{Z}_p[x]$ is an nth pseudo-power when $f'(x) \in n\mathbf{Z}_p[x]$.
   (*a*) Show that the following polynomials are nth pseudo-powers: x^n, $f(x)^n$ (f any polynomial), T_n (Chebyshev polynomial of the first kind; cf. previous exercise).
   (*b*) Using the mean value theorem, prove that if f is an nth pseudo-power, then

$$a \equiv b \pmod{p\mathbf{Z}_p} \implies f(a) \equiv f(b) \pmod{pn\mathbf{Z}_p}.$$

   (*c*) Suppose $(f_n)_{n\geq 0}$ is a sequence of polynomials with $\deg f_n = n$ and satisfying the congruences

$$f_{pn}(x) \equiv f_n(x^p) \pmod{pn\mathbf{Z}_p}.$$

   Show that f_n is an nth pseudo-power. Deduce that for $m \in \mathbf{N}, a \in \mathbf{Z}_p$, the sequence f_{mp^ν} has a limit for $\nu \to \infty$.

9. Define a sequence of polynomials inductively by the conditions

$$p_0 = 1, \quad p_n = \text{primitive of } p_{n-1} \text{ such that } \int_0^1 p_n(x)dx = 0 \quad (n \geq 1).$$

   The first one is $p_1(x) = x - \frac{1}{2}$.
   (*a*) Prove that $p_n(x) = B_n(x)/n!$, where $B_n(x)$ denotes the nth Bernoulli polynomial.
   (*b*) Prove that $p_n(1) = p_n(0)$ ($n \neq 1$) and compute the Fourier series expansions of the 1-periodic functions f_n extending $p_n|_{[0,1]}$,

$$f_n(x) = -\sum_{m\neq 0} \frac{e^{2\pi imx}}{(2\pi im)^n} \quad (n \geq 1).$$

   For even $n = 2k \geq 2$ there is absolute convergence, and

$$f_{2k}(0) = -\frac{2}{(2\pi i)^{2k}} \sum_{m\geq 1} \frac{1}{m^{2k}} = -\frac{2}{(2\pi i)^{2k}}\zeta(2k).$$

10. For any prime p, prove the following congruence for the Bernoulli numbers:

$$2n(b_{pn} - b_n) \equiv 0 \pmod{pn\mathbf{Z}_p} \quad (n \geq 1).$$

   (*Hint.* Use the congruence $j^{pn} \equiv j^n \pmod{pn\mathbf{Z}_p}$ (exercise 8), hence a similar congruence for the sums of powers $S_{pn}(p)$ and $S_p(p)$, and conclude by Proposition in (5.5).)

11. For each $m \geq 1$, show that the numerator of

$$2 + \frac{2^2}{2} + \frac{2^3}{3} + \cdots + \frac{2^n}{n}$$

   is divisible by 2^m when n is large enough.

(*a*) Check the preceding assertion experimentally for a few values of $n \geq 2$.

(*b*) Prove the general statement by consideration of the logarithm $1 + \mathbf{M}_2 \to \mathbf{C}_2$ and the expansion of $\log(1-2) = \log(-1) = 0$.

12. Show that all continuous homomorphisms $f : \mathbf{Z}_p^\times \to \mathbf{Q}_p^\times$ have the following form:

$$f(\zeta u) = \zeta^\nu u^x \quad (\zeta \in \mu_{p-1},\ u \in 1 + p\mathbf{Z}_p)$$

for some $\nu \in \mathbf{Z}/(p-1)\mathbf{Z}$ and $x \in \mathbf{Z}_p$.

13. Prove that an infinite product $\prod_{n\geq 0}(1+a_n)$, where all $a_n \neq -1$, converges for any sequence $(a_n)_{n\geq 0}$ converging to 0 in $\mathbf{C}_p$.
(*Hint.* Use $|\log(1+a_n)| = |a_n|$ if $|a_n|$ is small.)

14. Let I be an ordered set, $(E_i)_{i\in I}$ a family of sets (or groups, rings,...), and let $\varphi_{ij} : E_j \to E_i$ be maps (resp. homomorphisms,...) given for $i < j \in I$, subject to the transitivity conditions

$$\varphi_{ij} = \varphi_{i\ell} \circ \varphi_{\ell j} : E_j \to E_\ell \to E_i \quad (i < \ell < j).$$

Assume that I contains a countable cofinal sequence $S : i_0 < i_1 < \cdots$ and consider the projective system $(E_{i_n}, \varphi_{i_{n+1},i_n})_{n\geq 0}$ with projective limit $\varprojlim_S E_i$. Show that if T is another countable cofinal sequence in I and $\varprojlim_T E_i$ is similarly defined, there is a canonical isomorphism $\varprojlim_S E_i \cong \varprojlim_T E_i$ (use the universal property of projective limits). Provided that I has a countable cofinal subset, we may define $\varprojlim E_i$ by choosing such a sequence S and putting $\varprojlim E_i := \varprojlim_S E_i$. For example, let A be the maximal subring of a complete ultrametric field K and consider the ideals

$$I_r = B_{<r}(K) \subset A \quad (0 < r < 1).$$

Establish the following isomorphism:

$$A\{X\} \xrightarrow{\sim} \varprojlim (A/I_r)[X]$$

(limit when $r \searrow 0$).

6

Analytic Functions and Elements

A powerful method for defining functions is provided by power series (we have seen two examples in Chapter V: exp and log). This method is here developed systematically, and we come back to a more thorough study of formal power series. As is classically known, uniform limits of polynomials in a complex disk lead to analytic functions.

Another class of special functions is supplied by rational functions, namely quotients of polynomials: The simplest being the linear fractional transformations. We also study them in this chapter, especially since their uniform limits in the p-adic domain lead to the "analytic elements" in the sense of Krasner. Indeed, in ultrametric analysis, the sole consideration of balls is not sufficient and in particular not adapted to analytic continuation.

In this chapter the field K will still denote a complete *extension of $\mathbf{Q}_p$ in $\mathbf{C}_p$ (or in Ω_p) often with* dense *valuation. The results that also require K to be algebraically closed will be simply formulated for the field $\mathbf{C}_p$ (they are also valid for Ω_p).*

1. Power Series

1.1. Formal Power Series

Formal power series have already appeared repeatedly (with integral coefficients in (I.4.8), with coefficients in a field in (IV.5), (V.2)). We now study them more systematically.

Let $A \neq \{0\}$ be a commutative ring with a unit element 1. The formal power series ring $A[[X]]$ consists of sequences $(a_n)_{n\geq 0}$ of elements of A, with addition and multiplication respectively defined by

$$(a_n)_{n\geq 0} + (b_n)_{n\geq 0} = (a_n + b_n)_{n\geq 0},$$

$$(a_n)_{n\geq 0} \cdot (b_n)_{n\geq 0} = \left(\sum_{0\leq i\leq n} a_i b_{n-i} \right)_{n\geq 0}.$$

Instead of the sequence notation $(a_n)_{n\geq 0}$ we shall prefer to use the notation $f = f(X) = \sum_{n\geq 0} a_n X^n$ for a formal power series. The formal power series ring $A[[X]]$ contains the polynomial ring $A[X]$, and since $1 \in A$, we have $X^n \in A[[X]]$ $(n \geq 0)$.

Let us show that this formal power series ring constitutes a *completion* of the polynomial ring.

Definition. *Let $f(X) = \sum_{n\geq 0} a_n X^n$ be a nonzero power series. Its* order *is the integer*

$$\omega = \omega(f) = \min\{n \in \mathbf{N} : a_n \neq 0\}.$$

This order is the index of the first nonzero coefficient of $f(X)$. We shall also adopt the convention $\omega(0) = \infty$ with the usual rules

$$\infty \geq n, \quad \infty + n = n + \infty = \infty \quad (n \geq 0).$$

The following relation is then obvious:

$$\omega(f + g) \geq \min(\omega(f), \omega(g)),$$

with an equal sign if the orders are different. Moreover,

$$\omega(X^n f) = n + \omega(f)$$

shows that

$$\{f(X) : \omega(f) \geq n\} = X^n A[[X]]$$

is the principal ideal generated by X^n in the formal power series ring. Since

$$A[[X]]/X^n A[[X]] = A[X]/(X^n),$$

we also have

$$A[[X]] = \lim_{\longleftarrow} A[[X]]/X^n A[[X]] = \lim_{\longleftarrow} A[X]/(X^n)$$

(with obvious identifications), and the ring $A[[X]]$ appears as a completion of the ring $A[X]$ for the metrizable topology admitting the ideals (X^n) as a fundamental system of neighborhoods of 0.

When the ring A has no zero divisor, we have, moreover,

$$\omega(fg) = \omega(f) + \omega(g).$$

Taking $f = g$ we infer $\omega(f^2) = 2\omega(f)$ and $\omega(f^n) = n\,\omega(f)$ $(n \geq 0)$ by induction. In particular, we see that if A is an integral domain, so is the formal power series ring $A[[X]]$. If we iterate the construction, $A[[X]][[Y]] = A[[X, Y]]$ is also an integral domain. We have obtained the following result.

Lemma. *Let A be an integral domain and n a positive integer. Then the formal power series ring $A[[X_1, \ldots, X_n]]$ is also an integral domain.* ∎

Definition. *The formal derivation D of the ring $A[[X]]$ is the additive map defined by*

$$D\left(\sum_{n\geq 0} a_n X^n\right) = \sum_{n\geq 0} n a_n X^{n-1} = \sum_{n\geq 1} n a_n X^{n-1}.$$

It satisfies

$$D(fg) = D(f)g + f D(g) \quad (f, g \in A[[X]]).$$

Since $\ker D \supset A$, we see in particular that

$$D(af) = aD(f) \quad (a \in A,\ f \in A[[X]]),$$

namely, the derivation D is A-linear. Since

$$\omega(D(f)) \geq \omega(f) - 1,$$

it is also continuous for the previously defined topology.

If we iterate this derivation D we obtain

$$D^k(X^n) = n(n-1)\cdots(n-k+1)X^{n-k},$$

and since the product of k consecutive integers is divisible by $k!$, we can define

$$\frac{1}{k!}D^k : X^n \mapsto \binom{n}{k} X^{n-k}$$

even when the ring A does not contain inverses of the integral multiples of 1. Hence we define an A-linear map

$$\frac{1}{k!}D^k : A[[X]] \to A[[X]]$$

correspondingly. In spite of the fact that a formal power series does not define a function, we also use the notation $f(0)$ for the constant coefficient a_0 of f. Then if $f(X) = \sum_{n\geq 0} a_n X^n$, we have

$$a_k = \frac{1}{k!}D^k(f)(0) \quad (k \geq 0).$$

1.2. Convergent Power Series

Since we are assuming that the field K is *complete*, an ultrametric series converges when its general term tends to 0. If $r \geq 0$ denotes a real number such that $|a_n|r^n \to 0$, then $\sum_{n\geq 0} a_n x^n$ converges (at least) for $|x| \leq r$, and we get a function $B_{\leq r}(0) \to K$.

Definition. *The* radius of convergence *of a power series* $f = \sum_{n\geq 0} a_n X^n$ *having coefficients in the field* K *is the extended real number* $0 \leq r_f \leq \infty$ *defined by*

$$r_f = \sup\{r \geq 0 : |a_n|r^n \to 0\}.$$

Alternatively, we can consider the values of $r \geq 0$ for which $(|a_n|r^n)$ is *bounded*:

$$\sup\{r \geq 0 : |a_n|r^n \to 0\} \leq \sup\{r \geq 0 : (|a_n|r^n) \text{ bounded}\},$$

and conversely,

$$(|a_n|r^n) \text{ bounded} \Longrightarrow |a_n|s^n \to 0 \quad (s < r)$$

proves the other inequality, so that

$$r_f = \sup\{r \geq 0 : (|a_n|r^n) \text{ bounded}\}.$$

It is possible to compute this radius of convergence as in the classical complex case by means of *Hadamard's formula.*

Proposition 1. *The radius of convergence of* $f = \sum_{n\geq 0} a_n X^n$ *is*

$$r_f = \frac{1}{\overline{\lim}_{n\geq 0} |a_n|^{1/n}} = \frac{1}{\limsup_{n\to\infty} |a_n|^{1/n}}.$$

Proof. Define r_f by the Hadamard formula. If $|x| > r_f$ (this can happen only if $r_f < \infty$!), we have

$$\lim_{n\to\infty} \sup_{k\geq n} |x||a_k|^{1/k} = |x| \cdot \lim_{n\to\infty} \sup_{k\geq n} |a_k|^{1/k} = |x| \cdot \frac{1}{r_f} > 1.$$

Hence the decreasing sequence $\sup_{k\geq n} |x||a_k|^{1/k}$ is greater than 1, and for infinitely many values of $k \geq 0$ we have $|a_k||x|^k > 1$, namely, the general term $a_k x^k$ of the series does not tend to zero: The series $\sum a_k x^k$ diverges. Conversely, if $|x| < r_f$ (this can happen only if $r_f > 0$!) we can choose $|x| < r < r_f$, and from

$$\lim_{n\to\infty} \sup_{k\geq n} r|a_k|^{1/k} = r \cdot \lim_{n\to\infty} \sup_{k\geq n} |a_k|^{1/k} < 1$$

we infer that for some large N

$$\sup_{k\geq N} r|a_k|^{1/k} < 1.$$

Hence $|a_k|r^k < 1$ for all $k \geq N$ and

$$|a_k x^k| = |a_k| r^k \left(\frac{|x|}{r}\right)^k < \frac{|x|^k}{r^k} \searrow 0 \quad (k \to \infty).$$

This shows that the general term of the series $\sum a_k x^k$ tends to zero, and the series converges. ∎

The letter x will here be used for a variable element of $B_{<r_f}$, while the capital X denotes the indeterminate. When $r_f = 0$, the power series converges only for $x = 0$. Hence we shall mainly be interested in power series f for which $r_f > 0$.

Definition. *A* convergent power series *is a formal power series* f *with* $r_f > 0$.

Comments. (1) Let $f(X) \in K[[X]]$ be a convergent power series. If $K' \subset \Omega_p$ is a complete extension of K, then f can be evaluated at any point $x \in B_{<r_f}(K')$. The convergent power series $f(X)$ defines in this way a *continuous* function (still denoted by f)

$$f : B_{<r_f}(K') \to K'$$

because it is a uniform limit of continuous polynomial functions

$$f_N : x \mapsto \sum_{0 \leq n \leq N} a_n x^n.$$

Usually, we shall simply write the condition $|x| < r_f$, assuming implicitly that the element x is taken in K, $\mathbf{C}_p$, or even Ω_p.

(2) If $r_f > 1$, f is a restricted series, and by Theorem 1 in (V.2.4), it is strictly differentiable, with a derivative given by numerical evaluation of the formal derivative: $f'(x) = (Df)(x)$ $(|x| \leq 1)$. A similar result holds for any convergent power series: If $|x_0| < r_f$, then the restricted power series $g(X) = f(x_0 X)$ has the preceding property, and we conclude that

$$f'(x) = (Df)(x) \quad (|x| < r_f).$$

(3) Observe that a radius of convergence $r_f > 0$ does not necessarily belong to $|K^\times|$. If $0 < r_f \notin |K^\times|$, then the sphere $|x| = r_f$ is empty: $B_{\leq r_f} = B_{<r_f}$. These subtleties disappear when we take x in the universal field Ω_p, since $|\Omega_p| = \mathbf{R}_{\geq 0}$. Either the series converges at all points of the sphere $\{x \in \Omega_p : |x| = r_f\}$ or it diverges at all points of this sphere.

Examples. (1) The radius of convergence of the series $\sum_{n \geq 0} X^n$ is $r_f = 1$. This series diverges at all points of the sphere $|x| = 1$, since its general term does not tend to 0.

(2) More generally, any series $f = \sum_{n\geq 0} a_n X^n \in \mathbf{Z}_p[[X]]$ has a radius of convergence $r_f \geq 1$. Interesting examples are supplied by the expansions

$$f_a(X) = \sum_{n\geq 0} \binom{a}{n} X^n \in \mathbf{Z}_p[[X]]$$

for fixed $a \in \mathbf{Z}_p$. We also denote by $(1+X)^a$ the formal power series $f_a(X)$.

(3) When a series $\sum_{n\geq 0} a_n X^n$ converges on the sphere $|x| = 1$, it is a restricted series (V.2) and $r_f \geq 1$. Here is an example with $r_f = 1$. Consider $\sum_{n\geq 0} p^n X^{p^n}$, which obviously converges when $|x| = 1$. Since $|p^n|^{1/p^n} = |p|^{n/p^n} \to 1$, we have $r_f = 1$ by Hadamard's formula.

(4) The radius of convergence of a series can be 1 even when the coefficients are unbounded or when $|a_n| \to \infty$. The series $\sum_{n\geq 0} X^{p^n}/p^n$ illustrates this possibility. As in the previous example $r_f = 1$, since

$$|1/p^n|^{1/p^n} = |p|^{-n/p^n} \to 1.$$

This series converges only if $|x| < 1$: It obviously diverges if $|x| = 1$.

Proposition 2. *Let f and g be two convergent power series. Their product fg (computed formally) is a convergent power series, and more precisely, the radius of convergence of fg is greater than or equal to* $\min(r_f, r_g)$. *Moreover, the numerical evaluation of the power series fg can be made according to the usual rule*

$$(fg)(x) = f(x)g(x) \quad (|x| < \min(r_f, r_g)).$$

Proof. All statements are consequences of (V.2.2). ∎

Corollary 1. *Let $r > 0$. The set of power series $f = \sum a_n X^n$ such that $|a_n|r^n \to 0$ is a ring, and for each $x \in B_{\leq r}$ the evaluation map $f \mapsto f(x)$ is a homomorphism of this ring into the base field K.* ∎

Corollary 2. *For any* polynomial *f, the radius of convergence of the composite $f \circ g$ is $\geq r_g$ and*

$$(f \circ g)(x) = f(g(x)) \quad (|x| < r_g).$$

Proof. If $|x| < r_g$, taking $f = g$ in the preceding proposition, we obtain $g^2(x) = g(x)^2$ and by induction $g^n(x) = g(x)^n$ $(n \geq 0)$. Taking linear combinations of these equalities, we deduce

$$(f \circ g)(x) = f(g(x)) \quad (|x| < r_g)$$

for any *polynomial* f. ∎

The possibility of evaluating a composition $f \circ g$ according to the same rule will be established for general power series in (1.5).

Proposition 3. *The radius of convergence of $f = \sum_{n\geq 0} a_n X^n$ and of its derivative $Df = \sum_{n\geq 1} na_n X^{n-1}$ are the same: $r_f = r_{Df}$.*

PROOF. Let us prove this proposition when the field is either an extension of $\mathbf{Q}_p$ or an extension of $\mathbf{R}$ with the normalized absolute value. We know that

$$\tfrac{1}{n} \leq |n| \leq n \quad (n \in \mathbf{N})$$

and also

$$n^{\pm 1/n} \to 1 \quad (n \to \infty).$$

This proves

$$\overline{\lim}_{n\to\infty} |na_n|^{1/(n-1)} = \overline{\lim}_{n\to\infty} |na_n|^{1/n} = \overline{\lim}_{n\to\infty} |a_n|^{1/n},$$

which concludes the proof. ∎

Although f and Df always have the same radius of convergence, their behaviors on the sphere $|x| = r_f$ may differ. For example, the radius of convergence of the series $f = \sum_{n\geq 0} x^{p^n}$ is $r_f = 1$. This series diverges on the unit sphere, but the derivative $Df = \sum_{n\geq 0} p^n x^{p^n-1}$ converges at all points of the unit sphere.

Example. The series

$$\log(1+X) = \sum_{n\geq 1} (-1)^{n-1} X^n/n,$$

$$\frac{1}{(1+X)} = \sum_{n\geq 1} (-1)^{n-1} X^{n-1} = \sum_{n\geq 0} (-1)^n X^n$$

have the same radius of convergence, since the second one is the derivative of the first. Obviously, the radius of convergence of the second one is $r = 1$; hence the radius of convergence of the logarithmic series is also 1 (compare with (V.4.1)). Direct inspection (V.4.1) shows that the series $\log(1+X)$ diverges when $|x| = 1$, while the series $f(X) = \exp X$ diverges on the sphere $|x| = r_p$.

1.3. *Formal Substitutions*

In this section we study the composition of power series $f(X),\ g(X) \in K[[X]]$. In order to be able to substitute $X = g(Y)$ in the power series $f(X)$, it is essential to assume that the order of $g(X)$ is positive. This assumption is represented by any of the following equivalent notations:

$$\omega(g) \geq 1, \quad g(0) = 0, \quad g(X) \in XK[[X]].$$

Then $\omega(g^n) \geq n$, and if $f(X) = \sum_{n\geq 0} a_n X^n$, then

$$f(g(Y)) = \sum_{n\geq 0} a_n (g(Y))^n = \sum_{n\geq 0} c_n Y^n$$

is well-defined, since the family $(a_n(g(Y))^n)_{n\geq 0}$ is *summable*: The determination of any coefficient c_n involves the computation of at most a finite number of $a_m(g(Y))^m$ ($m \leq n$) and their coefficients of index at most n in each of them. We thus define the composite power series by

$$(f \circ g)(Y) = \sum_{n\geq 0} c_n Y^n \in K[[Y]].$$

The substitution $X = g(Y)$ furnishes a homomorphism

$$f(X) \mapsto (f \circ g)(Y) = f(g(Y)) : K[[X]] \to K[[Y]]$$

sending 1 to 1 and *continuous* for the metrizable topology having the ideals $X^k K[[X]]$ as fundamental neighborhoods of 0, since

$$\omega(f) \geq k \Longrightarrow \omega(f \circ g) \geq k.$$

For a fixed power series g of positive order, the identity of formal power series

$$(f_1 f_2) \circ g = (f_1 \circ g)(f_2 \circ g)$$

is easily verified. Hence $f^2 \circ g = (f \circ g)^2$, and by induction

$$f^n \circ g = (f \circ g)^n \quad (n \geq 1).$$

Observe that the exponents are relative to multiplication and not to composition. Iteration of composition is represented by

$$g^{\circ(2)} = g \circ g, \quad g^{\circ(n)} = \underbrace{g \circ g \circ \cdots \circ g}_{n \text{ factors}}.$$

Also distinguish the multiplication identity $f = 1$ (constant formal power series) and the composition identity $g = X = \mathrm{id}$:

$$f \circ X = f, \quad X \circ g = g \quad (f \in K[[X]],\ g \in XK[[X]]).$$

For example $X^{\circ(n)} \circ f = f$, but $X^n \circ f = f^n$ ($n \geq 0$).

Proposition. *Let g and h be two formal power series with positive order. Then for any formal power series f we have*

$$(f \circ g) \circ h = f \circ (g \circ h).$$

Proof. Both sides are well-defined. They are equal when $f(X) = X^n$, since $f \circ g = g^n$ in this case (the observation made just before the proposition is relevant). Hence the statement of the proposition is true by linearity for any polynomial f. Finally, in the general case let $f(X) = \sum_{n\geq 0} a_n X^n$. Then

$$(f \circ g) \circ h = \left(\sum_{n\geq 0} a_n g^n\right) \circ h = \sum_{n\geq 0} a_n (g^n \circ h)$$
$$= \sum_{n\geq 0} a_n (g \circ h)^n = f \circ (g \circ h).$$

■

Theorem 1. *Let $f(X) = \sum_{n\geq 0} a_n X^n$ be a formal power series. The following properties are equivalent:*

(i) $\exists\, g \in K[[X]]$ with $g(0) = 0$ and $(f \circ g)(X) = X$.
(ii) $a_0 = f(0) = 0$ and $a_1 = f'(0) \neq 0$.

When they are satisfied, there is a unique formal power series g as required by (i), and this formal power series also satisfies $(g \circ f)(X) = X$.

PROOF. (*i*) ⇒ (*ii*) If $g(X) = \sum_{m\geq 1} b_m X^m$, then the identity $(f \circ g)(X) = X$ can be written more explicitly as

$$\sum_{n\geq 0} a_n g(X)^n = a_0 + a_1 b_1 X + X^2(\cdots) = X.$$

In particular, $a_0 = 0$ and $a_1 b_1 = 1$; hence $a_1 \neq 0$.

(*ii*) ⇒ (*i*) The equality $(f \circ g)(X) = X$ requires that $a_1 b_1 = 1$ and that the coefficient of X^n in $a_1 g(X) + \cdots + a_n g(X)^n$ vanishes (for $n \geq 2$) (indeed, the coefficient of X^n in $a_m g(X)^m$, whenever $m > n$, vanishes). This coefficient of X^n is determined by an expression

$$a_1 b_n + P_n(a_2, \ldots, a_n; b_1, \ldots, b_{n-1})$$

with known polynomials P_n having integral coefficients (not that it matters, but these polynomials are linear in the first variables a_i; cf. (V.4.2)). The hypothesis $a_1 \neq 0 \in K$ makes it possible to choose iteratively the coefficients b_n according to

$$b_n = -a_1^{-1} P_n(a_2, \ldots, a_n; b_1, \ldots, b_{n-1}) \quad (n \geq 2).$$

These choices furnish the required inverse formal power series g.

Finally, if f satisfies (*ii*) and g is chosen as in (*i*), then $b_0 = 0$ and $b_1 = 1/a_1 \neq 0$, so that we may apply (*i*) to g and choose a formal power series h with $(g \circ h)(X) = X$. The associativity of composition shows that

$$h(X) = \underbrace{(f \circ g)}_{\text{id}} \circ h(X) = f \circ \underbrace{(g \circ h)}_{\text{id}}(X) = f(X).$$

This proves $g \circ f(X) = g \circ h(X) = X$. ■

We still need a formula for the formal derivative of a composition. The identity

$$D(f\,g) = (Df)\,g + f\,Dg \quad (f, g \in K[[X]])$$

is well-known and easy to check. In particular, if $f = g$, we see that $D(g^2) = 2g\,Dg$. By induction

$$D(g^n) = ng^{n-1}\,Dg \quad (g \in K[[X]]) \quad (n \geq 1)$$

and by linearity

$$D(f \circ g)(Y) = Df(X)\,Dg(Y) = Df(g(Y))\,Dg(Y)$$

for all polynomials $f \in K[X]$.

Theorem 2 (Chain Rule). *Let f and g be two formal power series with $g(0) = 0$. Then the formal derivative of $f \circ g$ is given by*

$$D(f \circ g)(Y) = Df(X)\,Dg(Y) = Df(g(Y))\,Dg(Y).$$

Proof. Fix the power series g and let f vary in $K[[X]]$. Then

$$\omega(f) \geq k \Longrightarrow \omega(f \circ g) \geq k \Longrightarrow \omega(D(f \circ g)) \geq k - 1$$

as well as

$$\omega(f) \geq k \Longrightarrow \omega(Df) \geq k - 1 \Longrightarrow \omega\Big[Df(g(Y))\,Dg(Y)\Big] \geq k - 1.$$

The identity $D(f \circ g)(Y) = Df(g(Y))\,Dg(Y)$, valid on the dense subspace of polynomials $f \in K[X]$, extends by continuity to $f \in K[[X]]$. ∎

Application. Let us come back to the formal power series

$$e^X = 1 + X + \tfrac{1}{2!}X^2 + \cdots = \textstyle\sum_{n\geq 0} \frac{1}{n!}X^n$$

of order 0 and

$$\log(1 + X) = X - \tfrac{1}{2}X^2 \pm \cdots = \textstyle\sum_{n\geq 1} \frac{(-1)^{n-1}}{n}X^n$$

of order 1. Their formal derivatives are respectively

$$D(e^X) = 0 + 1 + X + \tfrac{1}{2!}X^2 + \cdots = e^X,$$

$$D(\log(1 + X)) = 1 - X + X^2 \mp \cdots = \sum_{n\geq 1}(-1)^{n-1}X^{n-1} = \frac{1}{1 + X}.$$

For the composition, let us introduce here the formal power series of order 1

$$e(X) = e^X - 1 = \textstyle\sum_{n\geq 1} \frac{1}{n!}X^n, \quad D\,e(X) = D(e^X) = e^X.$$

The composite

$$\log(e^X) = \log(1 + e(X)) = X + \sum_{k \geq 2} c_k X^k$$

is well-defined, and its formal derivative is

$$D(\log(e^X)) = \begin{cases} \dfrac{1}{1 + e(X)} \cdot e^X = 1, \\ 1 + \sum_{k \geq 2} k c_k X^{k-1}. \end{cases}$$

Comparison of these two expansions gives

$$0 = k c_k \in \mathbf{Q}, \quad c_k = 0 \quad (k \geq 2),$$

and this proves

$$\log(e^X) = \log(1 + e(X)) = X.$$

The formal power series $e(X)$ is the inverse for composition of $\log(1 + X)$: By the last assertion of Theorem 1 we also have $e(X) \circ \log(1 + X) = X$, namely

$$\exp \circ \log(1 + X) - 1 = X,$$

or equivalently

$$e^{\log(1+X)} = 1 + X.$$

1.4. The Growth Modulus

Let f be a nonzero convergent power series with coefficients in the field K. For $|x| < r_f$ we have $f(x) = \sum_{n \geq 0} a_n x^n$ and hence

$$|f(x)| \leq \max_{n \geq 0} |a_n x^n|.$$

Although the sphere $|x| = r$ is not compact, f is bounded on this sphere:

$$|f(x)| \leq |a_m| r^m \quad (|x| = r),$$

for some $m \geq 0$. Let us say that a monomial $|a_m x^m|$ is *dominant* on a sphere $|x| = r \; (< r_f)$ when

$$|a_n| r^n < |a_m| r^m \quad \text{for all } n \neq m.$$

In this case, this monomial is responsible for the absolute value of f,

$$|f(x)| = |a_m x^m| = |a_m| r^m \quad (|x| = r),$$

which is *constant* on the sphere. When r is small enough, there is always a dominant monomial: If $m = \omega(f)$ is the order of f, we have $|f(x)| = |a_m x^m|$ for all sufficiently small values of $|x|$.

Definitions. (1) *The* growth modulus *of a convergent power series f is defined by*

$$M_r f = M_r(f) = \max_{n \geq 0} |a_n| r^n \quad (0 \leq r < r_f)$$

so that $r \mapsto M_r f$ is a positive increasing real function on the interval $[0, r_f) \subset \mathbf{R}$.

(2) *We say that $r \in [0, r_f)$ is a* regular radius *for f if the equality $M_r f = |a_n| r^n$ holds for one index $n = n(r) \geq 0$ only. The monomial $|a_n| r^n$ or $a_n x^n$ is called the* dominant monomial *for that radius.*

(3) *When there are (at least) two distinct indices $i \neq j$ such that $M_r f = |a_i| r^i = |a_j| r^j$, we say that r is a* critical radius *and the monomials $|a_i| r^i = M_r f$ are called* competing *monomials.*

By definition

$$|f(x)| \leq M_r(f) \quad \text{if } |x| = r < r_f,$$

and this inequality is an *equality* $|f(x)| = M_r(f)$ for all *regular radii* r. If $a_0 \neq 0$, then $r = 0$ is regular and $|f(x)| = |a_0|$ for small $|x|$. The positive critical radii satisfy $r^{i-j} = |a_j/a_i| \in |K^\times|$: They are roots of absolute values of elements of K. A critical radius of a power series with coefficients in K is the absolute value of an algebraic element (over K).

When the coefficients $a_n \in K$ are given, it is easy to sketch the curves $r \mapsto |a_n| r^n$ $(n \geq 0)$ and their upper bound $M_r f$ on the given interval. This upper bound is a *continuous convex* curve. Let us show that it is continuously differentiable except at a discrete set of points of the interval $[0, r_f)$.

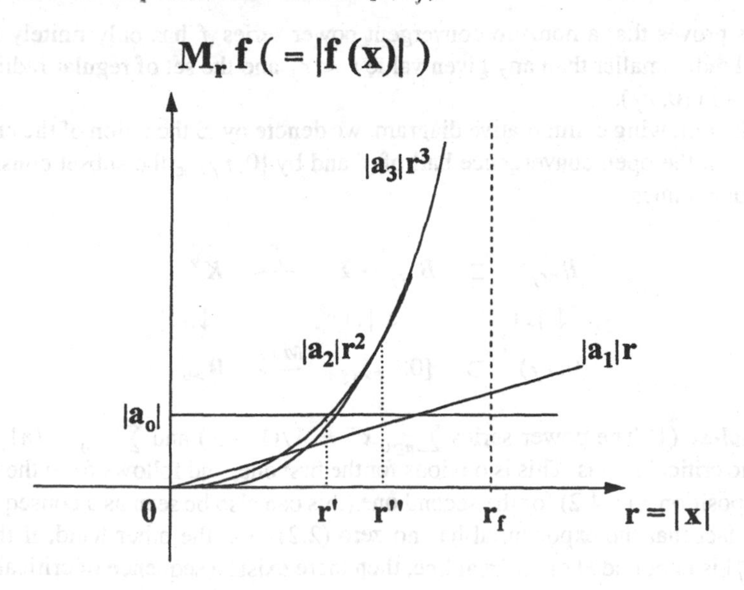

The growth modulus: $r \mapsto M_r f$

Classical Lemma. *Let $c_n \geq 0$ and $0 < R \leq \infty$ be such that for every $r < R$, $c_n r^n \to 0$ (as $n \to \infty$). Then*

$$r \mapsto M(r) = \sup_{n \geq 0} c_n r^n = \max_{n \geq 0} c_n r^n$$

is a continuous convex function on the interval $I = [0, R)$ that is smooth except on a discrete subset $\Delta = \{r' < r'' < r''' < \cdots\} \subset I$. *Between two consecutive values of Δ, M coincides with a single monomial $c_m r^m$.*

Proof. Let $0 < r < R$. Since $c_n r^n \to 0$ ($n \to \infty$), there is an integer $m \geq 0$ with

$$c_m r^m = \max_{n \geq 0} c_n r^n = M(r).$$

If $N > m$ and $0 < s < r$, then

$$c_N r^N \leq c_m r^m \Longrightarrow \frac{c_N}{c_m} \cdot r^{N-m} \leq 1 \Longrightarrow \frac{c_N}{c_m} \cdot s^{N-m} < 1 \Longrightarrow c_N s^N < c_m s^m.$$

Hence only finitely many monomials, namely those for which $N < m$, can compete with $c_m s^m$ for $s < r$. The critical radii $s < r$ are among the finite set of solutions of

$$s^{j-i} = \frac{c_i}{c_j} \quad (0 \leq i < j \leq m).$$

The set Δ is either finite—possibly empty—or consists of an increasing sequence converging to R. ∎

This proves that a nonzero convergent power series f has only finitely many critical radii smaller than any given value $r < r_f$ and the set of regular radii of f is *dense* in $[0, r_f)$.

In the following commutative diagram, we denote by Σ the union of the critical spheres in the open convergence ball of f and by $[0, r_f)_{\text{reg}}$ the subset consisting of regular values.

$$\begin{array}{ccccc} B_{<r_f} & \supset & B_{<r_f} - \Sigma & \xrightarrow{f} & K^\times \\ \downarrow |.| & & \downarrow |.| & & \downarrow |.| \\ [0, r_f) & \supset & [0, r_f)_{\text{reg}} & \xrightarrow{M_r f} & \mathbf{R}_{>0} \end{array}$$

Examples. (1) The power series $\sum_{n\geq 0} x^n = 1/(1-x)$ and $\sum_{n\geq 0} x^n/n! = e^x$ have no critical radius. This is obvious for the first one, and follows from the proof of Proposition 1 (V.4.2) for the second one (this can also be seen as a consequence of the fact that the exponential has no zero (2.2)). On the other hand, if the set $\{|a_n| r_f^n\}$ is unbounded on the real line, then there exists a sequence of critical radii $r_i \nearrow r_f$ (exercise).

(2) If f is a restricted power series, then $M_1(f)$ is the Gauss norm of f. This suggests that the maps $f \mapsto M_r(f)$ are *norms* on a suitable subspace of $K[[X]]$ depending on the value $r > 0$. This is indeed the case.

Proposition 1. *When $|K^\times|$ is dense in $\mathbf{R}_{>0}$ and $f \in K\{X\}$, the Gauss norm of f and the sup norm of the function defined by f on the unit ball A of K coincide.*

In other words (cf. (V.2.1)), the canonical homomorphism

$$K\{X\} \to \mathcal{C}(A; K)$$

is an isometric embedding.

PROOF. The Gauss norm of f is $M_1 f$, and the inequality

$$\sup_{|x|\le 1} |f(x)| \le M_1 f$$

holds in general. With our assumption, we can choose a sequence $x_n \in K$ with $|x_n|$ regular and $|x_n| \nearrow 1$; hence we have

$$M_1 f = \sup_{r\nearrow 1} M_r f \overset{!}{\le} \sup_{x\in A} |f(x)|. \qquad \blacksquare$$

Proposition 2. *When $r > 0$ is fixed, $f \mapsto M_r(f)$ is an ultrametric norm on the subspace consisting of formal power series $f(X) = \sum a_n X^n$ such that $|a_n|r^n \to 0\ (n \to \infty)$. This norm is multiplicative, i.e., $M_r(fg) = M_r(f)M_r(g)$ when f and g belong to this subspace.*

PROOF. If $f \ne 0$, then one a_n at least is nonzero, and $M_r(f) \ge |a_n|r^n > 0$, since $r > 0$. Hence M_r is a norm on the subspace considered. Moreover, the equality

$$M_r(fg) = M_r(f)M_r(g)$$

is true if r is a regular radius for f, g, and fg, since it is the common value (V.2.2)

$$|fg(x)| = |f(x)||g(x)| \quad (|x| = r,\ x \in \Omega_p).$$

The general result follows by density of regular values and continuity of the maps

$$r \mapsto M_r(f), \quad r \mapsto M_r(g), \quad r \mapsto M_r(fg). \qquad \blacksquare$$

In the classical case, a complex function with an infinite radius of convergence is an *entire function*. The only entire functions that are bounded on $\mathbf{C}$ are the constants. This is the theorem of Liouville. There is an analogous result in p-adic analysis.

Theorem. *Let the power series $f \in K[[X]]$ have infinite radius of convergence. If the function $|f|$ is bounded on K and $|K^\times|$ is dense, then f is a constant. More generally, if $|f(x)| \leq C|x|^N$ for some $C > 0$, $N \in \mathbf{N}$, and all $x \in K$ with $|x| \geq c$, then f is a polynomial of degree less than or equal to N.*

PROOF. It will suffice to prove the second, more general, statement. Write $f(x) = \sum a_n x^n$ as usual. We have

$$|a_n| r^n \leq M_r f \stackrel{!}{=} |f(x)|_{|x|=r} \leq C r^N,$$

provided that $r \geq c$ is a regular radius of f. By the lemma and since $|K^\times|$ is dense in $\mathbf{R}_{>0}$, this happens at least for a sequence of values $r_j = |x_j| \to \infty$, $x_j \in K$,

$$|a_n| \leq C r_j^{N-n}.$$

Letting $j \to \infty$, we get $a_n = 0$ for all $n > N$. This proves that f is a polynomial of degree at most N, as claimed. ∎

There are many entire functions that are bounded on $\mathbf{Q}_p$, just as there are many entire functions bounded on $\mathbf{R}$ (e.g., polynomials in $\sin x$ and $\cos x$).

1.5. Substitution of Convergent Power Series

Let $f(X) = \sum_{n\geq 0} a_n X^n$, $g(X) = \sum_{m\geq 1} b_m X^m$ be two convergent power series with $g(0) = 0$. The formal power series $(f \circ g)(X) = \sum_{k\geq 0} c_k X^k$ will turn out to be convergent, too, and we intend to prove the *validity of the numerical evaluation*

$$(f \circ g)(x) = f(g(x)),$$
$$\sum_{k\geq 0} c_k x^k = \sum_{n\geq 0} a_n \left(\Sigma_{m\geq 1} b_m x^m\right)^n$$

when $|x|$ is suitably small.

In order to be able to substitute the value $X = g(x)$ in the formal power series $f(X)$, it is necessary to assume that this is small: $|g(x)| < r_f$ will do. But even if $g(x) = 0$, namely, x on a critical sphere of g, $|x|$ might be too big to allow substitution in $f \circ g$. Recall that critical spheres occur when several monomials are of competing size. The circumstance $g(x) = 0$ does not prevent a few individual monomials to be large, and thus have an influence after rearrangement of these terms. On the other hand, the power series $f \circ g$ converges in a ball, and it would be unreasonable to expect to be able to take advantage of the single fact that "$g(x)$ small," i.e., x close to a root of g, and hence x on a critical sphere of g, while $\sup |g|$ on this critical sphere is $M_r(g)$ $(r = |x|)$. This explains the reason for the hypotheses made in the following theorem.

Theorem. *Let f and g be two convergent power series with $g(0) = 0$. If $|x| < r_g$ and $M_{|x|}(g) < r_f$, then $r_{f\circ g} > |x|$ and the numerical evaluation of the* *composite $f \circ g$ can be made according to*

$$(f \circ g)(x) = f(g(x)).$$

PROOF. Assume that $x \in K$ (or Ω_p) satisfies the assumptions and define $r = |x|$. Then recall that if $f(X) = \sum_{n\geq 0} a_n X^n$ and $g(Y) = \sum_{n\geq 1} b_n Y^n$, the formal power series $(f \circ g)(Y) = \sum_{k\geq 0} c_k Y^k$ is obtained by grouping equal powers in the expansion of $\sum_{n\geq 0} a_n g(Y)^n$ (this is a double series). Define the polynomials

$$f_N(X) = \sum_{0\leq n\leq N} a_n X^n.$$

The substitution

$$(f_N \circ g)(x) = f_N(g(x))$$

is valid if $|x| < r_g$ by Corollary 2 of Proposition in (1.2). Let $y = g(x)$. Since $|y| = |g(x)| \leq M_r(g) < r_f$, we have $f_N(y) \to f(y)$, and here is a diagram summing up the situation:

$$\begin{array}{ccc} f_N(g(x)) & \longrightarrow & f(g(x)) \quad (N \to \infty) \\ \| & \leftarrow & \text{polynomial case for } f_N \text{ (1.2)} \\ (f_N \circ g)(x) & \stackrel{?}{\longrightarrow} & (f \circ g)(x) \quad (N \to \infty). \end{array}$$

Introduce

$$(f_N \circ g)(Y) = \sum_{k\geq 0} c'_k(N) Y^k,$$
$$((f - f_N) \circ g)(Y) = \sum_{k>N} c''_k(N) Y^k,$$

so that the coefficients c_k of $f \circ g$ are

$$c_k = c'_k(N) + c''_k(N), \quad c_k = c'_k(N) \quad (k \leq N).$$

Recall that

$$(f \circ g)(Y) - (f_N \circ g)(Y) = \sum_{n>N} a_n g(Y)^n = \sum_{k>N} c''_k(N) Y^k$$

is obtained by grouping the monomials having the same degree. Any monomial of $g(Y)^n$ is a sum of products of n monomials of $g(Y)$. When we evaluate it at a point y with $M_{|y|} g = \rho$, the ultrametric inequality shows that its absolute value is less than or equal to ρ^n. This is where we use the assumption $M_r(g) < r_f$ in its full force: Choose $y \in \mathbf{C}_p$ with

(*a*) $|x| = r < |y| \stackrel{!}{<} r_g$, so that $g(y)$ is well-defined,

(*b*) $|g(y)| \leq M_{|y|} g = \rho \stackrel{!}{<} r_f$, so that $f(g(y))$ is well-defined (this is possible by continuity of $t \mapsto M_t g$ and $M_r(g) < r_f$). Our previous observation gives

$$|c_k''(N)y^k| \leq \sup_{n>N} |a_n|\rho^n \to 0 \quad (N \to \infty)$$

because $\rho < r_f$. This shows that the sequence $(c_k y^k)_{k\geq 0}$ lies in the closure of the set of sequences $(c_k'(N)y^k)_{k\geq 0}$ $(N \geq 1)$. But for each N, the sequence $c_k'(N)y^k \to 0$ $(k \to \infty)$ and the space c_0 of sequences tending to 0 is complete (IV.4.1). This proves

$$c_k y^k \to 0, \quad r_{f\circ g} \geq |y| > |x|,$$

and also

$$(f \circ g)(y) - (f_N \circ g)(y) \to 0. \qquad \blacksquare$$

Example. Take $K = \mathbf{Q}_p$ and consider the formal power series

$$f(X) = \sum_{n\geq 0} X^n = \frac{1}{1-X}, \qquad g(Y) = Y - Y^p.$$

Take a root $\zeta \in \mu_{p-1}$. Then $g(\zeta) = 0$, so that $f(g(\zeta)) = f(0) = 1$ is well-defined. But $r_{f\circ g} = 1$, and the power series of $f \circ g$ is not convergent on the unit sphere, so that $(f \circ g)(\zeta)$ is not defined. Here for $r = |\zeta| = 1$, $M_r(g) = 1$ is not less than $r_f = 1$, and the substitution is not allowed (cf. exercises for the case $p = 2$). This example also shows that for fields K having a discrete valuation, the condition on balls $g(B_{<r}(K)) \subset B_{<r_f}$ is not sufficient to allow substitution: If $(1 < r < p)$, $y \in \mathbf{Z}_p = B_{<r}(0; \mathbf{Q}_p)$, we have $y \equiv y^p \pmod{p\mathbf{Z}_p}$ hence $|y - y^p| \leq |p| < 1$ and thus $g(B_{<r}(\mathbf{Q}_p)) = g(B_{\leq 1}(\mathbf{Q}_p)) \subset B_{<1}$. But although f converges in the open unit ball, we cannot find a power series representing the composite $f \circ g$ in the ball $B_{\leq 1}$, since the rational function

$$\frac{1}{1 - y + y^p}$$

has poles at the roots of $1 - y + y^p = 0$. These poles are located on the unit sphere of a finite extension of $\mathbf{Q}_p$, and no power series can represent this rational function on the sphere $r = 1$ (cf. exercises).

Another quite interesting example where the composition $(f \circ g)(x)$ is well-defined but different from $f(g(x))$ will appear in (VII.2.4).

Application. As proved in (1.3), the formal power series $f(X) = \log(1 + X)$ and $g(X) = e(X) = \exp X - 1$ are inverses of each other. By (1.5),

$$\log(e^x) = x \quad (|x| < r_p),$$

since $M_{|x|}(e(\cdot)) = |x|$ and $|x| = r < r_p < r_{\log} = 1$. Similarly,

$$\exp\log(1+x) = 1+x \quad (|x| < r_p),$$

since $M_{|x|}(\log(1+\cdot)) = |x|$ for $|x| = r < r_p$, and r_p is the radius of convergence of the exponential. This is a second, independent, proof of the fact that exp and log are inverse isometries in the open ball $B_{<r_p}(K)$ (Proposition 2 in (V.4.2)).

1.6. The Valuation Polygon and its Dual

The study of

$$M_r = M_r(f) = \sup_{n\geq 0} |a_n| r^n$$

is best made using logarithms. We shall use Greek letters for these logarithms:

$$\begin{aligned} \rho &= \log r < \rho_f = \log r_f, \\ \alpha_n &= \log |a_n|, \\ \mu_\rho &= \log M_r = \sup_{n\geq 0} (n\rho + \alpha_n). \end{aligned}$$

It is convenient to choose the log to the base p in order to have $\log p = 1$ and

$$\alpha_n = \log |a_n| = -\mathrm{ord}_p(a_n) = -v_n.$$

The function μ_ρ is a *convex function* as the sup envelope of affine linear functions. It is a piecewise linear function, since the critical radii (and their logarithms) occur on a discrete subset. Its opposite

$$-\mu_\rho = \inf_{n\geq 0}(v_n - n\rho)$$

is a *concave function*. When $|x| = r = p^\rho$ is a regular radius, we have

$$-\mu_\rho = \mathrm{ord}_p f(x).$$

Definition. *The function*

$$\rho \mapsto h_\rho := \inf_{n\geq 0}(v_n - n\rho) \quad (-\infty < \rho < \rho_f),$$

or its graph, is the valuation polygon *of the power series* f.

(*a*) Let $a_i X^i$ be the dominant monomial between two consecutive critical radii, say ${}'r < r < r'$. Then

$$h_\rho = -\mu_\rho = \inf_{n\geq 0}(v_n - n\rho) = v_i - i\rho$$

is affine linear in the corresponding interval

$$'\rho = \log{}'r < \rho < \rho' = \log r'.$$

This gives a side of the valuation polygon. The valuation polygon, or the graph of $\rho \mapsto h_\rho := \inf_{n\geq 0}(v_n - n\rho)$, is the boundary of the convex intersection of lower half-planes determined by the lines of equations

$$\Delta_n^* : \quad \rho \mapsto v_n - n\rho.$$

The slope of Δ_n^* is $-n$, and this line passes through the point $(0, v_n)$. The segment of Δ_i^* above the interval $['\rho, \rho']$ is a portion of the boundary — a side — of this convex region.

(*b*) If the dominant monomial just beyond the critical radius r' is $a_j X^j$, then $i < j$ are the extreme indices for competition of the monomials

$$\begin{aligned} |a_i| r'^i &= |a_j| r'^j, \\ (r')^{j-i} &= |a_i/a_j| = p^{-v_i+v_j}, \\ \rho' = \log r' &= \frac{v_j - v_i}{j-i}. \end{aligned}$$

(*c*) From the definition $h_\rho = \inf_{n\geq 0}(v_n - n\rho)$ $(-\infty < \rho < \rho_f)$ we infer successively

$$\begin{aligned} h_\rho &\leq v_n - n\rho \quad (\rho,\ n \geq 0), \\ h_\rho + n\rho &\leq v_n \quad (\rho,\ n \geq 0), \\ \sup_\rho (h_\rho + n\rho) &\leq v_n \quad (n \geq 0). \end{aligned}$$

The function

$$n \mapsto \sup_\rho (h_\rho + n\rho)$$

is a piecewise linear convex function whose graph gives the boundary of the convex intersection of upper half-planes containing all points $P_n = (n, v_n)$. The line of equation

$$\Delta_\rho : \quad n \mapsto h_\rho + n\rho$$

has slope ρ and passes through the point $(0, h_\rho)$. This gives a method for computing $h_\rho = -\mu_\rho$ for a *fixed* value of ρ. The value $v_n - n\rho$ is geometrically the height above the origin of a straight line of slope ρ going through the point $P_n = (n, v_n)$. One can draw the graph of the function $n \mapsto \mathrm{ord}_p a_n$ of an integer variable — consisting of the points P_n — and look for the lowest line of slope ρ going through these points. The height above the origin of this lowest line gives the value h_ρ. Letting now the slope ρ vary, this construction furnishes the desired convex hull of the points P_n (or of the graph of $n \mapsto v_n = \mathrm{ord}_p\, a_n$).

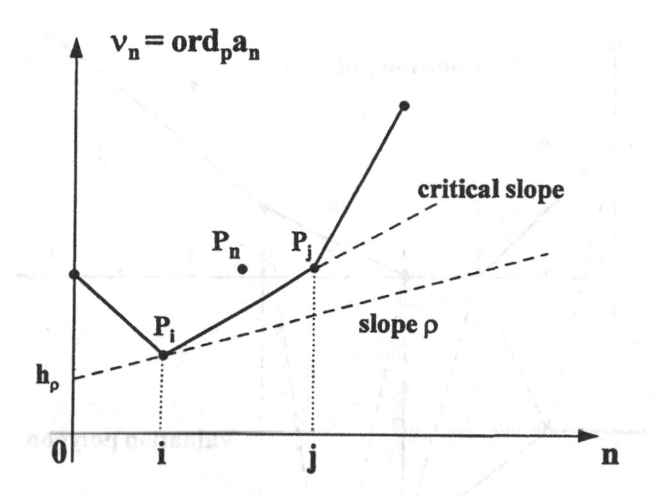

The Newton polygon: convex envelope of the P_n

Definition. *The function*

$$n \mapsto \sup_{\rho}(\nu_n + n\rho) \quad (-\infty < \rho < \rho_f),$$

or its graph, is the Newton polygon *of the power series* f.

The Newton polygon is the boundary of the sup convex envelope of the points $P_n = (n, \nu_n)$ $(n \geq 0)$.

A few conventions are useful at this point. When a coefficient a_n vanishes, its valuation $\nu_n = \infty$, and the corresponding point P_n is *at infinity* above all other ones. For example, the Newton polygon has a first vertical side at $m = \text{ord}(f)$, least integer m with $a_m \neq 0$. If f is a polynomial, it also has a last vertical side at $n = \deg(f)$ (since all P_n's are at ∞ when $n > \deg f$).

The two polygons constructed are *duals* of each other. The sides of one correspond to vertices of the other. For example, a side of the Newton polygon corresponds precisely to a slope of a lowest contact line going through two distinct points P_n. This situation occurs when two monomials have competing maximal absolute values, namely when this slope ρ is the logarithm of a critical radius: The valuation polygon has a vertex, and the graph of $r \mapsto M_r f$ exhibits an angle at the corresponding value of the radius r. More formally, from

$$h_\rho = \inf_{n \geq 0}(\nu_n - n\rho) \quad (\rho < \rho_f)$$

we infer

$$h_\rho \leq \nu_n - n\rho \quad (\rho < \rho_f,\ n \geq 0)$$

with equality for at least one index n. Equivalently,

$$h_\rho + n\rho \leq \nu_n \quad (\rho < \rho_f,\ n \geq 0),$$

$$\sup_{\rho}(h_\rho + n\rho) \leq \nu_n \quad (n \geq 0).$$

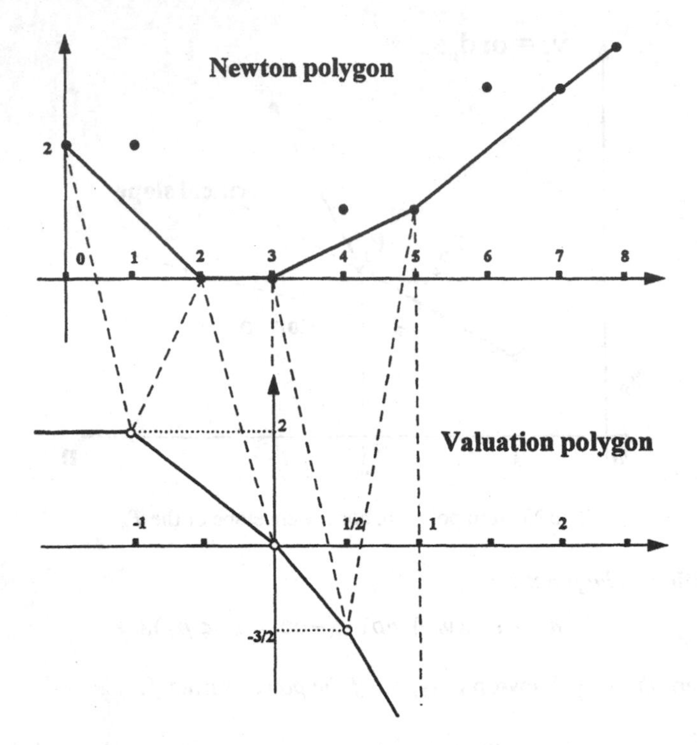

Duality of the Newton polygon and the valuation polygon

The affine lines

$$\Delta_\rho : t \mapsto h_\rho + t\rho,$$

which are below all P_n, have a sup that is the Newton polygon. Dually, the affine lines

$$\Delta_n^* : \rho \mapsto -(n\rho + \alpha_n) = v_n - n\rho$$

have an inf that is the valuation polygon.

This notion of duality is developed in CONVEXITY THEORY. It has numerous applications:

differential geometry (contact transformations),
variational calculus, classical mechanics, ...

Algorithm. Here is an efficient procedure to find the critical radii of a power series $f(X) = \sum_{n\geq 0} a_n X^n$. Let $v_n := \operatorname{ord}_p a_n$ and plot the points $P_n = (n, v_n)$. Determine the convex envelope of this set of points (and of $P_\infty = (0, \infty)$). The vertices of this convex envelope correspond to dominant monomials, those responsible for $|f| = M_r$ between two critical radii, and endpoints of sides correspond to competing monomials (responsible for a critical radius) with extremal indices.

If P_i and P_j $(i < j)$ are two endpoints of a side of the Newton polygon, the slope ρ' of this side

$$\rho' = \frac{\nu_j - \nu_i}{j - i} = \log r'$$

corresponds to the critical radius r' for which the two monomials $a_i X^i$ and $a_j X^j$ are the extreme competing monomials $|a_i|(r')^i = |a_j|(r')^j$,

$$(r')^{j-i} = |a_i/a_j| = p^{\nu_j - \nu_i}, \quad r' = p^{\rho'}.$$

For $\rho \nearrow \rho'$, the point P_i is the only contact point of the line Δ_ρ of slope ρ defining the Newton polygon

$$h_\rho = \nu_i - i\rho \quad (\rho \nearrow \rho').$$

For $\rho \searrow \rho'$, the point P_j plays a similar role:

$$h_\rho = \nu_j - j\rho \quad (\rho \searrow \rho').$$

Example 1. Consider an Eisenstein polynomial (II.4.2)

$$f(X) = X^n + a_{n-1}X^{n-1} + \cdots + a_0 \in \mathbf{Z}[X],$$

where $p \mid a_i$ $(0 \le i < n)$ and a_0 is not divisible by p^2. These assumptions mean that

$$\mathrm{ord}_p\, a_0 = 1, \quad \mathrm{ord}_p\, a_i \ge 1 \quad (1 \le i < n),$$

so that the Newton polygon of f can be drawn (see the figure).

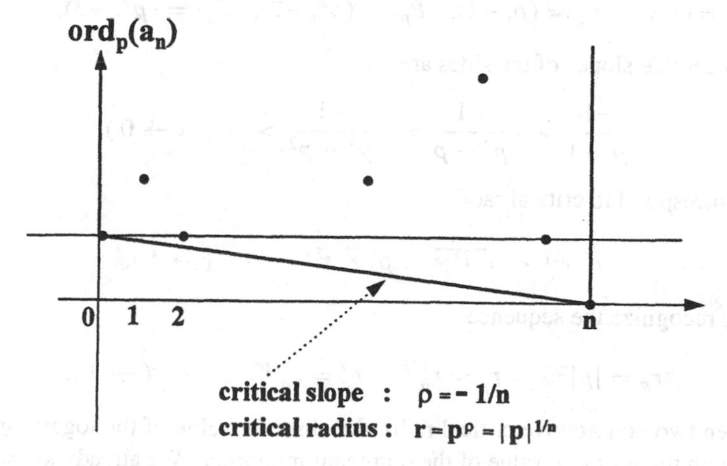

The Newton polygon of an Eisenstein polynomial

Example 2. Let us treat the case of the power series

$$f(X) = \log(1+X) = \sum_{n\geq 1} \tfrac{(-1)^{n-1}}{n} X^n.$$

We have

$$\mathrm{ord}_p\, a_n = 0 \text{ if } 1 \leq n < p, \quad \mathrm{ord}_p\, a_p = -1,$$

and

$$-1 \leq \mathrm{ord}_p\, a_n \leq 0 \text{ if } p < n < p^2, \quad \dots .$$

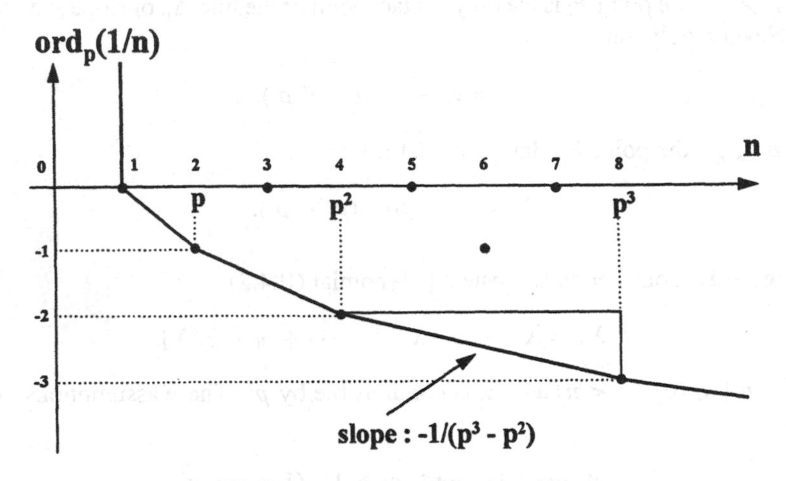

The Newton polygon of the logarithm

The vertices of the Newton polygon are the points

$$P_1 = (1,0), \quad P_p = (p,-1), \quad P_{p^2} = (p^2,-2), \quad P_{p^3} = (p^3,-3), \quad \dots .$$

The successive slopes of the sides are

$$\frac{-1}{p-1} > \frac{-1}{p^2-p} > \frac{-1}{p^3-p^2} > \cdots \quad (\to 0).$$

They correspond to critical radii

$$p^{-\frac{1}{p-1}} < p^{-\frac{1}{p^2-p}} < p^{-\frac{1}{p^3-p^2}} < \cdots \quad (\to 1),$$

and we recognize the sequence

$$r_p = |p|^{\frac{1}{p-1}} < r'_p = r_p^{1/p} < r''_p = r_p^{1/p^2} < \cdots \quad (\to 1).$$

Between two consecutive critical radii, the absolute value of the logarithm coincides with the absolute value of the dominant monomial. We already know that

$$|\log(1+x)| = |x| \quad (0 \leq |x| < r_p),$$

— the isometry domain of log (inverted by exp) — and see further that

$$r_p < |\log(1+x)| = \left|\frac{x^p}{p}\right| = p|x|^p < pr_p \quad (r_p < |x| < r'_p),$$

where $r'_p = r_p^{1/p}$ is the next critical radius. Quite generally,

$$p^{j-1}r_p < |\log(1+x)| = \left|\frac{x^{p^j}}{p^j}\right| = p^j|x|^{p^j} < p^j r_p$$

for

$$r_p^{1/p^{j-1}} < |x| < r_p^{1/p^j}.$$

Here we see how $|\log(1+x)|$ increases: We already knew by (V.4.4) that it can be arbitrarily large, since $\log : 1+\mathbf{M}_p \to \mathbf{C}_p$ is surjective. On the other hand, the zeros of $\log(1+x)$ can occur only when $|x|$ is equal to a critical radius. This gives an independent proof of (II.4.4) for the estimates of $|\zeta - 1|$ when $\zeta \in \mu_{p^\infty}$.

1.7. *Laurent Series*

Let us show how the preceding considerations extend to Laurent series. Let

$$f = \sum_{n\in\mathbf{Z}} a_n X^n = \sum_{-\infty}^{\infty} a_n X^n$$

be such a series with coefficients in the field K. Thus we consider this series as a sum of two formal power series

$$f = f^- + f^+ = \sum_{n<0} a_n X^n + \sum_{n\geq 0} a_n X^n$$

with $f^+ \in K[[X]]$ as before, and $f^- = \sum_{n<0} a_n X^n \in K[[X^{-1}]]$ has zero constant term. Convergence requires $|x| < r_f^+ = r_{f^+} = 1/\overline{\lim}_{n\to\infty} |a_n|^{1/n}$ for the first one and similarly $|x^{-1}| < 1/\overline{\lim}_{n\to\infty} |a_{-n}|^{1/n}$ for the second one. Let us define

$$r_f^- := \overline{\lim}_{n\to\infty} |a_{-n}|^{1/n}$$

and let us assume $r_f^- < r_f^+$, so that we have a common open annulus of convergence $r_f^- < |x| < r_f^+$ and hence a continuous (strictly differentiable) function — still denoted by f — in this annulus of K (or in any complete extension of K).

The absolute value of this function f is bounded on a sphere $|x| = r$ (where $r_f^- < r < r_f^+$),

$$|f(x)| \leq \sup_{-\infty<n<\infty} |a_n| r^n \quad (|x| = r),$$

and is even constant on this sphere, provided that a single monomial dominates all the others. In this case we say that it is a *regular radius*. We again define the *growth modulus* of f,

$$M_r f = M_r(f) := \sup_{-\infty<n<\infty} |a_n| r^n \quad (r_f^- < r < r_f^+),$$

as a positive real function on the interval $(r_f^-, r_f^+) \subset \mathbf{R}$. A *critical radius* r is a value $r_f^- < r < r_f^+$ such that for two monomials (at least)

$$M_r f = |a_i| r^i = |a_j| r^j \quad (-\infty < i < j < \infty).$$

The critical radii make up a discrete subset of the interval (r_f^-, r_f^+), and regular radii are dense in this interval.

The growth modulus $r \mapsto M_r f$ of a Laurent series is a *convex function* but is not necessarily increasing. Taking the log to the base p, define $\rho = \log r$, $\mu_\rho f = \log M_r f$. Then

$$\rho \mapsto h_\rho = \mu_\rho f \quad (\log r_f^- < \rho < \log r_f^+)$$

is a concave piecewise linear function (inf envelope of affine linear ones). It is the *valuation polygon* of the Laurent series f. All these facts are established exactly as in the case of power series.

Laurent series can also be multiplied in a common annulus of convergence. Let us indeed start with the case of Laurent polynomials. If

$$p = \sum_{\text{finite}} a_n X^n, \quad q = \sum_{\text{finite}} b_n X^n \in K[X, X^{-1}],$$

then their product is the polynomial $pq = \sum c_n X^n$ having coefficients

$$c_n = \sum_{k+l=n} a_k b_l, \quad |c_n| \leq \sup_{k+l=n} |a_k b_l| \quad (n \in \mathbf{Z}).$$

With the Gauss norms of p and q (sup norms on the coefficients) we have

$$|c_n| \leq \|p\| \cdot \|q\| \quad (n \in \mathbf{Z})$$

and consequently

$$\|pq\| \leq \|p\| \cdot \|q\|.$$

The product operation is (uniformly) continuous: It extends continuously to the completion $K\{X, X^{-1}\}$ with the same inequality. This completion consists of Laurent series $\sum_{-\infty<n<\infty} a_n X^n$, where $|a_n| \to 0$ for both limits $n \to \infty$ and $n \to -\infty$: These are called *restricted Laurent series*.

In particular, the powers of a restricted Laurent series are again restricted Laurent series, and if $f \in K\{X, X^{-1}\}$, then $x^m f^n \in K\{X, X^{-1}\}$ for all $m \in \mathbf{Z}$ and $n \in \mathbf{N}$. More generally, if f is a convergent Laurent series and $r_f^- < |a| < r_f^+$, then $g(X) = f(aX)$ is in $K\{X, X^{-1}\}$, and the same results are established for convergent Laurent series instead of restricted ones.

2. Zeros of Power Series

2.1. Finiteness of Zeros on Spheres

Let K be a complete extension of $\mathbf{Q}_p$ (in $\mathbf{C}_p$ or Ω_p),

$$K \supset A \supset M, \quad A/M = k: \text{ residue field.}$$

Select a nonzero convergent power series $f(X) = \sum_{n\geq 0} a_n X^n \in K[[X]]: r_f > 0$. If $f(a) = 0$ for some $a \in K^\times$, $|a| < r_f$, then $r = |a|$ is a critical radius: Indeed,

$$|f(a)| = 0 < M_r f := \sup_{n\geq 0} |a_n| r^n \neq 0$$

(cf. (1.4): $r = 0$ is critical precisely when $a_0 = f(0) = 0$).

We have already obtained an illustration of this fact in the study (1.6) of $|\log(1+x)|$ for $|x| < 1$: the zeros of log occur on the critical spheres, centered at 1, containing pth-power roots of unity (V.4.2).

Proposition. *Let $f \in A\{X\}$ be a restricted power series. Let $a \in A$. Then there is a formal power series g such that*

$$f(X) = f(a) + (X - a)g(X).$$

Moreover, $g \in A\{X\}$ and $r_g \geq r_f$.

PROOF. Replace f by $f_1(X) = f(aX)$, and hence $f_1 \in A\{X\}$ (if $|a| < 1$ we even have $r_{f_1} = r_f/|a| > 1$). Hence we only have to consider the typical case $a = 1$. We write $f = \sum_{n\geq 0} a_n X^n$ ($a_n \in A$, $|a_n| \to 0$), and we have to find $g = \sum_{n\geq 0} b_n X^n$ with

$$f(X) = f(1) + (X - 1)g(X).$$

Comparing coefficients, we find the conditions

$$a_0 = f(1) - b_0, \quad a_n = b_{n-1} - b_n \quad (n \geq 1),$$

or $b_0 = f(1) - a_0 = \sum_{i>0} a_i$, $b_n = b_{n-1} - a_n$. By induction we see that

$$b_n = \sum_{i>n} a_i \in A \quad (n \geq 0).$$

Hence $|b_n| \to 0$ and $g \in A\{X\}$, as desired. If $r_f = 1$, we are done. If $r_f > 1$, take any $r > 1$, $r < r_f$, so that $|a_i| r^i \to 0$. Hence there is a constant $c > 0$ such that

$$|a_i| r^i \leq c, \quad |a_i| \leq c/r^i \quad (i \geq 0).$$

Hence $|b_n| \leq \sup_{i>n} |a_i| \leq c \sup_{i>n} 1/r^i = c/r^{n+1}$,

$$|b_n| r^n \leq c/r \quad (n \geq 0).$$

Since the sequence $(|b_n|r^n)_{n\geq 0}$ is bounded, $r_g \geq r$. Letting r increase to r_f, we see that $r_g \geq \sup_{1<r<r_f} r = r_f$ (compare with Theorem 1 in (V.2.4)). ∎

Theorem (Strassman). *A nonzero restricted power series $f \in A\{X\}$ has only finitely many zeros in A.*

PROOF. (1) *Zeros on the unit sphere.* Assume $f = \sum_{n\geq 0} a_n X^n \neq 0$ and define

$$\mu := \min\{n : |a_n| = \sup_i |a_i|\} \leq \nu := \sup\{n : |a_n| = \sup_i |a_i|\},$$

so that $\mu \leq \nu$. If $\mu = \nu$, then f has no zero on the unit sphere. We are going to show more precisely that f has *at most* $\nu - \mu$ *zeros* (counting multiplicities) on the unit sphere. Suppose $\nu \geq 1$ and $f(a) = 0$ for some $a \in A^\times$, namely $|a| = 1$. Write

$$f = (X - a)g, \quad g \in A\{X\}.$$

By the definition of the extreme indices μ and ν, when we reduce the coefficients mod M,

$$\tilde{f}(X) = (X - \tilde{a})\tilde{g}(X) \in k[X],$$

we find that

$$\deg \tilde{f} = 1 + \deg \tilde{g}, \quad \omega(\tilde{f}) = \omega(\tilde{g}),$$

since $\tilde{a} \neq 0$ (ω denotes the order as in (1.1): first index of a nonzero coefficient),

$$\nu = 1 + \nu_g, \quad \mu = \mu_g.$$

This proves that

$$\nu_g - \mu_g = (\nu - \mu) - 1 < \nu - \mu.$$

But any zero $b \neq a$ of f is also a zero of g:

$$0 = f(b) = (b - a)g(b) \Longrightarrow g(b) = 0.$$

For example, if $\nu = \mu + 1$, we arrive at $\nu_g = \mu_g$, so that g cannot vanish on $A^\times$. In this case, f has only one zero in $A^\times$, and $\nu - \mu = 1$. If $\nu > \mu + 1$, we can repeat the procedure for g. In this way, we see that after at most $\nu - \mu$ steps, the last function $h \in A\{X\}$ obtained will satisfy $\nu_h = \mu_h$, hence will not vanish on the unit sphere $A^\times$. This process leads to a factorization

$$f = P \cdot h, \quad P \text{ polynomial}, \quad h \in A\{X\},$$

and h does not vanish on $A^\times$.

(2) *Zeros in M.* If $f(a) = 0$ for some $a \in M$, namely $|a| < 1$, consider $f_a(X) = f(aX)$, for which $r_{f_a} = r_f/|a| > r_f \geq 1$. By the first step, f_a has a finite

number of zeros on the unit sphere: f has a finite number of zeros on the critical sphere of radius $r = |a|$. Since f has only finitely many critical radii $r < 1$, the conclusion follows. ∎

The proof has shown more precisely that the number of zeros of f in A (counting multiplicities) is bounded by the telescoping sum of differences $\nu - \mu$ of exponents of critical monomials (corresponding to the critical radii less than 1). Hence we have obtained the following result.

Corollary. *Let $f = \sum_{n\geq 0} a_n X^n \in K[[X]]$ be a nonzero convergent power series and assume that $r < r_f$ is a critical radius of f. Let also*

$$\mu = \min\{n : |a_n| r^n = M_r f\} < \nu = \max\{n : |a_n| r^n = M_r f\}$$

be the extreme indices of the monomials of maximal absolute value. Then, counting multiplicities,

f has at most $\nu - \mu$ zeros in $S_r(K)$,
f has at most ν zeros in the closed ball $B_{\leq r}(K)$,
f has at most μ zeros in the open ball $B_{<r}(K)$. ∎

Remark. With the previous notation we have

$$M_s f = \begin{cases} |a_\mu| s^\mu & \text{for } r - \varepsilon < s < r, \\ |a_\nu| s^\nu & \text{for } r < s < r + \varepsilon \end{cases}$$

for small enough $\varepsilon > 0$. Taking logarithms (to the base p),

$$\mu_\sigma f = \begin{cases} \mu\sigma - \mathrm{ord}_p a_\mu & \text{for } \sigma \nearrow \rho, \\ \nu\sigma - \mathrm{ord}_p a_\nu & \text{for } \sigma \searrow \rho, \end{cases}$$

and $\nu - \mu$ appears as a difference of slopes of the valuation polygon at the corresponding vertex.

2.2. Existence of Zeros

We keep the same notation as in the preceding section.

Theorem 1. *Let K be a complete and* algebraically closed *extension of $\mathbf{Q}_p$ and $f = \sum a_n X^n \in K[[X]]$ a nonzero convergent power series. If f has a critical radius $r < r_f$, then f has a zero on the critical sphere of radius r in K. More precisely, if $\mu < \nu$ are the extreme indices for which $|a_n| r^n = M_r f$, then f has exactly $\nu - \mu$ zeros (counting multiplicities) on the critical sphere $|x| = r$ of K: There is a polynomial $P \in K[X]$ of degree $\nu - \mu$ and a convergent power* *series* $g \in K[[X]]$ *with*

$$f = P \cdot g, \quad r_g \geq r_f, \quad g \text{ does not vanish on } S_r(K).$$

PROOF. The result is trivial if $r = 0$, so we assume $r > 0$ from now on. Recall that $|a_\mu|r^\mu = |a_\nu|r^\nu$, $r^{\nu-\mu} = |a_\mu/a_\nu| \in |K^\times|$. Since K is algebraically closed, there is an element $a \in K$ with $|a| = r$. Replace f by $f_a(X) = f(aX)$ having $r = 1$ as critical radius. This converts f into a series having a radius of convergence $r_{f_a} = r_f/|a| > 1$, and in particular, $f_a \in K\{X\}$. We can similarly replace f by the multiple f/a_ν and assume $|a_\mu| = |a_\nu| = M_r f = 1$ (and $a_\nu = 1$). To sum up, it is sufficient to study the normalized situation

$$\begin{aligned}
&r = 1 < r_f \text{ is a critical radius of } f \in A\{X\} \subset K\{X\},\\
&|a_\mu| = |a_\nu| = M_r f = 1, \quad |a_n| \leq 1 \quad (n \geq 0),\\
&|a_n| < 1 \text{ for } n < \mu \text{ and also for } n > \nu.
\end{aligned}$$

We are going to show that (counting multiplicities)

$$f \text{ has precisely } \nu - \mu \text{ zeros on the critical sphere } S_r(K).$$

For $|x| \neq 1$ close to 1, the absolute value of $f(x)$ is given by

$$|f(x)| = \begin{cases} |a_\mu x^\mu| = |x|^\mu & \text{if} \quad |x| \nearrow 1 \quad \text{say } 1 - \varepsilon < |x| < 1, \\ |a_\nu x^\nu| = |x|^\nu & \text{if} \quad |x| \searrow 1 \quad \text{say } 1 < |x| < 1 + \varepsilon. \end{cases} \tag{$*$}$$

(The first estimate is valid when $|x|$ is larger than the largest critical radius less than 1, and similarly, the second one is valid when $|x|$ is smaller than the smallest critical radius greater than 1.)

First step: *Truncation.* For any index $\tau > 0$ define the polynomial $P_\tau = \sum_{n \leq \tau} a_n x^n$ (of degree $\leq \tau$) and the remainder $g_\tau = \sum_{n > \tau} a_n x^n$. We have $f = P_\tau + g_\tau$, and if $\tau \geq \nu$, then

$$M_1 g_\tau = \max_{n > \tau} |a_n| < 1 = M_1 f = M_1 P_\tau.$$

By continuity of the functions $r \mapsto M_r g_\tau$ and $r \mapsto M_r f$ we infer

$$\left| \sum_{n > \tau} a_n x^n \right| \leq M_r g_\tau < M_r f \quad (|x| = r \text{ close to } 1).$$

If r is regular for both g_τ and f (which is the case if $r \neq 1$ is close to 1), we have

$$|g_\tau(x)| = M_r g_\tau < M_r f = |f(x)| \quad (|x| = r \neq 1, \ r \text{ close to } 1).$$

Consequently, if $\tau \geq \nu$, then

$$|f(x)| = |P_\tau(x) + g_\tau(x)| = |P_\tau(x)|$$

for the same values of $|x| = r$. Choose $\tau \geq \nu$, so that $a_\tau \neq 0$: $\deg P_\tau = \tau$. Since K is algebraically closed, we can factorize this polynomial:

$$P_\tau(X) = a_\tau \prod_\xi (X - \xi).$$

More precisely, consider the partition of these roots into three subsets,

$$\begin{aligned} \Lambda = \Lambda_\tau &: \quad \text{roots } \xi \text{ with } \quad |\xi| < 1, \\ \Lambda' = \Lambda'_\tau &: \quad \text{roots } \xi \text{ with } \quad |\xi| > 1, \\ \Delta = \Delta_\tau &: \quad \text{roots } \xi \text{ with } \quad |\xi| = 1. \end{aligned}$$

Here is a table of the absolute values $|x - \xi|$, depending on ξ and $|x|$ close to 1:

	$\Lambda : \lvert\xi\rvert < 1$	$\Delta : \lvert\xi\rvert = 1$	$\Lambda' : \lvert\xi\rvert > 1$
$\lvert x\rvert \nearrow 1$	$\lvert x\rvert$	1	$\lvert\xi\rvert$
$\lvert x\rvert = 1$	1	$\lvert x - \xi\rvert$	$\lvert\xi\rvert$
$\lvert x\rvert \searrow 1$	$\lvert x\rvert$	$\lvert x\rvert$	$\lvert\xi\rvert$

In the middle column, we see that when $|x|$ crosses the value 1, then $|x - \xi|$ ($\xi \in \Delta$) varies from 1 to $|x|$. The number of roots $\xi \in \Delta$ — taking multiplicities into account — is responsible for the variation of growth of $|P_\tau|$. We have

$$|P_\tau(x)| = |a_\tau| \prod_\Lambda |x - \xi| \prod_\Delta |x - \xi| \prod_{\Lambda'} |x - \xi|$$

(the factors of this product are repeated as many times as the respective multiplicities require). Considering separately the cases $|x| < 1$ and $|x| > 1$, we have

$$|P_\tau(x)| = \begin{cases} |a_\tau||x|^{\#\Lambda} \cdot \quad 1 \quad \cdot \prod_{\Lambda'} |\xi| & \text{if} \quad |x| = r \nearrow 1, \\ |a_\tau||x|^{\#\Lambda} \cdot |x|^{\#\Delta} \cdot \prod_{\Lambda'} |\xi| & \text{if} \quad |x| = r \searrow 1 \end{cases} \tag{1}$$

(where multiplicities are taken into account in the exponents — the same notational abuse is made below). Recall (∗)

$$|f(x)| = M_r f = \begin{cases} |a_\mu||x|^\mu = |x|^\mu & \text{for } |x| = r \nearrow 1, \\ |a_\nu||x|^\nu = |x|^\nu & \text{for } |x| = r \searrow 1. \end{cases} \tag{2}$$

Comparing the first lines of (1) and (2), we infer

$$\mu = \#\Lambda, \quad |a_\tau| \prod_{\Lambda'} |\xi| = 1.$$

Observe that if $\tau > \nu$, then $|a_\tau| < 1$, so that N is not empty! Comparing now the second lines, we get

$$\delta := \#\Delta = \nu - \mu$$

independently from the index of truncation τ (recall that this takes into account the multiplicities of the roots $|\xi| = 1$ occurring in P_τ and is thus greater than or equal to the cardinality of this set of roots). Since $|a_\tau| \prod_{\Lambda'} |\xi| = 1$, we can now write

$$|P_\tau(x)| = \prod_{\Delta} |x - \xi| \quad \text{for } |x| = 1, \tag{3}$$

namely, the absolute value of $P_\tau(x)$ on the critical (unit) sphere is the product of the distances of x to the roots $\xi \in \Delta$.

Second step: *Convergence*. Let us compare *two successive truncations*: if $P_\tau \neq f$, then there is $\tau' > \tau$ with

$$P_{\tau'}(x) = \sum_{n \le \tau'} a_n x^n = P_\tau(x) + a_{\tau'} x^{\tau'}, \quad a_{\tau'} \neq 0.$$

By the first step, the roots of the polynomial $P_{\tau'}$ on the unit sphere constitute a set Δ' having the same number of elements (counting multiplicities) $\delta = \nu - \mu$ as Δ, and

$$|P_{\tau'}(x)| = \prod_{\Delta'} |x - \xi'| \quad \text{for } |x| = 1.$$

In particular, if we take a root $\xi \in \Delta$ of P_τ, we have

$$\prod_{\Delta'} |\xi - \xi'| = |P_{\tau'}(\xi)| = |P_{\tau'}(\xi) - P_\tau(\xi)| = |a_{\tau'} \xi^{\tau'}| = |a_{\tau'}|.$$

Hence for one root $\xi' \in \Delta'$ at least, we have

$$|\xi' - \xi| \le |a_{\tau'}|^{1/\delta}.$$

When f is not a polynomial, we can consider the infinite sequence of successive truncations of f, which are polynomials of degrees equal to their index

$$\tau < \tau_1 = \tau' < \tau_2 = \tau'' < \cdots.$$

Their sets of roots on the unit sphere

$$\Delta_0 = \Delta, \quad \Delta_1 = \Delta', \quad \Delta_2, \quad \ldots \subset \{x \in K : |x| = 1\}$$

have the same cardinality δ. Let us choose and fix a root $\xi = \xi_0 \in \Delta_0$. We have seen that we can choose a root $\xi_1 \in \Delta_1$ such that

$$|\xi_1 - \xi_0| \le |a_{\tau_1}|^{1/\delta}$$

and then a root $\xi_2 \in \Delta_2$ such that

$$|\xi_2 - \xi_1| \leq |a_{\tau_2}|^{1/\delta}, \quad \text{etc.}$$

Since $|a_n| \to 0$, this construction furnishes a Cauchy sequence $(\xi_i)_{i\geq 0}$ on the unit sphere of the complete field K. Let us call ξ_∞ its limit. By construction,

$$f(\xi_m) = \sum_{i>\tau_m} a_i \xi_m^i,$$

$$|f(\xi_m)| = \left|\sum_{i>\tau_m} a_i \xi_m^i\right| \leq \max_{i>\tau_m} |a_i| \to 0,$$

$$f(\xi_\infty) = f(\lim_{m\to\infty} \xi_m) = \lim_{m\to\infty} f(\xi_m) = 0.$$

This proves the *existence* of a root $a = \xi_\infty$ of f in the unit sphere of K. Writing $f = (X - a)g$, if $\nu - \mu > 1$, we can repeat the construction of a root of g. Eventually, we arrive at the precise statement of the theorem. ■

For example, if a convergent power series $f \in \mathbf{C}_p[[X]]$ has no zero in the open ball $B_{<r_f}$ of $\mathbf{C}_p$, then it has no critical radius. This is the case for $f(X) = e^X$, as was mentioned in (1.4) (before the Liouville theorem).

Corollary. *Let $f \in K[[X]]$ be a convergent power series having no zero in some closed ball $|x| \leq r$ ($< r_f$) of K^a. Then $1/f$ is given by a convergent power series with $r_{1/f} \geq r$. If f has no zero in an open ball $|x| < r'$ ($\leq r_f$) of K^a, then $1/f$ is given by a convergent power series with $r_{1/f} \geq r'$.*

PROOF. Let $f = \sum_{n\geq 0} a_n X^n$. Since $a_0 = f(0) \neq 0$, we may replace f by $f/f(0)$ and assume $a_0 = 1$. Define $g = \sum_{n\geq 1} a_n X^n$, so that $f = 1 + g$, $r_g = r_f$. The formal power series $1/f \in 1 + XK[[X]]$ is obtained by formal substitution (1.3)

$$\frac{1}{f} = \frac{1}{1+Y} \circ g(X),$$

since $\omega(g) \geq 1$. For the estimate of the radius of convergence of this power series, we may replace K by K^a, and hence assume that K is algebraically closed. By the theorem, f has no critical radius less than or equal to r, and if $f = \sum_{n\geq 0} a_n X^n$, then $|a_0| > |a_n| r^n$ for all $n \geq 1$. This shows that $M_r g = \max_{n\geq 1} |a_n| r^n < |a_0| = 1$. Numerical evaluation of the above composition (1.5) is valid when

$$|x| < r_g = r_f, \quad M_{|x|} g < r_{1/(1+Y)} = 1,$$

which is the case for $|x| \leq r$, since

$$M_{|x|} g \leq M_r g < 1.$$

The same reference (1.5) also proves that $r_{1/f} \geq r$. The second statement is obtained by letting $r \nearrow r'$. ■

When K is not algebraically closed, we can still give the following factorization result (in the following statement, μ and ν have the same meaning as before).

Theorem 2. *Let K be a complete extension of $\mathbf{Q}_p$ in Ω_p and $f \in K[[X]]$ a nonzero convergent power series.*

(a) If $f(a) = 0$ for some $a \in \Omega_p$, $|a| < r_f$, then a is algebraic over K.
(b) If $r = 1$ is a critical radius of f, then there is a factorization
$f = cP \cdot Q \cdot g_1$, $c \in K^\times$, P, $Q \in A[X]$ monic polynomials,
P of degree $\nu - \mu$, $|P(0)| = 1$, Q of degree μ, $Q \equiv X^\mu \pmod{M}$,
$g_1 \in 1 + XM\{X\}$ $(r_g \geq r_f)$ has no zero in the closed unit ball
of K^a. These conditions characterize uniquely this factorization.

Proof. (*a*) If f has a zero a on the sphere $|x| = r$ in $\mathbf{C}_p$ (or Ω_p), then r is a critical radius, and the preceding theorem shows that f has $\nu - \mu$ roots in Ω_p (counting multiplicities). If σ is a K-automorphism of Ω_p, it is continuous and isometric (III.3.2):

$$f(a^\sigma) = f(a)^\sigma = 0, \quad |a^\sigma| = |a| = r.$$

Hence a has a finite number of conjugates contained in the finite set of roots of f on the sphere $|x| = r$ of Ω_p. By Galois theory, this proves that a is algebraic over K. The same argument shows that the product $P = \prod_\xi (X - \xi) \in K^a[X]$ extended over all roots of f having absolute value r (all multiplicities counted) has coefficients fixed by all K-automorphisms of K^a and hence coefficients in K. This is a monic polynomial $P \in K[X]$ of degree $\nu - \mu$.

(*b*) Define $P = \prod_\xi (X - \xi) \in K[X]$, the product over all roots of f having absolute value $r = 1$ (taking into account multiplicities). Hence P is a monic polynomial of degree $\nu - \mu$, and $P(0) = \pm \prod \xi$ is a unit. Let similarly $Q = \prod_\xi (X - \xi) \in K^a[X]$ be the product corresponding to the roots of f in the open unit ball $|\xi| < 1$, i.e., $\xi \in M$. Then Q is a monic polynomial of degree μ having its coefficients in M except for the leading one. As before, Galois theory shows that $Q \in K[X]$. Now, $f = PQg$ with a convergent power series g, $r_g \geq r_f > 1$, having no zero in the closed unit ball of K^a, hence no critical radius. If $g = \sum_{n \geq 0} b_i X^i$, we have $|b_0| > |b_i x^i|$ for all $i > 1$, $|x| \leq 1$, since there is no critical radius in the unit ball. Hence $|b_0| > |b_i|$, and taking $c = b_0 \neq 0$, we see that ($b_i \to 0$ since $r_g > 1$)

$$f = cPQg_1, \quad g_1 = 1 + \sum_{i>0} (b_i/c) X^i \in 1 + XM\{X\}.$$

For uniqueness, observe that in a factorization $f = cP \cdot Q \cdot g_1$ with $g_1 \in 1 + XM\{X\}$ $(r_g \geq r_f)$, hence g_1 having no zero in the closed unit ball of K^a, the polynomial cPQ is a *multiple* of the product of the linear factors corresponding to the roots of f in the closed unit ball (counting multiplicities). If the degree of PQ is ν, this monic polynomial is the product $\prod_\xi (X - \xi) \in K^a[X]$ extended over all roots $|\xi| \leq 1$. If $P = \prod_\xi (X - \xi)$ is a product extended over a subset of roots, the condition $|P(0)| = |\prod_\xi \xi| = 1$ implies that the factors correspond to roots $|\xi| = 1$. The degree of P being $\nu - \mu$ by assumption, this product contains all the linear factors of f corresponding to the roots on the unit sphere. Consequently, Q is the product of the linear factors of f corresponding to the roots in the open unit ball. ∎

Remarks. (1) Under the assumptions of (*b*), if f has its coefficients in A, and not all in M, then the constant c is a unit, and by reduction mod M, the equality $f = cPQg_1$ leads to $\tilde{f} = \tilde{c}\tilde{P}\tilde{Q}\tilde{g_1} = \tilde{c}\tilde{P}X^\mu$, since $\tilde{g_1} = 1$. Since $\tilde{P}$ is a monic polynomial of degree $\nu - \mu$ with constant term $P(0) \in A^\times$, $\widetilde{P(0)} \neq 0$, we recognize the significance of μ and ν as the extreme indices of monomials of maximal absolute value for $|x| = 1$. In the normalized form $a_\nu = 1$; hence $\tilde{c} = 1$.

(2) This theorem is a version of the *Weierstrass preparation theorem*, which was initially proved for rings of germs of holomorphic functions in several complex variables. It has now several formulations in purely algebraic terms.

2.3. Entire Functions

Definition. *An* entire function *is a function f given by a formal power series $f \in K[[X]]$ having infinite radius of convergence: $r_f = \infty$.*

Before studying the entire functions more closely, let us prove the following elementary result.

Lemma. *For any sequence $(a_n)_{n\geq 0}$ in a complete ultrametric field K with $a_n \to 1$, the products $p_N := \prod_{n<N} a_n$ converge to a limit denoted by $\prod_{n\geq 0} a_n = \prod_{n=0}^{\infty} a_n$. More generally, if $(a_n)_{n\geq 0}$ is a sequence of K-valued functions defined on some set S, and if $a_n \to 1$ uniformly on S, the partial products p_N converge uniformly to $\prod_{n\geq 0} a_n$.*

PROOF. By assumption $|a_n| = 1$ for large n. Hence the partial products remain bounded, say $|p_N| \leq C$ $(N \geq 0)$. By definition,

$$p_{N+1} - p_N = (a_N - 1)p_N,$$
$$|p_{N+1} - p_N| \leq C|a_N - 1| \to 0 \quad (N \to \infty).$$

This proves that the sequence of partial products p_N is a Cauchy sequence. It converges in the complete field K. The second statement follows immediately from the first one. ∎

The exponential is an example of a function with no zero:

$$e^x \cdot e^{-x} = e^0 = 1 \quad \Longrightarrow \quad e^x \neq 0$$

(this is true for the complex exponential and for the p-adic exponential: Only the homomorphism property is used!). Although it is an example of an entire function in complex analysis, the finite radius of convergence of the p-adic exponential prevents this function from being entire in this context. In fact, any entire function having no zero in complex analysis is of the form $f(z) = e^{g(z)}$ for some entire function g, but the only entire functions in p-adic analysis having no zero in an algebraically closed field are the constants. This leads to an easy determination of entire functions in p-adic analysis. The main results are contained in the next statement.

Theorem. *Let $f \in K[[X]]$ be a formal power series with $r_f = \infty$.*

(a) If f does not vanish in K^a, then f is a nonzero constant.
(b) If f has only finitely many zeros in K^a, then it is a polynomial.
(c) If $0 \neq f \in \mathbf{C}_p[[X]]$, the following conditions are equivalent:
  (i) f has infinitely many zeros.
  (ii) f has a sequence of critical radii $\to \infty$.
  (iii) The growth of $|f|$ is not bounded by a polynomial in $|x|$,
  (iv) f is given by a convergent infinite product $f(x) = Cx^m \cdot \prod(1 - x/\xi)$ the product taken over nonzero roots of f, counting multiplicities, and $m = \mathrm{ord}_0 f$.

PROOF. (a) If f does not vanish, then $a_0 = f(0) \neq 0$ and $|f(x)| = |a_0|$ whenever $|x|$ is smaller than the first critical radius. Since f does not vanish, there is no critical radius; hence all $a_n = 0$ for $n \geq 1$. This proves that $f = a_0$ is constant.

(b) After division of f by the monic polynomial having the same roots as f, we are brought back to the first case.

(c) The equivalence $(i) \Longleftrightarrow (ii)$ is a consequence of the finiteness of zeros on each critical sphere. The equivalence $(ii) \Longleftrightarrow (iii)$ is Liouville's theorem (1.4). Finally, $(iv) \Longrightarrow (i)$ is clear, and we now show $(ii) \Longrightarrow (iv)$. By assumption $f \neq 0$, and if its order is $m \geq 0$, we can write

$$f(x) = \sum_{n \geq m} a_n x^n = Cx^m(1 + \sum_{n \geq 1} a'_n x^n)$$

with $C = a_m$ (and $a'_n = a_{m+n}/a_m$). Without loss of generality we may now assume that f is given by an expansion $f(x) = 1 + \sum_{n \geq 1} a_n x^n$. In this case $|f(x)| = 1$ for small x, namely for $|x| < r_0$ (r_0 denoting the first critical radius of f). Just beyond this critical radius, we shall have $|f(x)| = |x|^N$ if there are precisely N zeros (counting multiplicities) of f on the critical sphere $|x| = r_0$. Let us write

$$f(x) = p_0(x) \cdot f_1(x), \quad p_0(x) = \prod_{|\xi| = r_0,\, f(\xi) = 0} \left(1 - \frac{x}{\xi}\right).$$

The same procedure can obviously be iterated on each successive critical sphere and furnishes a factorization

$$f(x) = p_n(x) \cdot f_{n+1}(x), \quad p_n(x) = \prod_{|\xi| \leq r_n,\, f(\xi) = 0} \left(1 - \frac{x}{\xi}\right).$$

This construction makes it obvious that for fixed x, the terms $(1 - x/\xi)$ of the product tend to 1 and this convergence is uniform in x in a ball $B_{\leq R}$ (provided that $R < \infty$). This is the infinite product representation of f. It also shows that for each given sequence (ξ_i) with $|\xi_i| \to \infty$ there is an entire function having the ξ_i's as zeros (with correct multiplicities, and no other root): The corresponding infinite product converges uniformly on all bounded sets (its general term tends to 1 uniformly on bounded sets). Observe finally that if an infinite product of the form $\prod(1 - x/\xi)^\nu$ has the same zeros ($\neq 0$) as a power series f, the quotient

$$f(x) \Big/ \prod(1 - x/\xi)^\nu = g(x)$$

has no further zero $x \neq 0$. This function has no positive critical radius and can only be a monomial $c_m x^m$, m being the multiplicity of the zero at the origin. This concludes the proof of the theorem. ∎

2.4. Rolle's Theorem

Rolle's theorem for differentiable functions of a real variable is valid for scalar functions only. Here it is:

If $f : [a, b] \to \mathbf{R}$ $(a < b)$ is continuous and differentiable on the open interval (a, b), then there exists $a < c < b$ with

$$f'(c) = \frac{f(b) - f(a)}{b - a}.$$

The *mean value theorem with an intermediate point* follows from it. The preceding equality can be written $f(b) - f(a) = (b - a)f'(c)$ or with $a = t$, $b = t + h$, $c = t + \theta h$, as a limited expansion of the first order:

$$f(t + h) = f(t) + h \cdot f'(t + \theta h) \quad (0 < \theta < 1).$$

We give here the p-adic versions. Let us start with an easy observation.

Proposition. *Let $f \in K[[X]]$ be a convergent power series. For $\xi \in K$ with $|\xi| < r_f$, there is a unique convergent $f_\xi \in K[[X]]$ with $f(x) = f_\xi(x - \xi)$ for small $|x|$. Moreover, f_ξ has the same radius of convergence as f, and the preceding equality holds for $|x| < r_f$.*

If $f = \sum_{n\geq 0} a_n X^n$, this means that we can expand around ξ,

$$\sum_{n\geq 0} a_n x^n = \sum_{n\geq 0} a_n(\xi)(x - \xi)^n,$$

with no gain (no loss either) in convergence: $|x| < r_f \iff |x - \xi| < r_f$.

Proof. For a polynomial $f \in A[X]$, this is the Taylor formula for the expansion around the point $\xi \in A$:

$$\sum_{n\leq d} a_n X^n = \sum_{n\leq d} a_n(X - \xi + \xi)^n = \sum_{n\leq d} a_n(\xi)(X - \xi)^n = f_\xi(X - \xi),$$

with a polynomial $f_\xi \in A[X]$. This proves that $\sup_n |a_n(\xi)| \leq 1$ when $\sup_n |a_n| \leq 1$, so that $f \mapsto f_\xi$ diminishes the Gauss norms. The same is true for the converse isomorphism. We conclude that

$$f \mapsto f_\xi : K[X] \to K[X] \quad \text{is isometric.}$$

This isometry has a unique isometric extension to the completion $K\{X\}$: We still denote it by $f \mapsto f_\xi$. Now, if $f \in K[[X]]$ has $r_f > 0$, we may apply the preceding result to any $g = f(\alpha X)$ where $\alpha \in K^a$, $|\alpha| < r_f$, since $g \in K^a\{X\}$ in this case. This shows that the radius of convergence of f_ξ is greater than or equal to r_f, but as before, the inverse isometry proves the converse inequality and nothing is gained. ■

Theorem. *Let $f \in \mathbf{C}_p[[X]]$ have convergence radius $r_f > 1$. Then*

(a) if f has two distinct zeros $a \neq b$ in $B_{\leq 1}$ satisfying $|a - b| \leq r_p$, then f' has a zero in $B_{\leq 1}$;
(b) if f has two distinct zeros $a \neq b$ in $B_{<1}$ satisfying $|a - b| < r_p$, then f' has a zero in $B_{<1}$.

Proof. By the preceding proposition, we can replace f by its expansion centered at the point b. Thus, we may assume $a \neq b = 0$, $|a| \leq r_p$ (resp. $|a| < r_p$): $f(X) = \sum_{n \geq 1} a_n X^n$ and $a_1 \neq 0$ (otherwise, $f'(0) = 0$, and we are done). We can also assume that $|a| = r_c$ is the smallest positive critical radius. Hence there is an integer $n > 1$ such that

$$|a_1| r_c = |a_n| r_c^n,$$

whence

$$\left|\frac{a_1}{a_n}\right| = r_c^{n-1} \leq r_p^{n-1} \quad (\text{resp. } < r_p^{n-1}).$$

If $\nu = \text{ord}_p n$, say $n = p^\nu m$, m prime to p, we have

$$\frac{n-1}{p-1} = \frac{p^\nu m - 1}{p-1} \geq \frac{p^\nu - 1}{p-1} = p^{\nu-1} + \cdots + p + 1 \geq \nu$$

(with equality only for $m = 1$ and $\nu = 1$: $n = p$). Hence

$$\left|\frac{a_1}{a_n}\right| \leq r_p^{n-1} = |p|^{\frac{n-1}{p-1}} \leq |p|^\nu = |n|,$$

so that $|a_1| \leq |na_n|$ and $|a_1| = |na_n| r'^{n-1}$ for some $r' \leq 1$. Recalling that $r_{f'} = r_f > 1$, we see that the power series f' admits a critical radius $r' \leq 1$. By (2.2) f' has a zero in the closed unit ball. In the case (*b*), $|a_1| < |na_n|$ proves $r' < 1$, and the zero is in the open unit ball. ■

Example. Let $f = x^p - px$ and choose a root $p^{1/(p-1)} \in \mathbf{C}_p$. The zeros of f are 0 and $\mu_{p-1} \cdot p^{1/(p-1)}$. Hence two distinct roots are at a distance r_p. The zeros of f' are also the zeros of $f'/p = x^{p-1} - 1$, i.e., the elements of μ_{p-1} on the unit sphere.

Corollary. *Let $f \in \mathbf{C}_p[[X]]$ with $r_f > 1$. For each pair of points $a,\ b \in \mathbf{A}_p$ such that $|a-b| \le r_p$, there is a point $\xi \in \mathbf{A}_p$ such that*

$$f(b) - f(a) = (b-a)f'(\xi).$$

If $a,\ b \in \mathbf{M}_p$ and $|a-b| < r_p$, there is a point $\xi \in \mathbf{M}_p$ such that

$$f(b) - f(a) = (b-a)f'(\xi).$$

PROOF. As in the classical case, consider the function

$$\phi(x) = \begin{vmatrix} f(a) & f(x) & f(b) \\ a & x & b \\ 1 & 1 & 1 \end{vmatrix},$$

which vanishes at $x = a$ and $x = b$. Its derivative

$$\phi'(x) = \begin{vmatrix} f(a) & f'(x) & f(b) \\ a & 1 & b \\ 1 & 0 & 1 \end{vmatrix}$$

vanishes in $\mathbf{A}_p$ (resp. $\mathbf{M}_p$). ∎

2.5. *The Maximum Principle*

The preceding theory concerning critical radii — and particularly the existence of zeros on critical spheres — has important consequences for the study of power series.

Proposition. *Let $r < r_f$ be a critical radius of $f \in \mathbf{C}_p[[X]]$. Then $|f|$ takes all values between 0 and $M_r f$ in $|\mathbf{C}_p|$. More precisely, for each $y \in \mathbf{C}_p$ with $|y| < M_r f$, there is a solution $x \in \mathbf{C}_p$ of the equation $f(x) = y$ with $|x| = r$. If $|y| = M_r f$, the same equation also has a root of absolute value r, provided that $|y - f(0)| = M_r f$.*

PROOF. Consider the formal power series

$$f(X) - y = (a_0 - y) + \sum_{n \ge 1} a_n X^n \quad (a_0 = f(0)).$$

If $|y| < M_r f$, then $f - y$ has the same dominant monomials as f, and r is still a critical radius of $f - y$: This function vanishes on the corresponding sphere. If $|y| = M_r f$, the assumption made ensures that the formal power series

$f - y = (a_0 - y) + \sum_{n\geq 1} a_n X^n$ also admits the critical radius r and hence vanishes on the corresponding critical sphere of $\mathbf{C}_p$. This proves that

$$M_r f = \sup_{|x|<r} |f(x)| = \sup_{|x|\leq r} |f(x)| = \max_{|x|=r} |f(x)|$$

when r is critical. ■

Corollary. *Let $r < r_f$. Then*

$$M_r f = \sup_{|x|<r} |f(x)| = \sup_{|x|\leq r} |f(x)|.$$

Moreover, if $r \in |\mathbf{C}_p^\times|$ is a rational power of p, then

$$M_r f = \max_{|x|\leq r} |f(x)| = \max_{|x|=r} |f(x)|.$$

PROOF. For every $r < r_f$,

$$|f(x)| \leq M_{|x|} f \leq M_r f \quad (|x| \leq r)$$

implies that

$$\sup_{|x|<r} |f(x)| \leq \sup_{|x|\leq r} |f(x)| \leq M_r f.$$

Conversely, we can find a sequence (x_n) in $\mathbf{C}_p$ such that:

$$|x_n| = r_n \text{ is regular for all } n \text{ and } r_n \nearrow r.$$

Hence

$$|f(x)| = M_{r_n} f \nearrow M_r f$$

implies that

$$\sup_{|x|<r} |f(x)| \geq \sup_n |f(x_n)| = M_r f.$$

Finally, if r is regular, $|f(x)| = M_r f$ is constant on the sphere $|x| = r$, while if r is critical, $M_r f = \max_{|x|=r} |f(x)|$ follows from the proposition. ■

2.6. *Extension to Laurent Series*

Instead of Taylor series, we can work with convergent Laurent series

$$f = \sum_{-\infty}^{\infty} a_n X^n \in K[[X, X^{-1}]]$$

as in (1.7). Existence of zeros on critical spheres $r_f^- < |x| = r < r_f^+$ is ensured, provided that the field K is *algebraically closed* (typically, if $K = \mathbf{C}_p$).

For example, let r be a critical radius of f. As in (2.5), an equation $f(x) = y \in \mathbf{C}_p$ will have a solution $x \in \mathbf{C}_p$ with $|x| = r$, provided that

$$\text{either } |y| < M_r f \text{ or } |y| = M_r f \text{ and } |y - a_0| = M_r f.$$

This shows that

$$\sup_{|x|<r} |f(x)| = \max_{|x|=r} |f(x)| = M_r f \quad (r_f^- < r < r_f^+).$$

In the case of a Laurent series f, the function $r \mapsto M_r f$ is not necessarily increasing, but it is always a convex function on the interval (r_f^-, r_f^+). A consequence of this observation is the *maximum principle for annuli*:

$$\text{If } r_f^- < r_1 \leq r_2 < r_f^+ \quad (r_i \in |K^\times|), \text{ then}$$

$$\sup\nolimits_{r_1 \leq |x| \leq r_2} |f(x)| = \max\nolimits_{r_1 \leq |x| \leq r_2} |f(x)| = \max(M_{r_1} f\, ,\ M_{r_2} f).$$

We have also seen that

$$\rho = \log r \mapsto \mu_\rho f = \log M_r f$$

is a convex function on the interval $(\log r_f^-, \log r_f^+)$ (cf. (1.7)). Let us show that it is — as in complex analysis — a formal consequence of the maximum principle for all functions $x^m f^n$ ($m \in \mathbf{Z},\ n \in \mathbf{N}$) (given by convergent Laurent series by (1.7)) in annuli $r_f^- < r_1 \leq r_2 < r_f^+$.

Hadamard's Three-Circle Theorem. *Assume that $f \in K[[X, X^{-1}]]$ is a convergent Laurent series, so that f is also a function defined on an annulus $r^- < |x| < r^+$ of K. Then*

$$\rho = \log r \mapsto \mu_\rho f = \log M_r f = \log \max_{|x|=r} |f(x)|$$

is a convex function.

Proof. Let $(r^- < r_1 < r < r_2 < r^+)$, $M_i = M_{r_i} f$, so that

$$M_r f \leq \max(M_1, M_2)$$

by the maximum principle. Apply this inequality to $x^m f^n$ ($m \in \mathbf{Z},\ n \geq 0$),

$$r^m M_r^n \leq \max(r_1^m M_1^n, r_2^m M_2^n),$$

and taking nth roots

$$r^{m/n} M_r \leq \max(r_1^{m/n} M_1, r_2^{m/n} M_2).$$

If $K = \mathbf{C}_p$, we can choose the rational number $\alpha = m/n$ such that $r_1^\alpha M_1 = r_2^\alpha M_2$ (if $K = \Omega_p$ or $\mathbf{C}$, we can take a sequence of rational numbers m_k/n_k converging to the real root α of $r_1^\alpha M_1 = r_2^\alpha M_2$). With this choice for α (using continuity if

$K = \Omega_p$ or $\mathbf{C}$), we can write

$$r^\alpha M_r \leq r_1^\alpha M_1 = r_2^\alpha M_2 = (r_1^\alpha M_1)^s \cdot (r_2^\alpha M_2)^t$$

if $s + t = 1$. Since $\rho = \log r$ is a convex combination of the $\rho_i = \log r_i$, we can choose $s \geq 0, t \geq 0$ with $s + t = 1$ and $r = r_1^s \cdot r_2^t$. With this choice $r^\alpha = r_1^{\alpha s} \cdot r_2^{\alpha t}$, and the obtained inequality simplifies into

$$M_r \leq M_1^s \cdot M_2^t.$$

With $\mu_\rho = \log M_r$ and $\mu_i = \log M_i$ ($i = 1, 2$), we now get the announced convexity property

$$\mu_\rho \leq s\mu_1 + t\mu_2 \quad (\rho = s\rho_1 + t\rho_2). \qquad \blacksquare$$

Definition. *Let f be a Laurent series with $r_f^- = 0$. We say that the origin is an* isolated singularity. *Three cases can occur:*

(1) *f is a Taylor series ($a_n = 0$ for all $n < 0$):*
The origin is a removable singularity.
(2) *f has finitely many coefficients $a_n \neq 0$ for $n < 0$:*
The origin is a pole.
(3) *f has infinitely many coefficients $a_n \neq 0$ for $n < 0$:*
The origin is an essential singularity.

If the origin is a pole of f, its order is the smallest integer $m \geq 0$ such that $x^m f$ is a Taylor series (has a removable singularity at the origin). In the case of an essential singularity, the analogue of a classical result of Picard is valid.

Proposition. *Let f have an essential singularity (at the origin). Then there are infinitely many critical radii $r_i \searrow 0$, and for each $\varepsilon > 0$, $y \in \mathbf{C}_p$, the equation $f(x) = y$ has infinitely many solutions $0 < |x| < \varepsilon$.* $\blacksquare$

Proposition. *Let f be a Laurent series with $r_f^- = 0$ and $r_f^+ = \infty$. Then*

(a) *if f has no zero in $\mathbf{C}_p^\times$, f is a single monomial;*
(b) *if f has only finitely many zeros in $\mathbf{C}_p^\times$,*
then f is a polynomial in X and X^{-1};
(c) *f is given by a Weierstrass product*
$f(X) = CX^m \cdot \prod_{|\xi|\geq 1}(1 - X/\xi)^{\nu_\xi} \cdot \prod_{|\xi|<1}(1 - \xi/X)^{\nu_\xi}$
extended over the roots ξ of $f = 0$,
ν_ξ denoting the multiplicity of the root ξ. $\blacksquare$

Example: Theta Functions. Choose an element $q \in \mathbf{C}_p^\times$ with $0 < |q| < 1$. Consider the product

$$\Theta_1(X) = \prod_{n\geq 0}(1 - q^n X) \prod_{n>0}\left(1 - \frac{q^n}{X}\right),$$

which converges in the annulus $0 < |x| = r < \infty$. Obviously,

$$\Theta_1(q^{-1}X) = -\frac{X}{q} \cdot \Theta_1(X).$$

If we define more generally $\Theta_a(X) = \Theta_1(a^{-1}X)$ ($a \in \mathbf{C}_p^\times$), then we have

$$\Theta_a(q^{-1}X) = -\frac{X}{aq} \cdot \Theta_a(X).$$

Products of such *theta functions* satisfy functional equations of the form

$$\Theta(q^{-1}X) = C(-X)^d \cdot \Theta(X).$$

These functions are used for the construction of the Tate elliptic curves.

3. Rational Functions

Functions defined by convergent power series expansions are defined in a ball. Unfortunately, as we have seen in (2.4), it is impossible to obtain an analytic extension of such a function by looking at the expansions at different points of the ball of convergence: The radius of convergence does not change, so the ball of convergence is the same. Any point of a ball is a center of the ball, and there is no way of defining "points near the edge."

On the other hand, we like to consider rational functions (quotients of polynomial functions) as analytic functions outside their set of poles (zeros of their denominators). These functions can be expanded in power series in each ball containing none of their poles. More generally, uniform limits of rational functions will play a role similar to the analytic functions in complex analysis: They are the "analytic elements" introduced by Krasner.

3.1. Linear Fractional Transformations

A *linear fractional transformation* is a rational function

$$x \longmapsto f(x) = \frac{ax+b}{cx+d}$$

where $ad-bc \neq 0$. The coefficients are taken from a field K, and a linear fractional transformation defines a map

$$f : K \cup \{\infty\} \to K \cup \{\infty\}.$$

The space $K \cup \{\infty\} = \mathbf{P}^1(K)$ is the projective line over K: Its elements are the homogeneous lines in K^2, represented by quotients

$$\frac{x}{y} = [x : y] = \text{class of pairs proportional to } (x, y).$$

When $c = 0$ (a and $d \neq 0$), we get an *affine linear map*

$$x \mapsto \frac{a}{d}x + \frac{b}{d} = a'x + b' \quad (a' \neq 0).$$

When $c \neq 0$,

$$f(x) = \frac{ax+b}{cx+d} = \frac{1}{c}\left[a + \left(b - \frac{ad}{c}\right)\frac{1}{x + d/c}\right].$$

Typical examples of linear fractional transformations are

(*a*) *translations* $x \mapsto x + b$,
(*b*) *dilatations* (or *homotheties*) $x \mapsto ax$,
(*c*) *inversion* $x \mapsto 1/x$.

The preceding formula shows that these particular linear fractional transformations generate the group of all linear fractional transformations.

A good description of linear fractional transformations is supplied by 2×2 matrices: To each such invertible matrix we associate the linear fractional transformation having for coefficients the entries of the matrix

$$g = \begin{pmatrix} a & b \\ c & d \end{pmatrix} \longmapsto f_g(x) = \frac{ax+b}{cx+d}.$$

Composition of linear fractional transformations corresponds to matrix multiplication: The above correspondence is a homomorphism from the group $\mathrm{Gl}_2(K)$ of invertible 2×2 matrices with entries in K to the group of automorphisms of the projective line $\mathbf{P}^1(K) = K \cup \{\infty\}$. The kernel of this homomorphism consists of the nonzero multiples of the identity matrix I_2 (scalar matrices) $\begin{pmatrix} a & 0 \\ 0 & a \end{pmatrix} = a \cdot I_2$ ($a \neq 0$), namely the center of $\mathrm{Gl}_2(K)$. Hence there is an isomorphism

$$\mathrm{PGl}_2(K) = \mathrm{Gl}_2(K)/(K^\times I_2) \xrightarrow{\sim} \mathrm{Aut}\,(\mathbf{P}^1(K)).$$

Here are representative matrices for the three types of linear fractional transformations listed above:

(*a*) The matrix $\begin{pmatrix} 1 & b \\ 0 & 1 \end{pmatrix}$ produces the *translation* $x \mapsto x + b$.

(*b*) The matrix $\begin{pmatrix} a & 0 \\ 0 & 1 \end{pmatrix}$ produces the *dilatation* (or *homothety*) $x \mapsto ax$.

(*c*) The matrix $\begin{pmatrix} 0 & 1 \\ 1 & 0 \end{pmatrix}$ produces the *inversion* $x \mapsto 1/x$.

Proposition. *Let K be an ultrametric field. Then the image of a ball of K under a linear fractional transformation is either a ball or the complement of a ball.*

PROOF. Affine linear transformations send an open (resp. dressed) ball to an open (resp. dressed) ball. The formula

$$f(x) = \frac{ax+b}{cx+d} = \frac{1}{c}\left[a + (b - ad/c)\frac{1}{x+d/c}\right]$$

(when $c \neq 0$) shows that a linear fractional transformation that is not an affine linear map is nevertheless composed of such transformations *and* of an inversion. It is thus sufficient to prove the statement for the inversion. Consider, for example, the ball $B_{<r}(a)$ and its image B' by inversion. If the origin belongs to $B_{<r}(a)$, then this ball coincides with $B_{<r} = B_{<r}(0)$ and its inverse is the set defined by $|x| > 1/r$, i.e., the complement of the ball $B_{\leq 1/r}(a)$. Otherwise, $|a| \geq r$, and for $x \in B_{<r}(a)$,

$$|x-a| < r, \quad |x| = |a + (x-a)| = |a|,$$

so that

$$\left|\frac{1}{x} - \frac{1}{a}\right| = \left|\frac{a-x}{xa}\right| = \frac{|a-x|}{|a|^2} < \frac{r}{|a|^2}.$$

This proves that the image of the ball $B_{<r}(a)$ under inversion is contained in the ball $B_{<r/|a|^2}(1/a)$. Since the same argument shows that the image under inversion of the second ball must be contained in the first one, we conclude that the inversion is a bijection between these balls. A completely similar proof holds for closed balls instead of open ones (alternatively, one can use the relation $B_{\leq r}(a) = \bigcap_{s>r} B_{<s}(a)$ between closed and open balls.) Hence we have

$$\begin{aligned} f(B_{<r}(a)) &= B_{<r/|a|^2}(1/a) \quad \text{if } |a| \geq r, \\ f(B_{\leq r}(a)) &= B_{\leq r/|a|^2}(1/a) \quad \text{if } |a| > r, \\ f(\{x : |x-a| = r\}) &= \{y : |y - 1/a| = r/|a|^2\} \quad \text{if } |a| > r. \end{aligned}$$

∎

The image of a ball is called a *generalized ball*: A complement of a ball of K is identified to a ball of $\mathbf{P}^1(K)$ containing the point at infinity; note, however, that in general two such generalized balls satisfy no inclusion relation. The analogy with the classical complex case is striking. Indeed, recall that in $\mathbf{C}$ a linear fractional transformation preserves the family of generalized circles (circles and straight lines) and the family of generalized disks (disks, half planes, and complements of disks).

3.2. Rational Functions

Let us review a few elementary algebraic facts concerning rational functions having coefficients in an *algebraically closed field* K. Let $f \in K(x)$ be a rational function and write $f = g/h$ with two relatively prime polynomials g and h, h being monic. We say that f is *regular* at the point $a \in K$ if $h(a) \neq 0$. In this case, $f(a) = g(a)/h(a)$ is well-defined, and the numerator of the function $f - f(a)$

vanishes, and hence is divisible by $x - a$. This shows that if f is regular at a, we can write

$$f = f(a) + (x-a)f_{(1)}(x), \quad f_{(1)} \text{ rational, regular at } a.$$

Iterating this construction for $f_{(1)}$, we obtain a second-order limited expansion of f. By induction, we see that for any integer $m \geq 1$,

$$f = a_0 + a_1(x-a) + \cdots + a_{m-1}(x-a)^{m-1} + (x-a)^m f_{(m)},$$

where $f_{(m)}$ is rational, regular at the point a.

We say that $f \neq 0$ has a *pole of order* $m > 0$ if the denominator h has a zero of order m at the point a (hence $g(a) \neq 0$, since g is prime to h). In this case, we write $h(x) = (x-a)^m h_1(x)$, $h_1(a) \neq 0$; hence

$$f = \frac{g}{(x-a)^m h_1} = \frac{1}{(x-a)^m} f_1, \quad f_1 = \frac{g}{h_1} \text{ regular at } a.$$

If we write the above expansion for f_1, we obtain

$$\begin{aligned} f &= \frac{1}{(x-a)^m}\left(a_0 + a_1(x-a) + \cdots + a_{m-1}(x-a)^{m-1} + (x-a)^m \tilde{f}\right) \\ &= \frac{a_0}{(x-a)^m} + \frac{a_1}{(x-a)^{m-1}} + \cdots + \frac{a_{m-1}}{x-a} + \tilde{f} = P_a\left(\frac{1}{x-a}\right) + \tilde{f} \end{aligned}$$

with a polynomial P_a of degree m and zero constant term. The rational function

$$P_a\left(\frac{1}{x-a}\right) = \frac{a_0}{(x-a)^m} + \cdots + \frac{a_{m-1}}{x-a}$$

is the *principal part* of f at the pole a. It is uniquely characterized by the properties

$$\begin{cases} P_a \text{ is a polynomial with zero constant term,} \\ f - P_a\left(\dfrac{1}{x-a}\right) \text{ is regular at } a\,. \end{cases}$$

The order of $f \neq 0$ at the point $a \in K$ is by definition

$$\mathrm{ord}_a f = \mathrm{ord}_a g - \mathrm{ord}_a h \in \mathbf{Z}.$$

This integer is positive if f (is regular and) vanishes at a, negative $= -m$ if f has a pole of order m at a.

Consider the finite set $\{\alpha_i\} = \{a \in K : h(a) = 0\}$ of poles of $f = g/h$ and the respective principal parts $P_i\left(\frac{1}{x-\alpha_i}\right)$ of f. Then $f - \sum_i P_i\left(\frac{1}{x-\alpha_i}\right)$ is a rational function that is regular everywhere: It is a polynomial, and we have obtained the decomposition

$$f = \sum_i P_i\left(\frac{1}{x-\alpha_i}\right) + P_\infty \quad (P_\infty \in K[x]).$$

One way to obtain this decomposition is to start with the Euclidean division algorithm for polynomials,

$$g = P_\infty h + g_1, \quad \deg g_1 < \deg g.$$

Then

$$f = \frac{g}{h} = \frac{g_1}{h} + P_\infty \quad (P_\infty \in K[x]).$$

If $\deg g \geq \deg h$, then $\deg P_\infty = \deg g - \deg h$; otherwise, $P_\infty = 0$. The well-known partial fractions expansion for the first term leads to

$$\frac{g_1}{h} = \sum_i P_i\left(\frac{1}{x-\alpha_i}\right).$$

The particular rational functions

$$\frac{1}{(x-a)^m}, \quad x^n \quad (a \in K,\ m > 0,\ n \geq 0)$$

generate the K-vector space $K(x)$. Since they are also independent, they make a vector space basis of $K(x)$.

If K is a *valued field*, we have

$$\left|P_a\left(\frac{1}{x-a}\right)\right| \to 0 \quad (|x| \to \infty),$$

$$\left|P_a\left(\frac{1}{x-a}\right)\right| \to \infty \quad (|x| \to a).$$

With the previous notation

$$f(x) = \underbrace{\sum_{h(a)=0} P_a\left(\frac{1}{x-a}\right)}_{\to 0\ |x| \to \infty} + P_\infty(x),$$

we see that

$$\begin{aligned} |f(x)| \to 0 \text{ when } |x| \to \infty \quad &\Longleftrightarrow \quad P_\infty(x) = 0,\\ |f(x)| \text{ is bounded when } |x| \to \infty \quad &\Longleftrightarrow \quad P_\infty(x) \text{ is constant},\\ |f(x)| \to \infty \text{ when } |x| \to \infty \quad &\Longleftrightarrow \quad \deg P_\infty(x) > 0. \end{aligned}$$

Let us now specialize to $K = \mathbf{C}_p$, algebraically closed and complete. We can use the binomial series expansion: For $a \neq 0$, $m \geq 1$,

$$\begin{aligned} \frac{1}{(x-a)^m} &= \frac{1}{(-a)^m}\left(1 - \frac{x}{a}\right)^{-m}\\ &= \sum_{n\geq 0} (-1)^{m+n}\binom{-m}{n}\frac{x^n}{a^{m+n}} \quad (|x| < |a|) \end{aligned}$$

(a Taylor series),

$$\frac{1}{(x-a)^m} = \frac{1}{x^m}\left(1-\frac{a}{x}\right)^{-m}$$
$$= \sum_{n\geq 0}(-1)^n\binom{-m}{n}\frac{a^n}{x^{m+n}} \qquad (|x|>|a|)$$

(a Laurent series). In particular, in any region $r_1 < |x| < r_2$ containing no pole of f, we can choose the first type of expansion for the principal parts corresponding to poles $|\alpha_i| \geq r_2$ and the second type for the principal parts corresponding to poles $|\alpha_i| \leq r_1$. We obtain *Laurent series* expansions (1.7) (and (2.6)).

Proposition. *Let $f \in \mathbf{C}_p(x)$ be a nonzero rational function and $\{\alpha_i\}$ its set of poles. Then f admits three types of Laurent series expansions:*

(a) $\sum_{-m\leq n<\infty} a_n x^n \quad (0 < |x| < \min\{|\alpha_i| : \alpha_i \neq 0\})$,
(b) $\sum_{-\infty<n<\infty} a_n x^n \quad (\max\{|\alpha_i| : |\alpha_i| \leq r\} < |x| < \min\{|\alpha_i| : |\alpha_i| > r\})$,
(c) $\sum_{-\infty<n<N} a_n x^n \quad (|x| > \max\{|\alpha_i|\})$.

Proof. In the first case *(a)*, if f has a pole at the origin, then m is its order. Let us consider only the case *(b)*. If $r > 0$ is fixed, group the principal parts corresponding to poles in the closed ball $|x| \leq r$. For each individual monomial in these principal parts, a multiple of some $1/(x-\alpha_i)^m$, choose the Laurent expansion that converges for $|x| > |\alpha_i|$. Any linear combination of these expansions converges at least for $|x| > \max\{|\alpha_i| : |\alpha_i| \leq r\}$. Group similarly the principal parts corresponding to the poles $|\alpha_i| > r$, and choose the Taylor series for the corresponding monomials $1/(x-\alpha_i)^m$: Their linear combination converges at least for $|x| < \min\{|\alpha_i| : |\alpha_i| > r\}$. Adding these two contributions, we get a Laurent series as announced. Observe that since a Laurent series defines a continuous sum in its open annulus of convergence, this region cannot contain a pole of the sum, whence the precise estimate for the radii limiting its region of convergence. ■

3.3. The Growth Modulus for Rational Functions

Let us say that a radius $r > 0$ is *regular* for a rational function $f = g/h \in \mathbf{C}_p(x)$ (g and h relatively prime polynomials) when it is regular for both g and h, hence when g and h do not vanish on the sphere $|x| = r$ of $\mathbf{C}_p$. Hence, when $|x| = r$ is regular for $f = g/h$,

$$|g(x)| = M_r g, \quad |h(x)| = M_r h, \quad \text{and} \quad |f(x)| = \frac{M_r g}{M_r h}.$$

Lemma. *Let $f = g/h \in \mathbf{C}_p(x)$ and define $\widetilde{M}_r f := M_r g / M_r h$ for real $r > 0$. This expression is well-defined independently from the particular representation of f as a fraction g/h, and $r \mapsto \widetilde{M}_r f$ is continuous on $\mathbf{R}_{>0}$. For each regular*

$r > 0$, $|f(x)| = \widetilde{M}_r f$ on the sphere $|x| = r$. In each region where f has a Laurent series expansion, $\widetilde{M}_r f$ coincides with the growth modulus as defined in (2.6).

PROOF. If $f = g/h = g_1/h_1$, then $gh_1 = g_1 h$, $|g(x)h_1(x)| = |g_1(x)h(x)|$ implies

$$M_r g \cdot M_r h_1 = M_r g_1 \cdot M_r h$$

first for the regular values r (of g, h, g_1, and h_1, i.e., of the product $g h g_1 h_1$), and hence also for all values $r > 0$ by continuity. This proves that $M_r g / M_r h = M_r g_1 / M_r h_1$. The other assertions of the lemma are obvious. ∎

Considering the previous results, we shall simply denote by $M_r f = M_r g / M_r h$ the growth modulus of a rational function $f = g/h$.

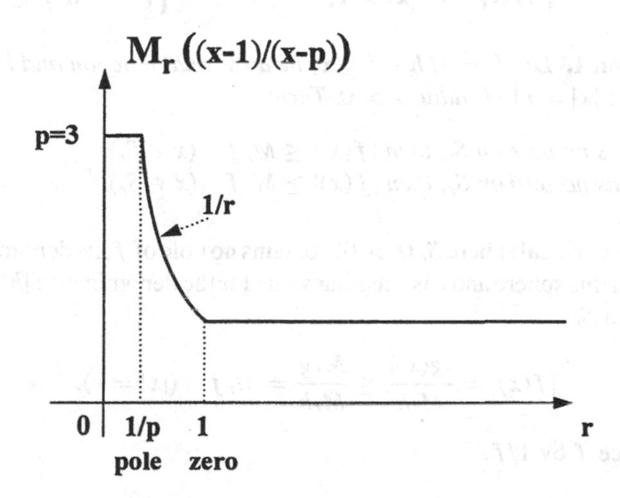

Growth modulus for a linear fractional transformation

Observe that for a nonzero rational function f, $M_r f > 0$ for $r > 0$. When a radius $r > 0$ is not regular for the rational function $f = g/h$, we say that it is a *critical radius*: f has some poles and/or zeros on the sphere $|x| = r$. Denote as before by $\{\alpha_i\}$ the poles of f and introduce its set $\{\beta_j\}$ of zeros.

Theorem. *Let $f = g/h \in \mathbf{C}_p(x)$ be a rational function. Then we have:*

(a) If f is regular at the origin, then $r \mapsto M_r f$ is convex increasing on the interval $0 \le r \le \min\{|\alpha_i|\}$.
(b) If f has no pole in the region $r_1 < |x| < r_2$, then $r \mapsto M_r f$ is convex in the interval $r_1 \le r \le r_2$.
(c) If $\deg g < \deg h$, then $r \mapsto M_r f$ is decreasing for $r \ge \max\{|\alpha_i|\}$.
(d) $|f(x)| = M_r f = c\, r^{\deg g - \deg h}$ for $|x| = r > \max\{|\alpha_i|, |\beta_j|\}$.

PROOF. (*a*) and (*b*) follow from the lemma and (1.4). For (*c*) and (*d*), observe that for any polynomial P of degree d, $M_r P = |a_d| r^d$ when r is bigger than the absolute value of all roots of P. Apply this to $P = h$ to obtain (*c*), and to both g and h to obtain (*d*). ■

Example. Consider the rational function

$$f(x) = \frac{x}{1 - x^2} \in \mathbf{C}_p(x),$$

which has a simple zero at the origin and two poles on the unit sphere. For $|x| < 1$ we have $|1 - x^2| = 1$, while $|1 - x^2| = |x|^2$ for $|x| > 1$. Hence

$$|f(x)| = \begin{cases} |x| & \text{if } |x| < 1, \\ 1/|x| & \text{if } |x| > 1, \end{cases} \qquad M_r f = \begin{cases} r & \text{if } r \leq 1, \\ r^{-1} & \text{if } r \geq 1. \end{cases}$$

Proposition 1. *Let $f = g/h \in \mathbf{C}_p(x)$ be a rational function and let S_r be the sphere $\{x : |x| = r\}$ of radius $r > 0$. Then:*

(*a*) *If f has no pole on S_r, then $|f(x)| \leq M_r f \quad (x \in S_r)$.*
(*b*) *If f has no zero on S_r, then $|f(x)| \geq M_r f \quad (x \in S_r)$.*

PROOF. (*a*) If a critical sphere S_r ($r > 0$) contains no pole of f, its denominator does not vanish on this sphere and r is a regular value for the denominator: $|h(x)| = M_r h$ is constant on S_r,

$$|f(x)| = \frac{|g(x)|}{M_r h} \leq \frac{M_r g}{M_r h} = M_r f \quad (|x| = r).$$

(*b*) Replace f by $1/f$. ■

If f has a zero $\beta \in S_r$, then by continuity, $|f|$ takes arbitrarily small values in a neighborhood $B_{<\varepsilon}(\beta)$ of β. Such a neighborhood is contained in the sphere S_r as soon as $\varepsilon \leq |\beta|$. Hence $|f|$ takes arbitrarily small values on the sphere S_r. The same holds for $1/f$ if f has a pole $\alpha \in S_r$. If f has both zeros and poles in S_r, this shows that $|f|$ takes arbitrarily small and large values on this sphere. This will be made more precise in the next propositions.

Proposition 2. *Let $f = g/h \in \mathbf{C}_p(x)$, S_r as before and consider an open ball $D = B_{<r}(a)$ of maximal radius in the sphere S_r (hence $|a| = r$). If f has no pole in D, then*

$$M_r f = \sup_{x \in D} |f(x)| := \|f\|_D.$$

PROOF. For $s > r = |a|$, the spheres S_s and $S_s(a) = \{x : |x - a| = s\}$ coincide. Hence $M_s f = M_{s,a} f$ (growth modulus with respect to the center a): This is obvious for regular values of s and by continuity also for all values $s \geq r$. This proves

$$M_r f = M_{r,a} f.$$

Since f is regular in the ball D, its growth modulus $M_{t,a} f$ (with respect to the center a) is an increasing function of $t < r$. By the maximum principle (2.5) for balls,

$$M_{t,a} f = \sup_{|x-a| \leq t} |f(x)| \quad (t < r),$$

and by continuity of the function $t \mapsto M_{t,a} f$,

$$\begin{aligned} M_{r,a} f &= \sup_{t<r} M_{t,a} f = \sup_{t<r} \sup_{|x-a| \leq t} |f(x)| \\ &= \sup_{|x-a|<r} |f(x)| = \|f\|_D. \end{aligned}$$ ■

Observe that since we work with the field $\mathbf{C}_p$, having an infinite residue field, a sphere S_r of positive radius is a disjoint union of infinitely many open balls of maximal radius r, so that it is always possible to choose a ball $D = B_{<r}(a)$ without pole of f as in Proposition 2.

Proposition 3. *Let $f = g/h \in \mathbf{C}_p(x)$ be a rational function, S_r as before. Then:*

(a) If f has no pole on S_r, then $M_r f = \sup_{|x|=r} |f(x)|$.
(b) If f has no zero on S_r, then $M_r f = \inf_{|x|=r} |f(x)|$.
(c) If f has both zeros and poles on S_r, then $|f(x)|$ assumes all values of $|\mathbf{C}_p|$ on $x \in S_r$.

Proof. Observe that if f has no pole and no zero in S_r, then r is regular and $|f(x)| = M_r f$ is constant on S_r. Now (a) follows from Propositions 1 and 2. For (b), replace f by $1/f$ and apply the previous result.

(c) Choose a pole $\alpha \in S_r$ and a zero $\beta \in S_r$ with

$$|\alpha - \beta| = \min |\alpha_i - \beta_j| := \delta$$

(minimum taken over the zeros β_j and poles α_i in S_r). Then

$$M_{s,\alpha} f = M_{s,\beta} f \quad (s > \delta),$$

since the spheres of radius $s > \delta$ and centers α (resp. β) coincide. By continuity,

$$M_{\delta,\alpha} f = M_{\delta,\beta} f := M.$$

Now, for each $y \in \mathbf{C}_p$, $|y| < M$, $f - y$ has a critical radius $r < \delta$ and $f(x) = y$ has a solution $x \in B_{<\delta}(\beta) \subset S_r$. Similarly, for each $y \in \mathbf{C}_p$, $|y| > M$, $f(x) = y$

has a solution $x \in B_{<\delta}(\alpha) \subset S_r$ (consider $1/f$). Finally, as in (2.6), f also assumes some values $y = f(x)$ where $|y| = M$. ∎

3.4. Rational Mittag-Leffler Decompositions

Recall that we denote the complement of a set $B \subset \mathbf{C}_p$ by $B^c = \mathbf{C}_p - B$.

Proposition 1. *Let* $0 \neq f = g/h \in \mathbf{C}_p(x)$. *Assume* $\deg g < \deg h$ *and that* f *has all its poles in a ball* $B = B_{<\sigma}$ *for some* $\sigma > 0$. *Then for any subset* D *disjoint from* B, $\|f\|_D \leq \|f\|_{B^c} = M_\sigma f$. *If* $D = B_{<\sigma}(a)$ *is a maximal open ball in the sphere* $|x| = \sigma$, *then*

$$\|f\|_D = \|f\|_{B^c} = M_\sigma f.$$

PROOF. Since f has all its poles α_i in B, we have $\sigma_p = \max_i |\alpha_i| < \sigma$ and

$$|f(x)| \leq M_{|x|} f \quad (|x| > \sigma_p).$$

On the other hand, since $\deg g < \deg h$, the growth modulus $M_r f$ decreases for $r \geq \sigma_p$,

$$|f(x)| \leq M_\sigma f \quad (|x| \geq \sigma),$$

and for $D \subset B^c$ we have

$$\|f\|_D \leq \|f\|_{B^c} \leq M_\sigma f.$$

Taking a sequence $x_n \in B^c$ with regular $r_n = |x_n| \searrow \sigma$, so that $|f(x_n)| = M_{r_n} f \nearrow M_\sigma f$, we see that $\|f\|_{B^c} \geq \sup_n |f(x_n)| = M_\sigma f$. Finally, if $D = B_{<\sigma}(a)$ is a maximal open ball in the sphere $|x| = \sigma$, then Proposition 2 of (3.3) shows that $\|f\|_D = M_\sigma f$, since f has no pole on $|x| = \sigma$. ∎

Observation. The last step of the preceding proof, $\|f\|_D = M_\sigma f$, requires only that we find a sequence (x_n) in D with regular $r_n = |x_n - a| \nearrow \sigma$, so that

$$|f(x_n)| = M_{r_n,a} f \to M_{\sigma,a} f = M_\sigma f,$$

$$\|f\|_D \geq \sup |f(x_n)| \geq \lim |f(x_n)| = M_\sigma f.$$

This will be essential for generalizations (cf. Proposition 2 in (4.2)).

If $f = g/h \in \mathbf{C}_p(x)$ is a rational function and $\sigma > 0$, we can group the principal parts corresponding to the poles α_i of f in the open ball $B = B_{<\sigma}$:

$$f_B(x) = \sum_{|\alpha_i| < \sigma} P_i \left(\frac{1}{x - \alpha_i} \right).$$

We can apply Proposition 1 to f_B, since this function has all its poles in B. The growth modulus $M_r f_B$ is decreasing for $r \geq \sigma$:

$$\|f_B\|_D = \sup_{x \in D} |f_B(x)| \leq M_\sigma f_B \quad (D \subset B^c).$$

Proposition 2. *Let $f \in \mathbf{C}_p(x)$ be a rational function and let f_B be the sum of the principal parts of f corresponding to its poles in $B = B_{<\sigma}$. If D is a maximal open ball $B_{<\sigma}(a)$ in the sphere $|x| = \sigma$ and D contains no pole of f, then*

$$\|f_B\|_D = M_\sigma f_B \leq M_\sigma f = \|f\|_D.$$

PROOF. We may assume $f_B \neq 0$ and let us introduce $f_0 := f - f_B \in \mathbf{C}_p(x)$, which is regular in B (but may have poles in the sphere $|x| = \sigma$). Hence

$$r \mapsto M_r f_0 \quad \text{is increasing (may be constant) for } r \leq \sigma.$$

On the other hand, $M_r f_B$ decreases (strictly) beyond

$$\sigma_p := \max\{|\alpha| : \alpha \text{ is a pole of } f \text{ in } B\} < \sigma.$$

There is at most one crossing point of $M_r f_0$ and $M_r f_B$ in the interval (σ_p, σ). Hence $M_r f_0 \neq M_r f_B$ with at most one exception $r \in (\sigma_p, \sigma)$. For all regular values (all except finitely many), these M_r represent absolute values of the corresponding functions where in a sum, the strongest wins:

$$M_r f_B \leq \max(M_r f_B, M_r f_0) \overset{!}{=} M_r(f_B + f_0) = M_r f \quad (\sigma_p(f_B) < r \nearrow \sigma).$$

Taking an increasing sequence of regular values $r_n \nearrow \sigma$, we conclude that

$$M_\sigma f_B \leq M_\sigma f.$$

Finally, by Proposition 2 of (3.3) we have $M_\sigma f_B = \|f_B\|_D$, $M_\sigma f = \|f\|_D$. ∎

Let us go one step further and group the poles of f in a finite number of balls.

Theorem. *Let $B_i = B_{<\sigma_i}(a_i)$ $(1 \leq i \leq \ell)$ be a finite set of disjoint open balls in the closed ball $B_{\leq r}$ and define $D = B_{\leq r} - \coprod_{1 \leq i \leq \ell} B_i$. Let $f \in \mathbf{C}_p(x)$ be a rational function regular in D, $f_i = f_{B_i}$ the sum of the principal parts of f corresponding to its poles in B_i $(1 \leq i \leq \ell)$. In the canonical decomposition*

$$f = f_0 + \sum_{1 \leq i \leq \ell} f_i = \sum_{0 \leq i \leq \ell} f_i,$$

where f_0 is regular in $B_{\leq r}$, we have

$$\|f\|_D = \max_{0 \leq i \leq \ell} \|f_i\|_D.$$

PROOF. As in (3.3) (Proposition 2) we can select open balls D_i of maximal radius σ_i on the spheres $|x - a_i| = \sigma_i$ $(1 \le i \le \ell)$ and containing no pole of f.

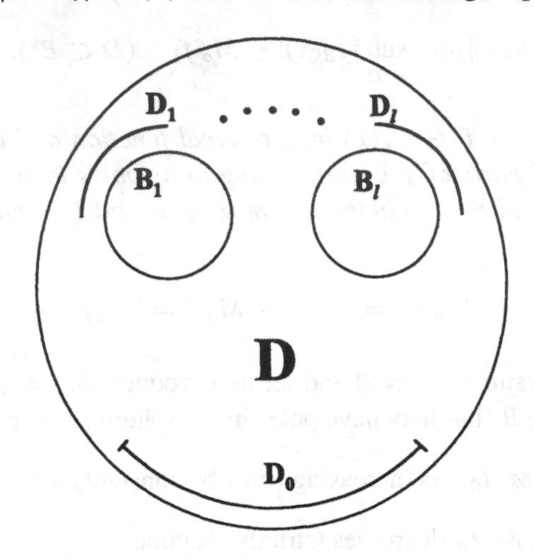

Mittag-Leffler decomposition of a rational function

By Proposition 1, the inclusions $D_i \subset D \subset B_i^c$ lead to the same sup norms

$$\|f_i\|_{D_i} = \|f_i\|_D = \|f_i\|_{B_i^c} \ (= M_{\sigma_i, a_i} f_i).$$

By Proposition 2, we have $\|f_i\|_{D_i} \le \|f\|_{D_i}$ $(i \ge 1)$, hence

$$\|f_i\|_D = \|f_i\|_{D_i} \le \|f\|_{D_i} \le \|f\|_D \quad (1 \le i \le \ell).$$

Now the competitivity principle (II.1.2) in $f - f_0 - \sum_{1 \le i \le \ell} f_i = 0$ shows that we have

$$\|f\|_D = \max_{0 \le i \le \ell} \|f_i\|_D.$$

■

Note that for the regular part f_0 of f in $B_{\le r}$ we have

$$\|f_0\|_D = \sup_{|x| \le r} |f_0(x)| = \max_{|x|=r} |f_0(x)| = \sup_{|x-b|<r} |f_0(x)|$$

for any $|b| = r$. On the other hand, if $\deg g < \deg h$, $f = g/h \to 0$ $(|x| \to \infty)$, we find that $f_0 = 0$, so that we also have

$$\|f\|_D = \sup_{1 \le i \le \ell} \|f_i\|_D$$

for the unbounded domain $D = \mathbf{C}_p - \coprod_{1 \le i \le \ell} B_i$.

Let us introduce the following notation for any subset D of $\mathbf{C}_p$:

$R(D)$: ring of rational functions having no pole in D,
$R_0(D) \subset R(D)$: subring defined by $|f(x)| \to 0$ when $|x| \to \infty$.

For any open ball B and any D disjoint from B, we can look at

$$R(D \cup B) \xrightarrow{\text{res}} R(D) \to R_0(B^c).$$

The first map is the restriction $f \mapsto f|_D$ (injective as soon as D is infinite), while the second is $f \mapsto f_B$ (principal part of f in B). This second map is surjective and admits a *section* $f_B \mapsto f_B|_D$ also given by restriction. If a rational function is regular on $D \cup B$ and on B^c, i.e., regular everywhere, then it is a polynomial. If moreover it tends to 0 (when $|x| \to \infty$), then it is 0:

$$R(D \cup B) \cap R_0(B^c) = \{0\}.$$

Hence the preceding sequence is a short exact sequence: It splits

$$f = f_0 + f_B \leftrightarrow (f_0, f_B) \ : \ R(D) \cong R(D \cup B) \oplus R_0(B^c).$$

More generally, with the notation and assumptions of the theorem,

$$0 \longrightarrow R(B_{\leq r}) \xrightarrow{\text{res}} R(D) \longrightarrow \bigoplus_{1 \leq i \leq \ell} R_0(B_i^c) \longrightarrow 0$$

is a split short exact sequence (the case $\ell = 1$ is as in the previous example). The section of $f \mapsto (f_i)_{1 \leq i \leq \ell}$ furnishing the splitting is $(f_i)_{1 \leq i \leq \ell} \mapsto \sum_{1 \leq i \leq \ell} f_i|_D$. Indeed, the difference $f - \sum_{1 \leq i \leq \ell} f_i|_D$ extends to the regular $f_0 \in R(B_{\leq r})$. All these maps are linear and contracting; hence

$$R(D) \cong R(B_{\leq r}) \oplus \bigoplus_{1 \leq i \leq \ell} R_0(B_i^c)$$

is an isomorphism of the normed space $R(D)$ with a direct sum of normed spaces over $\mathbf{C}_p$ (IV.4.1).

3.5. *Rational Motzkin Factorizations*

It is easy to give a *product decomposition* for rational functions quite similar to the sum decomposition given in the preceding section. Let $B \subset \mathbf{C}_p$ be an open ball and $f = g/h \in \mathbf{C}_p(x)$ a rational function having all its zeros and poles in B:

$$S = \{a \in \mathbf{C}_p : g(a)h(a) = 0\} \subset B.$$

Hence $f = c \prod_{a \in S}(x - a)^{\mu_a}$ ($\mu_a \in \mathbf{Z}$, $c \in \mathbf{C}_p^\times$). Choose b in B and write

$$f = c(x - b)^{\Sigma \mu_a} \prod_{a \in S} \left(\frac{x - a}{x - b} \right)^{\mu_a} = c(x - b)^{\Sigma \mu_a} h(x).$$

Observe that

$$\frac{|x - a|}{|x - b|} = 1 \quad (x \notin B).$$

More precisely,

$$\frac{x-a}{x-b} = 1 + \frac{b-a}{x-b}, \qquad \left|\frac{b-a}{x-b}\right| < 1 \quad (x \notin B),$$

$$\left\|\frac{x-a}{x-b} - 1\right\|_{B^c} < 1 \ (a \in S), \qquad \left\|\prod_{a\in S}\left(\frac{x-a}{x-b}\right)^{\mu_a} - 1\right\|_{B^c} < 1.$$

This gives a factorization

$$f(x) = c(x-b)^m h(x),$$

where

$$m = (\text{number of zeros} - \text{number of poles}) \text{ of } f \text{ in } B,$$

and

$$\|h-1\|_{B^c} < 1, \quad h(x) \to 1 \ (|x| \to \infty).$$

If we take another center $\tilde{b}$ in B, we shall have

$$f(x) = c(x-\tilde{b})^m \tilde{h}(x)$$

with

$$\tilde{h} = \left(\frac{x-b}{x-\tilde{b}}\right)^m h.$$

In particular, we see that c and m are independent of the choice of center of B (but h depends on this choice). The integer m is called *index of f relative to the ball B*. Asymptotically,

$$f(x) \sim c(x-b)^m \quad (|x| \to \infty).$$

We can formulate a more general result when the zeros are located in a finite union of balls. Let $r > 0$ and let

$$B_i = B_{<\sigma_i}(b_i) \subset B_{\le r} \quad (1 \le i \le \ell)$$

be a finite set of disjoint open balls contained in $B_{\le r}$ $(0 < \sigma_i \le r, |b_i| \le r)$. Consider the domain

$$D = B_{\le r} - \coprod_{1\le i\le \ell} B_i.$$

The next three propositions concern rational functions f that are units in the ring $R(D)$: Neither f nor $1/f$ has poles in D, i.e., f has neither zero nor pole in D.

Proposition 1. *Any $f \in R(D)^\times$ can be uniquely factorized as*

$$f = f_0 \cdot \prod_{1\le i\le \ell} f_i \quad \textit{(Motzkin factorization)},$$

where $f_0 \in R(B_{\leq r})^\times$ *and for* $1 \leq i \leq \ell$

$$f_i = (x - b_i)^{m_i} h_i \in R(B_i^c)^\times, \quad \|h_i - 1\|_{B_i^c} < 1, \quad h_i(x) \to 1 \ (|x| \to \infty).$$

PROOF. The only possibility consists in collecting the zeros and poles of f in B_i and defining f_i as the product of the corresponding factors $(x - a)^{\mu_a}$ ($a \in B_i$, $\mu_a \in \mathbf{Z}$ positive for zeros and negative for poles of f). With $f_0 = f / \prod_{1 \leq i \leq \ell} f_i$ all requirements are satisfied. ∎

As before, the difference m_i between the number of zeros and poles of f in B_i (taking multiplicities into account) is the *index of f with respect to B_i*. With any choice of center b_i of B_i, the Motzkin factor f_i of f relative to B_i satisfies

$$\|h_i - 1\|_D = \left\| \frac{f_i}{(x - b_i)^{m_i}} - 1 \right\|_D \leq \left\| \frac{f_i}{(x - b_i)^{m_i}} - 1 \right\|_{B_i^c} < 1.$$

Proposition 2. *Assume* $\|f - 1\|_D < 1$. *Then* f *has as many zeros as poles in each ball* $B_i \subset D^c$.

PROOF. The assumption implies $|f(x) - 1| < 1$ for all $x \in D$, hence $|f(x)| = 1$ is constant in D. Consider a ball B_i and consider the growth modulus centered at $b_i \in B_i$. Without loss of generality, we may assume $i = 1$, $b_1 = 0$, since b_1 is also a center of the ball $B_{\leq r}$. Since D contains a maximal open ball D_1 of the sphere $|x| = \sigma_1 := \sigma$ having no pole of $f - 1$ (in fact infinitely many such balls), Proposition 2 of (3.3) shows that

$$M_\sigma(f - 1) = \|f - 1\|_{D_1} \leq \|f - 1\|_D < 1.$$

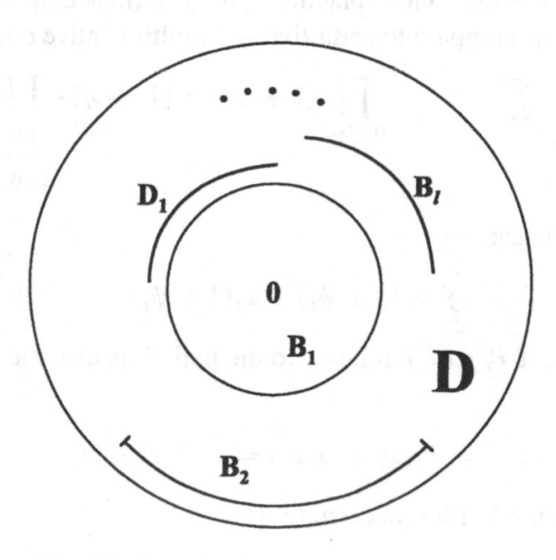

Motzkin factorization of a rational function

By continuity of the growth modulus, we have $M_t(f-1) < 1$ for all t close to σ. For regular values of $t < \sigma$ close to σ we have

$$|f(x) - 1| = M_t(f-1) < 1 \Longrightarrow |f(x)| = 1 \quad (|x| = t).$$

Hence $M_t f = 1$ for $t \nearrow \sigma$. But our study of Laurent series has shown (3.3) that for $t < \sigma$ close to σ, $M_t f = t^m$, where m is the difference between the number of zeros and poles of f in B_1. Hence $m = 0$, as asserted. ∎

Under the assumptions of the preceding proposition, if f has some zeros in B_i, it also has some poles in this ball, and we can look at the principal part $P_i f$ of f in B_i (3.4).

Proposition 3. *If $\|f-1\|_D < 1$, then the principal part $P_i f$ of f relative to the ball B_i and the Motzkin factor $f_i = h_i$ defined in Proposition 1 are related by*

$$\|P_i f\|_D = \|f_i - 1\|_D \quad (1 \le i \le \ell).$$

Proof. Let S denote the set of zeros and poles of f, and let $f = f_0 \cdot \prod_{1\le i\le \ell} f_i$ be the Motzkin factorization of f. By Proposition 2, we have

$$f_i = \prod_{a\in S\cap B_i} \left(\frac{x-a}{x-b_i}\right)^{\mu_a} = \prod_{a\in S\cap B_i} (x-a)^{\mu_a} := 1 + \omega_i.$$

Hence $\omega_i = f_i - 1$ is a rational function that is regular outside B_i and tends to 0 when $|x| \to \infty$: ω_i is a sum of principal parts of poles in B_i. Moreover,

$$\|\omega_i\|_D \le \|\omega_i\|_{B_i^c} < 1.$$

Similarly, $f_0 = c(1+\omega_0)$, and replacing f by f/c (hence f_0 by f_0/c) we may assume $c = 1$. Let us compare the additive and multiplicative decompositions

$$f = P_0 f + \sum_{1\le i\le \ell} P_i f = \prod_{0\le j\le \ell} (1+\omega_j) = (1+\omega_i)\cdot \underbrace{\prod_{j\ne i}(1+\omega_j)}_{1+\psi_i \text{ regular in } B_i}$$

with $\|\psi_i\|_D < 1$. Hence

$$f = (1+\psi_i) + \omega_i(1+\psi_i),$$

and the principal part $P_i f$ of f relative to the ball B_i is also the principal part of $\omega_i(1+\psi_i)$:

$$P_i f = P_i(\omega_i + \omega_i\psi_i) = \omega_i + P_i(\omega_i\psi_i). \qquad (*)$$

By the rational Mittag-Leffler theorem (3.4),

$$\|P_i(\omega_i\psi_i)\|_D \le \|\omega_i\psi_i\|_D < \|\omega_i\|_D,$$

and in $(*)$ the first term is dominant:

$$\|P_i f\|_D = \|\omega_i\|_D = \|f_i - 1\|_D. \qquad \blacksquare$$

3.6. Multiplicative Norms on $K(X)$

When $r > 0$ is fixed, the growth modulus $M_r(f)$ (1.4) defines an absolute value, in other words a multiplicative norm, on the field $K(X)$ of rational functions having coefficients in an extension K of $\mathbf{Q}_p$. Other norms of the same type are obtained if we consider $a \in K$ and consider the growth modulus centered at the point a: For a rational function f regular at a, having a power series expansion

$$f(x) = \sum_{n\geq 0} a_n(a)(x-a)^n \quad (|x-a| < r_f),$$

$$M_{r,a}(f) := \sup_{n\geq 0} |a_n(a)| r^n \quad (r < r_f).$$

When $|K|$ is dense in $\mathbf{R}_{\geq 0}$, the maximum principle (2.5) shows that

$$M_{r,a} f = \sup_{|x-a|<r} |f(x)| = \sup_{|x-a|\leq r} |f(x)| \quad (f \in K[X]);$$

hence $M_{r,a} f = \|f\|_B$ is the sup norm on the ball $B = B_{\leq r}(a)$ when f is a polynomial. Since a multiplicative norm on $K(X)$ is completely determined by its values on $K[X]$, we deduce that the following properties are equivalent:

$$(i)\ M_{r,a} = M_{r,b}; \quad (ii)\ B_{\leq r}(a) = B_{\leq r}(b); \quad (iii)\ |a-b| \leq r.$$

When the field K is *algebraically closed*, a multiplicative norm is completely determined by its values on linear functions. As we have seen, for a linear function

$$X - \xi = (a - \xi) + (X - a)$$

we have

$$M_{r,a}(X - \xi) = \sup(|\xi - a|, r) \quad (r \geq 0).$$

When ξ varies in K we have $M_{r,a}(X-\xi) \geq r$ and hence

$$\inf_{\xi\in K} M_{r,a}(X-\xi) \geq r.$$

In fact, this inequality is an equality: Take $\xi \in K$ with $|\xi - a| \leq r$.

Proposition. *Let K be an algebraically closed, spherically complete extension of $\mathbf{Q}_p$. Then any absolute value ψ on the field of rational functions $K(X)$ that extends the absolute value of K is of the form $M_{r,a}$ for some $a \in K$ and $r > 0$.*

Proof. By (II.1.6), the absolute value ψ is ultrametric. We are looking for a ball $B = B_{\leq r}(a)$ leading to

$$\psi(X - \xi) = M_{r,a}(X - \xi) = \|X - \xi\|_B \quad (\xi \in K).$$

Let us consider

$$r = \inf_{\xi \in K} \psi(X - \xi).$$

(1) If this is a minimum, take $a \in K$ such that $\psi(X - a) = r$. Then

$$r \le \psi(X - \xi) \le \sup(\underbrace{\psi(X - a)}_{=r}, \underbrace{\psi(a - \xi)}_{=|\xi - a|}).$$

If $|\xi - a| \neq r$, this is an equality

$$\psi(X - \xi) = \sup(r, |\xi - a|) = M_{r,a}(X - \xi).$$

If $|\xi - a| = r$, the preceding inequality gives

$$r \le \psi(X - \xi) \le \sup(r, r) = r;$$

hence $\psi(X - \xi) = r = \sup(r, |\xi - a|) = M_{r,a}(X - \xi)$. This proves $\psi = M_{r,a}$.

(2) In general, take a sequence

$$\psi(X - a_n) = r_n \searrow r \quad (n \ge 0).$$

As we have seen,

$$r \le \psi(X - \xi) \le \sup(\underbrace{\psi(X - a_n)}_{=r_n}, \underbrace{\psi(a_n - \xi)}_{=|\xi - a_n|}) = M_{r_n,a_n}(X - \xi).$$

Hence

$$r \le \psi(X - \xi) \le \liminf_{n\to\infty} M_{r_n,a_n}(X - \xi) \quad (\xi \in K),$$

$$\psi \le \liminf_{n\to\infty} M_{r_n,a_n}.$$

If $\psi(X - \xi) > r$, then $\psi(X - \xi) > r_n$ for all large n, and by (1),

$$r < \psi(X - \xi) = \sup(r_n, |\xi - a_n|) = M_{r_n,a_n}(X - \xi) \quad (n \ge N)$$

proves that

$$\psi(X - \xi) = \liminf_{n\to\infty} M_{r_n,a_n}(X - \xi).$$

If there is a $\xi \in K$ with $\psi(X - \xi) = r$, we are brought back to the first case already treated.

(3) Let us study $\liminf_{n\to\infty} M_{r_n,a_n}(X - \xi)$. Consider the inequality

$$r_{n+1} = \psi(X - a_{n+1}) \le \sup(r_n, |a_{n+1} - a_n|) = M_{r_n,a_n}(X - a_{n+1}).$$

If $r_n \neq |a_{n+1} - a_n|$, then $r_{n+1} = \sup(r_n, |a_{n+1} - a_n|) \ge r_n$. Since we suppose on the contrary $r_{n+1} < r_n$, there is a competition $r_n = |a_{n+1} - a_n|$. Define the sequence of balls $B_n = B_{\le r_n}(a_n)$ $(n \ge 0)$. We have just proved that

$$B_{n+1} \subset B_n \quad (n \ge 0).$$

Since the field K is *spherically complete* by assumption,

$$B = \bigcap_{n \geq 0} B_n \neq \emptyset,$$

and any element a in this intersection is a possible center of the ball $B = B_{\leq r}(a)$:

$$\liminf_{n \to \infty} M_{r_n, a_n}(X - \xi) = \lim_{n \to \infty} \|X - \xi\|_{B_n} = \|X - \xi\|_B = M_{r,a}(X - \xi).$$

This concludes the proof. ∎

4. Analytic Elements

Since analytic continuation cannot be achieved by means of Taylor expansions in p-adic analysis (cf. (1.2)), another procedure has to be devised. It was Krasner's idea to mimic the Runge theorem of complex analysis: A holomorphic function f defined in a domain D of the complex plane $\mathbf{C}$ can be uniformly approximated by means of rational functions. More precisely, for each compact subset $C \subset D$, choose $A = \{a_i\}_{i \in I}$ with one point in each connected component of the complement of C in the Riemann sphere. Then f can be uniformly approximated on C by rational functions having all their poles in the set A.

We shall adopt this point of view here, and we start by a discussion of the domains of $\mathbf{C}_p$ in which the idea of Krasner can be carried out. For simplicity, we shall limit ourselves to bounded analytic elements.

4.1. Enveloping Balls and Infraconnected Sets

Let D be any nonempty subset of the field $\mathbf{C}_p$. Its diameter is defined by

$$\delta = \delta(D) = \sup_{x,y \in D} |x - y| = \sup_{x \in D} |x - a| \leq \infty \quad (a \in D).$$

The closed ball

$$B_D := B_{\leq \delta}(a) \quad (a \in D)$$

is called the *enveloping ball* of D (if D is unbounded — i.e., $\delta = \infty$ — we take $B_D = \mathbf{C}_p$). It is the intersection of all closed balls containing D and hence the *smallest closed ball* containing D. When D is closed and bounded,

$$a \in \mathbf{C}_p - D \Longrightarrow r = d(a, D) = \inf_{x \in D} |a - x| > 0,$$

and the open ball $B_{<r}(a)$ is a *maximal* open ball in the complement of D. Each maximal open ball of $B_D - D$ is called a *hole* of D. The preceding observations show that any closed bounded subset D has a representation

$$D = B_D - \coprod_i B_i, \quad \text{where } B_D = D \cup \coprod_i B_i$$

with (possibly infinitely many) holes $B_i = B_{<r_i}(a_i)$.

Examples. (1) Let $D = B_{<1}$ be the open unit ball. Then $B_D = B_{\leq 1}$, and the holes of D are all the open balls $B_{<1}(a)$ contained in the unit sphere ($|a| = 1$).

(2) If $0 < r \notin |\mathbf{C}_p|$, we have $B_{<r}(a) = B_{\leq r}(a)$, and this set coincides with its enveloping ball (it has no hole).

We shall be interested in a special class of closed bounded subsets.

Definition. *A subset $D \subset \mathbf{C}_p$ is called* infraconnected *if its diameter δ is positive and for each $a \in D$*

$$\{|x - a| : x \in D\} \textit{ is dense in } [0, \delta] \subset \mathbf{R}_{\geq 0}.$$

In other words, D is infraconnected when for all pairs of distinct points $a \neq b \in D$, all annuli

$$\{x \in \mathbf{C}_p : r_1 < |x - a| < r_2\} \quad (0 < r_1 < r_2 < |b - a|)$$

meet D. In particular, if D is infraconnected, it has infinitely many elements.

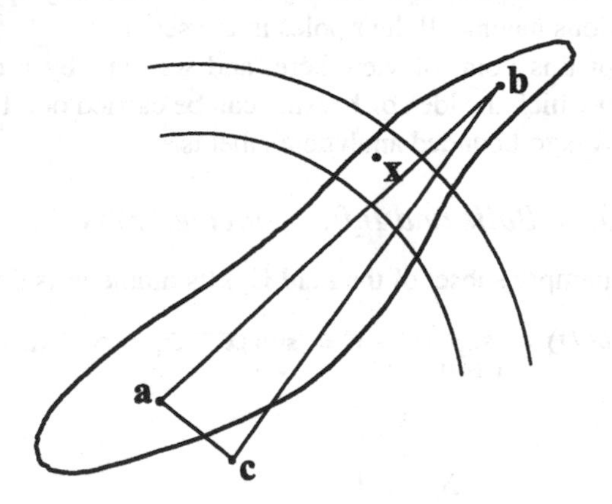

Infraconnected sets

Lemma. *Let D be an infraconnected set. Then for each $c \in \mathbf{C}_p$,*

$$I_c = \{|x - c| : x \in D\}$$

is dense in an interval of $\mathbf{R}$.

Proof. (1) If $c = a \in D$, by definition $\{|x - a| : x \in D\}$ is dense in the interval $[0, \delta]$, where δ is the diameter of D.

(2) If $c \notin D$ is at a distance $r \leq \delta$ of D, we have to prove

$$I_c = \{|x - c| : x \in D\} \text{ is dense in the interval } [r, \delta].$$

There is nothing to prove if $r = \delta$. Otherwise, choose $a \in D$ with $|c - a| < \delta$. Since $\delta = \sup_{b \in D} |a - b|$, we can choose $b \in D$ with $|c - a| < |a - b| \leq \delta$. Hence

$$r \leq |c - a| < |a - b| = |c - b| \leq \delta$$

(make a picture !). The annuli of inner radius $r_1 > |a - c|$ having center a or c coincide. If we select any outer radius $r_2 > r_1$, $r_2 < |c - b| = |c - a|$, the corresponding annulus meets D, since this subset is infraconnected.

(3) Finally, if D is bounded and $c \in \mathbf{C}_p$ is not in the enveloping ball of D, D is contained in the sphere $|x - c| = d(c, D) > \delta$ centered at c and I_c is an interval reduced to a point. ■

Examples. (1) Let $0 < r_1 < r_2 < \infty$. Then the annulus $r_1 < |x - a| < r_2$ is infraconnected. But the complement of this annulus is not infraconnected. The complement of a sphere in a ball is infraconnected. For example, the subset $|x| \neq 1/p$ of the unit ball $B_{\leq 1}$ is infraconnected.

(2) The compact subset $\mathbf{Z}_p \subset \mathbf{C}_p$ is not infraconnected.

(3) Let $(B_i)_{1 \leq i \leq \ell}$ be a finite family of disjoint open balls contained in $B_{\leq 1}$. Then

$$D = B_{\leq 1} - \coprod_{1 \leq i \leq \ell} B_i$$

is infraconnected, and the holes of D are the open balls B_i (a more general class of examples will be given below).

Proposition. *Let $(B_i)_{i \geq 0}$ be a sequence of disjoint open balls contained in the closed unit ball $B_{\leq 1}$. If the sequence of radii r_i tends to 0, then*

$$D = B_{\leq 1} - \coprod_{i \geq 0} B_i$$

is an infraconnected set. Its enveloping ball is $B_D = B_{\leq 1}$.

PROOF. Let us recall the following fact (systematically used in the proof):

> *Any nonempty sphere of radius $0 < \rho \leq 1$ is a union of infinitely many open balls of equal radii $\leq \rho$.*

Let us order the radii of the balls B_i in strictly decreasing order

$$r' > r'' > r''' > \cdots \searrow 0.$$

By assumption, there are only finitely many balls B_i of any given radius in this sequence. There exists an open ball $B' \subset B_{\leq 1}$ of radius r' and disjoint from the finite number of balls B_n having radius greater than or equal to r'. In this ball B' we can find an open ball B'' of radius r'' disjoint from the finite number of balls B_n having radius greater than or equal to r'' (and $< r'$), etc. We can construct a sequence of clopen balls $B' \supset B'' \supset B''' \supset \cdots$ having radii (diameter) approaching 0. Since the field $\mathbf{C}_p$ is complete, the intersection $B' \cap B'' \cap B''' \cdots$ is a common point and is not in the union of the sequence $(B_i)_{i \geq 0}$. This construction shows that $B_{\leq 1} - \coprod B_i$ is nonempty and has infinitely many points. The proof that it is infraconnected follows from the observation that at each step of the above construction, we can choose the ball $B^{(i)}$ in any given nonempty sphere of prescribed radius $\geq r^{(i)}$. ■

4.2. *Analytic Elements*

As in (3.4), let $R(D)$ denote the ring of rational functions having no pole in D:

$$R(D) = \{f = g/h : g,\ h \in \mathbf{C}_p[X],\ h \text{ having no zero in } D\}.$$

Definition. *Let D be a closed subset of $\mathbf{C}_p$. A function $f : D \to \mathbf{C}_p$ is an* analytic element *if it is a uniform limit of a sequence of rational functions $f_n \in R(D)$.*

The analytic elements on D make up a vector space $H(D)$, which is a *uniform completion* of $R(D)$. However, note that in general an $f \in R(D)$ can be an unbounded function on D, so that $R(D)$ is not a metric space. Let us start with the important case where it *is* a metric space (in (4.3) we shall show how to treat the other case).

Proposition 1. *When $D \subset \mathbf{C}_p$ is a closed and bounded subset, each $f \in R(D)$ is bounded on D, and $H(D)$ is the closure of $R(D)$ in the Banach algebra $C_b(D)$ for the sup norm.*

PROOF. Recall (3.2): The functions

$$x^n, \quad \frac{1}{(x-a)^m} \quad (n \geq 0,\ a \notin D,\ m \geq 1)$$

constitute a basis of the vector space $R(D)$. When D is bounded, the functions x^n $(n \geq 0)$ are bounded on D. Moreover, when D is closed and $a \notin D$, the distance $\inf_{x \in D} |x - a|$ is positive, so that the functions $1/(x-a)^m$ $(m \geq 1)$ are also bounded on D. This proves that all rational functions having no pole on D define bounded continuous functions $D \to \mathbf{C}_p$, and the same is true for the analytic elements (IV.2.1). Since the closure of a subalgebra of $C_b(D)$ is also a subalgebra, the statement follows. ■

Corollary. *The product of two analytic elements on a closed and bounded set D is an analytic element.* ■

We can now generalize Proposition 1 in (3.4) for infraconnected sets.

Proposition 2. *Let $D \subset \mathbf{C}_p$ be a closed, bounded, and infraconnected set. Assume $0 \in B_D$ and let $0 \leq d(0, D) \leq r \leq \delta(D)$. Then*

$$M_r f \leq \|f\|_D \quad (f \in R(D)).$$

If the sphere $|x| = r$ meets D, we have, more precisely,

$$M_r f \leq \|f\|_{S_r \cap D} \leq \|f\|_D \quad (f \in R(D)).$$

PROOF. Let $\sigma = d(0, D)$, $\delta = \delta(D)$, so that $\{|x| : x \in D\}$ is dense in the interval $[\sigma, \delta]$. If $f \in R(D)$, let us show that there exists a sequence $x_n \in D$ with

$$|f(x_n)| = M_{|x_n|} f \to M_r f \quad (n \to \infty),$$

so that

$$\|f\|_D \geq \sup |f(x_n)| \geq \lim |f(x_n)| = M_r f.$$

First case: D does not meet the sphere $S_r = \{|x| = r\}$. Since D is infraconnected and $r \in \overline{\{|x| : x \in D\}}$, we can find points $x_n \in D$ with $|x_n| \to r$ monotonically. All except finitely many (that we may discard) are regular, and we have finished in this case.

Second case: There is a point $a \in D \cap S_r$ (observe that in this case $D \subset S_r$ may well happen for $r = \delta$!). We have $M_{s,a} f = M_s f$ for $s > r$, since the spheres of radius s and respective centers a or 0 coincide. By continuity we also have $M_{r,a} f = M_r f$. By density of the values $|x - a|$ ($x \in D$ infraconnected) in the interval $[0, \delta]$ we can find a sequence $x_n \in D$ such that $|x_n - a| = r_n$ is regular for f (with respect to the center a), $r_n \nearrow r$: $|x_n| = |a| = r$. Hence $x_n \in S_r \cap D$,

$$|f(x_n)| \to M_{r,a} f = M_r f,$$

$$\|f\|_{S_r \cap D} \geq \sup_n |f(x_n)| \geq M_r f.$$

The proof is complete. ■

Corollary. *Let $c \in B_D$ where D is a closed bounded infraconnected set, and r in the interval $J_c :=$ closure of $\{|x - c| : x \in D\}$. Then the growth modulus $M_{r,c}$ (centered at the point c) has a continuous extension to $H(D)$. More precisely, $f \mapsto M_{r,c} f$ is a contracting map*

$$|M_{r,c} f - M_{r,c} g| \leq M_{r,c}(f - g) \leq \|f - g\|_D \quad (f, g \in R(D)).$$

PROOF. Take a sequence (x_n) as in the proof of the above proposition (working for both f and g) and let $n \to \infty$ in the inequality

$$\left| |f(x_n)| - |g(x_n)| \right| \le |f(x_n) - g(x_n)| \le \|f - g\|_D. \qquad \blacksquare$$

Definition. *Let D be closed, bounded, and infraconnected:*

$$c \in B_D, \quad \sigma = d(c, D) \le r \le \delta(D).$$

The growth modulus $M_{r,c}$ *is defined on $H(D)$ by continuous extension of*

$$f \mapsto M_{r,c} f : R(D) \to \mathbf{R}_{\ge 0}.$$

For fixed r and c, the growth modulus is a *seminorm* on $H(D)$. Beware of the fact that for complicated infraconnected sets D there can be nonzero analytic elements f on D with $M_{r,c} f = 0$: This can happen only when r is an extremity of the interval $[\sigma, \delta]$. For this more specialized topic involving a discussion of T-filters, we refer to the recent book by A. Escassut.

4.3. Back to the Tate Algebra

A power series $f(x) = \sum_{n \ge 0} a_n x^n$ with $0 < r_f < \infty$ does not necessarily define an analytic element on $B_{<r_f}$. If it does, this sum is bounded on the closed and bounded ball $B_{<r_f}$ (Proposition 1 in (4.2)). In the typical case $r_f = 1$, if the $|a_n|$ are unbounded, there are infinitely many critical radii less than 1, and the sum is unbounded: It is not an analytic element on $B_{<1} = \mathbf{M}_p$. When the $|a_n|$ are bounded, both cases can happen. The series $\sum_{n \ge 0} x^n = 1/(1 - x)$ has bounded $|a_n|$ $(= 1)$ and is an analytic element on $\mathbf{M}_p$ (indeed a rational function with a single pole at $1 \notin \mathbf{M}_p$). It is more difficult to give an example of a power series with bounded coefficients that is not an analytic element on $\mathbf{M}_p$ (a criterion will be given in (4.6)). When $|a_n| \to 0$, the sum f is a uniform limit of polynomials (partial sums) on $\mathbf{A}_p$, and we get an analytic element on the closed unit ball. This simple observation shows that a convergent power series defines analytic elements on all balls $B_{\le r}$ $(r < r_f)$.

Theorem. *The space $H(\mathbf{A}_p)$ of analytic elements on the closed unit ball coincides with the Tate algebra $\mathbf{C}_p\{x\}$ with its norm: for $f = \sum_{n \ge 0} a_n x^n$,*

$$\|f\| = \sup_{|x| \le 1} |f(x)| = \sup_{n \ge 0} |a_n|.$$

PROOF. When $|a| > 1$ the series expansion (3.2)

$$\frac{1}{(x - a)^m} = \frac{1}{(-a)^m} \left(1 - \frac{x}{a}\right)^{-m} = \sum_{n \ge 0} (-1)^{m+n} \binom{-m}{n} \frac{x^n}{a^{m+n}}$$

converges for $|x| < |a|$ and a fortiori for $|x| \le 1$: Its coefficients tend to 0:

$$\left|\frac{(-1)^{m+n}}{a^{n+m}}\binom{-m}{n}\right| \le \frac{1}{|a|^{n+m}} \to 0 \quad (n \to \infty).$$

This shows that the space of rational functions without a pole on $\mathbf{A}_p$ is a subspace of the Banach algebra $\mathbf{C}_p\{x\}$. This subspace contains the dense subspace consisting of the polynomials

$$\mathbf{C}_p[x] \subset R(\mathbf{A}_p) \subset \mathbf{C}_p\{x\},$$

and the closure is $H(\mathbf{A}_p) = \mathbf{C}_p\{x\}$. ∎

To be able to speak of analytic elements on the complement of a ball (which is unbounded) we now approach the case of *unbounded* domains D, and hence $R(D)$ is *not* a metric space. Let us introduce the vector subspaces

$$R_b(D) := \{f \in R(D) : f \text{ bounded on } D\},$$

consisting of the rational functions $f = g/h$, $\deg g \le \deg h$, having no pole in D,

$$R_0(D) := \{f \in R(D) : f \to 0 \ (|x| \to \infty)\} \subset R_b(D)$$

consisting of the rational functions $f = g/h$, $\deg g < \deg h$, having no pole in D. The Euclidean division algorithm shows more precisely that

$$\begin{aligned} R(D) &= R_0(D) \oplus \mathbf{C}_p[x] \\ &= \underbrace{R_0(D) \oplus \mathbf{C}_p}_{=R_b(D)} \oplus x\mathbf{C}_p[x]. \end{aligned}$$

A fundamental system of neighborhoods of an f_0 in $R(D)$ is given by

$$V_\varepsilon(f_0) = \{f \in R(D) : \sup_{x \in D} |f(x) - f_0(x)| < \varepsilon\} \quad (\varepsilon > 0).$$

In particular, if f_0 is bounded, then $V_\varepsilon(f_0) \subset R_b(D)$, namely:

$$V_\varepsilon(f_0) \cap x\mathbf{C}_p[x] = \{0\}.$$

This proves that the topology induced by uniform convergence on $x\mathbf{C}_p[x]$ is the discrete one:

$$R(D) = \underbrace{R_b(D)}_{\text{normed space}} \oplus \underbrace{x\mathbf{C}_p[x]}_{\text{uniformly discrete}}.$$

By completion we get

$$H(D) = \underbrace{H_b(D)}_{\text{Banach space}} \oplus \underbrace{x\mathbf{C}_p[x]}_{\text{uniformly discrete}}.$$

We can also write

$$H(D) = H_0(D) \oplus \mathbf{C}_p \oplus x\mathbf{C}_p[x]$$

and group the last two factors

$$H(D) = H_0(D) \oplus \mathbf{C}_p[x],$$

but the uniform structure on the last factor is not the discrete one.

When D is unbounded, we shall only use *bounded* analytic elements and thus work in the *Banach algebra* $H_b(D) = H_0(D) \oplus \mathbf{C}_p$. We note that $H_0(D)$ is a (maximal) ideal in this algebra with quotient $H_b(D)/H_0(D) \cong \mathbf{C}_p$ (a field).

Let us now treat explicitly the case of the complement of open balls.

Proposition. *The bounded analytic elements on* $\mathbf{C}_p - \mathbf{M}_p = \{|x| \geq 1\}$ *are the formal restricted power series in* $1/x$*:*

$$H_b(\{|x| \geq 1\}) = \mathbf{C}_p\{1/x\} \supset H_0(\{|x| \geq 1\}) = x^{-1}\mathbf{C}_p\{1/x\}.$$

PROOF. The inversion $x \mapsto y = 1/x$ transforms bounded rational functions having no pole in $|x| \geq 1$ into rational functions having no pole in $|y| \leq 1$,

$$R_b(\{|x| \geq 1\}) = R(\{|y| \leq 1\}),$$

and the completion is

$$H_b(\{|x| \geq 1\}) = H(\{|y| \leq 1\}) = \mathbf{C}_p\{y\}$$

by the preceding theorem. ∎

The comments made prior to the proposition prove that the analytic elements on $\{|x| \geq 1\}$ are given by Laurent series having only finitely many nonzero terms $a_n x^n$ with $n > 0$:

$$H(\{|x| \geq 1\}) = H_b(\{|x| \geq 1\}) \oplus x\mathbf{C}_p[x] = \mathbf{C}_p\{1/x\} \oplus x\mathbf{C}_p[x].$$

More generally, if $B = B_{<r}(a)$ is an open ball, then

$$H(B^c) = H_b(B^c) \oplus (x-a)\mathbf{C}_p[x-a],$$

and

$$H_b(B^c) \subset \mathbf{C}_p[[(x-a)^{-1}]]$$

is the subspace consisting of the formal power series

$$f(x) = \sum_{n\geq 0} a_n/(x-a)^n \quad \text{such that } |a_n|/r^n \to 0 \quad (n \to \infty).$$

Similarly,

$$H_0(B^c) \subset (x-a)^{-1}\mathbf{C}_p[[(x-a)^{-1}]]$$

is the subspace consisting of the formal power series

$$f(x) = \sum_{n\geq 1} a_n/(x-a)^n \quad \text{such that } |a_n|/r^n \to 0 \quad (n \to \infty).$$

4.4. *The Amice-Fresnel Theorem*

Let $B = 1 + \mathbf{M}_p \subset \mathbf{C}_p$ be the open ball of radius 1 and center 1. We are going to give a useful description of the space $H_0(B^c)$ of generalized principal parts relative to the hole B: These are the analytic elements in the complement of B that tend to zero at infinity. By (4.3) we know that these analytic elements $f \in H_0(B^c)$ are given by restricted power series in $1/(x-1)$ with zero constant term:

$$f(x) = \sum_{m\geq 0} \lambda_m \frac{1}{(x-1)^{m+1}} \quad (|\lambda_m| \to 0)$$

and

$$\|f\|_{B^c} = M_{1,1}f = \sup_{m\geq 0} |\lambda_m|.$$

Let us expand each term $(x-1)^{-m-1}$ according to the binomial formula

$$(x-1)^{-m-1} = (-1)^{m+1} \sum_{n\geq 0} \binom{-m-1}{n} (-x)^n.$$

Recall the elementary identity $(-1)^n\binom{-m-1}{n} = (-1)^m\binom{-n-1}{m}$ e.g., if $m > n$, then

$$\begin{aligned}(-1)^n\binom{-m-1}{n} &= (-1)^n \frac{(-m-1)\cdots(-m-n)}{n!} \cdot \frac{m(m-1)\cdots(n+1)}{(n+1)\cdots(m-1)m} \\ &= \frac{(n+1)(n+2)\cdots m(m+1)\cdots(m+n)}{m!} \\ &= (-1)^m\binom{-n-1}{m}\end{aligned}$$

(similar computations are valid when $n \geq m$). Grouping terms, we see that the coefficient of x^n in $f(x)$ is

$$a_n = \sum_{m\geq 0} (-1)^{m+1}(-1)^m \lambda_m \binom{-n-1}{m} = -\sum_{m\geq 0} \lambda_m \binom{-n-1}{m}.$$

Define $\varphi : \mathbf{Z}_p \to \mathbf{C}_p$ by the uniformly convergent series

$$\varphi(x) = -\sum_{m\geq 0} \lambda_m \binom{-x-1}{m}$$

(this is a Mahler series in $y = -x-1$ (IV.2.3)). The inequality $\|\varphi\|_{\mathbf{Z}_p} \leq \max |\lambda_m|$ is obvious, and conversely, since λ_m is a linear combination with integral coefficients of $\varphi(-1), \ldots, \varphi(-m-1)$, we also have

$$\max |\lambda_m| \leq \|\varphi\|_{\mathbf{Z}_p}$$

(recall Theorem 1 in (IV.2.4) for Mahler series). This proves that

$$\|\varphi\|_{\mathbf{Z}_p} = \sup_{m \geq 0} |\lambda_m| = \|f\|_{B^c}.$$

By definition, φ is a continuous extension of $n \mapsto a_n$ to $\mathbf{Z}_p$. Reversing the operations, we have proved the following result of Y. Amice and J. Fresnel.

Theorem. *Let $f = \sum_{n \geq 0} a_n x^n \in \mathbf{C}_p[[x]]$ be a convergent power series with $r_f \geq 1$ and denote by B the open ball $1 + \mathbf{M}_p$. Then the following properties are equivalent:*

(i) The sequence $n \mapsto a_n$ has a continuous extension $\varphi : \mathbf{Z}_p \to \mathbf{C}_p$.
(ii) f is the restriction of an analytic element of $H_0(B^c)$.

4.5. *The p-adic Mittag-Leffler Theorem*

For a simple region $D = B_{\leq r} - \coprod_{1 \leq i \leq \ell} B_i$ where the $B_i = B_{<\sigma_i}(a_i)$ are disjoint open balls in $B_{\leq r}$, namely holes of D (notations and assumptions of (3.4)), the rational Mittag-Leffler theorem leads by completion to a simple decomposition of analytic elements in D:

$$H(D) \xrightarrow{\sim} H(B_{\leq r}) \oplus \bigoplus_{1 \leq i \leq \ell} H_0(B_i^c).$$

Explicitly, this means that each $f \in H(D)$ can be uniquely written as

$$f = f_0 + \sum_i f_i \quad (f_0 \in H(B_{\leq r}); \quad f_i \in H_0(B_i^c),\ 1 \leq i \leq \ell),$$

namely with *generalized principal parts* f_i of f, regular outside B_i, or equivalently, having all their singularities in the hole B_i of D. If we choose a center a_i in the hole B_i, such a generalized principal part $f_i \in H_0(B_i^c)$ is given by a Laurent expansion (Corollary in (4.3))

$$f_i(x) = \sum_{n \geq 1} \frac{a_n}{(x - a_i)^n}, \qquad \frac{|a_n|}{\sigma_i^n} \to 0.$$

Let us turn to a closed, bounded, and *infraconnected* domain $D \subset \mathbf{C}_p$.

Proposition. *Let D be closed, bounded, and infraconnected, $f \in R(D)$. Let also $B = B_{<\sigma}$ be a hole of D. If f_B denotes the sum of the principal parts* *attached to the poles of f in B, and $f_0 = f - f_B$, then*

$$\|f_B\|_D \le M_\sigma f_B \le M_\sigma f \le \|f\|_D,$$
$$\|f\|_D = \max(\|f_B\|_D, \|f_0\|_D).$$

PROOF. If $f \in R(D)$ is a rational function without a pole in D, $B = B_{<\sigma}$ is a hole of D, and f_B is the sum of the principal parts of f corresponding to its poles in B, then we have $f = f_B + f_0$ with a regular $f_0 \in R(D \cup B)$, $f_B(x) \to 0$ $(x \to \infty)$, and

$$\|f_B\|_D \le M_\sigma f_B \quad \text{by (3.4) Proposition 1,}$$
$$M_\sigma f_B \le M_\sigma f \quad \text{by (3.4) Proposition 2,}$$
$$M_\sigma f \le \|f\|_D \quad \text{by (4.2) Proposition 2.}$$

This proves $\|f_B\|_D \le \|f\|_D$, and the competitivity principle (II.1.2) in

$$f - f_B - f_0 = 0$$

leads to

$$\|f\|_D = \max(\|f_B\|_D, \|f_0\|_D).$$

■

This means that we have an isomorphism

$$R(D) \xrightarrow{\sim} R(D \cup B) \oplus R_0(B^c)$$

of normed spaces.

Let now $(B_i)_{i \in I}$ be the family of holes of D, so that $B_D = D \cup \coprod_I B_i$ is the enveloping ball of D. We also have a split short exact sequence

$$0 \to R(B_D) \xrightarrow{\text{res}} R(D) \to \bigoplus_i R_0(B_i^c) \to 0$$

with linear contracting maps of normed spaces. The surjective map is $f \mapsto (f_i)_{i \in I}$, where f_i denotes the sum of the principal parts of f at its poles in B_i: When $f \in R(D)$ is given, finitely many f_i are nonzero. The map

$$(f_i)_{i \in I} \mapsto \sum_i f_i|_D : \bigoplus_i R_0(B_i^c) \to R(D)$$

is a splitting: $f - \sum_i f_i|_D$ is the restriction of a rational $f_0 \in R(B_D)$ and $f(x) = f_0(x) + \sum_i f_i(x)$ $(x \in D)$. The central term is the normed direct sum of the extreme ones, and

$$R(D) \xrightarrow{\sim} R(B_D) \oplus \bigoplus_i R_0(B_i^c)$$

is an isometry of normed spaces. By completion, we obtain the following general result.

Theorem. *Let D be a closed, bounded, infraconnected set, $(B_i)_{i\in I}$ its family of holes. Then there is a Banach direct sum decomposition*

$$H(D) \xrightarrow{\sim} H(B_D)\widehat{\bigoplus}_{i\in I} H_0(B_i^c).$$

Each $f \in H(D)$ can be uniquely expressed as a sum

$$f = f_0 + \sum_{i\in I} f_i, \quad \|f\|_D = \max(\|f_0\|, \sup_i \|f_i\|),$$

where

> *f_0 is an analytic element on the enveloping ball of D,*
> *f_i are analytic elements on B_i^c with $f_i(x) \to 0$ $(x \to \infty)$,*
> *$\|f_i\| = \|f_i\|_{B_i^c} = \|f_i\|_D \to 0$ $(i \to \infty)$.* ∎

In particular, in the summable family $(f_i)_{i\in I}$ of generalized principal parts of an analytic element $f \in H(D)$, at most countably many f_i are nonzero (IV.4.1).

4.6. The Christol-Robba Theorem

A formal power series $f = \sum_{n\geq 0} a_n x^n \in \mathbf{C}_p[[x]]$ having bounded coefficients converges at least for $|x| < 1$: $r_f \geq 1$ but it does not always define an analytic element on the open ball $\mathbf{M}_p$. Let us determine the space $H(\mathbf{M}_p)$ of analytic elements on this ball.

Since $H(\mathbf{M}_p) \subset H(B_{\leq r})$ for all $r < 1$, it follows that any $f \in H(\mathbf{M}_p)$ is given by a power series $\sum_{n\geq 0} a_n x^n$ such that $|a_n|r^n \to 0$ $(n \to \infty)$ for all $r < 1$, hence by a power series having a radius of convergence $r_f \geq 1$. On the other hand, $\mathbf{M}_p$ is closed and bounded; hence $H(\mathbf{M}_p) \subset C_b(\mathbf{M}_p)$ (Proposition 1 in (4.2)).

Lemma. *The subspace of $\mathbf{C}_p[[x]]$ consisting of the convergent power series f having a radius of convergence $r_f \geq 1$ and a bounded sum in $\mathbf{M}_p$ coincides with the space ℓ^∞ of formal power series having bounded coefficients: The map*

$$(a_n)_{n\geq 0} \mapsto \sum_{n\geq 0} a_n x^n \;:\; \ell^\infty \to \mathbf{C}_p[[x]] \cap C_b(\mathbf{M}_p)$$

is an isometric isomorphism.

Proof. If $(a_n)_{n\geq 0}$ is a bounded sequence,

$$|f(x)| = \left|\sum_{n\geq 0} a_n x^n\right| \leq \sup_{n\geq 0} |a_n| \quad (|x| < 1),$$

hence $\|f\| \le \sup_{n\ge0} |a_n|$ with the sup norm of f on $\mathbf{M}_p$. Conversely, assume that the power series $\sum_{n\ge0} a_n x^n$ converges for $|x| < 1$ and has a bounded sum in this ball. For a *regular* $|x| = r < 1$ we have

$$|a_n r^n| \le M_r f \overset{!}{=} |f(x)| \le \|f\| \quad (n \ge 0).$$

Taking a sequence of regular $r \nearrow 1$, we infer $|a_n| \le \|f\|$ $(n \ge 0)$, and consequently $\sup_{n\ge0} |a_n| \le \|f\|$. ∎

The preceding proof works for any field K having a *dense valuation*: Compare with (V.2.1), where the residue field k was assumed to be infinite.

The lemma shows that we have an isometric embedding

$$f \mapsto (a_n)_{n\ge0} : H(\mathbf{M}_p) \to \ell^\infty = \ell^\infty(\mathbf{C}_p).$$

The following theorem characterizes the image: It gives a criterion for a formal power series with bounded coefficients to define an analytic element $f \in H(\mathbf{M}_p)$.

Theorem (Christol-Robba). *Let $f = \sum_{n\ge0} a_n x^n \in \mathbf{C}_p[[x]]$ be a formal power series with bounded coefficients. Define $p_\nu = p^\nu(p^\nu - 1)$ $(\nu \ge 1)$. Then f defines an analytic element on $\mathbf{M}_p$ precisely when the following condition* (CR) *holds:*

> *For each $\varepsilon > 0$ there exist ν and $N \ge 0$*
> *such that $|a_{n+p_\nu} - a_n| \le \varepsilon \quad (n \ge N)$.*

PROOF. The proof is based on the Mittag-Leffler theorem (4.5) for the bounded and closed infraconnected set

$$D = \mathbf{M}_p = B_{\le1} - \coprod_{\zeta\in\mu_{(p)}} B_{<1}(\zeta) \subset B_D = B_{\le1} = \mathbf{A}_p.$$

The condition is necessary. Let us write the Mittag-Leffler decomposition of the space of analytic elements $H(\mathbf{M}_p)$ on the open unit ball as

$$H(\mathbf{M}_p) \cong H(\mathbf{A}_p) \oplus \widehat{\bigoplus}_\zeta H_0((\zeta + \mathbf{M}_p)^c)$$

with a sum parametrized by $\zeta \in \mu_{(p)}$. The space $H(\mathbf{A}_p)$ is the Tate algebra with normal basis $(x^i)_{i\ge0}$, and $H_0((\zeta + \mathbf{M}_p)^c)$ has normal basis $1/(x-\zeta)^{m+1}$ $(m \ge 0)$. This proves that the family of functions

$$x^n, \quad \frac{1}{(x-\zeta)^{m+1}} \quad (n \ge 0,\ \zeta \in \mu_{(p)},\ m \ge 0)$$

constitutes a normal basis of $H(\mathbf{M}_p)$. Let us show that each basis element satisfies the condition given in the theorem. This is obvious for the powers x^n $(n \ge 0)$. On the other hand, the rational function

$$f(x) = \frac{1}{(x-\zeta)^{m+1}} \quad (\zeta \in \mu_{(p)},\ m \geq 0)$$

(having a pole at the point $\zeta \notin \mathbf{M}_p$) can be expanded according to the binomial formula

$$\begin{aligned} f(x) &= \frac{1}{(-\zeta)^{m+1}} \sum_{n\geq 0} \binom{-m-1}{n} (-1)^n \zeta^{-n} x^n \\ &= \frac{1}{(-\zeta)^{m+1}} \sum_{n\geq 0} (-1)^m \binom{-1-n}{m} \zeta^{-n} x^n \end{aligned}$$

by the elementary identity on binomial coefficients recalled in (4.4). We have obtained

$$f(x) = -\zeta^{-m-1} \sum_{n\geq 0} \binom{-1-n}{m} \zeta^{-n} x^n.$$

Let us estimate the difference between two coefficients as in the condition (CR):

$$\binom{-1-n-p_\nu}{m} \zeta^{-n-p_\nu} - \binom{-1-n}{m} \zeta^{-n} =$$
$$\zeta^{-n} \left[\binom{-1-n-p_\nu}{m} \zeta^{-p_\nu} - \binom{-1-n}{m} \right].$$

Now, since since p_ν is a multiple of $p^\nu - 1$ and $\zeta^{p^\nu - 1} = 1$, we have $\zeta^{-p_\nu} = 1$ for ν large enough (depending on ζ). On the other hand, $x \mapsto \binom{x}{m}$ is uniformly continuous on $\mathbf{Z}_p$, so that, uniformly in n,

$$\binom{-1-n-p_\nu}{m} - \binom{-1-n}{m} \text{ is small if } p_\nu \text{ is small in } \mathbf{Z}_p,$$

which is the case for large ν, since p_ν is a multiple of p^ν:

$$\zeta^{-n} \left[\binom{-1-n-p_\nu}{m} \zeta^{-p_\nu} - \binom{-1-n}{m} \right] =$$
$$\zeta^{-n} \left[\binom{-1-n-p_\nu}{m} - \binom{-1-n}{m} \right] \leq \varepsilon \quad (\nu \geq \nu_\zeta).$$

Finally, the conclusion will be reached as soon as we observe that (CR) characterizes a *closed* subspace of ℓ^∞. Let $(\bar{a}_n)$ be a sequence in the closure of the space satisfying (CR). If $\varepsilon > 0$ is given, we can first find a sequence (a_n) satisfying (CR) and with $|\bar{a}_n - a_n| \le \varepsilon$ $(n \ge 0)$. Then

$$|\bar{a}_{n+p_\nu} - \bar{a}_n| \le \max(|\bar{a}_{n+p_\nu} - a_{n+p_\nu}|,\ |a_{n+p_\nu} - a_n|,\ |a_n - \bar{a}_n|)$$
$$\le \max(\varepsilon,\ |a_{n+p_\nu} - a_n|).$$

This is less than or equal to ε when n and ν are large enough, by assumption on the sequence (a_n). Hence $(\bar{a}_n)$ still satisfies (CR).

The condition is sufficient. Fix a positive integer N and consider the rational function

$$g_N(x) = \sum_{n<N} a_n x^n + \frac{\sum_{N\le n<N+p_\nu} a_n x^n}{1 - x^{p_\nu}}$$

having all its poles in the set of roots of unity (on the unit sphere). We have

$$f(x) - g_N(x) = \sum_{n\ge N} a_n x^n - \frac{\sum_{N\le n<N+p_\nu} a_n x^n}{1 - x^{p_\nu}}.$$

The numerator is

$$\begin{aligned}
&(1 - x^{p_\nu})(f(x) - g_N(x)) \\
&= \sum_{n\ge N} a_n x^n - x^{p_\nu} \sum_{n\ge N} a_n x^n - \sum_{N\le n<N+p_\nu} a_n x^n \\
&= \sum_{n\ge N+p_\nu} a_n x^n - x^{p_\nu} \sum_{n\ge N} a_n x^n = \sum_{n\ge N} (a_{n+p_\nu} - a_n) x^{n+p_\nu}.
\end{aligned}$$

We have obtained

$$f(x) - g_N(x) = \frac{\sum_{n\ge N} (a_{n+p_\nu} - a_n) x^{n+p_\nu}}{1 - x^{p_\nu}},$$

and since $|1 - x^{p_\nu}| = 1$ for $|x| < 1$, we have

$$|f(x) - g_N(x)| = \left| \sum_{n\ge N} (a_{n+p_\nu} - a_n) x^{n+p_\nu} \right| \le \sup_{n\ge N} |a_{n+p_\nu} - a_n| \quad (|x| < 1).$$

With the postulated condition, $\|f - g_N\| \le \varepsilon$. This proves that f is a uniform limit on $|x| < 1$ of rational functions g_N having no pole in this ball. ∎

Examples. Here are three power series with bounded coefficients (hence a bounded sum in $|x| < 1$) that do not define an analytic element on $\mathbf{M}_p$:

- $\sum_{n\ge 0} x^{p^n}$ (follows from the Christol-Robba condition),
- $\exp \pi x$ $(|\pi| = r_p)$ (exercise),
- $(1 + x)^{1/m}$ $(m > 1$ not multiple of $p)$ (book by A. Escassut).

Analytic elements

Formal power series	$\mathbf{C}_p[[x]]$	Sequences $(a_n)_{n\geq 0}$	$\prod_{n\geq 0}\mathbf{C}_p$
power series converging in $\lvert x\rvert<1$		$\limsup \lvert a_n\rvert^{1/n}\leq 1$ ($r_f\geq 1$)	
power series bounded in $\lvert x\rvert<1$		$(a_n)_{n\geq 0}$ bounded sequence	ℓ^∞
analytic elements in $\lvert x\rvert<1$	$H(\mathbf{M}_p)$	Christol-Robba condition (4.6)	
analytic elements in $\lvert x\rvert\leq 1$	$H(\mathbf{A}_p)=\mathbf{C}_p\{x\}$	$a_n\to 0$ ($n\to\infty$)	c_0
polynomials	$\mathbf{C}_p[x]$	$a_n\neq 0$ for finitely many n's	$\mathbf{C}_p^{(\mathbf{N})}$

4.7. *Analyticity of Mahler Series*

Theorem. *Let $f:\mathbf{Z}_p\to\mathbf{C}_p$ be a continuous function with Mahler series*

$$f(x)=\sum_{k\geq 0}c_k\binom{x}{k}.$$

Then f is the restriction of an analytic element $\tilde{f}\in\mathbf{C}_p\{x\}$ iff $\lvert c_k/k!\rvert\to 0$.

PROOF. Consider the triangular change of basis of the space of polynomials given by

$$(x)_k=\sum_{n\leq k}(-1)^{k-n}\begin{bmatrix}k\\n\end{bmatrix}x^n\quad(k\geq 0)$$

where the coefficients are positive integers: Stirling numbers of the first kind. Conversely,

$$x^n=\sum_{k\leq n}\begin{Bmatrix}n\\k\end{Bmatrix}(x)_k,$$

where the coefficients are positive integers: Stirling numbers of the second kind. Hence if $f=\sum_n a_nx^n=\sum_k b_k\,(x)_k$ is a polynomial, we have $\sup_n\lvert a_n\rvert=\sup_k\lvert b_k\rvert$, and this isometry $(a_n)\mapsto(b_k)$ extends to an isometric embedding of the completion

$$\mathbf{C}_p\{x\} \longrightarrow C(\mathbf{Z}_p;\mathbf{C}_p),$$
$$\sum_{n\geq 0} a_n x^n \mapsto \sum_{k\geq 0} b_k\,(x)_k = \sum_{k\geq 0} c_k\binom{x}{k},$$

where $b_k = c_k/k!$. The assertion follows. ■

Corollary. *If the Mahler coefficients c_k of a continuous function f satisfy $|c_k| \leq cr^k$ $(k \geq k_0)$ for some $c > 0$ and $r < r_p$, then f is the restriction of an analytic element $\tilde{f}$ on the closed unit ball $\mathbf{A}_p$ of $\mathbf{C}_p$.*

Proof. Under the assumption, $|c_k/k!| \leq c\,r^k/|k!|$. Since the general term of the exponential series e^x tends to 0 when $|x| = r < r_p$, we have $r^k/|k!| \to 0$. The conclusion follows by the Theorem in (4.3). ■

Example. Choose and fix an element $t \in \mathbf{C}_p$ with $|t| < r_p$. According to the preceding corollary, the Mahler series

$$f(x) = \sum_{k\geq 0} t^k\binom{x}{k}$$

is the restriction of an analytic element $\tilde{f} \in H(\mathbf{A}_p) = \mathbf{C}_p\{x\}$. *I claim that the analytic element in question is*

$$\tilde{f}(x) = e^{x\log(1+t)}.$$

Proof. The assumption $|t| < r_p$ indeed implies (Proposition 1 in (V.4.2))

$$|\log(1+t)| = |t| < r_p,$$

so that the series expansion

$$f_1(x) = \sum_{n\geq 0} \frac{(x\log(1+t))^n}{n!}$$

converges for $|x| \leq 1$. Now, for integers $m \geq 0$,

$$f_1(m) = \sum_{n\geq 0} \frac{(m\log(1+t))^n}{n!}$$
$$= e^{m\log(1+t)} = \left(e^{\log(1+t)}\right)^m = (1+t)^m = f(m)$$

(we have used the identities $e^{x+y} = e^x e^y$; hence $e^{mx} = (e^x)^m$ and $e^{\log(1+t)} = 1+t$). By continuity and density,

$$f_1(x) = f(x) = \tilde{f}(x) \quad (x \in \mathbf{Z}_p).$$

Since $\tilde{f} - f_1$ is given by a power series expansion, it vanishes identically in its convergence disk (a nonzero power series expansion has a discrete sequence of critical radii with finitely many zeros on each critical sphere). ■

Since

$$f(x) = \sum_{k \geq 0} t^k \binom{x}{k} = (1+t)^x \quad (x \in \mathbf{Z}_p),$$

it is also clear that for larger values of $t \in \mathbf{M}_p \subset \mathbf{C}_p$ we can still define $(1+t)^x$ for smaller values of $x \in \mathbf{C}_p$. Recall that

$$p^{j-1} r_p < |\log(1+t)| = \left| \frac{t^{p^j}}{p^j} \right| = p^j |t|^{p^j} < p^j r_p$$

for

$$r_p^{1/p^{j-1}} < |t| < r_p^{1/p^j}$$

(cf. (1.6)). Hence

$$|t| \leq r_p^{1/p^j} \Longrightarrow |\log(1+t)| \leq p^j r_p,$$

and it is enough to assume $|x| < 1/p^j$ to have a convergent series expansion for

$$(1+t)^x = e^{x \log(1+t)}.$$

Still, for $|t| \leq r_p^{1/p^j}$ and $|x - n| < 1/p^j$, we can define

$$(1+t)^x = (1+t)^n \cdot (1+t)^{x-n} = (1+t)^n \cdot e^{(x-n)\log(1+t)}.$$

For these values of the parameter $t \in \mathbf{C}_p$, the function $x \mapsto (1+t)^x$ is a *locally analytic function* defined in the neighborhood

$$V_j = \bigcup_{a \in \mathbf{Z}_p} B_{<1/p^j}(a) = \mathbf{Z}_p + B_{1/p^j}$$

of $\mathbf{Z}_p$ in $\mathbf{C}_p$.

Remark. The identity

$$e^{x \log(1+t)} = (1+t)^x$$

leads to

$$\sum_{n \geq 0} \frac{[x \log(1+t)]^n}{n!} = \sum_{k \geq 0} t^k \binom{x}{k},$$

hence to

$$\sum_{n\geq 0}\frac{[x\log(1+t)]^n}{n!}=\sum_{k\geq 0}\frac{t^k}{k!}(x)_k=\sum_{k\geq 0}\frac{t^k}{k!}\sum_{n\leq k}(-1)^{k-n}\begin{bmatrix}k\\n\end{bmatrix}x^n.$$

Identifying the coefficients of x^n we get the well-known classical identity

$$\frac{[\log(1+t)]^n}{n!}=\sum_{k\geq n}(-1)^{k-n}\begin{bmatrix}k\\n\end{bmatrix}\frac{t^k}{k!},$$

where the coefficients are again the Stirling numbers of the first kind.

4.8. The Motzkin Theorem

Let D be a closed, bounded, and infraconnected set, $f \in H(D)^\times$ an invertible analytic element on D. Assume that $B = B_{<\sigma}(a)$ is a hole of D (maximal open ball in the complement of D). A *Motzkin factorization* of f relative to the hole B is a product decomposition $f = gf_{(B)}$, where

(1) $g \in H(D \cup B)^\times$,
(2) $f_{(B)} = (x-a)^m h \quad (m \in \mathbf{Z}, h \in H(B^c)^\times)$,
$h(x) \to 1 \ (x \to \infty)$, and $\|h-1\|_{B^c} < 1$.

Remarks. (1) We have seen in (4.3) that an analytic element on B^c admits a convergent Laurent series expansion. If it is not zero, we can write it as

$$\sum_{j\leq m} a_j(x-a)^j = a_m(x-a)^m \cdot h(x).$$

Here $h(x) = 1 + b_1/(x-a) + \cdots$ $(a_m b_1 = a_{m-1}, \ldots)$ is invertible if it does not vanish on B^c, i.e., if it has no critical radius greater than or equal to σ. Since $h(x) - 1 \to 0 \ (x \to \infty)$, $M_r h \ (r \geq \sigma)$ decreases, $M_r h \searrow 0 \ (r \to \infty)$, and

$$\|h-1\|_{B^c} = M_{r,a}(h-1) = \max_{j\geq 1} |b_j|/r^j \overset{!}{<} 1.$$

(2) When $\|f-1\|_D < 1$, then f is invertible. In fact, since D is closed and bounded, $H(D)$ is a closed subalgebra (Banach subalgebra) of $C_b(D)$ (4.2). Hence the geometric series expansion

$$\frac{1}{f}=\frac{1}{1-(1-f)}=\sum_{n\geq 0}(1-f)^n$$

converges (in norm) in $H(D)$. More generally, if $f \in H(D)$ and $\|f - f(a)\| < |f(a)|$ for some $a \in D$, then $f/f(a)$ (hence also f) is invertible.

(3) The *existence* of a Motzkin factorization (with respect to a hole $B = B_{<\sigma}$) requires $M_\sigma f > 0$: The growth modulus function is multiplicative:

$$M_\sigma f = M_\sigma g \cdot M_\sigma(f_{(B)}) > 0.$$

This condition is also sufficient, as we are going to prove.

Theorem. *Let D be a closed, bounded, and infraconnected set, $B = B_{<\sigma}$ a hole of D. Then each $f \in H(D)$ satisfying $\|f - 1\|_D < 1$ admits a unique Motzkin factorization (with index $m = 0$)*

$$f = g \cdot h, \quad \|h - 1\|_{B^c} < 1, \quad h(x) \to 1 \ (x \to \infty).$$

PROOF. Let $(f_n)_{n\geq 0}$ be a sequence of $R(D)$ converging uniformly to f. Since

$$\|f_n - 1\|_D \leq \max(\|f_n - f\|_D, \|f - 1\|_D) < 1$$

for large n, we can disregard the first few values of n and assume $\|f_n - 1\|_D < 1$ for all n. By the rational Motzkin factorization result (3.5) we can write

$$f_n = g_n \cdot h_n \quad (h_n = (f_n)_{(B)} \to 1).$$

Now set $P(f) = f_B$, the principal part of a rational function f with respect to the hole B (as in the Mittag-Leffler decomposition) and $Q(f) = f_{(B)}$ the Motzkin factor of f relative to the same hole. Obviously, $P(f) = P(f - 1)$ and $Q(f_n/f_m) = Qf_n/Qf_m$ (the Motzkin factorization of rational functions is simply obtained by gathering the linear terms corresponding to the zeros and poles in B). The norm estimate in (3.5) for the function f_n/f_m leads to

$$\|Q(f_n/f_m) - 1\|_D = \|h_n/h_m - 1\|_D = \|P(f_n/f_m)\|_D.$$

By the proposition in (4.5),

$$\begin{aligned} \|h_n/h_m - 1\|_D = \|P(f_n/f_m)\|_D &= \|P(f_n/f_m - 1)\|_D \\ \leq M_\sigma\left(\frac{f_n - f_m}{f_m}\right) &\leq \left\|\frac{f_n - f_m}{f_m}\right\|_D. \end{aligned}$$

Hence we have

$$\begin{aligned} \|h_n/h_m - 1\|_D = \|h_n/h_m - 1\|_{B^c} &= M_\sigma\left(\frac{h_n - h_m}{h_m}\right) \\ &\leq M_\sigma\left(\frac{f_n - f_m}{f_m}\right) \end{aligned}$$

and by multiplicativity of the growth modulus

$$\frac{M_\sigma(h_n - h_m)}{M_\sigma h_m} \leq \frac{M_\sigma(f_n - f_m)}{M_\sigma f_m}.$$

But since $|f_m - 1| < 1$, we have $|f_m| = 1$ (on D) and $M_\sigma f_m = 1$. The same is true for h_m and f. We have obtained

$$\|h_n - h_m\|_{B^c} = M_\sigma(h_n - h_m) \leq M_\sigma(f_n - f_m) \leq \|f_n - f_m\|_D, \qquad (*)$$

which proves the uniform convergence of the singular (Motzkin) factors h_n to $h \in H(B^c)$. Since $\|h - 1\|_{B^c} < 1$, we even have $h \in H(B^c)^\times$. Moreover, we have a uniform convergence

$$\frac{1}{h_n} \to \frac{1}{h} \in H(B^c) \quad (n \to \infty),$$

which implies a convergence

$$g_n = \frac{f_n}{h_n} \to \frac{f}{h} = g$$

in $H(D)$. The maximum principle on the ball B and Proposition 2 in (4.2) give

$$\|g_n - g_m\|_B \le M_\sigma(g_n - g_m) \le \|g_n - g_m\|_D,$$

hence

$$\|g_n - g_m\|_{D \cup B} \le \|g_n - g_m\|_D \to 0$$

when $m, n \to \infty$, so that $g_n \to g \in H(D \cup B)$. Since

$$\frac{1}{g_n} = \frac{h_n}{f_n} \to \frac{h}{f} = \frac{1}{g}$$

uniformly (first on D, but also on $D \cup B$ by the maximum principle), the function g is a unit of $H(D \cup B)$ and does not vanish in $D \cup B$.

Let us prove *uniqueness*. If $f = g_1 h_1 = g_2 h_2$ are two decompositions, then

$$\frac{g_1}{g_2} \cdot \frac{h_1}{h_2} = 1$$

is a Motzkin factorization of $f = 1$, and it is sufficient to prove the uniqueness in this simple case. Assume that $gh = 1$ is a Motzkin factorization and choose

$$g_n \to g \quad (g_n \in R(D \cup B)^\times), \quad h_n \to h \quad (h_n \in R(B^c)^\times).$$

Then the inequality $(*)$ for the rational functions h_n and 1 gives

$$\|h_n - 1\|_{B^c} \le \|g_n h_n - 1\|_D \to 0 \quad (n \to \infty).$$

This proves $h_n \to 1$ $(n \to \infty)$, $h = \lim_{n\to\infty} h_n = 1$, and $g = 1$. ■

Exercises for Chapter 6

1. (*a*) Show that $K[X][[Y]] \neq K[[Y]][X]$ (consider $\sum_{i<j} X^i Y^j$).
   (*b*) Give a description of the fraction field $K((X))$ of $K[[X]]$ using Laurent series of order $> -\infty$.

2. Using the definition of the product of formal power series prove the identity

$$D(fg) = g\,Df + f\,Dg \quad (f,\ g \in K[[X]])$$

for the formal derivative of a product.

3. (*a*) Let $f,\ g \in K[[X]]$ be two convergent power series with $g(0) = 0$. Assume that the value group $|K^\times|$ is dense in $\mathbf{R}_{>0}$. Prove that the numerical evaluation result $(f \circ g)(x) = f(g(x))$ is valid in the ball $|x| \le r$ of K, provided that $r < r_g$ and $g(B_{\le r}) \subset B_{<r_f}$.
   (*b*) Show that the radius of convergence of a composite $f \circ g$ satisfies

$$r_{f\circ g} \ge \min\left(r_g, \sup\{r : M_r g < r_f\}\right)$$

   and $M_r(f \circ g) = M_{M_r g}(f)$.

4. Let $f(X) \in K[[X]]$ be a convergent power series and fix $a \in K$, $|a| < r_f$. Let $X = a + Y$ and $f(X) = f(a+Y) = f_a(Y)$ (this substitution $g(Y) = a + Y$ is not a substitution of formal power series of the type considered in the theory, since $\omega(g) = 0$). Show that the following double series is summable when X is replaced by an element $x \in K$, $|x| < r_f$:

$$\sum_{0\le m\le n} a_n \binom{n}{m} a^m Y^{n-m}.$$

   Reorder its terms to obtain another proof of the proposition in (VI.2.4):

$$f(a) + Df(a)Y + \frac{1}{2!}D^2 f(a)Y^2 + \cdots.$$

   Deduce that the radius of convergence of g is at least equal to r_f. Interchanging the roles of f and g, conclude that $r_f = r_g$. [By contrast to the classical case, it is impossible to obtain an analytic continuation of f using a Taylor series centered at a different point.]

5. Let p be an odd prime. There is a sequence $(a_n)_{n\ge 0}$ in $\mathbf{C}_p$ with

$$\tfrac{1}{2^{n+1}} < |a_n| < \tfrac{1}{2^n} \quad (n \ge 0).$$

   What is the radius of convergence of the power series $f = \sum_{n\ge 0} a_n X^n$? Show that the sphere $|x| = r_f$ in $\mathbf{C}_p$ is empty (the corresponding closed and open balls coincide). What can one say of the convergence of f on the sphere $|x| = r_f$ in Ω_p?

6. Take $K = \mathbf{Q}_p$ and consider the formal power series

$$f(X) = \sum_{n\ge 0} X^n = \frac{1}{1-X}, \qquad g(Y) = Y - Y^p.$$

   Find the power series representing the composite $f \circ g$, which is the rational function

$$\frac{1}{1 - X + X^p}$$

   when $p = 2$. In this case, the two roots ξ, η of $1 - X + X^2 = 0$ are easily determined. Give the power series expansion of $(1 - X + X^2)^{-1}$ explicitly using the partial fraction decomposition

$$\frac{1}{1 - X + X^2} = \frac{a}{X - \xi} + \frac{b}{X - \eta}.$$

   The coefficients of the corresponding power series are periodic mod 6. (*Hint.* Note that ξ and η are 6th roots of unity.)

7. Show that

$$\tan x = \frac{\sin x}{\cos x} \quad (|x| < r_p),$$

$$\arctan x = x - \frac{x^3}{3} + \frac{x^5}{5} \mp \cdots \quad (|x| < 1),$$

are inverse functions for $|x| < r_p$. L. van Hamme has suggested the following extension. Choose $i \in \mathbf{C}_p$ with $i^2 = -1$ and use the Iwasawa logarithm to define

$$\arctan x = \frac{1}{2i}\mathrm{Log}\frac{1+ix}{1-ix} \quad (x \neq \pm i).$$

Using

$$2i(\arctan x - \arctan a) = \mathrm{Log}\frac{1+ix}{1+ia} - \mathrm{Log}\frac{1-ix}{1-ia}$$

prove that if a point $a \in \mathbf{C}_p$ is selected, then $\arctan x$ is given by a series expansion valid in the ball

$$|x - a| < \min(|a - i|, |a + i|).$$

Prove that

$$\arctan 1 = 0, \qquad \lim_{|x|\to\infty} \arctan x = 0.$$

8. (*a*) Let $f(X) = \sum_{n\geq 0} a_n X^n$ be a convergent power series and assume that the set $\{|a_n|r_f^n\}$ is unbounded. Prove that there exists an infinite sequence of critical radii $r_i \nearrow r_f$.
(*b*) Let $(a_n)_{n\geq 0}$ be a sequence in $\mathbf{C}_p$ with

$$|a_0| < |a_1| < \cdots < |a_n| < \cdots < 1.$$

Prove that the formal power series $f(X) = \sum_{n\geq 0} a_n X^n$ defines a bounded function in $|x| < r_f$ with infinitely many critical radii converging to r_f.

9. Prove that for any ultrametric field K, $1 + XK[[X]]$ is a multiplicative group, and for $r > 0$,

$$G_r = \left\{\sum_{n\geq 0} a_n X^n : |a_n| \leq 1/r^n \ (n \geq 0)\right\}$$

is a multiplicative subgroup of $1 + XK[[X]]$.
(*Hint.* For $r = 1$, the subgroup G_1 is simply $1 + XA[[X]]$; use dilatations to get the general statement.)

10. Prove the Liouville theorem in the case K has a discrete valuation but an infinite residue field.

11. (*a*) Show that $r > 0$ is a regular radius for $f_1, f_2, \ldots, f_n$ iff it is regular for $f_1 f_2 \cdots f_n$.
(*b*) Let $f, g \in K[[X]]$ be two convergent power series. Assume $f(x_n) = g(x_n)$ for a convergent sequence $x_n \to x_\infty$ ($x_n \neq x_\infty$ for all $n \geq 0$), where $|x_\infty| < \min(r_f, r_g)$. Show that $f = g$.
(*c*) Formulate and prove a statement analogous to (*b*) for Laurent series.

12. Let $K = \mathbf{Q}_p(\mu_{p^\infty})$. This is a totally ramified extension of $\mathbf{Q}_p$, and $|K^\times|$ is dense (the residue field of K is $\mathbf{F}_p$). Give an example of a polynomial f for which $\|f\|_{\mathrm{Gauss}} = \sup_{x\leq 1} |f(x)|$ is not a maximum.

13. Let f denote the Taylor series at the origin of

$$\frac{X^4 + (p^2-1)X^2 - p^2}{1 - pX}.$$

What is the canonical factorization $f = cPQg_1$ (Theorem 2 in VI.2.2) of this formal power series? Draw the Newton polygon of f.

14. Show that the formal power series $\sum_{n\geq 1} p^{n!} X^n$ defines an entire function. What is the location of its zeros? Give the form of an infinite product that represents this function.

15. Give the Laurent expansions of the rational function

$$\frac{x}{(x-1)(x-p)}$$

valid in the region $|x| > 1$. Same question for the region $1/p < |x| < 1$.

16. Let $f(X) = \sum_{n\geq 0} a_n X^n$ be a formal power series with coefficients $|a_n| \leq 1$. Consider the map $g : \mathbf{M}_p - \{0\} \to \mathbf{C}_p - \mathbf{A}_p$ defined by $y = g(x) = f(x) + 1/x$. Prove that g is bijective.
(*Hint.* To show that g is surjective, proceed as follows. For given y with $|y| > 1$ we are looking for a solution x of $x = y^{-1}(1 + xf(x))$. Show that the sequence defined inductively by $x_0 = 0$, $x_{n+1} = y^{-1}(1 + x_n f(x_n))$ is a Cauchy sequence with $|x_n| = 1/|y| < 1$, and hence it converges in the open unit ball $\mathbf{M}_p$.)

17. What are the critical radii of the polynomial

$$x^{n+1} + x^n + \cdots + x^2 + \tfrac{1}{p^2}x + 1 \quad (n \geq 1)\,?$$

How many roots are there on each critical sphere?

18. Let $f : \mathbf{R}_{>0} \to \mathbf{R}_{>0}$ be a C^2-function and define $\log f(r) = \varphi(\rho)$, where $\rho = \log r$. Show that φ may be convex even when f is not convex (consider $f(r) = \sqrt{r}$).
(*a*) Consider the functions $f_\alpha(r) = r^\alpha$ ($\alpha \in \mathbf{R}$) and discuss the convexity of f_α and the corresponding φ_α.
(*b*) Prove $r^2 f''(r)/f(r) = \varphi''(\rho) + \varphi'(\rho)(\varphi'(\rho) - 1)$ and deduce that if φ is convex and φ' does not take values in $(0, 1)$, then f is convex.

19. What is the Newton polygon of the polynomial

$$px^3 - (4p^2+1)x^2 + (4p^3+4p)x - 4p^2\,?$$

(*a*) Compute the absolute value of the zeros of this polynomial.
(*b*) Factorize the polynomial and compare with the result obtained in (*a*).

20. (*a*) Show that the logarithm $\log : 1 + B_{\leq r_p} \to B_{\leq r_p}$ is surjective. More precisely, show that for each $|y| = r_p$ there are exactly p preimages x_i with $|x_i| = r_p$ and $\log(1 + x_i) = y$ (but if $|y| < r_p$, there are only $p - 1$ preimages).
(*b*) Draw the valuation polygon of the formal power series of $\log(1 + X)$.

21. What are the Newton polygons of the Chebyshev polynomials T_3, T_6, T_9 (any prime p) (cf. exercises for Chapter V). Same question for $T_1, T_p, T_{p^2}, \ldots, T_{p^n}$.

22. Let $f = g/h \in \mathbf{C}_p(x)$ be a rational function. Assume that f has a zero and a pole on a sphere $|x| = r$. Show that $|f|$ assumes all values (in $p^{\mathbf{Q}} = |\mathbf{C}_p^\times|$) on this sphere.

23. Generalize the mean value theorem (as given in (2.4)) to the case of a parametrized curve $t \mapsto (f(t), g(t), h(t))$ by considering the determinant

$$\Phi(t) = \begin{vmatrix} f(a) & f(t) & f(b) \\ g(a) & g(t) & g(b) \\ h(a) & h(t) & h(b) \end{vmatrix},$$

which vanishes for the two values $t = a$ and $t = b$.

24. (a) Draw the graph of the growth modulus $M_r f$ of the following rational functions:

$$\frac{1-X}{1+X}, \quad \left(\frac{1-X}{1+X}\right)^n, \quad \frac{1-X^n}{1+X^m}.$$

Can one guess the location of the zeros and poles of a rational function by the sole observation of the graph of $M_r f$? Sketch the graph of the growth modulus $M_{r,1} f$, centered at the point 1, for the same functions.

(b) Draw the graph of the growth modulus $M_r f$ of

$$\frac{X}{1-X} \frac{1-pX}{1-p^2X}.$$

25. Give the principal parts Pf of the rational functions

$$f = \frac{x^2+x+1}{x^2}, \quad \frac{2x^2-x-1}{2x^2}$$

at the origin. Take $\sigma > 1$ and consider a region $D := \{\sigma \leq |x| \leq r\}$. Compare $\|Pf\|_D$ and $\|f-1\|_D$.

26. With the notation of (3.5), let $f = c\prod_{a\in S}(x-a)^{\mu_a}$. Prove that the principal part $P_i(f'/f)$ with respect to some ball B_i is f_i'/f_i, where f_i is the Motzkin factor of f relative to B_i. If $\|f-1\|_D < 1$, use (3.4) and prove $\left\|f'\right\|_D = \max_{0\leq i\leq \ell} \left\|f_i'\right\|_D$.

27. Fix $r > 0$ and choose $c \in \Omega_p - \mathbf{C}_p$. Then

$$f \mapsto M_{r,c}(f)$$

is a multiplicative norm on $\mathbf{C}_p(X)$. If $\delta = \operatorname{dist}(c, \mathbf{C}_p)$, show that

$$\inf_{a\in\mathbf{C}_p} M_{r,c}(X-a) = \delta.$$

In particular, the general inequality $\inf_{a\in\mathbf{C}_p} M_{r,c}(X-a) \geq r$ can be a strict inequality.

28. Show that the union of two infraconnected subsets having a nonempty intersection is infraconnected.

29. The subsets of $\mathbf{C}_p$

$$B_{\leq 1/\varepsilon}(0) \cap \{|x-a| \geq \varepsilon : \text{ for all } a \in \mathbf{Q}_p\}$$

are infraconnected. What are their enveloping balls? What are their holes? Conclude that $\mathbf{C}_p - \mathbf{Q}_p$ is a union of an increasing sequence of bounded infraconnected sets, each of them having finitely many holes.

30. (*a*) Let B be a dressed ball and D a closed and bounded subset of $\mathbf{C}_p$ so that

$$\|f\|_B = \|f\|_D \quad \text{for all polynomials } f.$$

Prove that $D \subset B$ and $B = B_D$ is the enveloping ball of D.
(*b*) Let D be closed and bounded. Prove $\|f\|_{B_D} = \|f\|_D$ for $f \in H(B_D)$.
(*Hint.* Choose $g \in R(B_D)$ such that $\|f - g\|_{B_D} < \|f\|_D$.)

31. Let $D = B_{\leq r} - \coprod_{1 \leq i \leq \ell} B_i$ and $B \subset D$ any nonempty open ball. Show that the restriction map $H(D) \to H(B)$ is injective.
(*Hint.* Use the Mittag-Leffler decomposition to show that each $f \in H(D)$ is described by a power series expansion in B and hence has isolated zeros if nonzero.)

32. The simple domains of the form $D = B_{\leq r} - \coprod_{1 \leq i \leq \ell} B_i$ can be patched together. For example, if the hole B_1 of D has radius r_1, show that any $D_1 = B_{\leq r_1} - \coprod_{1 \leq j \leq \ell_1} B'_j$ has a nonempty intersection with D and the union $D \cup D_1$ is again a simple domain of the same form.

33. Choose a sequence $r_n \nearrow 1$ ($r_n < 1$) and let $S_n = \{|x| = r_n\}$. Show that

$$D = B_{<1} - \coprod_n S_n \subset \mathbf{C}_p$$

is infraconnected. Moreover, if $B \subset D$ is any nonempty open open ball, show that the restriction map $H(D) \to H(B)$ is injective.
(Motzkin calls *analytic* a set D having this property. If $D' = D \cup B_{<1}(1)$, it can be shown that the restriction $H(D') \to H(B_{<1}(1)$ is *not* injective; hence D' is not an analytic set; this is again the phenomenon of T-filters.)

34. Find domains of uniform convergence for the following sequences of rational functions:

$$\frac{x^n}{1-x^n}, \quad \frac{x^n}{1-x^{2n}}, \quad \frac{x^{2n}}{1-x^n}.$$

35. Let $0 < \varepsilon < 1$ and $D = \{x \in \mathbf{C}_p : |x| \notin [1-\varepsilon, 1+\varepsilon]\}$. Consider the sequence of functions

$$f_n(x) = \frac{1}{1-x^n} \in R(D).$$

This sequence converges in $H(D)$: The limit $f \neq 0$ is a zero divisor (a nontrivial idempotent). Conclude that for any $D \in \mathbf{C}_p$ having at least two points, and which is not infraconnected, $H(D)$ is not an integral domain.
(*Hint.* Choose $a \neq b \in D$ and an annulus $\{0 < r_1 < |x-a| < r_2 < |b-a|\}$ that does not meet D and use a sequence similar to the one above. It can be shown that infraconnectedness is also a sufficient condition for $H(D)$ to be an integral domain.)

36. Show that the series $\sum_{n\geq 0} p^n/(x-n)$ converges for $x \in \mathbf{C}_p - \mathbf{Z}_p$ and that it defines an analytic element on the subsets $D_n = \{x \in \mathbf{C}_p : |x-a| \geq 1/n \ \forall\, a \in \mathbf{Z}_p\}$.

37. Show that the series

$$\sum_{n\geq 0}\left(\frac{1}{x-p^n}-\frac{1}{x}\right)$$

converges for $x \notin \{0\} \cup p^{\mathbf{N}}$ and defines an analytic element in the complement D_ε of any (finite) union of balls of the form $\bigcup_{n\geq 0} B_{<\varepsilon}(p^n)$ $(\varepsilon > 0)$.

38. Show that the exponential is not an analytic element on its convergence ball. (*Hint.* Let π be a root of $X^{p-1} + p = 0$, so that the convergence ball of $f(X) = e^{\pi x}$ is 1. Then use the Christol-Robba criterion (VI.4.6).)

39. Let $p_\nu = p^\nu(p^\nu - 1)$ and $\varphi_\nu(x) = 1 - x^{p_\nu}$. Check the following assertions:
(*a*) $\varphi_\nu(x) \to 0$ $(\nu \to \infty)$ for any $|x| = 1$.
(*b*) $|\varphi_\nu(x)| = 1$, if $|x| < 1$, $\quad M_1(\varphi_\nu) = 1$.
(*c*) $\sup_{|x|=1} |\varphi_\nu(x)| = 1$.

40. (*a*) Let $D \subset \mathbf{C}_p$ be closed, bounded, and infraconnected. Assume $f \in H(D)$ and $\|f-1\|_D < 1$. Prove that $\log f \in H(D)$.
(*b*) If $f \in R(D)^\times$, $\|f-1\|_D < 1$, and $f = \prod_i f_i$ is the Motzkin factorization of f (with respect to a finite family of open balls B_i as in (3.5)), show that $\log f = \sum_i \log f_i$ is the Mittag-Leffler decomposition of $\log f$.

41. Let $0 < r < 1$ and $(a_n)_{n\geq 1}$ a sequence on the unit sphere with $|a_n - a_m| = 1$ whenever $n \neq m$. Define

$$B_0 = B_{<r}(0), \quad B_n = B_{<r}(a_n), \quad D = \mathbf{C}_p - \cup_{n\geq 0} B_n.$$

Choose a sequence $\lambda_n \to 0$ and consider the rational functions

$$g_n(x) = 1 + \frac{\lambda_n}{x-a_n}.$$

Show that $\|g_n - 1\|_{B_n^c} \leq |\lambda_n|/r \to 0$. Conclude that $\prod_{n\geq 1} g_n$ converges uniformly on D, $x^2\prod_{n\geq 1} g_n \in H(D)^\times$, but the sequence $f_N = x^2\prod_{1\leq n\leq N} g_n$ is not uniformly convergent on D.
(*Hint.* Observe that $f_{N+1} - f_N$ is not bounded on D due to the presence of the unbounded factor x^2.)

7

Special Functions, Congruences

The applications given in this chapter concern congruences.

They rely on the first two sections of the preceding chapter (convergence of power series, growth modulus, critical radii). The more technical notion of analytic element developed in the last two sections of Chapter VI is not used here.

1. The Gamma Function Γ_p

The special functions of classical analysis are defined by a variety of methods: series expansions, differential equations, parametric integrals, functional equations, etc. We have seen in (V.4) that the power series method is well adapted to the definition of the exponential and logarithm in a suitable ball of $\mathbf{C}_p$.

Here is another method adapted to p-adic analysis. Let f be a classical function defined on some interval $[a,\infty) \subset \mathbf{R}$ with rational values $f(n) \in \mathbf{Q}$ on the integers $n \geq a$, we may look for a continuous function $\mathbf{Z}_p \to \mathbf{C}_p$ extending $n \mapsto f(n)$. By the density of $\mathbf{Z} \cap [a,\infty)$ in $\mathbf{Z}_p$, there is at most one such interpolation. Of course, this possibility requires arithmetic properties of the sequence of values $f(n)$ and the method works only in particular cases. A suitable modification of the function $n \mapsto n!$ will lead to an analogue of the classical gamma function. Another successful example of this method (not treated here) is the Riemann zeta function, using its values at the negative integers.

To simplify our considerations, we assume first that the prime p is *odd*: $p \geq 3$. The case $p = 2$ is treated later in (1.7).

1.1. Definition

The function $n \mapsto n!$ cannot be extended by continuity on $\mathbf{Z}_p$. Indeed, let us look for a continuous function

$$f : \mathbf{Z}_p \to \mathbf{Q}_p$$

satisfying $f(n) = nf(n-1)$ for all integers $n \geq 1$. By continuity and density the same relation will hold for all $n \in \mathbf{Z}_p$. Iterating it, we get

$$f(n) = n(n-1)(n-2)\cdots p^m f(p^m - 1)$$

for all integers $n > p^m$, where p^m is a fixed power of p. Since f is continuous on the compact space $\mathbf{Z}_p$, it is bounded and there is a constant $C > 0$ such that $|f(x)| \leq C$ $(x \in \mathbf{Z}_p)$. The preceding factorization also shows that

$$|f(n)| \leq |p|^m \cdot C$$

for all integers $n > p^m$. But these integers make up a dense subset of $\mathbf{Z}_p$; hence

$$\|f\|_\infty \leq |p|^m \cdot C.$$

Since the integer $m \geq 1$ is arbitrary, the only possibility is $\|f\|_\infty = 0$. (The single consideration of the case $m = 1$ is sufficient: Taking $C = \|f\|_\infty$ we get $\|f\|_\infty \leq |p|\,\|f\|_\infty$, $(1 - 1/p)\|f\|_\infty \leq 0$.) The only continuous function f on $\mathbf{Z}_p$ satisfying the functional equation $f(n) = nf(n-1)$ for all integers $n \geq 1$ is $f = 0$.

The trouble obviously comes from the multiples of p in the factorial $n!$: Let us omit them and consider a *restricted factorial* $n!^*$

$$n!^* := \prod_{1 \leq j \leq n,\ p \nmid j} j.$$

The key to the construction of the p-adic gamma function lies in a generalization of the classical Wilson congruence

$$(p-1)! \equiv -1 \bmod p.$$

Proposition. *Let a and* $\nu \geq 1$ *be two integers. Then*

$$\prod_{a \leq j < a + p^\nu,\ p \nmid j} j \equiv -1 \pmod{p^\nu}.$$

PROOF. The integers $a \leq j < a + p^\nu$ make up a complete set of representatives of the quotient $\mathbf{Z}/p^\nu\mathbf{Z}$. Those that are not multiples of p represent the invertible elements, namely the elements of the unit group $G = (\mathbf{Z}/p^\nu\mathbf{Z})^\times$. Grouping each element $g \in G$ with its inverse g^{-1} we obtain compensations in the product *except when* $g = g^{-1}$. But

$$g = g^{-1} \Longleftrightarrow g^2 = 1.$$

In the *ring* $\mathbf{Z}/p^\nu\mathbf{Z}$ we can write

$$g^2 = 1 \Longleftrightarrow g^2 - 1 = 0 \Longleftrightarrow (g-1)(g+1) = 0.$$

The elements in question are $g = \pm 1$, or both $g - 1$ and $g + 1$ are zero divisors. The second case corresponds to

$$p \text{ divides both } g - 1 \text{ and } g + 1.$$

Obviously, this second case can happen only when p divides $2 = (g+1)-(g-1)$, i.e., $p = 2$. Since we are considering only the case p odd, this does not occur, and the proposition is proved. ∎

The proposition implies that the products

$$f(n) := (-1)^n \prod_{1 \le j < n,\ p \nmid j} j \quad (n \ge 2)$$

satisfy $f(a) \equiv f(a + p^\nu) \pmod{p^\nu}$. More generally, they also satisfy

$$f(a) \equiv f(a + mp^\nu) \pmod{p^\nu} \quad (m \in \mathbf{N}).$$

The function $a \mapsto f(a) : \mathbf{N} - \{0, 1\} \to \mathbf{Z}$ is uniformly continuous for the p-adic topology, hence has a unique continuous extension $\mathbf{Z}_p \to \mathbf{Z}_p$.

Definition. *The Morita* p-adic gamma function *is the continuous function*

$$\Gamma_p : \mathbf{Z}_p \to \mathbf{Z}_p$$

that extends

$$f(n) := (-1)^n \prod_{1 \le j < n,\ p \nmid j} j \quad (n \ge 2).$$

Observe that by construction, this p-adic gamma function takes its values in the clopen subset $\mathbf{Z}_p^\times$ of $\mathbf{Z}_p$.

Since its definition depends on the prime p, this function is denoted by Γ_p. (But as with the functions log and exp, we might simply denote it by Γ when the prime p is fixed and there is no risk of confusion.)

1.2. *Basic Properties*

We have

$$\Gamma_p(2) = 1, \quad \Gamma_p(3) = -2,$$

$$\Gamma_p(n+1) = \begin{cases} n! & \text{if } n \text{ odd}, n \le p-1, \\ -n! & \text{if } n \text{ even}, n \le p-1. \end{cases}$$

From the definition it also follows that $\Gamma_p(n) \in \mathbf{Z}_p^\times$ is given by

$$\Gamma_p(n+1) = \frac{(-1)^{n+1} n!}{\prod_{1 \le kp \le n} kp} = \frac{(-1)^{n+1} n!}{[n/p]!\, p^{[n/p]}}$$

when the integer n is greater than or equal to 2. Still, by its definition, we have

$$\Gamma_p(n+1) = \begin{cases} -n\Gamma_p(n) & \text{if } n \text{ is not multiple of } p, \\ -\Gamma_p(n) & \text{if } n \text{ is multiple of } p, \end{cases}$$

and by continuity, more generally,

$$\Gamma_p(x+1) = \begin{cases} -x\Gamma_p(x) & \text{if } x \in \mathbf{Z}_p^\times, \\ -\Gamma_p(x) & \text{if } x \in p\mathbf{Z}_p. \end{cases}$$

It is convenient to introduce a function h_p:

$$h_p(x) = \begin{cases} -x & \text{if } x \in \mathbf{Z}_p^\times \quad (|x| = 1), \\ -1 & \text{if } x \in p\mathbf{Z}_p \quad (|x| < 1), \end{cases}$$

in order to be able to write the functional equation

$$\Gamma_p(x+1) = h_p(x) \cdot \Gamma_p(x) \quad (x \in \mathbf{Z}_p).$$

This functional equation can be used backwards to compute the values $\Gamma_p(1)$ and $\Gamma_p(0)$ from $\Gamma_p(2) = 1$. In particular, we check that $\Gamma_p(0) = 1$. This normalization also follows by continuity: By the proposition in (1.1) (with $a = 0$),

$$\Gamma_p(p^n) = -\prod_{1 \le j < p^n,\, p \nmid j} j \equiv +1 \pmod{p^n};$$

hence $\Gamma_p(p^n) \to 1$ as $n \to \infty$.

Theorem. *For an odd prime p, the p-adic gamma function $\Gamma_p : \mathbf{Z}_p \to \mathbf{Q}_p$ is continuous. Its image is contained in $\mathbf{Z}_p^\times$. Moreover:*

(1) $\Gamma_p(0) = 1$, $\Gamma_p(1) = -1$, $\Gamma_p(2) = 1$,
$\Gamma_p(n+1) = (-1)^{n+1} n! \quad (1 \le n < p)$.
(2) $|\Gamma_p(x)| = 1$.
(3) $|\Gamma_p(x) - \Gamma_p(y)| \le |x - y|, \quad |\Gamma_p(x) - 1| \le |x|$.
(4) $\Gamma_p(x+1) = h_p(x)\Gamma_p(x)$.
(5) $\Gamma_p(x) \cdot \Gamma_p(1-x) = (-1)^{R(x)}$,
where $R(x) \in \{1, 2, \ldots, p\}, \quad R(x) \equiv x \pmod{p}$.

As we shall see in (1.7), the property (3) has to be suitably modified for $p = 2$. The exponent in (5) is also different in this case.

PROOF. (3) follows from $\Gamma(a+mp^\nu) \equiv \Gamma(a) \pmod{p^\nu}$ ($a \in \mathbf{Z}$, $m \in \mathbf{N}$) by continuity.

There only remains to prove (5). Put $f(x) = \Gamma_p(x) \cdot \Gamma_p(1-x)$. We have

$$f(x+1) = h_p(x)\Gamma_p(x) \cdot \Gamma_p(-x),$$

and since $\Gamma_p(-x) = \Gamma_p(1-x)/h_p(-x)$,

$$f(x+1) = \varepsilon(x) \cdot f(x), \quad \varepsilon(x) = h_p(x)/h_p(-x).$$

Now,

$$\varepsilon(x) = \begin{cases} -1 & \text{if } |x| = 1, \\ +1 & \text{if } |x| < 1. \end{cases}$$

Take for x an integer n and iterate,

$$f(n+1) = \varepsilon(n) \cdot f(n) = \cdots = (-1)^{\#} f(1),$$

with an exponent # equal to the number of integers $j \le n$ prime to p. Since the number of integers $j \le n$ divisible by p is $[n/p]$, this exponent # is $n - [n/p]$. Hence

$$f(n+1) = \Gamma_p(n+1) \cdot \Gamma_p(-n) = (-1)^{n-[n/p]} \cdot \underbrace{\Gamma_p(1)\Gamma_p(0)}_{-1} = (-1)^{n+1-[n/p]}.$$

To find the formula given in (5), let us take $x = m = n+1$ (integer), whence $1 - m = -n$ and

$$\Gamma_p(m) \cdot \Gamma_p(1-m) = \Gamma_p(n+1) \cdot \Gamma_p(-n) = (-1)^{n+1-[n/p]}.$$

With the expansion of the integer n in base p,

$$n = n_0 + n_1 p + \cdots = n_0 + p\left[\frac{n}{p}\right]$$

we infer

$$n - \left[\frac{n}{p}\right] = n_0 + (p-1)\left[\frac{n}{p}\right].$$

Since we assume p odd — hence $p-1$ even — this proves that $n - [n/p]$ has the same parity as n_0:

$$(-1)^{n+1-[n/p]} = (-1)^{n_0+1}.$$

Since $m = n+1 \equiv n_0+1 \pmod{p}$ and n_0+1 is in the correct range $\{1, 2, \ldots, p\}$, we have $R(m) = n_0 + 1$, and the formula is proved for integral values $x = m$ of the variable. By density and continuity, it remains true for all $x \in \mathbf{Z}_p$. ∎

Comment. The classical Γ-function satisfies the Legendre relation

$$\Gamma(z)\Gamma(1-z) = \frac{\pi}{\sin \pi z},$$

which implies for $z = \frac{1}{2}$

$$\Gamma(\tfrac{1}{2})^2 = \pi, \quad \Gamma(\tfrac{1}{2}) = \sqrt{\pi}.$$

Hence we can say that in $\mathbf{Q}_p$, an analogue of the number π could be taken as $\Gamma_p(\frac{1}{2})^2 = (-1)^{(p+1)/2}$. In particular, if $p \equiv 1 \pmod 4$, $\Gamma_p(\frac{1}{2}) = \sqrt{-1}$ is a *canonical* square root of -1 in $\mathbf{Q}_p$. This canonical imaginary unit can be identified easily. In the case $p \equiv 1 \pmod 4$, the Wilson congruence

$$(p-1)! \equiv \left(\frac{p-1}{2}\right)!^2 \equiv -1 \pmod p$$

shows that $\left(\frac{p-1}{2}\right)! \bmod p$ is a square root of -1. Since $(p+1)/2 \equiv \frac{1}{2} \pmod p$, the point (3) of the above theorem gives

$$\Gamma_p(\tfrac{1}{2}) \equiv \Gamma_p\left(\frac{p+1}{2}\right) = (-1)^{(p+1)/2}\left(\frac{p-1}{2}\right)! \equiv -\left(\frac{p-1}{2}\right)! \pmod p.$$

1.3. *The Gauss Multiplication Formula*

The classical gamma function satisfies the identity

$$\prod_{0 \le j < m} \Gamma\left(z + \frac{j}{m}\right) = (2\pi)^{(m-1)/2} m^{(1-2mz)/2} \cdot \Gamma(mz) \quad (m \ge 2),$$

which is the *Gauss multiplication formula.* It is remarkable that Γ_p satisfies a similar relation.

Proposition. *Let $m \ge 1$ be an integer prime to p. Then*

$$\prod_{0 \le j < m} \Gamma_p\left(x + \frac{j}{m}\right) = \varepsilon_m \cdot m^{1-R(mx)} \cdot (m^{p-1})^{s(mx)} \cdot \Gamma_p(mx),$$

where

$$\varepsilon_m = \prod_{0 \le j < m} \Gamma_p\left(\frac{j}{m}\right),$$

$$R(y) \in \{1, \dots, p\}, \quad R(y) \equiv y \bmod p,$$

$$s(y) = \frac{R(y) - y}{p} \in \mathbf{Z}_p.$$

PROOF. Let

$$f(x) = f_m(x) = \prod_{0 \le j < m} \Gamma_p\left(x + \frac{j}{m}\right),$$

$$G(x) = G_m(x) = f(x)/\Gamma_p(mx).$$

We have to compute the Gaussian factor $G(x)$. Start with

$$\begin{aligned} G(x+1/m) &= \frac{1}{\Gamma_p(mx+1)} \prod_{1 \le j \le m} \Gamma_p(x + j/m) \\ &= \frac{1}{h_p(mx)\Gamma_p(mx)} \cdot \frac{\Gamma_p(x+1)}{\Gamma_p(x)} \prod_{0 \le j < m} \Gamma_p(x + j/m) \\ &= \frac{h_p(x)}{h_p(mx)} G(x). \end{aligned}$$

Consider the locally constant function

$$\lambda(x) = \frac{h_p(x)}{h_p(mx)} = \begin{cases} -x/(-mx) = 1/m & \text{if } |x| = 1, \\ -1/(-1) = 1 & \text{if } |x| < 1 \end{cases}$$

(since $(m, p) = 1$). This multiplier is useful to compute the successive values

$$\begin{aligned} G(1/m) &= \lambda(0) \cdot G(0), \\ G(2/m) &= \lambda(0)\lambda(1/m) \cdot G(0), \ \dots \\ G(j/m) &= \prod_{0 \le i < j} \lambda(i/m) \cdot G(0). \end{aligned}$$

Since $(m, p) = 1$, we have

$$\prod_{0 \le i < j} \lambda(i/m) = (1/m)^{\#}$$

with an exponent

$$\begin{aligned} \# &= \#\{i \text{ prime to } p,\ 0 < i < j\} \\ &= j - 1 - \left[\frac{j-1}{p}\right]. \end{aligned}$$

Let us find a convenient form for this exponent. Start with the p-adic expansion of the integer $j - 1$:

$$j - 1 = (j-1)_0 + p\left[\frac{j-1}{p}\right],$$

$$j = \underbrace{(j-1)_0 + 1}_{=:R(j)} + p\left[\frac{j-1}{p}\right],$$

where $R(j) \equiv j \pmod{p}$ is in the correct range $\{1, \ldots, p\}$. This proves

$$j - 1 - \left[\frac{j-1}{p}\right] = R(j) - 1 + (p-1)\left[\frac{j-1}{p}\right],$$

and hence

$$\prod_{0 \le i < j} \lambda(i/m) = m^{1-R(j)}(m^{p-1})^{s(j)}$$

with

$$s(j) = -\left[\frac{j-1}{p}\right] = -\frac{j - R(j)}{p},$$

an expression that admits a continuous extension to $\mathbf{Z}_p$

$$s(x) = \frac{R(x) - x}{p} \quad (x \in \mathbf{Z}_p).$$

We have proved

$$\begin{aligned} G(j/m) &= \prod_{0 \le i < j} \lambda(i/m) \cdot G(0) \\ &= m^{1-R(j)}(m^{p-1})^{s(j)} \cdot G(0), \end{aligned}$$

and

$$G(x) = m^{1-R(mx)}(m^{p-1})^{s(mx)} \cdot G(0)$$

for integral $x = j/m$ ($j = mx > 0$ multiple of m). By continuity, the last formula will also hold for all $x \in \mathbf{Z}_p$. This proves the expected formula

$$\prod_{0 \le j < m} \Gamma_p(x + j/m) = \varepsilon_m \cdot m^{1-R(mx)}(m^{p-1})^{s(mx)} \Gamma_p(mx)$$

with $\varepsilon_m = G(0)$. ■

Finally, let us observe that $\varepsilon_m = G(0)$ is always a fourth root of unity.

Lemma. *We have $\varepsilon_m^4 = 1$. In fact, $\varepsilon_m^2 = 1$ except when $p \equiv 1 \pmod 4$ and m is even, in which case $\varepsilon_m^2 = -1$.*

Proof. When m is odd,

$$\varepsilon_m = \Gamma_p(\tfrac{1}{m}) \cdots \Gamma_p(\tfrac{m-1}{m})$$

since $\Gamma_p(0) = 1$. Now we can group pairs

$$\Gamma_p\left(\frac{j}{m}\right)\Gamma_p\left(\frac{m-j}{m}\right) = \pm 1$$

(by the analogue of the Legendre relation). In this case $\varepsilon_m = \pm 1$. Assume now m even. The same grouping leaves the middle term $\Gamma_p(\frac{1}{2})$ solitary:

$$\varepsilon_m = \pm\Gamma_p(\tfrac{1}{2}), \quad \varepsilon_m^2 = \Gamma_p(\tfrac{1}{2})^2.$$

As we have seen in (1.2), $\Gamma_p(\frac{1}{2})^2 = -1$ when $p \equiv 1 \pmod 4$, so that ε_m is a square root of -1 and

$$\varepsilon_m = G(0) = \prod_{0 \le j < m} \Gamma_p(\tfrac{j}{m})$$

is a root of unity of order 4. ∎

1.4. The Mahler Expansion

If f is a continuous function on $\mathbf{Z}_p$, it can be represented by a Mahler series

$$f(x) = \sum_{k \ge 0} a_k \binom{x}{k}, \quad a_k = (\nabla^k f)(0).$$

We have shown (Comment 2 in (IV.1.1)) that these coefficients are linked to the values of f by the identity of formal power series

$$\sum_{k \ge 0} a_k \frac{X^k}{k!} = e^{-X} \cdot \sum_{n \ge 0} f(n) \frac{X^n}{n!}.$$

Proposition. *Let $\Gamma_p(x+1) = \sum_{k \ge 0} a_k \binom{x}{k}$ be the Mahler series of Γ_p. Then its coefficients satisfy the following identity:*

$$\sum_{k \ge 0} (-1)^{k+1} a_k \frac{x^k}{k!} = \frac{1-x^p}{1-x} \exp\left(x + \frac{x^p}{p}\right).$$

PROOF. Let us compute $e^{-x}\varphi(x)$, where $\varphi(x) = \sum_{n \ge 0} \Gamma_p(n+1) x^n/n!$. For this purpose, we make a partial summation over the cosets mod p:

$$\varphi(x) = \sum_{0 \le j < p} \sum_{m \ge 0} \Gamma_p(mp+j+1) \frac{x^{mp+j}}{(mp+j)!}.$$

Here, we can use

$$\Gamma_p(n+1) = \frac{(-1)^{n+1} n!}{[n/p]!\, p^{[n/p]}}$$

for $n = mp + j$, $[n/p] = m$ and get

$$\Gamma_p(mp+j+1) = (-1)^{mp+j+1} \cdot \frac{(mp+j)!}{m!\, p^m}.$$

$$\begin{aligned}\varphi(x) &= \sum_{0\le j<p}\sum_{m\ge 0}(-1)^{mp+j+1}\frac{x^{mp+j}}{m!p^m}\\ &= \sum_{m\ge 0}(-1)^{mp+1}\left(\frac{x^p}{p}\right)^m\frac{1}{m!}\sum_{0\le j<p}(-1)^j x^j\\ &= -\exp\left(\frac{(-x)^p}{p}\right)\sum_{0\le j<p}(-1)^j x^j\\ &= -\exp\left(\frac{(-x)^p}{p}\right)\frac{1-(-x)^p}{1-(-x)}\end{aligned}$$

Finally,

$$e^{-x}\varphi(x) = -\exp\left(-x+\frac{(-x)^p}{p}\right)\frac{1-(-x)^p}{1-(-x)} = \sum_{k\ge 0} a_k\frac{x^k}{k!},$$

whence the desired formula. ∎

1.5. *The Power Series Expansion of* $\log\Gamma_p$

We shall use the following formula (V.5.3) for the Volkenborn integral:

$$S(f')(x) = \int_{\mathbf{Z}_p}[f(x+y)-f(y)]\,dy \quad (x\in\mathbf{Z}_p)$$

with the function

$$f(x) = \begin{cases} x\mathrm{Log}\,x - x & \text{if } |x|=1,\\ 0 & \text{if } |x|<1,\end{cases}$$

$$\begin{aligned} f'(x) &= \begin{cases}\mathrm{Log}\,x & \text{if } |x|=1\\ 0 & \text{if } |x|<1\end{cases}\\ &= \mathrm{Log}\,h_p(x).\end{aligned}$$

Here, Log denotes the Iwasawa logarithm (V.4.5): It vanishes on roots of unity, so that $\mathrm{Log}\,(-x)=\mathrm{Log}\,x$. This implies that the function f is odd, so that $\int_{\mathbf{Z}_p} f(t)\,dt=0$ (Corollary of Proposition 4 in (V.5.3)), a fact that we are going to use presently. On the other hand h_p still denotes the function occurring in the functional equation

$$\Gamma_p(x+1) = h_p(x)\Gamma_p(x);$$

hence

$$\nabla\mathrm{Log}\,\Gamma_p(x) = \mathrm{Log}\,\Gamma_p(x+1) - \mathrm{Log}\,\Gamma_p(x) = \mathrm{Log}\,h_p(x).$$

Since $S\nabla f = f - f(0)$ and $\operatorname{Log}\Gamma_p(0) = \operatorname{Log} 1 = 0$, we infer

$$\operatorname{Log}\Gamma_p(x) = S \operatorname{Log} h_p(x).$$

The above formula for the Volkenborn integral with $f' = \operatorname{Log} h_p$ is now

$$\operatorname{Log}\Gamma_p(x) = S(\operatorname{Log} h_p)(x) = S(f')(x) = \int_{\mathbf{Z}_p} f(x+y)\,dy$$

$$= \int_{\mathbf{Z}_p^\times} [(x+y)\operatorname{Log}(x+y) - (x+y)]\,dy.$$

For the computations, we come back to

$$\operatorname{Log}(x+y) = \operatorname{Log} y + \operatorname{Log}(1 + x/y),$$

and $\operatorname{Log}(1 + x/y) = \log(1 + x/y)$ is given by the series expansion if $|x/y| < 1$ (e.g., $x \in p\mathbf{Z}_p$ and $|y| = 1$). Since (1.2) $|\Gamma_p(x) - 1| \le |x|$, we also have

$$|\operatorname{Log}\Gamma_p(px)| = |\Gamma_p(px) - 1| \le |px|,$$

since $|px| < r_p$.

Theorem. *For $x \in p\mathbf{Z}_p$, we have*

$$\operatorname{Log}\Gamma_p(x) = \lambda_0 x - \sum_{m\ge 1} \frac{\lambda_m}{2m(2m+1)} x^{2m+1},$$

where

$$\lambda_0 = \int_{\mathbf{Z}_p^\times} \operatorname{Log} t\,dt, \quad \lambda_m = \int_{\mathbf{Z}_p^\times} t^{-2m}\,dt \quad (m \ge 1).$$

The radius of convergence of the power series is 1, *and this provides a continuation of* $\operatorname{Log}\Gamma_p$ *in the open unit ball* $\mathbf{M}_p \subset \mathbf{C}_p$.

PROOF. The preliminary considerations already prove that

$$\operatorname{Log}\Gamma_p(x) = \int_{\mathbf{Z}_p^\times} \left[(x+y)\operatorname{Log} y - x - y + (x+y)\sum_{n\ge 1}(-1)^{n-1}\frac{x^n}{ny^n}\right] dy,$$

which is equal to

$$x \underbrace{\int_{\mathbf{Z}_p^\times} \operatorname{Log} y\,dy}_{=\lambda_0} + \underbrace{\int_{\mathbf{Z}_p^\times} (y\operatorname{Log} y - y)\,dy}_{=0:\text{previous comment}} + \int_{\mathbf{Z}_p^\times} \left(-x + (x+y)\sum\cdots\right) dy.$$

Here is the elementary computation for the series appearing under the integral sign:

$$\begin{aligned}
(x+y)\sum_{n\geq 1}(-1)^{n-1}\frac{x^n}{ny^n} &= \sum_{n\geq 1}(-1)^{n-1}\frac{x^{n+1}}{ny^n} + \sum_{n\geq 1}(-1)^{n-1}\frac{x^n}{ny^{n-1}} \\
&= \sum_{n\geq 1}(-1)^{n-1}\frac{x^{n+1}}{ny^n} + x - \sum_{n\geq 1}\frac{(-1)^{n-1}x^{n+1}}{(n+1)y^n} \\
&= x + \sum_{n\geq 1}(-1)^{n-1}\left(\frac{1}{n} - \frac{1}{n+1}\right)\frac{x^{n+1}}{y^n} \\
&= x + \sum_{n\geq 1}(-1)^{n-1}\cdot\frac{1}{n(n+1)}\cdot\frac{x^{n+1}}{y^n}.
\end{aligned}$$

Now, for odd $n \geq 1$, the function equal to 0 on $p\mathbf{Z}_p$ and to $1/y^n$ on $\mathbf{Z}_p^\times$ is odd with a vanishing derivative at the origin. This shows that $\int_{\mathbf{Z}_p^\times} y^{-n}\,dy = 0$ for odd $n \geq 1$ (Corollary at the end of (V.5.3)). There remain only the even terms

$$\operatorname{Log}\Gamma_p(x) = \lambda_0\,x - \sum_{m\geq 1}\frac{\lambda_m}{2m(2m+1)}x^{2m+1}$$

with $\lambda_0 = \int_{\mathbf{Z}_p^\times}\operatorname{Log} t\,dt$, and $\lambda_m = \int_{\mathbf{Z}_p^\times} t^{-2m}\,dt \quad (m \geq 1)$ as asserted. We can deduce an estimate of these coefficients. If f_n denotes the function equal to 0 on $p\mathbf{Z}_p$ and to x^{-2n} on $\mathbf{Z}_p^\times$, we have $\|f_n\|_1 = 1$ (cf. (V.1.5)), as follows from

$$\Phi f_n(x,y) = \frac{x^{-2n} - y^{-2n}}{x-y} = \frac{y^{2n} - x^{2n}}{x^{2n}y^{2n}(x-y)} \qquad (|x| = |y| = 1),$$

$$|\Phi f_n(x,y)| = \left|\frac{y^{2n} - x^{2n}}{x-y}\right| = |y^{2n-1} + \cdots + x^{2n-1}| \leq 1 \quad (|x| = |y| = 1).$$

This proves $|\lambda_n| \leq p$ (Proposition 1 in (V.5.1)) and $p\lambda_n \in \mathbf{Z}_p$ $(n \geq 1)$. The isometric property of the logarithm on $1 + p\mathbf{Z}_p$ makes it easy to prove that the norm $\|.\|_1$ (V.1.5) of the function equal to 0 on $p\mathbf{Z}_p$ and to Log on units is 1: This proves $|\lambda_0| \leq p$ also. But we can prove a more precise result directly (cf. below). We have seen (Proposition 3 in (VI.1.2)) that the radius of convergence of a power series f is the same as the one for its derivative f' and hence also for f''. Let us apply it to

$$f(x) = \lambda_0 x - \sum_{n\geq 1}\frac{\lambda_n}{2n(2n+1)}x^{2n+1}$$

and

$$f''(x) = -\sum_{n\geq 1} \lambda_n x^{2n-1}.$$

Since the coefficients λ_n are bounded, we infer $r_{f''} \geq 1$. Finally, $r_{f''} \leq 1$ comes from the fact that $|\lambda_n| \not\to 0$: We show this in the next lemma. ∎

Recall that the Bernoulli numbers (V.5.4) are given by the Volkenborn integral

$$b_k = \int_{\mathbf{Z}_p} x^k \, dx \quad (k \geq 0).$$

Lemma. *For $n \geq 1$ we have $\lambda_n \equiv b_{2n} \pmod{\mathbf{Z}_p}$. Moreover, $|\lambda_0| \leq 1$ and $|\lambda_n| = p$ for all integers $n \geq 1$, such that $2n$ is a multiple of $p-1$.*

PROOF. (1) We have

$$\lambda_0 = (\log \Gamma_p)'(x)|_{x=0} = \frac{\Gamma'_p(0)}{\Gamma_p(0)} = \Gamma'_p(0),$$

and since we have seen in (1.2) that $\Gamma_p(p^n) \equiv 1 \pmod{p^n}$ we infer

$$\frac{\Gamma_p(p^n) - 1}{p^n} = \frac{\Gamma_p(p^n) - \Gamma_p(0)}{p^n} \in \mathbf{Z},$$

whence

$$\Gamma'_p(0) = \lim_{n\to\infty} \frac{\Gamma_p(p^n) - \Gamma_p(0)}{p^n} \in \mathbf{Z}_p$$

and $|\lambda_0| \leq 1$.

(2) The units of the ring $\mathbf{Z}/p^m\mathbf{Z}$ are represented by the integers $0 \leq j < p^m$ that are prime to p. The involution $u \mapsto u^{-1}$ on these units shows

$$\sum_{1\leq j<p^m,\, p\nmid j} j^{-2n} \equiv \sum_{1\leq j<p^m,\, p\nmid j} j^{2n} \pmod{p^m}.$$

Dividing by p^m and letting $m \to \infty$, we obtain by definition (V.5.1) (adapted to a function vanishing outside $\mathbf{Z}_p^\times$)

$$\lambda_n = \int_{\mathbf{Z}_p^\times} t^{-2n} \, dt \equiv \int_{\mathbf{Z}_p^\times} t^{2n} \, dt \pmod{\mathbf{Z}_p}.$$

(3) Start with

$$\int_{\mathbf{Z}_p^\times} t^{2n} \, dt = \int_{\mathbf{Z}_p} t^{2n} \, dt - \int_{p\mathbf{Z}_p} t^{2n} \, dt.$$

Let us compute explicitly the second integral:

$$\begin{aligned}
\int_{p\mathbf{Z}_p} t^{2n}\,dt &= \lim_{n\to\infty} \frac{1}{p^n} \sum_{1\le j<p^n,\ p|j} j^{2n} \\
&= \frac{1}{p} \lim_{n\to\infty} \frac{1}{p^{n-1}} \sum_{1\le m<p^{n-1}} m^{2n} p^{2n} \\
&= \frac{1}{p} \cdot p^{2n} \int_{\mathbf{Z}_p} t^{2n}\,dt = p^{2n-1} b_{2n}.
\end{aligned}$$

(Observe that this computation proves that in the Volkenborn integral over $p\mathbf{Z}_p$ we could have replaced formally t by ps with $d(ps) = |p|ds = (1/p)ds$!) We have obtained

$$\begin{aligned}
\int_{\mathbf{Z}_p^\times} t^{2n}\,dt &= \int_{\mathbf{Z}_p} t^{2n}\,dt - p^{2n-1} \int_{\mathbf{Z}_p} s^{2n}\,ds \\
&= (1 - p^{2n-1}) b_{2n} \equiv b_{2n} \pmod{\mathbf{Z}_p} \quad (n \ge 1).
\end{aligned}$$

The last assertion of the lemma follows now from the Clausen-von Staudt theorem (V.5.5). ■

The lemma and hence also the theorem are completely proved. Let us summarize two formulas that follow immediately from the theorem (and its proof).

Corollary. *We have*

$$\frac{\Gamma_p'(x)}{\Gamma_p(x)} = \int_{\mathbf{Z}_p^\times} \operatorname{Log}(x+t)\,dt, \quad (\operatorname{Log}\Gamma_p)''(x) = \int_{\mathbf{Z}_p^\times} \frac{1}{x+t}\,dt.$$

PROOF. Everything follows from the previous proof and Proposition 3 of (V.5.3) (justification of derivation under the integral sign). Observe that the expansion

$$\frac{1}{x+t} = \frac{1}{t} \cdot \frac{1}{(1+x/t)} = \frac{1}{t} \sum_{n\ge 0} (-1)^n \frac{x^n}{t^n}$$

can be integrated termwise:

$$\int_{\mathbf{Z}_p^\times} \frac{1}{x+t}\,dt = \sum_{n\ge 0} (-1)^n x^n \int_{\mathbf{Z}_p^\times} t^{-n-1}\,dt.$$

Since the integrals of odd functions (with zero derivative at the origin vanish by the corollary of Proposition 4 in (V.5.3)), there remain only the even powers of t (corresponding to odd powers of x):

$$\int_{\mathbf{Z}_p^\times} \frac{1}{x+t}\,dt = -\sum_{m\ge 1} \lambda_m x^{2m-1}, \quad \lambda_m = \int_{\mathbf{Z}_p^\times} t^{-2m}\,dt.$$

This confirms our previous expression for the coefficients of $\operatorname{Log}\Gamma_p$. ■

1.6. The Kazandzidis Congruences

We have already given in (V.3.3) some congruences for the binomial coefficients

$$\binom{pn}{pk} \equiv \binom{n}{k} \pmod{pn\mathbf{Z}_p}.$$

It turns out that these congruences hold modulo higher powers of p.

Theorem (Kazandzidis). *For all primes $p \geq 5$ we have*

$$\binom{pn}{pk} \equiv \binom{n}{k} \pmod{p^3nk(n-k)\tbinom{n}{k}\mathbf{Z}_p}.$$

For $p = 3$ the same congruence holds only mod $3^2nk(n-k)\binom{n}{k}\mathbf{Z}_3$ *(namely one power of 3 fewer).*

The form of these congruences suggests that we should prove (when $p \geq 5$)

$$\binom{pn}{pk} \Big/ \binom{n}{k} \equiv 1 \pmod{p^3nk(n-k)\mathbf{Z}_p}.$$

It is clear that the left-hand side is a p-adic unit, and L. van Hamme had already observed that it can be expressed in terms of Γ_p (or in terms of a p-adic beta function) as follows:

$$\binom{pn}{pk} \Big/ \binom{n}{k} = \frac{\Gamma_p(pn)}{\Gamma_p(pk)\Gamma_p(pl)} \qquad (k + l = n).$$

The Kazandzidis congruence states that this unit belongs to a multiplicative subgroup $1 + p^r\mathbf{Z}_p \subset \mathbf{Z}_p^\times$ with a precisely determined integer $r > 0$. The preceding unit can be studied by means of the logarithm: We have indeed proved $|\log \xi| = |\xi - 1|$ if $|\xi - 1| < r_p$. On the other hand, we also have $|\Gamma_p(x) - 1| \leq |x|$, proving, for example,

$$\Gamma_p(px) \in 1 + p\mathbf{Z}_p \quad (x \in \mathbf{Z}_p).$$

Since we are assuming $p \geq 3$ we have $|p| < r_p$, and the isometric property of the logarithm is valid for $\xi = \Gamma_p(px)$, resp. $\xi = \Gamma_p(py)$ and $\xi = \Gamma_p(px + py)$ $(x, y \in \mathbf{Z}_p)$. Hence

$$\left|\frac{\Gamma_p(px + py)}{\Gamma_p(px)\Gamma_p(py)} - 1\right| = \left|\log \frac{\Gamma_p(px + py)}{\Gamma_p(px)\Gamma_p(py)}\right| \qquad (x,\ y \in \mathbf{Z}_p).$$

Let us introduce the restricted power series (all its coefficients are in $p\mathbf{Z}_p$ as we shall see)

$$f(x) = \log \Gamma_p(px) = \lambda_0 px - \sum_{n \geq 1} \frac{\lambda_n}{2n(2n+1)} p^{2n+1} x^{2n+1}.$$

This is an odd function (this is also a consequence of the Legendre relation, which implies $\Gamma_p(px)\cdot\Gamma_p(-px)=1$). We now have

$$\left|\log\frac{\Gamma_p(px+py)}{\Gamma_p(px)\Gamma_p(py)}\right| = |f(x+y)-f(x)-f(y)|.$$

As we have seen in (V.2.2), the linear term of $f(x+y)-f(x)-f(y)$ disappears, and $|f(x+y)-f(x)-f(y)|\le C\cdot|xy(x+y)|$, where the constant C is the sup of the absolute value of the coefficients (of index $n\ge 3$) of f. Here, we need to examine carefully these coefficients. The Kazandzidis congruences will follow from the next theorem. ∎

Theorem. *Let* $f(x)=\log\Gamma_p(px)$ $(x\in\mathbf{Z}_p)$. *Then*

(a) f *is given by a restricted series having all its coefficients in* $p\mathbf{Z}_p$,
(b) $|f(x+y)-f(x)-f(y)|\le|p^3xy(x+y)|$.

Proof. Let us start with

$$f(x)=\lambda_0 px-\sum_{n\ge1}\frac{\lambda_n}{2n(2n+1)}p^{2n+1}x^{2n+1}.$$

(a) The radius of convergence of f is $p>1$ (this function is obtained by a dilatation $x\mapsto px$ from the function considered in (1.5)) and hence the series for $f(x)$ is a restricted power series. We can write

$$f(x)=\lambda_0 px-p\sum_{n\ge1}p\lambda_n\cdot\frac{p^{2n-1}}{2n(2n+1)}x^{2n+1}.$$

We have seen that $\lambda_0\in\mathbf{Z}_p$ and $p\lambda_n\in\mathbf{Z}_p$ $(n\ge1)$. It is enough to observe that

$$\frac{p^{2n-1}}{2n(2n+1)}\in\mathbf{Z}_p\quad(n\ge1).$$

This is obvious for $n=1$ and $n=2$ and follows from the lemma below for $n\ge3$.

(b) Let us repeat the expression for $f(x+y)-f(x)-f(y)$ in the following form:

$$-\frac{\lambda_1}{2\cdot3}p^3((x+y)^3-x^3-y^3)-\frac{\lambda_2}{4\cdot5}p^5((x+y)^5-x^5-y^5)-\cdots.$$

The leading term in this expression of $f(x+y)-f(x)-f(y)$ is

$$\frac{\lambda_1}{2\cdot3}p^3\cdot3(x^2y+xy^2)\equiv\frac{b_2}{2}p^3xy(x+y)=\frac{1}{2^2\cdot3}p^3xy(x+y)\bmod\mathbf{Z}_p.$$

When the prime p is greater than 3, this term is in $p^3xy(x+y)\mathbf{Z}_p$, whereas it is only in $3^2xy(x+y)\mathbf{Z}_3$ when $p=3$. The next term is treated similarly:

$$\left|\frac{\lambda_2}{4\cdot5}p^5xy(x+y)(*)\right|\le\left|\frac{b_4}{4\cdot5}p^5xy(x+y)\right|=\left|\frac{1}{4\cdot5^2\cdot6}p^5xy(x+y)\right|.$$

When $p = 3$ there is a factor 3^4 making this term smaller than the first one. When $p = 5$ there is still a factor 5^3 of the same size as in the first term. When $p > 5$ the factor p^5 makes this term strictly smaller than the first one. The subsequent terms are treated in the next lemma. ∎

Lemma. *For $n \geq 3$ we have* $\left| \dfrac{p^{2n-3}}{2n(2n+1)} \right| \leq 1.$

PROOF. Let us estimate the p-adic order of the fraction:

$$\begin{aligned} 2n - 3 - \operatorname{ord}_p 2n(2n+1) &\geq 2n - 3 - \operatorname{ord}_p (2n+1)! \\ &= 2n - 3 - \frac{2n+1-S_p(2n+1)}{p-1} \\ &\geq 2n - 3 - \frac{2n+1-1}{p-1} \\ &\geq 2n - 3 - \frac{2n}{2} = n - 3 \geq 0. \end{aligned}$$

∎

1.7. *About* Γ_2

Let us show here how the Morita gamma function is defined for the prime $p = 2$.

Preliminary comment. Let G be a finite abelian group written additively and let

$$s = s(G) = \sum_{g \in G} g.$$

In this sum the pairs $\{g, -g\}$ consisting of *two distinct* elements contribute 0 to the sum, and we see that

$$s = \sum_{g=-g} g.$$

But $g = -g$ is equivalent to $2g = 0$, and

$$H = \{g \in G : 2g = 0\} \subset G$$

is a subgroup of G, isomorphic to a product of cyclic groups of order 2: H is of type $(2, 2, \ldots, 2)$. Moreover, we have seen that $s(G) = s(H)$. Now, the sum $s(H)$ is obviously invariant under *any automorphism* of the group H: The only case where $s(H)$ can be nonzero is thus

$$H \text{ cyclic with two elements,}$$

in which case $s(H) = 1$ is the nontrivial element of this group. Equivalently, $s \neq 0$ precisely when the 2-Sylow subgroup of G is cyclic and not trivial.

Proposition. *For $\nu \geq 3$ the kernel of the homomorphism*

$$x \mapsto x \bmod 4 : (\mathbf{Z}/2^\nu\mathbf{Z})^\times \to (\mathbf{Z}/4\mathbf{Z})^\times \cong \{\pm 1\}$$

is a cyclic group $C(2^{\nu-2})$ of order $2^{\nu-2}$ generated by the class of 5, and $(\mathbf{Z}/2^\nu\mathbf{Z})^\times$ is isomorphic to the direct product of $C(2^{\nu-2})$ and $\{\pm 1\}$.

Proof. Since the order of $(\mathbf{Z}/2^\nu\mathbf{Z})^\times$ is $2^{\nu-1}$, the kernel of the homomorphism onto $(\mathbf{Z}/4\mathbf{Z})^\times \cong \{\pm 1\}$ has order $2^{\nu-2}$. We shall prove that this kernel contains an element x of order $2^{\nu-2}$. Take $x = 1 + 4t$ (obviously in the kernel) and use the fourth form of the fundamental inequality (III.4.3) (Corollary at the end of (V.3.6))

$$(1+t)^n \equiv 1 + nt \pmod{pntR}$$

for $n = 2^k$ and $p = 2$. Replacing t by $4t$ ($t \in R$) we obtain

$$(1+4t)^{2^k} \equiv 1 + 2^k 4t \pmod{2 \cdot 2^k \cdot 4R},$$
$$(1+4t)^{2^k} \equiv 1 + 2^{k+2} t \pmod{2^{k+3}R}.$$

The element $1+4t$ has order $2^{\nu-2}$ precisely (and is a generator of the kernel) when $(1+4t)^{2^{\nu-3}} \not\equiv 1 \pmod{2^\nu}$:

$$(1+4t)^{2^{\nu-3}} \equiv 1 + 2^{\nu-1} t \not\equiv 1 \pmod{2^\nu R}.$$

As appears now, this will be the case exactly when t is odd, $|t| = 1$. This proves that the class of an integer $x = 1 + 4t$ is a generator of $C(2^{\nu-2})$ precisely when $x \not\equiv 1 \pmod 8$ and $x = 5 = 1 + 4$ is an eligible candidate! ■

Corollary *The product of all units of $\mathbf{Z}/2^\nu\mathbf{Z}$ is*

$$1\ (\nu = 1), \quad -1\ (\nu = 2), \quad 1\ (\nu \geq 3).$$

Proof. This follows from the preliminary observation, since

$$(\mathbf{Z}/2\mathbf{Z})^\times \cong \{1\}, \quad (\mathbf{Z}/4\mathbf{Z})^\times \cong \{\pm 1\},$$

whereas if $\nu \geq 3$, then

$$(\mathbf{Z}/2^\nu\mathbf{Z})^\times \cong \{\pm 1\} \times C(2^{\nu-2})$$

is a product of two nontrivial cyclic groups. ■

Now let us consider the following sequence:

$$f(1) = 1, \quad f(n) = \prod_{1 \leq j < n,\ j \text{ odd}} j \quad (n \geq 2).$$

Hence $f(2) = 1$ and

$$f(2n+1) = f(2n) \quad (n \geq 1).$$

Since $f(n)$ is odd for all $n \geq 1$, we infer (for the 2-adic absolute value!)

$$|f(n)| = 1, \quad |f(m) - f(n)| \leq |2| = \tfrac{1}{2} \quad (n, m \geq 1).$$

For $n = 2^\nu$ we have

$$f(1) = 1, \quad f(2) = 1, \quad f(4) = 3 \equiv -1 \pmod 4$$

and then

$$f(2^\nu) = \prod_{1 \leq j < 2^\nu,\, j \text{ odd}} j \equiv +1 \pmod{2^\nu} \quad (\nu \geq 3).$$

As in (1.1) we infer

$$|f(n + 2^\nu) - f(n)| = \left| f(n) \left(\prod_{n \leq j < n+2^\nu,\, j \text{ odd}} j - 1 \right) \right| \leq |2^\nu| \quad (\nu \geq 3)$$

and more generally

$$|f(m) - f(n)| \leq |m - n| \quad (m, n \geq 1,\ |m - n| \leq \tfrac{1}{8}).$$

This proves that the function f is uniformly continuous, and hence has a unique extension to $\mathbf{Z}_2 \to \mathbf{Z}_2^\times = 1 + 2\mathbf{Z}_2$, which we still denote by f:

$$|f(x) - f(y)| \leq |x - y| \quad (|x - y| \leq \tfrac{1}{8}).$$

Since $f(2^\nu) \equiv 1 \bmod 2^\nu$ $(\nu \geq 3)$ we deduce $f(0) = 1$.

Lemma. *We have* $|f(x + 4) - f(x)| = \frac{1}{2}$ $(x \in \mathbf{Z}_2)$.

Proof. Since the image of f is contained in $1 + 2\mathbf{Z}_2$ of diameter $\frac{1}{2}$, we have quite generally $|f(x) - f(y)| \leq \frac{1}{2}$. The relation

$$f(2n + 2) = (2n + 1) f(2n) = 2n f(2n) + f(2n)$$

shows that

$$|f(2n + 2) - f(2n)| = |2n f(2n)| = |2n| \quad (\leq |2| = \tfrac{1}{2}).$$

Similarly,

$$f(2n + 4) - f(2n) \equiv f(2n) \cdot 1 \cdot 3 - f(2n) \equiv -2 f(2n) \pmod 4,$$
$$|f(2n + 4) - f(2n)| = |2 f(2n)| = |2| \quad (= \tfrac{1}{2}).$$

Since we also have

$$f((2n + 1) + 4) - f(2n + 1) = f(2n + 4) - f(2n),$$

we may conclude that $|f(x + 4) - f(x)| = |2| = \frac{1}{2}$ $(x \in \mathbf{Z}_2)$. ∎

In order to have

$$\Gamma_p(0) = 1, \quad \Gamma_p(1) = -1, \quad \Gamma_p(2) = 1$$

for all primes (including $p = 2$), we decide to change the sign of $f(n)$ when n is odd. Thus we define

$$\Gamma_2(n) = (-1)^n f(n) = (-1)^n \prod_{1 \le j < n,\, j \text{ odd}} j \quad (n \ge 2).$$

The formula (Definition (1.1))

$$\Gamma_p(n) = (-1)^n \prod_{1 \le j < n,\, p \nmid j} j \quad (n \ge 2)$$

holds now for all primes p. By definition, we have

$$\Gamma_2(x) = \begin{cases} f(x) & \text{if } x \in 2\mathbf{Z}_2, \\ -f(x) & \text{if } x \in 1 + 2\mathbf{Z}_2. \end{cases}$$

Consequently, when x and y are in the same coset mod 2,

$$\Gamma_2(x) - \Gamma_2(y) = \pm(f(x) - f(y)),$$

and this shows that

$$|\Gamma_2(x) - \Gamma_2(y)| = |f(x) - f(y)| \quad (x \equiv y \pmod 2),$$

so that the inequalities obtained for f are still valid for Γ_2.

Observe that we have

$$\Gamma_2(x + 1) = h_2(x)\Gamma_2(x),$$

where

$$h_2(x) = \begin{cases} -x & \text{if } x \in 1 + 2\mathbf{Z}_2 \quad (|x| = 1), \\ -1 & \text{if } x \in 2\mathbf{Z}_2 \quad (|x| < 1), \end{cases}$$

in complete similarity with the odd-prime case (1.2).

2. The Artin-Hasse Exponential

The exponential series has a radius of convergence $r_p < 1$ because its coefficients $a_n = 1/n!$ have increasing powers of p in the denominator. It turns out that the Artin-Hasse power series

$$\exp\left(x + \tfrac{1}{p}x^p + \tfrac{1}{p^2}x^{p^2} + \cdots\right) = \sum_{n \ge 0} a_n x^n$$

has p-integral coefficients: $a_n \in \mathbf{Q} \cap \mathbf{Z}_p$. As a consequence, this power series has a radius of convergence equal to 1. Dwork has used this power series for the construction of pth roots of unity in $\mathbf{C}_p$ (similar to the construction of nth roots of unity in $\mathbf{C}$). The Dieudonné-Dwork criterion explains the integrality property of the Artin-Hasse power series, and Hazewinkel has found a deep generalization of this phenomenon. We shall present only the initial aspects of these theories.

2.1. Definition and Basic Properties

Let us start by reviewing a couple of elementary formulas concerning the *Möbius function.* Recall that for an integer $n \geq 1$ this function is defined by $\mu(1) = 1$ and

$\mu(n) = 0$ if n is divisible by a square $k^2 > 1$,
$\mu(p_1 p_2 \cdots p_m) = (-1)^m$ if the p_i are distinct primes.

Lemma. *We have*

$$\sum_{d|n} \mu(d) = 0, \quad \sum_{d|n} |\mu(d)| = 2^k \quad (n > 1),$$

where k is the number of distinct prime divisors of n.

PROOF. In fact, if $n = p_1^{\nu_1} \cdots p_k^{\nu_k}$ and $d \mid n$ is a divisor with $\mu(d) \neq 0$, then d is a product of a subset of primes p_i, and quite explicitly,

$$\begin{aligned}\sum_{d|n} \mu(d) &= 1 + \sum_i \mu(p_i) + \sum_{i,j} \mu(p_i p_j) + \cdots \\ &= 1 - k + \binom{k}{2} - \cdots + (-1)^k \\ &= \sum_{0 \leq i \leq k} (-1)^i \binom{k}{i} = (1-1)^k = 0.\end{aligned}$$

Similarly,

$$\sum_{d|n} |\mu(d)| = 1 + \sum_i |\mu(p_i)| + \sum_{i,j} |\mu(p_i p_j)| + \cdots = (1+1)^k = 2^k. \qquad \blacksquare$$

Proposition. *We have identities of formal power series*

$$\sum_{n \geq 1} -\frac{\mu(n)}{n} \log(1 - x^n) = x,$$

and for each prime p

$$\sum_{n \geq 1,\, p \nmid n} -\frac{\mu(n)}{n} \log(1 - x^n) = x + \frac{1}{p} x^p + \frac{1}{p^2} x^{p^2} + \cdots.$$

PROOF. Recall that

$$-\log(1-t) = \log\frac{1}{1-t} = \sum_{m\geq 1}\frac{t^m}{m}.$$

Hence

$$\sum_{n\geq 1} -\frac{\mu(n)}{n}\log(1-x^n) = \sum_{n\geq 1}\mu(n)\sum_{m\geq 1}\frac{x^{nm}}{nm}$$
$$= \sum_{N\geq 1}\frac{x^N}{N}\sum_{n|N}\mu(n) = x$$

by the first identity of the lemma. Similarly,

$$\sum_{n\geq 1,\, p\nmid n} -\frac{\mu(n)}{n}\log(1-x^n) = \sum_{n\geq 1,\, p\nmid n}\mu(n)\sum_{m\geq 1}\frac{x^{nm}}{nm}$$
$$= \sum_{N\geq 1}\frac{x^N}{N}\sum_{n|N,\, p\nmid n}\mu(n).$$

The conditions $n \mid N$ and n prime to p amount to $n \mid Np^{-\nu}$, where $\nu = \mathrm{ord}_p N$ (also denoted by $p^\nu \parallel N$). The corresponding sum vanishes (still by the first identity of the lemma) except if $Np^{-\nu} = 1$, namely $N = p^\nu$ ($\nu \geq 0$):

$$\sum_{n\geq 1,\, p\nmid n} -\frac{\mu(n)}{n}\log(1-x^n) = \sum_{N=p^\nu}\frac{x^N}{N} = x + \frac{1}{p}x^p + \frac{1}{p^2}x^{p^2} + \cdots. \qquad \blacksquare$$

Corollary. *We have formal power series identities:*

$$\exp(x) = \prod_{n\geq 1}(1-x^n)^{-\mu(n)/n},$$

$$\exp\left(x + \tfrac{1}{p}x^p + \tfrac{1}{p^2}x^{p^2} + \cdots\right) = \prod_{n\geq 1,\, p\nmid n}(1-x^n)^{-\mu(n)/n}. \qquad \blacksquare$$

Definition. *The* Artin-Hasse exponential *is the formal power series defined by*

$$E_p(x) = \exp\left(x + \tfrac{1}{p}x^p + \tfrac{1}{p^2}x^{p^2} + \cdots\right) = 1 + x + \cdots.$$

Since log and exp are inverse power series for composition (VI.1), we have

$$\log E_p(x) = x + \frac{1}{p}x^p + \frac{1}{p^2}x^{p^2} + \cdots,$$

and by the corollary,

$$E_p(x) = \prod_{n\geq 1,\, p\nmid n}(1-x^n)^{-\mu(n)/n}$$

is an identity of formal power series.

2.2. *Integrality of the Artin-Hasse exponential*

The power series $e^{x^p/p} = 1 + x^p/p + \cdots$ converges at least for $|x| < r_p$, since $|x^p/p| \le |x|$ for $|x| \le r_p$. Consider the product of the two power series

$$e^x = 1 + x + \cdots + \frac{x^p}{p!} + \cdots \quad \text{and} \quad e^{x^p/p} = 1 + \frac{x^p}{p} + \cdots.$$

Its first coefficients are

$$\exp x \cdot \exp \frac{x^p}{p} = 1 + x + \cdots + \frac{x^{p-1}}{(p-1)!} + x^p\left(\frac{1}{p!} + \frac{1}{p}\right) + \cdots.$$

The coefficient of x^p is

$$\frac{1 + (p-1)!}{p!}.$$

A miracle happens: The numerator is divisible by p — Wilson's theorem — so that the whole fraction is in $\mathbf{Z}_p$. More is true: All the coefficients in the product

$$\exp x \cdot \exp \frac{x^p}{p} \cdot \exp \frac{x^{p^2}}{p^2} \cdots = \prod_{j \ge 0} \exp \frac{x^{p^j}}{p^j} = \exp \sum_{j \ge 0} \frac{x^{p^j}}{p^j}$$

are p-integral, hence in $\mathbf{Z}_p$. As a consequence, this power series converges for $|x| < 1$.

The radius of convergence of the power series

$$h(x) = x + \frac{1}{p}x^p + \frac{1}{p^2}x^{p^2} + \cdots$$

is the same as for its derivative (Proposition 3 in (VI.1.2)):

$$h'(x) = 1 + x^{p-1} + x^{p^2-1} + \cdots,$$

namely $r_h = r_{h'} = 1$. The critical radii and the growth modulus of h are the same as for the logarithm $\log(1+x)$: Both series have the same dominant monomials. In particular, $E_p(x) = \exp h(x)$ is well-defined, it converges at least for $|x| < r_p$, and

$$|\log E_p(x)| = |h(x)| = |x| \quad (|x| < r_p).$$

(But $|h|$ is unbounded in the open unit ball $\mathbf{M}_p \subset \mathbf{C}_p$.) This proves that $E_p(x) = \exp h(x)$ is well-defined in the ball $|x| < r_p$.

Theorem. *The coefficients of the Artin-Hasse power series E_p are p-integral rational numbers, so that $E_p(x) \in 1 + x\mathbf{Z}_p[[x]]$. Moreover, the radius of convergence of this power series is $r_{E_p} = 1$, and*

$$|E_p(x)| = 1, \quad |E_p(x) - 1| = |x| \quad (|x| < 1).$$

PROOF. When p does not divide n, $-\mu(n)/n$ is equal to 0 or to $\pm 1/n$: The binomial series expansion of $(1-x^n)^{-\mu(n)/n}$ has its coefficients in $\mathbf{Z}_p$, and hence converges for $|x|<1$ (at least). The infinite product has coefficients in $\mathbf{Z}_p$ too. It also converges in the ball $|x|<1$ by the lemma in (VI.2.3). This proves $E_p(x) \in 1+x\mathbf{Z}_p[[x]]$, and in particular $r = r_{E_p} \geq 1$. Let us show that this radius of convergence is precisely 1. For this purpose, let us prove the identity of formal power series

$$E_p(x^p) = E_p(x)E_p(\zeta x) \cdots E_p(\zeta^{p-1}x)$$

where ζ is a primitive pth root of unity: $\zeta \neq 1 = \zeta^p$. The exponent in the product is indeed

$$\underbrace{(1+\zeta+\cdots+\zeta^{p-1})}_{=0}x + p\left(\tfrac{1}{p}x^p + \tfrac{1}{p^2}x^{p^2} + \cdots\right),$$

whence the identity. Now, each power series $E_p(\zeta^i x)$ has the same radius of convergence $r = r_{E_p}$, while the radius of convergence of $E_p(x^p)$ is $r^{1/p}$. By Proposition 2 in (VI.1.2), we obtain

$$r^{1/p} \geq \min(r, r, \ldots, r) = r,$$

namely $r \geq r^p$. This proves $r \leq 1$.[1] Now let

$$E_p(x) - 1 = x + \sum_{n\geq 2} a_n x^n \quad (a_n \in \mathbf{Z}_p).$$

Hence we have

$$|a_n x^n| \leq |x|^n \leq |x|^2 < |x| \quad (|x| < 1,\ n \geq 2),$$

and $|E_p(x) - 1| = |x|$ $(|x| < 1)$ since the strongest wins. ■

As we have already observed, the coefficient of x^p in the expansion of

$$e^{x+x^p/p} = e^x \cdot e^{x^p/p}$$

is p-integral. Let us show that this product furnishes a *transition* between exp and E_p, with an intermediate radius of convergence (a quantitative way of saying that it has fewer powers of p in the denominators of its coefficients than the exponential).

Proposition. *The radius of convergence of the power series* $f(x) = e^{x+x^p/p}$ *is*

$$r_f = r_p^{(2p-1)/p^2},$$

hence $r_p < r_f < 1$. *We have*

$$\left|\exp\left(x + \tfrac{1}{p}x^p\right)\right| = 1 \quad (|x| < r_f).$$

---

[1] or $r = \infty$, but look at exercise 9.

PROOF. (1) As formal power series, we have

$$E_p(x) = e^{x+x^p/p} \cdot \exp \sum_{j\geq 2} \frac{x^{p^j}}{p^j},$$

and conversely,

$$e^{x+x^p/p} = E_p(x) \cdot \exp\left(-\sum_{j\geq 2} \frac{x^{p^j}}{p^j}\right).$$

To prove that this product converges beyond $|x| = r_p$ and get an estimate of its radius of convergence, it is sufficient to show that the radius of convergence of its second factor is greater than r_p (Proposition 2 in (VI.1.2)). To get an estimate of the radius of convergence of

$$\exp\left(-\sum_{j\geq 2} \frac{x^{p^j}}{p^j}\right)$$

we use (VI.1.5). First, let us recall that

$$M_r \sum_{j\geq 2} \frac{x^{p^j}}{p^j} = \left|\frac{1}{p^2}\right| r^{p^2} \quad \text{for } 0 \leq r \leq r_p^{1/p^2} = r_p''$$

(the dominant monomial of the log series in the interval $r_p' \leq r \leq r_p''$ is x^{p^2}/p^2: Since the preceding monomials are absent in $\sum_{j\geq 2} x^{p^j}/p^j$, the first one is dominant up to r_p''). The numerical substitution of $g(x) = \sum_{j\geq 2} x^{p^j}/p^j$ in $f(x) = \exp x$ is allowed when

$$|x| < r_g = 1 \text{ and } M_{|x|} g < r_f = r_p.$$

The second condition is

$$|x|^{p^2}/|p^2| < |p|^{\frac{1}{p-1}}, \quad |x|^{p^2} < |p|^{2+\frac{1}{p-1}} = |p|^{\frac{2p-1}{p-1}} = r_p^{2p-1},$$

namely

$$|x| < r_p^{(2p-1)/p^2}.$$

Since

$$\tfrac{1}{p} < \tfrac{1}{p}\left(2 - \tfrac{1}{p}\right) = \tfrac{2p-1}{p^2} < \tfrac{2p}{p^2} \leq 1,$$

we see that

$$r_p < r_p^{(2p-1)/p^2} < 1,$$

and the numerical evaluation is valid in the region considered above. The radius of convergence of the composite is at least $r_p^{(2p-1)/p^2} > r_p$. In its ball of convergence, all factors in

$$e^{x+x^p/p} = E_p(x) \cdot \prod_{j\geq 2} \exp\left(-\frac{x^{p^j}}{p^j}\right)$$

have absolute value equal to 1, hence $|e^{x+x^p/p}| = 1$ ($|x| < r_p^{(2p-1)/p^2}$).

(2) To simplify the notation, let

$$\rho = \text{radius of convergence of } \frac{1}{E_p},$$

$$\rho_2 = \text{radius of convergence of } \exp\left(\frac{x^{p^2}}{p^2}\right),$$

$$\rho_3 = \text{radius of convergence of } \exp\left(\sum_{j\geq 3} \frac{x^{p^j}}{p^j}\right).$$

Since $|E_p(x) - 1| = |x|$ for $|x| < 1$, we have $\rho \geq 1$. More precisely, $1/E_p(x) = E_p(-x)$ if p is odd, proves that $\rho = 1$ in this case. Now, let us write

$$\exp\left(-\frac{x^{p^2}}{p^2}\right) = \frac{1}{E_p(x)} \cdot f(x) \cdot \exp\left(\sum_{j\geq 3} \frac{x^{p^j}}{p^j}\right).$$

This shows that

$$\rho_2 \geq \min(\rho, r_f, \rho_3)$$

(Proposition 2 in (VI.1.2)), and since $\rho_2 < \rho_3 < 1 \leq \rho$, we infer that $\rho_2 \geq r_f$. ∎

2.3. *The Dieudonné-Dwork Criterion*

Another proof of the p-integrality of the coefficients of the Artin-Hasse power series will now be given.

Let k be a field of characteristic p. The identity $x^p = x$ in k characterizes its prime field $\mathbf{F}_p$. In the polynomial ring $k[x]$, the identity $f(x)^p = f(x^p)$ characterizes polynomials f having coefficients in the prime field. For a polynomial f with integral coefficients, the congruence $f(x)^p \equiv f(x^p) \pmod{p}$ means that

$$f(x)^p - f(x^p) \in p\mathbf{Z}[x],$$

and it should therefore be written more precisely as $f(x)^p \equiv f(x^p) \pmod{p\mathbf{Z}[X]}$. For polynomials f with rational coefficients, it turns out that the same congruence characterizes the integrality of its coefficients. This principle also holds for power series. The extent to which the operations

first raising x to the power p and then applying f,
first computing $f(x)$ and then raising to the pth power

lead to similar results, is a measure of the integrality of the coefficients of f. A precise formulation of this principle can now be given.

Theorem (Dieudonné-Dwork). *Let $f(x) \in 1 + x\mathbf{Q}_p[[x]]$ be a formal power series. Then the following conditions are equivalent:*

(*i*) *The coefficients of f are in $\mathbf{Z}_p$.*
(*ii*) $f(x)^p/f(x^p) \in 1 + px\mathbf{Z}_p[[x]]$.

Proof. (*i*) $\Rightarrow$ (*ii*) If $f(x) \in 1 + x\mathbf{Z}_p[[x]]$, then $f(x)^p \equiv f(x^p) \pmod{p}$. Both series belong to $1 + x\mathbf{Z}_p[[x]]$, and $f(x^p) \in 1 + x\mathbf{Z}_p[[x]]$ is invertible, so that (*ii*) follows.

(*ii*) $\Rightarrow$ (*i*) Let us write $f(x) = \sum_{i \geq 0} a_i x^i \quad (a_0 = 1,\ a_i \in \mathbf{Q}_p)$ and assume

$$f(x)^p = f(x^p)\left(1 + p\sum_{j\geq 1} b_j x^j\right) \quad (b_j \in \mathbf{Z}_p). \tag{$*$}$$

We have $a_0 = 1$ and $a_1 = b_1 \in \mathbf{Z}_p$. Let us assume by induction that $a_i \in \mathbf{Z}_p$ for $i < n$ and let us compare the coefficients of x^n in both members of $(*)$. The coefficient of x^n in the left-hand side is the same as in

$$\left(\sum_{i\leq n} a_i x^i\right)^p = \sum_{i\leq n} a_i^p x^{ip} + p(\cdots).$$

The nonwritten monomials are products $a_{i_1}a_{i_2}\cdots a_{i_p}x^{i_1+i_2+\cdots+i_p}$ having at least two distinct indices i_j. It is enough to determine them mod $\mathbf{Z}_p$, and for this reason, all monomials not containing a_n will play no explicit role, since — by the induction assumption — they have coefficients in $\mathbf{Z}_p$. The only monomials containing a_n that are of interest for us have a single factor $a_n x^n$ and all other factors $a_0 = 1$ (all other monomials containing a_n lead to powers x^m, $m > n$). Hence we find that the coefficient of x^n in the left-hand side of $(*)$ is

$$\underbrace{a_i^p}_{\text{if } ip=n} + pa_n + \text{ terms in } p\mathbf{Z}_p.$$

With the convention $a_{n/p} = 0$ when n is not divisible by p (i.e., n/p not an integer), we may write this coefficient as

$$a_{n/p}^p + pa_n + \text{ terms in } p\mathbf{Z}_p.$$

The right-hand side of $(*)$ is

$$\sum_{i\leq n/p} a_i x^{pi} \cdot \left(1 + p\sum_{j\leq n} b_j x^j\right),$$

and the coefficient of x^n in this expression is

$$a_{n/p} + \text{ terms in } p\mathbf{Z}_p.$$

Since $n/p < n$, the induction hypothesis shows that $a_{n/p} \in \mathbf{Z}_p$, and hence $a_{n/p}^p \equiv a_{n/p} \pmod{p\mathbf{Z}_p}$. By comparison we infer $pa_n \in p\mathbf{Z}_p$ and $a_n \in \mathbf{Z}_p$. ∎

Application. Consider, for example, the Artin-Hasse power series E_p. As formal power series we have the following identities:

$$E_p(x)^p = \exp\left(p\sum_{j\geq 0}\frac{x^{p^j}}{p^j}\right) = \exp\left(px + x^p + \frac{x^{p^2}}{p} + \cdots\right) = e^{px}E_p(x^p).$$

Hence

$$\frac{E_p(x)^p}{E_p(x^p)} = e^{px} \in 1 + px\mathbf{Z}_p[[x]],$$

as we are just going to show. In other words, the p-integrality of the coefficients of the Artin-Hasse power series follows from the Dieudonné-Dwork criterion and the following observation.

Proposition. *We have*

$$e^{px} \in 1 + px\mathbf{Z}_p[[x]]$$

and even

$$e^{px} \in 1 + px\mathbf{Z}_p\{x\} \quad (p \text{ an odd prime}).$$

Proof. For $n \geq 1$ we have

$$\begin{aligned}\operatorname{ord}_p \frac{p^n}{n!} &= n - \frac{n - S_p(n)}{p-1}\\ &\geq n - \frac{n-1}{p-1} = \frac{p-2}{p-1}\cdot n + \frac{1}{p-1} \geq 1,\end{aligned}$$

hence the first result. For $p = 2$ there remains only $\operatorname{ord}_2(2^n/n!) \geq 1$ with equality precisely when $S_2(n) = 1$, namely when $n = 2^\nu$ is a power of 2. For $p \geq 3$ we see that

$$\operatorname{ord}_p \frac{p^n}{n!} \geq \frac{p-2}{p-1}\cdot n \to \infty \quad (n \to \infty),$$

and the second result follows. ∎

2.4. *The Dwork Exponential*

The roots of the equation $x + x^p/p = 0$ are 0, as well as the roots of $x^{p-1} + p = 0$. All the roots π of $x^{p-1} + p = 0$ have the same absolute value $|\pi| = r_p$. Since the radius of convergence of $\exp(x + x^p/p)$ is greater than r_p, we may evaluate this power series on any such root π. But crude substitution of π in $x + x^p/p$ gives 0, and $e^0 = 1$ is not the correct result for $e^{x+x^p/p}\big|_{x=\pi}$! In fact, the condition given in

(VI.1.5) for numerical substitution is not satisfied, since

$$M_{|\pi|}\left(x+\frac{x^p}{p}\right)=r_p=r_{\exp}.$$

In the classical, complex case, all roots of unity are special values of the exponential. It turns out that pth roots of unity can also be constructed analytically by means of a generalized exponential. Recall that if $1 \neq \zeta \in \mu_p$, then $|\zeta - 1| = r_p$ (II.4.4).

Proposition (Dwork). *Choose a root π of the equation $x^{p-1}+p=0$ and let ζ_π denote the result of the substitution $e^{x+x^p/p}\big|_{x=\pi}$. Then $\zeta_\pi \in \mu_p$ is the pth root of unity such that*

$$\zeta_\pi \equiv 1+\pi \pmod{\pi^2}.$$

PROOF. (1) We have

$$e^{x+x^p/p} = 1+x+x^2(\cdots) \equiv 1+x \pmod{x^2} \quad (x \text{ indeterminate}).$$

Let us show that we also have

$$e^{x+x^p/p}\big|_{x=\pi} \equiv 1+\pi \pmod{\pi^2}.$$

The sup norm of the function $e^{x+x^p/p}$ on its ball of convergence is 1, hence the coefficients a_n of its power series expansion $e^{x+x^p/p}=\sum_{n\geq 0} a_n x^n$ satisfy

$$|a_n| r_p^{\frac{2p-1}{p^2}n} \leq 1 \quad (n \geq 0)$$

(Lemma in (VI.4.6)). Hence

$$|a_n \pi^n| \leq r_p^{n-n\frac{2p-1}{p^2}}.$$

The exponent of r_p is

$$\frac{np^2-2np+n}{p^2} = n\left(\frac{p-1}{p}\right)^2.$$

We want to show that $|a_n\pi^n| < |\pi|$ $(n \geq 2)$. This is certainly the case when $(\frac{p-1}{p})^2 n > 1$. When $p \geq 5$, we have $(\frac{p-1}{p})^2 n \geq \frac{16}{25}n > 1$ for all $n \geq 2$ and we are done. When $p = 3$, we have $(\frac{p-1}{p})^2 n = \frac{4}{9}n > 1$ for all $n \geq 3$. We have to estimate a_3. But the coefficients a_n of $\exp(x+\frac{1}{3}x^3)$ are the same as those of the Artin-Hasse power series for $n \leq 8$, hence are 3-integers, and the conclusion follows. When $p = 2$, we have $(\frac{p-1}{p})^2 n = \frac{1}{4}n > 1$ for all $n \geq 5$. We have to estimate the coefficients a_n for $n \leq 4$. But (exercise)

$$e^{x+x^2/2}\Big|_{x=\pi=-2} = 1+\pi+\pi^2+\tfrac{2}{3}\pi^3+\tfrac{5}{12}\pi^4+\cdots,$$

and since $\frac{1}{4}\pi^2 = 1$ (is a 2-integer!), $\frac{5}{12}\pi^4 \equiv 0 \pmod{\pi^2}$ as we desired to show.

(2) As formal power series, we have

$$\left(e^{x+x^p/p}\right)^p = e^{p(x+x^p/p)} = e^{px+x^p} = e^{px} \cdot e^{x^p}.$$

In detail, let φ denote the polynomial $\varphi(x) = x^p$. Then $\varphi \circ \exp(x) = \exp px$ as formal power series, and hence with $h(x) = x + x^p/p$

$$\begin{aligned}(\exp(x + x^p/p))^p &= \varphi \circ \exp(x + x^p/p) = \varphi \circ (\exp \circ h)(x)\\ &= (\varphi \circ \exp) \circ h(x) = \exp(ph(x))\\ &= e^{px+x^p} = e^{px} \cdot e^{x^p}\end{aligned}$$

(since φ is a polynomial, no condition on the order of $\exp \circ h$ is required in Corollary 2 of Proposition 2 (VI.1.2)). Since $|\pi^p| = |p\pi| = |p|r_p < r_p$, the numerical evaluation of both exponentials is obtained by substitution (VI.1.5):

$$\zeta_\pi^p = \left(e^{x+x^p/p}\big|_{x=\pi}\right)^p = e^{p\pi} \cdot e^{\pi^p} = e^{p\pi} \cdot e^{-p\pi} = 1. \qquad \blacksquare$$

Let us renormalize the situation. Choose a root π of $x^{p-1} + p = 0$; hence $|\pi| = r_p$. Substitute $x = \pi y$, so that $e^{\pi y}$ converges whenever $|y| < 1$. The same substitution in $\exp(x + x^p/p)$ leads to a power series

$$\exp\left(\pi y + \frac{\pi^p y^p}{p}\right) = \exp\left(\pi y - \pi y^p\right) = \exp \pi(y - y^p)$$

converging at least for

$$|\pi y| < r_p^{(2p-1)/p^2}, \quad |y| < r_p^{(2p-1)/p^2-1}.$$

The exponent of r_p is

$$\frac{2p-1}{p^2} - 1 = \frac{2p-1-p^2}{p^2} = -\frac{(p-1)^2}{p^2}.$$

The power series $\exp \pi(y - y^p)$ converges at least for

$$|y| < |p|^{-(p-1)/(p^2)} = p^{(p-1)/(p^2)};$$

hence its radius of convergence is greater than or equal to $p^{(p-1)/(p^2)} > 1$.

Definition. *When π is a root of $x^{p-1} + p = 0$ in $\mathbf{Q}_p^a$, the* Dwork series *is the formal power series*

$$E_\pi(x) = \exp(\pi(x - x^p)) \in \mathbf{Q}_p(\pi)[[x]].$$

The radius of convergence of the Dwork series is $p^{(p-1)/(p^2)} > 1$.

Hence $E_\pi = f \circ g$, where $f(x) = e^x$ and $g(x) = \pi(x - x^p)$ has order 1 (sufficient to enable substitution). We shall be interested in the special values taken by this power series when its exponent vanishes:

$$x^p - x = 0 \Longleftrightarrow x = 0 \text{ or } x \in \mu_{p-1}.$$

As we have seen,

$$E_\pi(1) \equiv 1 + \pi \pmod{\pi^2}$$

is a generator of μ_p.

Theorem (Dwork). *Let π be a root of $x^{p-1} + p = 0$. Then $K = \mathbf{Q}_p(\pi)$ is a Galois extension of $\mathbf{Q}_p$. It is totally and tamely ramified of degree $p - 1$, and $K = \mathbf{Q}_p(\mu_p)$. More precisely:*

(a) The field K contains a unique pth root of unity $\zeta_\pi \in \mu_p$ such that

$$\zeta_\pi \equiv 1 + \pi \pmod{\pi^2}.$$

(b) The series $E_\pi(x)$ has a radius of convergence $p^{(p-1)/p^2} > 1$.
(c) For every $a \in \mathbf{Q}_p$ with $a^p = a$ we have

$$E_\pi(a) \in \mu_p, \quad E_\pi(a) \equiv 1 + a\pi \pmod{\pi^2},$$

so that $E_\pi(1) = \zeta_\pi$.

Proof. Nearly everything has already been proved. Observe that $X^{p-1} + p$ is an Eisenstein polynomial relative to the prime p and hence is irreducible over $\mathbf{Q}_p$ (II.4.2). If π and π' are two roots of this polynomial, then $(\pi'/\pi)^{p-1} = 1$, hence $\pi'/\pi \in \mu_{p-1} \subset \mathbf{Q}_p$. Thus the splitting field of $X^{p-1} + p$ over $\mathbf{Q}_p$ is obtained by adding a single root π of this polynomial to $\mathbf{Q}_p$. This proves that K is totally ramified of degree $p - 1$ over $\mathbf{Q}_p$ and hence tamely ramified. The uniqueness of a pth root of unity $\zeta_\pi \equiv 1 + \pi \pmod{\pi^2}$ follows from the simple observation that the distance between pth roots of unity is r_p (Example 2 in (II.4.2), and also (II.4.4)): Two distinct pth roots of unity are not congruent mod π^2. The other statements of the theorem follow easily from previous observations. ∎

Comments (1) If $1 \neq \zeta \in \mathbf{C}_p$ is a root of unity of order p, we have seen in (II.4.4) that $\xi = \zeta - 1$ is a root of

$$x^{p-1} + px(\cdots) + p = 0,$$

and hence $|\xi| = r_p = |p|^{1/(p-1)}$. We are now considering roots π of the simpler equation

$$x^{p-1} + p = 0.$$

Since π and ξ have the same absolute value, $\xi = \pi u$ for some u with $|u| = 1$:

$$\zeta - 1 = \pi u, \quad \zeta = 1 + \pi u \in \mu_p \subset \mathbf{C}_p.$$

If $a^{p-1} = 1$, say $a \equiv k \pmod p$ with $1 \le k < p$ (namely $k = a_0$ is the first digit in the p-adic expansion of a), then both $E_\pi(a)$ and $E_\pi(1)^k$ are pth roots of unity congruent to $1 + k\pi \pmod{\pi^2}$, and the theorem implies

$$E_\pi(a) = E_\pi(1)^k.$$

(2) The Dwork power series is a kind of exponential map: $E_\pi(0) = 1$ and

$$\begin{array}{ccccc} \mu_{p-1} & \subset & \{a : a^p = a\} & \xrightarrow{E_\pi} & \mu_p \\ \updownarrow & & \updownarrow & \nearrow & \\ \mathbf{F}_p^\times & \subset & \mathbf{F}_p & & \end{array}$$

(3) Let $f \ge 1$ and $E_\pi^f(x) = \exp \pi(x - x^{p^f})$, so that $E_\pi(x) = E_\pi^1(x)$. Then

$$\begin{aligned} E_\pi^f(x) &= \exp \pi(x - x^{p^f}) \\ &= \exp \pi(x - x^p) \cdot \exp \pi(x^p - x^{p^2}) \cdots \exp \pi(x^{p^{f-1}} - x^{p^f}) \\ &= E_\pi(x) E_\pi(x^p) \cdots E_\pi(x^{p^{f-1}}) \end{aligned}$$

converges at least when each factor converges. The most restrictive condition is given by the last one: Convergence of $E_\pi(x^{p^{f-1}})$ occurs if

$$\left| x^{p^{f-1}} \right| < p^{(p-1)/p^2}, \quad |x| < p^{(p-1)/(p^2 p^{f-1})}.$$

With $q = p^f$, we see that the radius of convergence of E_π^f is $p^{(p-1)/(pq)} > 1$.

f	e^x	$e^{x + \frac{x^p}{p}}$	E_p (Artin-Hasse) $\exp \sum_{j \ge 0} \frac{x^{p^j}}{p^j}$	E_π^f (Dwork) $e^{\pi(x - x^q)}$ $(q = p^f)$
r_f	$r_p = \lvert p \rvert^{\frac{1}{p-1}}$	$r_p^{\frac{2p-1}{p^2}}$	1	$p^{\frac{p-1}{pq}}$

Radii of convergence of some exponential series
(listed in increasing order)

2.5. *Gauss Sums*

Sums of roots of unity play an important role in number theory. Let us show how they can be used to prove that any quadratic extension of the rational field $\mathbf{Q}$ is contained in a *cyclotomic* one, i.e., in an extension generated by roots of unity. It is enough to show that the quadratic extensions $\mathbf{Q}(\sqrt{p})$ (p prime) are contained in a cyclotomic extension of $\mathbf{Q}$.

Let us choose a root of unity ζ of prime order $p \geq 3$ in an algebraically closed field K of characteristic 0. For example, take $K = \mathbf{C}$ and $\zeta = e^{2\pi i/p}$. Then the sum of roots of unity

$$S_p = \sum_{0<\nu<p} \left(\frac{\nu}{p}\right) \zeta^\nu$$

is the simplest example of a Gauss sum: Here — as in (I.6.6) — $\left(\frac{\nu}{p}\right) = \pm 1$ denotes the quadratic residue symbol of Legendre.

Proposition 1. *For an odd prime p, we have $S_p^2 = \pm p$.*

Proof. The square of the sum S_p is

$$S_p^2 = \sum_{0<\nu,\mu<p} \left(\frac{\nu}{p}\right)\left(\frac{\mu}{p}\right) \zeta^{\nu+\mu} = \sum_{0<\nu,\mu<p} \left(\frac{\nu\mu}{p}\right) \zeta^{\nu+\mu}.$$

For fixed $\mu \neq 0$, $\nu\mu$ goes through all nonzero classes mod p, and we can replace ν by $\nu\mu$ in the double sum:

$$S_p^2 = \sum_{\nu,\mu} \left(\frac{\nu\mu^2}{p}\right) \zeta^{(\nu+1)\mu} = \sum_{\nu,\mu} \left(\frac{\nu}{p}\right) \zeta^{(\nu+1)\mu}.$$

We consider separately the terms with $\nu = p - 1$:

$$\sum_{\mu} \left(\frac{-1}{p}\right) \zeta^0 = (p-1)\left(\frac{-1}{p}\right)$$

and for $\nu \neq p - 1$

$$\sum_{\nu \neq p-1} \left(\frac{\nu}{p}\right) \sum_{\mu \neq 0} \zeta^{\mu(\nu+1)}.$$

Recall that

$$\sum_{\mu \neq 0} \zeta^{\mu(\nu+1)} = \underbrace{\sum_{0 \leq \mu < p} \zeta^{\mu(\nu+1)}}_{=0 \text{ because } \nu+1 \neq 0} - 1 = -1.$$

Hence

$$S_p^2 = (p-1)\left(\frac{-1}{p}\right) - \sum_{\nu\neq -1}\left(\frac{\nu}{p}\right)$$
$$= p(-1/p) - \underbrace{\left(\frac{-1}{p}\right) - \sum_{\nu\neq -1}\left(\frac{\nu}{p}\right)}$$
$$= p(-1/p) - \underbrace{\sum_{0<\nu<p}\left(\frac{\nu}{p}\right)}_{=0}.$$

The announced formula is proved. ■

Corollary 1. *For a prime $p \geq 3$, the complex absolute value of S_p is*

$$|S_p|_{\mathbf{C}} = \sqrt{p}.$$ ■

Corollary 2. *For a prime $p \geq 3$, the quadratic extension $\mathbf{Q}(\sqrt{p})$ is contained in the cyclotomic field $\mathbf{Q}(\zeta, \sqrt{-1})$.* ■

Observe that if $p = 2$, we have $(1+\sqrt{-1})^2 = 2\sqrt{-1}$, so that $\sqrt{2} \in \mathbf{Q}(\sqrt[4]{-1})$ and the quadratic extension $\mathbf{Q}(\sqrt{2})$ is also contained in a cyclotomic one.

Comment. A theorem of Kronecker asserts that any Galois extension of the rational field $\mathbf{Q}$ with abelian Galois group is contained in a cyclotomic one. This is a deeper theorem, which has been widely generalized, and belongs now to *class field theory*.

The general form of Gauss sums in a field K containing a pth root of unity ζ is obtained as follows. The map $\nu \mapsto \zeta^\nu$, $\mathbf{F}_p \to K^\times$ is a group homomorphism:

$$\zeta^{\nu+\mu} = \zeta^\nu \cdot \zeta^\mu.$$

The map $\nu \mapsto \left(\frac{\nu}{p}\right)$, $\mathbf{F}_p^\times \to K^\times$ is a group homomorphism:

$$\left(\frac{\nu\mu}{p}\right) = \left(\frac{\nu}{p}\right)\left(\frac{\mu}{p}\right),$$

extended by $\left(\frac{0}{p}\right) = 0$. Replace $\mathbf{Z}/p\mathbf{Z} = \mathbf{F}_p$ by a finite field $\mathbf{F}_q$ (where $q = p^f$ is a power of p) and let more generally ψ and χ be two group homomorphisms

$$\psi : \mathbf{F}_q \to K^\times \quad \text{and} \quad \chi : \mathbf{F}_q^\times \to K^\times \text{ extended by } \chi(0) = 0.$$

According to tradition, we shall say that ψ is an *additive character* of $\mathbf{F}_q$ and χ a *multiplicative character* of $\mathbf{F}_q$. By definition, the Gauss sum attached to this pair of characters is the sum

$$G(\psi, \chi) = \sum_{\nu \in \mathbf{F}_q} \psi(\nu)\chi(\nu) = \sum_{\nu \in \mathbf{F}_q^\times} \psi(\nu)\chi(\nu).$$

In the next section we give the p-adic absolute value of Gauss sums. Now, let us show how to determine all additive characters of a finite field.

Proposition 2. *Let G be a group and K a field. Any set of distinct homomorphisms $G \to K^\times$ is linearly independent in the K-vector space of functions $G \to K$.*

Proof. Since linear independence of any family is a property of its finite subsets, it is enough to prove that all finite sets of distinct homomorphisms are linearly independent. We argue by induction on the number of homomorphisms ψ_i. Since homomorphisms are nonzero maps, the independence assertion is true for one homomorphism. Assume that $n-1$ distinct homomorphisms are always independent and consider n distinct homomorphisms ψ_i $(1 \le i \le n)$. Starting from a linear dependence relation

$$\alpha_1\psi_1(x) + \cdots + \alpha_n\psi_n(x) = 0 \quad (x \in G,\ \alpha_i \in K),$$

we multiply it by the value $\psi_1(a)$ (for some $a \in G$):

$$\alpha_1\psi_1(a)\psi_1(x) + \cdots + \alpha_n\psi_1(a)\psi_n(x) = 0 \quad (x \in G).$$

On the other hand, we may replace x by ax in the first equality, and since $\psi_i(ax) = \psi_i(a)\psi_i(x)$, we obtain

$$\alpha_1\psi_1(a)\psi_1(x) + \cdots + \alpha_n\psi_n(a)\psi_n(x) = 0 \quad (x \in G).$$

If we subtract the two relations obtained, the first term disappears, and we get a shorter relation:

$$\alpha_2(\psi_1(a) - \psi_2(a))\psi_2 + \cdots + \alpha_n(\psi_1(a) - \psi_n(a))\psi_n = 0.$$

By the induction assumption, all the coefficients of this new relation vanish. If we choose $a \in G$ such that $\psi_1(a) - \psi_n(a) \neq 0$ (this is possible since $\psi_1 \neq \psi_n$), we see that $\alpha_n = 0$. Using the induction assumption again, we get $\alpha_i = 0$ $(1 \le i < n)$. ■

Proposition 3. *Let F be a finite field and $\tau : F \to K^\times$ a nontrivial additive character. Then any other additive character ψ has the form $\psi(x) = \tau(ax)$ for some $a \in F$.*

Proof. The identity

$$\tau(a(x + y)) = \tau(ax + ay) = \tau(ax)\tau(ay)$$

shows that for any $a \in F$, $\tau_a(x) := \tau(ax)$ defines an additive character. Now, $a \mapsto \tau_a$ is a homomorphism

$$\tau_{a+b}(x) = \tau((a+b)x) = \tau(ax+bx) = \tau(ax)\tau(bx) = \tau_a(x)\tau_b(x).$$

It is injective, since τ is a nontrivial character:

$$\tau_a(x) = 1 \quad (x \in F) \implies a = 0.$$

The additive characters $(\tau_a)_{a\in F}$ constitute a basis of the F-vector space of functions $F \to K^\times$. Any additive character must be in this family by Proposition 1. ∎

As a consequence, we observe that the Gauss sums $G(\psi, \chi)$ can be computed easily from $G(\tau, \chi)$:

$$G(\psi, \chi) = G(\tau_a, \chi) = \sum_{x \in F^\times} \tau(ax)\chi(x).$$

If we assume $a \neq 0$ and replace x by $a^{-1}x$ in the sum, then we obtain

$$G(\psi, \chi) = \sum_{x \in F^\times} \tau(aa^{-1}x)\chi(a^{-1}x) = \sum_{x \in F^\times} \tau(x)\chi(a^{-1})\chi(x) = \chi(a^{-1})G(\tau, \chi).$$

2.6. *The Gross-Koblitz Formula*

Let us choose a primitive pth root of unity ζ_p in $\mathbf{C}_p$ and let $K = \mathbf{Q}_p(\zeta_p)$. Then $|\zeta_p - 1| = r_p$, and $\zeta_p - 1$ is a generator of the maximal ideal P of $R \subset K$. As we have seen in (2.4), there is a generator π of P uniquely characterized by

$$\pi^{p-1} = -p, \quad \pi \equiv \zeta_p - 1 \pmod{(\zeta_p - 1)^2}.$$

Conversely, if we choose a root $\pi \in \mathbf{C}_p$ of $\pi^{p-1} = -p$, the field $K = \mathbf{Q}_p(\pi)$ is a Galois extension (2.3) of $\mathbf{Q}_p$, it contains all roots of unity of order p, and the Dwork series furnishes $E_\pi(1) = \zeta_p$, the unique root of unity of order p satisfying

$$\zeta_p \equiv 1 + \pi \pmod{\pi^2}.$$

Since an additive character of the field $\mathbf{F}_p$ is uniquely determined by its value $\psi(1) \in \mu_p$, we choose the nontrivial additive character

$$\psi(1) = \zeta_\pi, \quad \psi(\nu) = \zeta_\pi^\nu \quad (\nu \in \mathbf{F}_p).$$

We can now consider Gauss sums of the form

$$G(\chi, \psi) = \sum_{x \in \mathbf{F}_p} \chi(x)\zeta_\pi^x \quad (\chi(0) = 0),$$

where χ is a multiplicative character of $\mathbf{F}_p$ with values in K. More precisely, the values of χ are roots of unity having order dividing $p-1$ (and 0):

$$G(\chi, \psi) \in \mathbf{Q}(\mu_p, \mu_{p-1}) = \mathbf{Q}(\mu_{p(p-1)}).$$

We shall consider Gauss sums of the form

$$\sum_{0\neq x\in \mathbf{F}_p} \omega(x)^{-a}\psi(x) = \sum_{x\in \mathbf{F}_p} \omega(x)^{-a}\zeta_\pi^x \quad (\omega(0)=0),$$

where $\omega(x) \in \mu_{p-1}$ denotes the unique root of unity in K having reduction x in the residue field R/P of K ((II.4.3) and (III.4.4)). Here, the integer a only counts mod $p-1$: It is better to take $\alpha \in \frac{1}{p-1}\mathbf{Z}/\mathbf{Z}$ and set

$$G_\alpha = -\sum_{x\in \mathbf{F}_p} \omega(x)^{-(p-1)\alpha}\zeta_\pi^x \quad (\omega(0)=0).$$

A reason for the choice of sign is that we now have $G_0 = 1$:

$$\sum_{0\le \nu<p} \zeta_\pi^\nu = 0 \Longrightarrow \sum_{0<\nu<p} \zeta_\pi^\nu = -1.$$

It is remarkable that these Gauss sums are linked to the Morita p-adic gamma function: When $\alpha = a/(p-1)$ $(0 \le a < p-1)$ we have explicitly

$$G_\alpha = \pi^a \Gamma_p\left(\frac{a}{p-1}\right).$$

This is a particular case of the Gross-Koblitz formula. Since the values of Γ_p are units of $\mathbf{Z}_p$, the preceding formula gives the exact order of G_α, and

$$|G_\alpha| = |\pi|^a = r_p^a = |p|^{\frac{a}{p-1}}.$$

Conversely, this case of the Gross-Koblitz formula shows that

$$\Gamma_p\left(\frac{a}{p-1}\right) \in \mathbf{Q}(\pi, \mu_{p(p-1)}),$$

and this is an algebraic value, since $\pi^{p-1} = -p$.

There is a more general formula. Let $\alpha \in \mathbf{Z}_{(p)} = \mathbf{Q} \cap \mathbf{Z}_p$ be a rational number with denominator N prime to p and choose a sufficiently high power $q = p^f$ of p so that the extension $\mathbf{F}_q$ of degree f of its prime field contains a root of unity of order N. We shall work in the tamely ramified extension

$$K = \mathbf{Q}_p(\pi, \mu_{q-1}) \subset \mathbf{C}_p$$

having ramification index $e = p-1$, residue degree f, and hence degree $n = ef$ over $\mathbf{Q}_p$. Considering $\alpha \in \frac{1}{N}\mathbf{Z}/\mathbf{Z} \subset \frac{1}{q-1}\mathbf{Z}/\mathbf{Z}$, we choose a representation

$$0 \le \langle\alpha\rangle = \frac{a}{q-1} < 1$$

of α and write the p-adic expansion of the numerator:

$$a = a_0 + a_1 p + \cdots + a_{f-1}p^{f-1} < q-1 < q = p^f.$$

Let $S_p(a) = \sum_{0\le j<f} a_j$ denote the sum of digits of a as in (V.3.1) and introduce the integers $a^{(i)}$ having the p-adic expansions obtained by cyclic permutations from the expansion of $a = a^{(0)}$:

$$\begin{aligned} a^{(1)} &= a_{f-1} + a_0 p + \cdots + a_{f-2}p^{f-1}, \\ a^{(2)} &= a_{f-2} + a_{f-1}p + \cdots + a_{f-3}p^{f-1}, \\ &\cdots \\ a^{(f-1)} &= a_1 + a_2 p + a_3 p^2 + \cdots + a_0 p^{f-1}. \end{aligned}$$

On the other hand, if the nontrivial additive character ψ of the prime field $\mathbf{F}_p$ is chosen as before, the composite of ψ with the trace

$$\mathrm{Tr} : \mathbf{F}_q \to \mathbf{F}_p,\ x \mapsto x + x^p + \cdots + x^{p^{f-1}}$$

is a nontrivial additive character of $\mathbf{F}_q$ (the trace is nontrivial, since the extension $\mathbf{F}_q/\mathbf{F}_p$ is separable: All extensions of finite fields are separable). Then we have the following general formula.

Theorem (Gross-Koblitz). *Let* $0 \le \alpha = \dfrac{a}{q-1} < 1$. *The value of the Gauss sum* G_α *is explicitly given by*

$$G_\alpha = -\sum_{0\ne x\in\mathbf{F}_q} \omega(x)^{-a}\psi(\mathrm{Tr}\,(x)) = \pi^{S_p(a)} \prod_{0\le j<f} \Gamma_p\left(\frac{a^{(j)}}{q-1}\right). \qquad \blacksquare$$

3. The Hazewinkel Theorem and Honda Congruences

3.1. *Additive Version of the Dieudonné-Dwork Quotient*

The power series

$$f(x) = \sum_{j\ge 0} \tfrac{1}{p^j} x^{p^j} = x + \tfrac{1}{p}x^p + \tfrac{1}{p^2}x^{p^2} + \cdots$$

does not have coefficients in $\mathbf{Z}_p$ (powers of p appear in the denominators). However, its exponential — the Artin-Hasse power series — has p-integral coefficients. This phenomenon will now be studied more closely. Observe that

$$f(x^p) = x^p + \tfrac{1}{p}x^{p^2} + \tfrac{1}{p^2}x^{p^3} + \cdots$$

so that

$$f(x) - \frac{f(x^p)}{p} = x$$

has integral coefficients! Let us introduce the operator

$$H_p f(x) = f(x) - \frac{f(x^p)}{p}$$

on formal power series. We have an identity of formal power series:

$$\exp pH_p f(x) = \exp(pf(x) - f(x^p))$$
$$= \frac{(\exp f(x))^p}{\exp f(x^p)}.$$

The expression $H_p f$ is an *additive version* of the Dieudonné-Dwork quotient $\varphi(x)^p/\varphi(x^p)$ (2.3), and we shall formulate criteria for p-integrality of some formal power series in terms of H_p.

Proposition. *Let f denote the formal power series $f(x) = \log(1+x)$. Then*

$$H_2 f(x) = \sum_{n \text{ odd}} (x^n - x^{2n})/n \in \mathbf{Z}_{(2)}[[x]],$$

$$H_p f(x) = \sum_{n\geq 1,\, p\nmid n} (-1)^{n-1} x^n/n \in \mathbf{Z}_{(p)}[[x]] \quad (p \text{ an odd prime}).$$

Hence for all primes p, $H_p(\log(1+x))$ has p-integral coefficients.

PROOF. We have

$$\log(1+x) = \sum_{n\geq 1} (-1)^{n-1}\frac{x^n}{n},$$

$$\frac{1}{p}\log(1+x^p) = \frac{1}{p}\sum_{n\geq 1}(-1)^{n-1}\frac{x^{pn}}{n}.$$

If the prime p is odd, we have $(-1)^n = (-1)^{pn}$, and the announced result follows in this case. When $p = 2$, let us write explicitly

$$\log(1+x) = \sum_{n \text{ odd}} \frac{x^n}{n} - \sum_{n \text{ even}} \frac{x^n}{n},$$

and

$$H_2(\log(1+x)) = \sum_{n \text{ odd}} \frac{x^n}{n} - \sum_{n=2m \text{ even}} (1 + (-1)^{m-1})\frac{x^{2m}}{2m}$$
$$= \sum_{n \text{ odd}} \frac{x^n}{n} - \sum_{m \text{ odd}} \frac{x^{2m}}{m}.$$

As announced, all coefficients are in $\mathbf{Z}_{(2)}$. ∎

3.2. *The Hazewinkel Maps*

Let us consider the following setting: Either

$$A = \mathbf{Z}_{(p)}[t] \subset B = \mathbf{Q}[t]$$

or

$$A = \mathbf{Z}_p[t] \subset B = \mathbf{Q}_p[t],$$

and σ is the $\mathbf{Q}$-linear map (resp. $\mathbf{Q}_p$-linear map)

$$\sigma : B \to B, \quad a(t) = \sum a_i t^i \mapsto a(t^p) = \sum a_i t^{pi}$$

extended to

$$\sigma_* : B[[x]] \to B[[x]], \quad \sum_{i \geq 0} a_i(t)x^i \mapsto \sum_{i \geq 0} a_i(t^p)x^i,$$

letting σ act on the coefficients only. Note that $(\sigma_* f)(x^n) = \sigma_*(f(x^n))$, so that we may unambiguously write this term $\sigma_* f(x^n)$.

Definition. *Any map $H_p : B[[x]] \to B[[x]]$ of the form*

$$f \mapsto H_p f = f(x) - \frac{1}{p} \sum_I \sigma_*^i f(x^{p^i}),$$

where $I \subset \mathbf{N}^* = \{1, 2, \ldots\}$ is a subset of indices, will be called a Hazewinkel map.

In the next three propositions, H_p denotes a Hazewinkel map.

Proposition 1. *Let $f = \sum_{m \geq 1} f_m x^m \in B[[x]]$, so that $f(0) = 0$. Then*

$$H_p f \in A[[x]] \Longrightarrow m f_m \in A \quad (m \geq 1).$$

Proof. The coefficients of $H_p f = \sum h_m x^m \in A[[x]]$ are given by

$$h_m = f_m - \frac{1}{p} \sum_I \sigma^i f_{m/p^i} \in A$$

with the convention $f_{m/p^i} = 0$ if $i > \operatorname{ord}_p(m)$, namely if m/p^i is not an integer. This series of identities starts with $h_m = f_m \in A$ when $(m, p) = 1$. We proceed by induction on the order ν of m, the case $\nu = 0$ having just been treated. When $p \mid m$, we have

$$m f_m - \frac{m}{p} \sum_I \sigma^i f_{m/p^i} = m h_m \in mA,$$

so that

$$m f_m \equiv \frac{m}{p} \sum_I \sigma^i f_{m/p^i} = \sum_I p^{i-1} \sigma^i \Big(\underbrace{\frac{m}{p^i} f_{m/p^i}}_{\in A \text{ by induction}} \Big) \pmod{mA}$$

and hence $m f_m \in A$ as expected. ∎

Remark. When $f = \sum f_m x^m \in B[[x]]$ and $H_p f \in A[[x]]$, we can write $f_m = a_m/m$ with coefficients $a_m \in A$, and the formal power series f is a *logarithmic series*

$$f(x) = \sum_{m\geq 1} f_m x^m = \sum_{m\geq 1} a_m \frac{x^m}{m}.$$

Proposition 2. *Let* $g = \sum_{m\geq 1} g_m x^m$ *and* $h = \sum_{m\geq 1} h_m x^m$ *be two formal power series with zero constant term. Then*

$$H_p(g \circ h) = H_p(g)(h) + \frac{1}{p} \sum_I \sum_{m\geq 1} (\sigma^i g_m) \cdot \left(h(x)^{p^i m} - (\sigma_*^i h(x^{p^i}))^m \right).$$

PROOF. By definition,

$$H_p(g \circ h) = g \circ h - \frac{1}{p} \sum_I \sigma_*^i (g \circ h)(x^{p^i}),$$

while

$$H_p(g)(h) = g(h) - \frac{1}{p} \sum_I \sigma_*^i g(h^{p^i}).$$

The first terms are the same and cancel by subtraction. Using the obvious relation

$$\sigma_*^i (g \circ h) = \sigma_*^i (g) \circ \sigma_*^i (h)$$

and the expansion $g = \sum_{m\geq 1} g_m x^m$ we get the announced result. ∎

In the special case $I = \{1\}$ (a single term in the index set I),

$$H_p f = f(x) - \frac{1}{p} \sigma_* f(x^p),$$

and we recover the additive version (3.1) of the Dieudonné-Dwork quotient in the case of constant coefficients (cf. generalization in (3.3) below). The following conditions are equivalent:

(i) $f_m - (1/p)\sigma f_{m/p} \in A$.
(ii) $a_m - \sigma a_{m/p} \in mA$.
(iii) $a_m \equiv \sigma a_{m/p} \pmod{mA}$.
(iv) $a_m(t) \equiv a_{m/p}(t^p) \pmod{p^\nu A[t]}$ $\quad (\nu = \mathrm{ord}_p\, m)$.

Definition. *A sequence* $(a_m)_{m\geq 1}$ *in* $A = \mathbf{Z}_{(p)}[t]$ *is a p*-Honda sequence *when it satisfies the following* Honda congruences*:*

$$a_m(t) \equiv a_{m/p}(t^p) \pmod{m\mathbf{Z}_{(p)}[t]} \quad \text{when } p \mid m.$$

In particular, a sequence $(a_m)_{m\geq 1}$ in $\mathbf{Z}_{(p)}$ is a p-Honda sequence when

$$a_m \equiv a_{m/p} \pmod{m\mathbf{Z}_{(p)}} \quad \text{when } p \mid m.$$

The paragraph preceding the definition proves the following result.

Proposition 3. *For a formal power series* $f(x) = \sum_{m\geq 1} a_m x^m/m \in B[[x]]$ *we have the equivalences*

(*i*) $H_p f = f(x) - \dfrac{1}{p}\sigma_* f(x^p)$ *has its coefficients in* $A \subset B$,

(*ii*) $(a_m)_{m\geq 1}$ *is a p-Honda sequence in A.*

Proposition 4. *Let A be a ring, I an ideal of A containing a prime p, and x and y two elements of A satisfying* $x \equiv y \pmod{I^r}$ *for some integer* $r \geq 1$. *Then*

$$p^\nu \mid m \quad \Longrightarrow \quad x^m \equiv y^m \pmod{I^{r+\nu}} \quad (\nu \in \mathbf{N}).$$

Proof. (1) Let us write $x = y + z$ with $z \in I^r$. Hence

$$x^p = (y+z)^p = y^p + zp(\cdots) + z^p$$

with

$$zp(\cdots) \in zI \subset I^r \cdot I = I^{r+1}$$

and

$$z^p \in I^{pr} \subset I^{2r} \subset I^{r+1}.$$

This establishes the case $m = p$ ($\nu = 1$) of the lemma.

(2) The case $m = p^\nu$ is treated by induction on ν, the basic step $\nu \mapsto \nu + 1$ being analogous to the first case already treated. Hence

$$x^{p^\nu} \equiv y^{p^\nu} \pmod{I)^{r+\nu}} \quad (t \geq 0).$$

(3) Finally, if we raise a congruence to the power $\ell = m/p^\nu$, it is preserved: If $x' \equiv y' \pmod{I)^s}$, say $x' = y' + z'$ with $z' \in I^s$, then

$$(x')^\ell = (y')^\ell + z'(\cdots) \in (y')^\ell + I^s. \qquad \blacksquare$$

This proposition shows that the sequence $((1+x)^m)_{m\geq 1}$ is a p-Honda sequence for any prime p. Let us state it explicitly.

Corollary. *Let* $m \geq 1$ *be an integer divisible by p. If* $\nu \geq 1$ *denotes its p-adic order, then*

$$(1+x)^m \equiv (1+x^p)^{m/p} \pmod{p^\nu \mathbf{Z}[x]}.$$

PROOF. Observe that $(1+x)^p \equiv 1+x^p \pmod{p}$ and apply the proposition. ■

3.3. The Hazewinkel Theorem

The particular form of the Hazewinkel theorem that we are going to state and prove has been specifically studied by various authors:

Barsky, Cartier, van Hamme, Honda, ..., Zuber

(neither exhaustive nor chronological... but in alphabetical order!). It has many applications. Let us first give the Dieudonné-Dwork theorem (2.3) in a more general form.

Theorem (Dieudonné-Dwork). *Let $f(x) \in 1+x\mathbf{Q}_p[t][[x]]$ be a formal power series. Then the following conditions are equivalent:*

(*i*) *The coefficients of f are in $\mathbf{Z}_p[t]$.*
(*ii*) $f(x)^p/\sigma_* f(x^p) \in 1+px\mathbf{Z}_p[t][[x]]$.

PROOF. As in (2.3): Only observe at the end of the implication (*ii*) ⇒ (*i*) that the coefficient of x^n in the left-hand side is now

$$a_{n/p}(t)^p + pa_n(t) + \text{ terms in } p\mathbf{Z}_p[t]$$

and in the right-hand side

$$\sigma_* a_{n/p}(t) = a_{n/p}(t^p) \equiv a_{n/p}(t)^p \pmod{p\mathbf{Z}_p[t]}.$$

The conclusion follows. ■

If we know a priori that the coefficients of f are rational, namely

$$f(x) \in 1+x\mathbf{Q}[t][[x]],$$

we get equivalent statements:

(i) *The coefficients of f are in $\mathbf{Z}_{(p)}[t]$,*
(ii) $f(x)^p/\sigma_* f(x^p) \in 1+px\mathbf{Z}_{(p)}[t][[x]]$

simply since $\mathbf{Z}_p \cap \mathbf{Q} = \mathbf{Z}_{(p)}$. Let us come back to the notation of (3.2):

$$\text{either } A = \mathbf{Z}_{(p)}[t] \subset B = \mathbf{Q}[t] \quad \text{or } A = \mathbf{Z}_p[t] \subset B = \mathbf{Q}_p[t]$$

and

$$\sigma : B \to B, \quad a(t) = \sum a_i t^i \mapsto a(t^p) = \sum a_i t^{pi}$$

is extended to

$$\sigma_* : B[[x]] \to B[[x]], \quad \sum_{i\geq 0} a_i(t)x^i \mapsto \sum_{i\geq 0} a_i(t^p)x^i$$

by letting σ act on the coefficients only.

Theorem. *For a formal power series $f(x) = \sum_{m\geq 1} a_m x^m/m \in B[[x]]$ we have equivalent statements:*

(i) $H_p f = f(x) - \frac{1}{p}\sigma_* f(x^p)$ has its coefficients in $A \subset B$.

(ii) $\varphi = e^f$ has coefficients in A.

PROOF. $(i) \Rightarrow (ii)$ Assume that $f(x) = \sum_{m\geq 1} f_m x^m = \sum_{m\geq 1} a_m x^m/m$ satisfies (i). By Proposition 3 in (3.2), (a_m) is a p-Honda sequence. Then $H_p(f)$ has coefficients in A and by the proposition in (2.3), $\exp pH_p(f)$ has p-integral coefficients

$$\begin{aligned}\exp pH_p(f) &= (\exp f)^p/\exp \sigma_* f(x^p) = (\exp f)^p/\sigma_* \exp f(x^p)\\ &= \varphi(x)^p/\sigma_*\varphi(x^p) \in 1 + px\mathbf{Z}_{(p)}[t][[x]].\end{aligned}$$

By the general form of the Dieudonné-Dwork criterion,

$$\varphi(x) = \exp(f) \in 1 + xA[[x]]$$

has p-integral coefficients.

The proof of the converse $(ii) \Rightarrow (i)$ is based on Proposition 4 in (3.2). Assume that $\varphi = \exp(f)$ has p-integral coefficients. Write

$$f(x) = \log\exp(f(x)) = \log(1 + (e^{f(x)} - 1)) = g(h(x)),$$

namely $f = g \circ h$ with $g(x) = \log(1+x)$ and $h(x) = e^{f(x)} - 1$. Proposition 2 in (3.2) can be applied to this composition:

$$H_p(f) = H_p(g \circ h) = H_p(g)(h) + \frac{1}{p}\sum_I \sum_{m\geq 1} \sigma^i g_m \left(h(x)^{p^i m} - (\sigma_*^i h(x^{p^i}))^m\right).$$

By (3.1) $H_p(g)$ has p-integral coefficients and by assumption, h has p-integral coefficients. There only remains to consider the second term, where g has constant coefficients (independent of t)

$$\sigma^i g_m = g_m = \pm 1/m.$$

Now, for all formal power series $h \in A[[x]]$ having p-integral coefficients, we have

$$h(x)^{p^i} \equiv \sigma_*^i h(x^{p^i}) \pmod{p}.$$

As Proposition 4 of (3.2) shows, this congruence is improved when raised to a power m:

$$h(x)^{p^i m} \equiv \sigma_*^i h(x^{p^i})^m \pmod{p)^{\nu+1} \quad (\nu = \mathrm{ord}_p m).$$

This proves

$$h(x)^{p^i m} - \sigma_*^i h(x^{p^i})^m \in pmA[[x]],$$

and the p-integrality of the remaining sum follows. ■

3.4. Applications to Classical Sequences

Proposition (Beukers). *Let M be a $d \times d$ matrix with integer coefficients. Define $a_n = \mathrm{Tr}\,(M^n)$. Then for any prime p, $(a_n)_{n\geq 1}$ is a p-Honda sequence*

$$a_m \equiv a_{m/p} \pmod{m\mathbf{Z}_{(p)}} \quad \textit{if } p \mid m.$$

PROOF. We have to prove that e^f has coefficients in $A = \mathbf{Z}_{(p)}$ where $f(x) = \sum_{m\geq 1} a_m x^m/m$. This logarithmic generating function is easily evaluated:

$$\begin{aligned} f(x) &= \sum_{m\geq 1} (\mathrm{Tr}\, M^m) \frac{x^m}{m} = \mathrm{Tr}\left(\sum_{m\geq 1} \frac{M^m x^m}{m}\right) \\ &= \mathrm{Tr}\,(-\log(1 - Mx)). \end{aligned}$$

Hence

$$\begin{aligned} \exp f(x) &= \exp \mathrm{Tr} \log(1 - Mx)^{-1} = \det \exp \log\left((1 - Mx)^{-1}\right) \\ &= \det(1 - Mx)^{-1} = 1/\det(1 - Mx) \end{aligned}$$

has its coefficients in A. ■

Corollary 1. *The Lucas sequence*

$$\ell_0 = 2, \quad \ell_1 = 1, \quad \ell_{n+1} = \ell_n + \ell_{n-1} \quad (n \geq 1),$$

is a p-Honda sequence for any prime p.

PROOF. Let $M = \begin{pmatrix} 1 & 1 \\ 1 & 0 \end{pmatrix} \in M_2(\mathbf{Z})$. The characteristic polynomial of M is $x^2 - x - 1$, hence $M^2 - M - I = 0$ (Hamilton-Cayley). We deduce

$$M^{n+2} = M^{n+1} + M^n \quad (n \geq 0).$$

Since $\mathrm{Tr}\, I_2 = 2$, $\mathrm{Tr}\, M = 1$, this proves that $\ell_n = \mathrm{Tr}\, M^n$ is the Lucas sequence. ■

Corollary 2. *The Perrin sequence*

$$a_0 = 3, \quad a_1 = 0, \quad a_2 = 2, \quad a_{n+2} = a_n + a_{n-1} \quad (n \geq 1),$$

is a p-Honda sequence for any prime p.

PROOF. Let $M = \begin{pmatrix} 0 & 1 & 1 \\ 1 & 0 & 0 \\ 0 & 1 & 0 \end{pmatrix} \in M_3(\mathbf{Z})$. The characteristic polynomial of M is $-x^3 + x + 1$, hence $M^3 - M - I = 0$ (Hamilton-Cayley). We deduce

$$M^{n+3} = M^{n+1} + M^n \quad (n \geq 0).$$

Since $\operatorname{Tr} I_3 = 3$, $\operatorname{Tr} M = 0$, and $\operatorname{Tr} M^2 = 2$, this proves that $a_n = \operatorname{Tr} M^n$ is the Perrin sequence. ∎

3.5. Applications to Legendre Polynomials

Let $(P_n)_{n\geq 0}$ denote the sequence of *Legendre polynomials*. This sequence can be defined by its generating function

$$\frac{1}{R} = \sum_{n\geq 0} P_n(t)x^n,$$

where $R^2 = 1 - 2xt + x^2$. Recall that these polynomials $P_n(t) \in \mathbf{Q}[t]$ satisfy

$$\deg P_n = n, \quad P_n(1) = 1 \quad (n \geq 0).$$

They can be computed according to the *Rodrigues formula*

$$P_n(t) = \frac{1}{2^n n!}\left(\frac{d}{dt}\right)^n (t^2 - 1)^n.$$

This formula shows that the coefficients of P_n are rational numbers with denominators powers of 2. More precisely, the coefficients of P_n belong to $(1/2^n)\mathbf{Z}$. They are p-integral for all *odd primes* p.

The following generating functions are well known (they can be checked by differentiation with respect to x):

$$\sum_{m\geq 1} P_{m-1}(t)\frac{x^m}{m} = \log\frac{x - t + R}{1 - t},$$

$$\sum_{m\geq 1} P_m(t)\frac{x^m}{m} = \log\frac{2}{1 - tx + R}.$$

Hence

$$\exp\left(\sum_{m\geq 1} P_{m-1}(t)\frac{x^m}{m}\right) = \frac{x - t + R}{1 - t},$$

$$\exp\left(\sum_{m\geq 1} P_m(t)\frac{x^m}{m}\right) = \frac{2}{1 - tx + R},$$

and for each odd prime p, we get two p-Honda sequences. Explicitly,

$$p \mid m \Longrightarrow P_{m-1}(t) \equiv P_{(m/p)-1}(t^p) \pmod{m\mathbf{Z}_{(p)}[t]},$$

$$p \mid m \Longrightarrow P_m(t) \equiv P_{m/p}(t^p) \pmod{m\mathbf{Z}_{(p)}[t]}.$$

For example, to check that

$$\frac{x-t+R}{1-t} \in \mathbf{Z}_{(p)}[t][[x]],$$

write

$$\begin{aligned}\frac{x-t+R}{1-t} &= 1+\frac{x-1}{1-t}+\frac{1-x}{1-t}\left(1+2x\frac{1-t}{(1-x)^2}\right)^{1/2}\\ &= 1+\frac{1-x}{1-t}(-1+\sqrt{\cdots})\\ &= 1+\frac{1-x}{1-t}\sum_{n\geq 1}\binom{1/2}{n}\frac{(1-t)^n}{(1-x)^{2n}}2^n x^n,\end{aligned}$$

so that the denominator $1-t$ disappears: All coefficients are in $\mathbf{Z}[\frac{1}{2}, t] \subset \mathbf{Z}_{(p)}[t]$. The integrality verification for the other generating function is similar and therefore left as an exercise. ∎

The change of variable $t = 1 + 2\tau$ clears the powers of 2 in the denominators, and congruences mod 2 (or mod 4) can also be established.

3.6. *Applications to Appell Systems of Polynomials*

Let $(A_n(t))_{n\geq 0}$ be an Appell family (IV.6.1) in $\mathbf{Z}_p[t]$: $A'_n = nA_{n-1}$ $(n \geq 1)$. The following result generalizes the corollary of Proposition 4 in (3.2).

Theorem (Zuber). *For an Appell family $(A_n(t))_{n\geq 0}$ in $\mathbf{Z}_p[t]$, the following conditions are equivalent:*

(i) There exists $a \in \mathbf{Z}_p$ such that
$A_n(a) \equiv A_{n/p}(a) \pmod{n\mathbf{Z}_p}$ $(n \geq 1,\ p \mid n)$.

(ii) There exists $a \in \mathbf{Z}_p$ such that $(A_n(a))_{n\geq 1}$ is a Honda sequence
$A_n(a) \equiv A_{n/p}(a^p) \pmod{n\mathbf{Z}_p}$ $(n \geq 1,\ p \mid n)$.

(iii) $(A_n)_{n\geq 1}$ is a Honda sequence of polynomials
$A_n(t) \equiv A_{n/p}(t^p) \pmod{n\mathbf{Z}_p[t]}$ $(n \geq 1,\ p \mid n)$.

PROOF. *(i)* ⇔ *(ii)* by the p-adic mean value theorem (V.3.2),

$$A'_{n/p} = \frac{n}{p}A_{(n/p)-1} \quad\Longrightarrow\quad \|A'_{n/p}\| \leq \frac{|n|}{|p|},$$

$$|a^p - a| \leq |p| \leq r_p \quad (a \in \mathbf{Z}_p).$$

Hence

$$|A_{n/p}(a^p) - A_{n/p}(a)| \le |p| \cdot \frac{|n|}{|p|} \le |n|,$$

and the equivalence follows.

(iii) ⇒ *(ii)* is obvious.

(i) ⇒ *(iii)* This uses (3.3): It is enough to show that

$$\exp \sum_{n\ge1} A_n(t)\frac{x^n}{n} = \sum_{k\ge0} q_k(t)x^k \quad (q_0 = 1) \tag{$*$}$$

has p-integral coefficients, namely $q_k(t) \in \mathbf{Z}_p[t]$.

(1) Let us compute the partial derivative $\partial/\partial t$ of the defining equation $(*)$:

$$\sum_{n\ge1} A_n' \frac{x^n}{n} \cdot \exp \sum \cdots = \sum_{k\ge0} q_k' x^k$$

or equivalently (using $A_n' = nA_{n-1}$),

$$\sum_{n\ge1} A_{n-1}x^n \cdot \sum_{m\ge0} q_m x^m = \sum_{k\ge0} q_k' x^k.$$

This gives

$$q_k' = \sum_{n+m=k, n\ge1} A_{n-1}q_m = A_0 q_{k-1} + A_1 q_{k-2} + \cdots + A_{k-1}. \tag{I}$$

(2) Let us compute the partial derivative $\partial/\partial x$ of the defining equation $(*)$:

$$\sum_{n\ge1} A_n x^{n-1} \cdot \exp \sum \cdots = \sum_{k\ge1} k q_k x^{k-1}$$

or equivalently,

$$\sum_{n\ge1} A_n x^{n-1} \cdot \sum_{m\ge0} q_m x^m = \sum_{k\ge1} k q_k x^{k-1}.$$

This gives

$$k q_k = \sum_{n+m=k, n\ge1} A_n q_m \quad (k \ge 1),$$

and

$$(k-1)q_{k-1} = \sum_{n+m=k-1,\, n\ge1} A_n q_m = A_1 q_{k-2} + \cdots + A_{k-1} \quad (k \ge 1). \tag{II}$$

(3) Comparing (I) with (II),

$$q_k' = A_0 q_{k-1} + (k-1)q_{k-1} = (k + A_0 - 1)q_{k-1} \quad (k \ge 1).$$

Iteration leads to

$$q_k'' = (k + A_0 - 1)(k + A_0 - 2)q_{k-2}, \quad \ldots,$$
$$q_k^{(\ell)} = (k + A_0 - 1)(k + A_0 - 2)\cdots(k + A_0 - \ell)q_{k-\ell} \quad (1 \le \ell \le k).$$

Now, the Taylor formula is

$$q_k(t) = \sum_{0\le j\le k} \binom{k-1+A_0}{j} q_{k-j}(a)(t-a)^j.$$

Since $A_0 \in \mathbf{Z}_p$, all binomial coefficients are in $\mathbf{Z}_p$. Moreover, by assumption, all $q_n(a) \in \mathbf{Z}_p$, since $(A_n(a))_{n\ge1}$ is a p-Honda sequence. We conclude that the polynomials $q_k(t)$ have p-integral coefficients. ∎

Exercises for Chapter 7

1. For $f,\ g \in \mathbf{Q}[x]$, prove that

$$f \equiv g \pmod{n\mathbf{Z}_p[x]} \iff f \equiv g \pmod{p^\nu \mathbf{Z}_{(p)}[x]}$$

(cf. Exercise 29 in Chapter I).

2. Let p be an odd prime. Show that the closure of the set of pairs $(n, n!^*)$ in $\mathbf{Z}_p \times \mathbf{Z}_p$ is the union of two graphs.
(*Hint.* Consider the graphs of $\pm\Gamma_p$.)

3. Find the limit $\lim_{n\to\infty} \Gamma_p(p^n)$. More precisely, can you evaluate

$$\lim_{n\to\infty} (\Gamma_p(p^n) - 1)/p^n?$$

4. Prove the congruence

$$(1+4t)^{2^k} \equiv 1 + 2^{k+2}t \pmod{2^{k+3}} \quad (t \in \mathbf{Z})$$

by induction on the integer $k \ge 0$.

5. More on the gamma function Γ_2.
(*a*) Check the formula

$$\Gamma_2(n) = \frac{(-1)^n(2[n/2])!}{(2^{[n/2]}[n/2]!)} \quad (n \ge 1).$$

(*b*) Prove $\Gamma_2(n+1)\Gamma_2(-n) = (-1)^{1+[(n+1)/2]} \quad (n \ge 1)$.
(*c*) Let $m \ge 1$ be an odd integer. Prove $\prod_{0\le k<m} \Gamma_2(k/m) = \pm 1$.

6. For any prime p and $0 \le a < p$, show that

$$n \mapsto (-1)^{pn} \frac{(a+pn)!}{p^n n!}$$

has a continuous extension to $\mathbf{Z}_p$ given by

$$x \mapsto (-1)^{a+1}\Gamma_p(a + px + 1) \in \mathbf{Z}_p^\times \subset \mathbf{Q}_p.$$

Prove the following generalization. Let $q = p^f$ $(f \geq 1)$ be a fixed power of p, and $0 \leq a < q$. Show that

$$m \mapsto (-1)^{(q-1)m} \frac{(a+qm)!}{(-p)^{\frac{q-1}{p-1}m} m!}$$

admits a continuous extension $\mathbf{Z}_p \to \mathbf{Q}_p$ given by

$$x \mapsto (-1)^{f+a}(-p)^{\mathrm{ord}_p a!} \prod_{0 \leq i < f} \Gamma_p([a/p^i] + p^{f-i}x + 1).$$

(*Hint.* Write a telescopic product with $n_0 = a + qm,\ n_f = m$

$$\begin{aligned}
\frac{(a+qm)!}{m!} &= \frac{n_0!}{n_1!} \frac{n_1!}{n_2!} \cdots \frac{n_{f-1}!}{n_f!} \\
&= \frac{(a_0 + pn_1)!}{n_1!} \frac{(a_1 + pn_2)!}{n_2!} \cdots \frac{(a_{f-1} + pm)!}{m!} \\
&= \pm p^{n_1}\Gamma_p(a_0 + pn_1 + 1) \cdot p^{n_2}\Gamma_p(n_1 + 1) \cdots p^m \Gamma_p(n_{f-1} + 1).
\end{aligned}$$

Observe that when the prime p is odd, $q - 1$ is even and $(-1)^{(q-1)m} = +1$: Hence this sign is relevant only if $p = 2$, in which case it is $(m) = (-1)^m$.)

7. With $\pi^{p-1} = -p$, prove

$$e^{\pi x} \in 1 + \pi x \mathbf{Z}_p[\pi][[x]].$$

(*Hint.* For $n \geq 1$, $\mathrm{ord}_p \dfrac{\pi^n}{n!} = \dfrac{n}{p-1} - \dfrac{n - S_p(n)}{p-1} = \dfrac{S_p(n)}{p-1} \geq \dfrac{1}{p-1}$.)

8. Compute the first coefficients of the Artin-Hasse exponential E_p for $p = 2,\ 3$ (and 5). In particular, show that

$$E_2(X) = 1 + X + X^2 + \tfrac{2}{3}X^3 + \tfrac{2}{3}X^4 + X^5(\cdots),$$

$$E_3(X) = 1 + X + X^2 + \tfrac{1}{2}X^3 + \tfrac{1}{24}X^4 + X^5(\cdots).$$

Compute the first coefficients of the Dwork exponential for $p = 2$, 3 (and 5).

9. Here is another proof of the fact that the radius of convergence r of the Artin-Hasse exponential E_p is smaller than or equal to 1.
   (*a*) Show that if this radius r were greater than 1, then the unit sphere would be a critical sphere of E_p, and E_p would have a zero $a \neq 1$ on this sphere.
   (*b*) Use the identity

$$E_p(x)E_p(\zeta x) \cdots E_p(\zeta^{p-1}x) = E_p(x^p)$$

   (where ζ is a primitive pth root of unity: $\zeta \neq 1 = \zeta^p$) to show that E_p would have infinitely many zeros on the unit sphere, thus contradicting (VI.2.1).

(c) When $p \neq 2$, give another proof of $r_{E_p} \leq 1$ based on the identity

$$E'_p(x)/E_p(x) = \sum_{j\geq 0} x^{p^j-1}.$$

(*Hint.* Use Propositions 2 and 3 in (VI.1.2), as well as $1/E_p(x) = E_p(-x)$ to show that $r_{1/E_p} = r_{E_p}$.)

10. When $\pi^{p-1} = -p$ and $\pi y = x$, then we have $e^{\pi(y-y^p)} = e^{x+x^p/p}$. Find the corresponding general expression for $e^{\pi(y-y^q)}$ ($q = p^f$, $f \geq 1$) in terms of $x = \pi y$.

11. Prove the following relations for the coefficients of Dwork's exponential $e^{\pi(x-x^q)} = \sum_{n\geq 0} A_n x^n$

$$nA_n = \pi A_{n-1} \; (1 \leq n < q), \quad nA_n = \pi(A_{n-1} - qA_{n-q}) \quad (n \geq q).$$

(*Hint.* Differentiate the above generating function.)

12. For $0 < \alpha = a/(p-1) < 1$ let G_α denote the Gauss sum $-\sum \zeta^{-a} E_\pi(\zeta)$. Use the Gross-Koblitz formula to prove $G_\alpha G_{1-\alpha} = \pm p$.

13. Let χ be a nontrivial multiplicative character $\mathbf{F}_p^\times \to \mathbf{C}^\times$, and consider the Gauss sum

$$G(\chi) = \sum_{\nu\in\mathbf{F}_p^\times} \chi(\nu)\zeta^\nu$$

(where $\zeta \neq 1$ is a pth root of unity in $\mathbf{C}$). Show that the complex absolute value of this Gauss sum is

$$|G(\chi)|_\mathbf{C} = \sqrt{p}.$$

(*Hint.* Prove $G(\chi)\overline{G(\chi)} = p$ exactly as in the proof of Proposition 1 in (VII.2.5). There, the Legendre symbol was a multiplicative character χ such that $\chi \neq 1 = \chi^2$. But here, χ^2 may be nontrivial.)

14. Let $\chi : \mathbf{F}_p^\times \to \mathbf{C}^\times$ be a nontrivial, complex-valued, multiplicative character of $\mathbf{F}_p$. As in the previous exercise, we consider the Gauss sums $G(\chi)$: $|G(\chi)|_\mathbf{C} = \sqrt{p}$. Show that the only case when $G(\chi) = \varepsilon\sqrt{p}$ for some root of unity ε, happens when $\chi = \left(\frac{\cdot}{p}\right)$.
(*Hint.* Let $\omega : \mathbf{F}_p^\times \to\subset \mathbf{C}^\times$ denote an injective homomorphism, considered as a complex-valued, multiplicative character of $\mathbf{F}_p$ (analogous to the Teichmüller character (III.4.4)). Show that any nontrivial multiplicative character $\chi : \mathbf{F}_p^\times \to \mathbf{C}^\times$ can be written uniquely $\chi = \omega^{-a}$ ($1 \leq a \leq p-2$). If $G(\omega^{-a})^2/p$ is a root of unity, use the Gross-Koblitz formula to show that $a = \frac{p-1}{2}$.)

15. Check the following formulas by differentiation with respect to x

$$\sum_{m\geq 1} P_{m-1}(t)\frac{x^m}{m} = \log\frac{x-t+R}{1-t},$$

$$\sum_{m\geq 1} P_m(t)\frac{x^m}{m} = \log\frac{2}{1-tx+R}.$$

16. Let $(B_n)_{n\geq 0}$ denote the sequence of the Bernoulli polynomials. If p is an odd prime, prove the following congruences:

$$B_{pn}(t) \equiv B_n(t^p) \pmod{n\mathbf{Z}_p} \quad (n \geq 1).$$

For $p = 2$, prove that a single power of 2 is lost, i.e.,

$$2B_{2n}(t) \equiv 2B_n(t^2) \pmod{n\mathbf{Z}_2} \quad (n \geq 1).$$

(*Hint.* Use (Chapter V, Exercise 10) and (VII.3.6) for the Appell sequence $\beta_n(t) = 2pB_n(t) \in \mathbf{Z}_p[t]$.)

Specific References for the Text

I.2

A. Cuoco: *Visualizing the p-adic integers*, Amer. Math. Monthly **98** (1991), pp. 355–364.

A. Robert: *Euclidean Models of p-adic Spaces*, Proc. of the 4th Int. Conf. (Nijmegen), M. Dekker (1997), pp. 349–361.

I.3.1

N. Bourbaki: *Topologie Générale*, Hermann (Paris 1974), Chap. IX §3.

I.6.8

F.K. Schmidt: *Mehrfach Perfekte Körper*, Mathematische Annalen **108** (1933), pp. 1–25 (cf. Th. V, p. 7).

I Exercise 1

These numbers were apparently already considered by Gergonne and Lucas in the nineteenth century and called *congruent numbers*:

R. Cuculière: *Jeux Mathématiques*, in *Pour la Science* (Juin 1986), pp. 10–15.

II.1.2

N. Bourbaki: *Topologie Générale*, Hermann (Paris 1974), Chap. IX §3.

II.4.6

A. Robert: *A Good Basis for Computing with Complex Numbers*, Elemente Math. **49** (1994), pp. 111–117.

It was pointed out to me by H. Brunotte (Düsseldorf) that the surjectivity proof given in this article has a gap. One can deduce it from

I. Kátai: *Number systems in imaginary quadratic fields*, Ann. Univ. Sci. Budapest, Sect. Comp. 1994, pp. 91–93, MR#95k:11134.

II.A

N. Bourbaki: *Algèbre commutative*, Hermann (Paris 1964), Chap. VI §9.

A. Weil: *Basic Number Theory*, Springer Verlag (3d ed. 1974), Chap. 1.

III.1.6

M. KRASNER: *Nombre des extensions d'un degré donné d'un corps p-adique*, in: Les tendances géométriques en algèbre et théorie des nombres, Colloques Internat. du CNRS N^0 143 (1966), pp. 143–169.

III.2.4

For spherical completeness, cf. Chapter 4 of A.C.M. VAN ROOIJ: *Non-Archimedean Functional Analysis*

III.2.5

B. DIARRA: *Ultraproduits ultramétriques de corps valués*, Ann. Sc. de l'Univ. de Clermont II, Série Math. **22** (1984), pp. 1–37.

III.3.3

S. LANG: *Algebra*, Addison-Wesley (2nd ed. 1984), p. 412 and 420, or (3d ed. corrected 1994), p. 347 and p. 482.

III Exercise 7

Trees have always played an important role in p-adic analysis. They were revived recently by Berkovitch, Escassut, Mainetti, and others in the context of circular filters, from which our version for balls is derived (Exercise 2 in Chapter II).

IV.1.3

J.F. ADAMS: *On the groups* $J(X)-III$, Topology, vol. 3 (1965), pp. 193–222.

IV.1.5

L. VAN HAMME: *Three generalizations of Mahler's expansion ...*, in p-adic Analysis, Proc. Trento 1989, ed. by F. Baldassari et al., Springer Verlag (1990), Lect. Notes #1454, pp. 356–361.

IV.4.6 For compact operators and compactoids

J.-P. SERRE: *Endomorphismes complètement continus des espaces de Banach p-adiques*, IHES $n^0$12 (1962), pp. 69–85.

A.C.M. VAN ROOIJ: *Non-Archimedean Functional Analysis* p. 133.

IV.5.5

L. VAN HAMME: *Continuous operators which commute with translations ...*, Proc. of the 1st Int. Conf. (Laredo), M. Dekker (1992), pp. 75–88.

G.-C. ROTA, D. KAHANER, A. ODLYZKO: *Finite Operator Calculus*, J. of Math. Anal. and Appl., vol. **42**, N^0 3, (1973) pp. 684–760.

S.M. ROMAN, G.-C. ROTA: *The Umbral Calculus*, Adv. in Math., vol. **27**, nb. 2 (1978) pp. 95–188.

V.3

A. ROBERT: *A note on the numerators of the Bernoulli numbers*, Expositiones Math. **9** (1991) pp. 189–191.

V.4.5

M.-C. SARMANT(-DURIX): *Prolongement de la fonction exponentielle en dehors de son cercle de convergence*, C.R. Acad. Sc. Paris (A), **269** (1969), pp. 123–125.

VI.1.4

For an example of entire function, bounded on $\mathbf{Q}_p$ see the book by W. Schikhof, *Ultrametric Calculus*, p. 126.

VI.2.2

S. Lang: *Algebra*, Addison-Wesley 2nd ed. p. 215, Th. 11.2 (or 3d ed. p. 208, Th. 9.2).

VI.2.3 Part (c) is already in

L. Schnirelmann: *Sur les fonctions dans les corps normés* ..., Bull. Acad. Sci. USSR Math., vol. 5/6 (1938) pp. 487–497.

VI.2.6

D. Husemöller: *Elliptic Curves*, Springer-Verlag, GTM **111** (1987) ISBN: 0–387–96371–5 (3–540–96371–5 Berlin),

J.H. Silverman: *Advanced Topics in The Arithmetic of Elliptic Curves*, Springer-Verlag, GTM **151** ISBN: 0–387–94325–0 (3–540–94325–0 Berlin).

VI.3

For real/complex analysis, my favorite is

W. Rudin: *Real and Complex Analysis*, 2nd. ed., McGraw-Hill, New-York (1974) ISBN: 0–07–054233–3.

[Runge p. 288; Mittag-Leffler p. 291; Hadamard three-circle p. 281; Picard for essential singularities p. 227.]

VI.3.6

B. Guénnebaud: *Sur une notion de spectre pour les algèbres normées ultramétriques* (Thèse, Univ. Poitiers 1973).

G. Garandel: *Les semi-normes multiplicatives sur les algèbres d'éléments analytiques* ..., Indag. Math. **37** n^0 4 (1975) pp. 327–341.

VI.4.7

M. Abramowitz, I. Stegun: *Handbook of Mathematical Functions*, Dover (1972) ISBN: 486–61272–4 (Chap. 24 p. 824).

VI.4.8

E. Motzkin: *La décomposition d'un élément analytique en facteurs singuliers*, Ann. Inst. Fourier **27**, fasc. 1 (1977), pp. 67–82.

VII.1.6

A. Robert, M. Zuber: *The Kazandzidis Supercongruences. A simple Proof and an Application.* Rend. Sem. Mat. Univ. Padova, vol. **94** (1995), pp. 235–243.

VII.2.5

B. Gross, N. Koblitz: *Gauss sums and the p-adic Γ-function*, Annals of Math. **109** (1979), pp. 569–581.

R.F. Coleman: *The Gross-Koblitz formula*, pp. 21–52 in Galois representations and arithmetic algebraic geometry, Papers from a Symposium held at Kyoto Univ. (1985) and Tokyo Univ. (1986), ed. by Y. Ihara, Adv. Studies in Pure Maths. **12**, North-Holland (1987), ISBN 0–444–70315–2.

S. Lang: *Cyclotomic Fields*, Springer-Verlag, GTM 59 (1978),

– . – : *Cyclotomic Fields II*, GTM 69 (1980).

VII.2.6

A simple proof of the Gross-Koblitz formula appears in A.M. Robert: The Gross-Koblitz Formula Revisited, Rend. Sem. Mat. Univ. Padova (to appear in 2001).

VII.3.2

Proposition 4 appears repeatedly in various forms

N. Bourbaki: *Algèbre commutative*, Chap. VI, lemme 1, p. 157 and Chap. IX, lemme 1, p. 2.

VII.3.3

M. Hazewinkel: *Formal Groups and Applications*, Academic Press, New-York (1978), 573 p. (contains over 500 references!).

VII.3.4

L. van Hamme: *The p-adic Moment Problem*, pp. 151–163 in p-adic Functional Analysis (Santiago-Chile 1992), Editorial Univ. de Santiago.

VII.3.5

T. Honda: *Two Congruence Properties of Legendre Polynomials*, Osaka Journal of Math., vol. **13** (1976), pp. 131–133.

A. Robert: *Polynômes de Legendre mod p*, in Univ. Blaise Pascal, Sém. d'analyse 1990-91, t. **6** (pp. 19.01–19.11).

–. – : *Polynômes de Legendre mod* 4,

C.R. Acad. Sc. Paris, t. **316**, Série I (1993), pp. 1235–1240.

VII.3.6

M. Zuber: *Propriétés de congruence de certaines familles de polynômes*, C.R. Acad. Sc. Paris, t. **315**, Série I (1992), pp. 869–872.

Bibliography

General

Y. Amice: *Les nombres p-adiques*, PUF, Collection Sup. "Le mathématicien" **14** (Paris 1975).

G. Bachmann: *Introduction to p-adic numbers and valuation theory*, Academic Press, New York (1964).

A. Escassut: *Analytic Elements in p-adic Analysis*, World Scientific (1995) ISBN: 981–02–2234–3.

F. Gouvea: *p-adic Numbers: An Introduction*, Springer-Verlag, Universitext (1993) ISBN : 0–387–56844–1 (3–540–56844–1 Berlin).

N. Koblitz: *p-adic Numbers, p-adic Analysis and Zeta-Functions*, Springer-Verlag, GTM **58** (1977, 1984) ISBN: 0–387–90274–0 (3–540–90274–0 Berlin).

K. Mahler: *p-adic Numbers and their Functions*, Cambridge University Press, Cambridge tract **76** (2nd ed. 1981).

A. Monna: *Analyse non-archimédienne*, Springer-Verlag, New York (1970).

A.C.M. van Rooij: *Non-Archimedean Functional Analysis*, Marcel Dekker, Pure and Appl. Math. **51** (1978).

W.H. Schikhof: *Ultrametric Calculus, An Introduction to p-adic Analysis*, Cambridge Studies in Adv. Math. 4, Cambridge Univ. Press (1984), ISBN: 0–521–24234–7.

J.-P. Serre: *Cours d'arithmétique*, PUF, Collection Sup. "Le mathématicien" **2** (Paris 1970), English translation: *A Course in Arithmetic*, Springer-Verlag (1973).

Advanced, More Specialized

S. BOSCH, U. GUNTZER, R. REMMERT: *Non-Archimedean Analysis, A Systematic Approach to Rigid Analytic Geometry*, Springer-Verlag, Grundlehren Nr. **261** (1984).

B. DWORK: *Lectures on p-adic Differential Equations*, Springer-Verlag, Grundlehren Nr. **253** (1982).

B. DWORK, G. GEROTTO, F.J. SULLIVAN: *An Introduction to G-functions*, Ann. of Math. Studies **133**, Princeton Univ. Press (1994), ISBN: 0–691–03681–0.

J. FRESNEL, M. VAN DER PUT: *Géométrie Analytique Rigide et Applications*, Birkhäuser, Progress in Math. **18** (1981).

L. GERRITZEN, M. VAN DER PUT: *Schottky Groups and Mumford Curves*, Springer-Verlag, L. N. in Math. **817** (1980).

Proceedings of Congresses in p-adic Functional Analysis

J. BAYOD ET AL. (editors): *p-adic Functional Analysis*, Proc. of the 1st Int. Conf. (1990, Laredo, Spain) Lect. Notes in pure and appl. math. **137**, Marcel Dekker (1992).

N. DE GRANDE-DE KIMPE ET AL. (editors): *p-adic Functional Analysis*, Proc. of the 2nd Int. Conf. (1992, Santiago, Chile) Editorial Universidad de Santiago (1994).

A. ESCASSUT ET AL. (editors): *p-adic Functional Analysis*, Proc. of the 3d Int. Conf. (June 1994, Clermont-Ferrand, France) Annales Math. Blaise Pascal, vol. 2, N^{0}1 (1995).

W.H. SCHIKHOF ET AL. (editors): *p-adic Functional Analysis*, Proc. of the 4th Int. Conf. (June 1996, Nijmegen, The Netherlands) Lect. Notes in pure and appl. math. **192**, Marcel Dekker (1997) ISBN: 0–8247–0038–4.

J. KAKOL ET AL. (editors): *p-adic Functional Analysis*, Proc. of the 5th Int. Conf. (June 1998, Poznan, Poland) Lect. Notes in pure and appl. math. **207**, Marcel Dekker (1999) ISBN: 0–8247–8254–2.

A.K. KATSARAS ET AL. (editors): *p-adic Functional Analysis*, Proc. of the 6th Int. Conf. (July 2000, Ioannina, Greece) (to appear).

Tables

Field	Units	Squares	Roots of unity	Number of quadratic extensions
$\mathbf{Q}_2$	$\mathbf{Z}_2^\times = 1 + 2\mathbf{Z}_2$	$1 + 8\mathbf{Z}_2$ index 4 in $\mathbf{Z}_2^\times$	$\mu_2 = \{\pm 1\}$	7
$\mathbf{Q}_p$ p odd prime	$\mathbf{Z}_p^\times \supset 1 + p\mathbf{Z}_p$ index p-1	index 2 in $\mathbf{Z}_p^\times$	μ_{p-1}	3

Field $\supset B_{\leq 1} \supset B_{<1}$	Residue field	Nonzero $\lvert . \rvert$	Properties
$\mathbf{Q}_p \supset \mathbf{Z}_p \supset p\mathbf{Z}_p$	$\mathbf{F}_p$	$p^{\mathbf{Z}}$	locally compact
$K \supset R \supset P = \pi R$	$\mathbf{F}_q \ (q = p^f)$	$\lvert\pi\rvert^{\mathbf{Z}} = p^{\frac{1}{e}\mathbf{Z}}$	$ef = \dim_{\mathbf{Q}_p} K < \infty$; locally compact
$\mathbf{Q}_p^a \supset A^a \supset M^a$	$k^a = \mathbf{F}_p^a = \mathbf{F}_{p^\infty}$	$p^{\mathbf{Q}}$	algebraically closed; not locally compact
$\mathbf{C}_p \supset \mathbf{A}_p \supset \mathbf{M}_p$	$\mathbf{F}_p^a = \mathbf{F}_{p^\infty}$	$p^{\mathbf{Q}}$	algebraically closed; complete
$\Omega_p \supset A_\Omega \supset M_\Omega$	k_Ω uncountable	$\mathbf{R}_{>0}$	algebraically closed; spherically complete

Umbral calculus

Delta operator (IV.5)	Basic sequence of polynomials (IV.5.2)	Related sequences (IV.6.1)
$D = d/dx$	$(x^n)_{n\geq 0}$	Appell sequences $Dp_n = np_{n-1}$
	$T \downarrow$ umbral operator	
δ	$(p_n)_{n\geq 0}$	Sheffer sequences $\delta s_n = ns_{n-1}$
$\tau_{-y}\delta$ (IV.5.5)	$\left(\frac{x}{x+ny} p_n(x+ny)\right)_{n\geq 0}$ translation principle	

Binomial identity: $p_n(x+y) =$ "$(p(x)+p(y))^n$,"
Appell sequences: $p_n(x+y) =$ "$(p(x)+y)^n$,"
Sheffer sequences: $s_n(x+y) =$ "$(s(x)+p(y))^n$,"

Analytic elements

Formal power series	$\mathbf{C}_p[[x]]$	Sequences $(a_n)_{n\geq 0}$	$\prod_{n\geq 0} \mathbf{C}_p$
power series converging in $\lvert x\rvert < 1$		$\limsup \lvert a_n\rvert^{1/n} \leq 1$ $(r_f \geq 1)$	
power series bounded in $\lvert x\rvert < 1$		$(a_n)_{n\geq 0}$ bounded sequence	ℓ^∞
analytic elements in $\lvert x\rvert < 1$	$H(\mathbf{M}_p)$	Christol-Robba condition (4.6)	
analytic elements in $\lvert x\rvert \leq 1$	$H(\mathbf{A}_p) = \mathbf{C}_p\{x\}$	$a_n \to 0$ $(n \to \infty)$	c_0
polynomials	$\mathbf{C}_p[x]$	$a_n \neq 0$ for finitely many n's	$\mathbf{C}_p^{(\mathbf{N})}$

Radius of convergence of some exponential series
(listed in increasing order)

f	e^x	$e^{x+\frac{x^p}{p}}$	E_p (Artin-Hasse) $\exp\sum_{j\geq 0}\frac{x^{p^j}}{p^j}$	E_π^f (Dwork) $e^{\pi(x-x^q)}$ $(q=p^f)$
r_f	$r_p = \lvert p\rvert^{\frac{1}{p-1}}$	$r_p^{\frac{2p-1}{p^2}}$	1	$p^{\frac{p-1}{pq}}$

Basic Principles of Ultrametric Analysis in an Abelian Group

(1) *The strongest wins*

$$|x| > |y| \Longrightarrow |x+y| = |x|.$$

(2) *Equilibrium: All triangles are isosceles (or equilateral)*

$$a+b+c=0,\ |c| < |b| \Longrightarrow |a| = |b|.$$

(3) *Competitivity*

$$a_1 + a_2 + \cdots + a_n = 0 \Longrightarrow$$
$$\textit{there is } i \neq j \textit{ such that } |a_i| = |a_j| = \max |a_k|.$$

(4) *A dream realized*

$$(a_n)_{n\geq 0} \textit{ is a Cauchy sequence} \Longleftrightarrow d(a_n, a_{n+1}) \to 0.$$

(5) *Another dream come true* (*in a* complete *group*)

$$\sum_{n\geq 0} a_n \textit{ converges} \Longleftrightarrow a_n \to 0.$$

When $\sum_{n\geq 0} a_n$ *converges*, $\sum_{n\geq 0} |a_n|$ may *diverge*, but

$$\Big|\sum_{n\geq 0} a_n\Big| \leq \sup |a_n| = \max |a_n|$$

and the infinite version of (3) is valid.

(6) *Stationarity of the absolute value*

$$a_n \to a \neq 0 \Longrightarrow \textit{there is } N \textit{ with } |a_n| = |a| \textit{ for } n \geq N.$$

Conventions, Notation, Terminology

We use the abbreviations
iff "if and only if," := "equal by definition," $\stackrel{!}{=}$ nontrivial equality.
■ is the "end of proof" (or "absence of proof") sign.
In a statement: (*i*), (*ii*), ... always denote *equivalent* properties.
In the table of contents, an asterisk * before a section indicates that it will not be used later and may be omitted in a first reading.

Set Theory

$\mathcal{P}(E)$ power set of E: Set of subsets of E; Ø: Empty set.
$A \subset B$ means "$x \in A \Longrightarrow x \in B$" hence: $A \subset E \Longleftrightarrow A \in \mathcal{P}(E)$.
(certain authors denote this inclusion by $\subseteq$).
When $A \subset B \subset E$, $B - A = B \setminus A$ denotes the complement of A in B,
$E - A = A^c$ is the complement of a subset $A \subset E$.
A subset of E having only one element is a *singleton set*: $x \in E \Longrightarrow \{x\} \in \mathcal{P}(E)$.
$\coprod$: Disjoint union symbol, partition of a set.
E^I: Set of families (or functions) $I \to E$.
$E^{(I)}$: Set of families $I \to E$ having components equal to the base point of E (the neutral element in a group G, the 0 in a ring A ...) *except for finitely many indices.*
Let $f : E \to F, x \mapsto f(x)$ be a map. Then
f is *injective* when $x \neq y \Longrightarrow f(x) \neq f(y)$, namely f is *one-to-one*,
or equivalently when $f(x) = f(y) \Longrightarrow x = y$,
f is *surjective* when $f(E) = F$ (namely f is *onto*),
f is *bijective* when it is *one-to-one and onto*.

The characteristic function of a subset $A \subset E$ is the function

$$\varphi(x) = \varphi_A(x) = \begin{cases} 1 & \text{if } x \in A, \\ 0 & \text{if } x \notin A. \end{cases}$$

Fundamental Sets of Numbers

$\mathbf{N} = \{0, 1, 2, \ldots, n, \ldots\} \subset \mathbf{Z} \subset \mathbf{Q} \subset \mathbf{R} \subset \mathbf{C}$, $\mathbf{N}^* = \{1, 2, \ldots, n, \ldots\} = \mathbf{N}_{>0}$.
When $p \in \{2, 3, 5, 7, 11, \ldots\}$ is a prime, $\mathbf{F}_p = \mathbf{Z}/p\mathbf{Z}$.
$p \mid n$ means p divides n, $p \nmid n$ means p does *not* divide n,
$p^\nu \parallel n$ means that p^ν is the highest power of p dividing n,
$\mathbf{R}_{>0} = \{x \in \mathbf{R} : x > 0\}$, $\mathbf{R}_{\geq 0} = \{x \in \mathbf{R} : x \geq 0\}$, $[a, b)$: interval $a \leq x < b$.
$\mathbf{Z}_{(p)} = \{a/b : a \in \mathbf{Z},\ b \geq 1, b \text{ prime to } p\} \subset \mathbf{Q}$,
$\mathbf{Z}[1/p] = \{ap^\nu : a \in \mathbf{Z},\ \nu \in \mathbf{Z}\} \subset \mathbf{Q}$.
When $a > 0$ and $S \subset \mathbf{R}$, $a^S = \{a^s : s \in S\} \subset \mathbf{R}_{>0}$, e.g., $p^{\mathbf{Z}} \subset p^{\mathbf{Q}} \subset \mathbf{R}_{>0}$.
$[x] \in \mathbf{Z}$ integral part of $x \in \mathbf{R}$: $[x] \leq x < [x] + 1$.
$\langle x \rangle$ fractional part of $x \in \mathbf{R}$: $x = [x] + \langle x \rangle$.
gcd: Greatest common divisor; lcm: Least common multiple.
δ_{ij}: Kronecker symbol ($= 1$ if $i = j$, $= 0$ otherwise).

Groups, Rings and Modules

$A^\times$: Multiplicative group of units (i.e., invertible elements) in a ring A.
$A[X]$: Polynomial ring in one indeterminate X and coefficients in the ring A,
a monic polynomial f is a polynomial having leading coefficient 1:
$X^n + a_{n-1}X^{n-1} + \cdots + a_0$ if $\deg f = n$.
$A[[X]]$: Formal power series ring.
$A\{X\}$: Restricted power series over a valued ring A
(Chapter V: Power series with coefficients $\to 0$) $A[X] \subset A\{X\} \subset A[[X]]$.
An *integral domain* is a commutative ring $A \neq \{0\}$ having no zero divisor.
$K = \operatorname{Frac} A$: Fraction field of an integral domain A. In particular,
$K(X) = \operatorname{Frac} A[X]$: Rational fractions,
$K((X)) = \operatorname{Frac} A[[X]]$ ($\supset K(X)$): Formal Laurent series ring.
$A[1/q]$: Partial fraction ring corresponding to denominators in $\{1, q, q^2, \ldots\}$,
where q is not a zero divisor in the ring A.
If G is an abelian group, then $\{g \in G : g^n = e \text{ for some integer } n \geq 1\}$
is the *torsion subgroup* of G: In particular,
$\mu(A)$ denotes the group of roots of unity in a commutative ring A,
$\mu = \mu(\mathbf{C}^\times) = \mu_{p^\infty} \times \mu_{(p)}$, where
μ_{p^∞}: pth-power roots of unity (p-Sylow subgroup of μ),
$\mu_{(p)}$: Roots of unity having order prime to p,
$\mu_n(A) = \{x \in A : x^n = 1\}$: nth roots of unity in the ring A.
A pair of homomorphisms $A \xrightarrow{f} B \xrightarrow{g} C$ is *exact* when $f(A) = \ker g$.
A short exact sequence (SES) is an exact pair with
f injective and g surjective; hence C is a quotient of B by $f(A) \cong A$,
written $0 \to A \xrightarrow{f} B \xrightarrow{g} C \to 0$ for additive groups
(replace 0 by 1 for multiplicative groups).

Fields, Extensions

Characteristic of a field K: Either 0 or the prime p such that $p \cdot 1_K = 0 \in K$, in which case the prime field $\mathbf{F}_p$ is contained in K.
For each prime p, the group $\mathbf{F}_p^\times$ is cyclic; when the prime p is odd, the squares in $\mathbf{F}_p^\times$ make up a subgroup of index two, kernel of the *Legendre symbol* $\left(\frac{a}{p}\right) = \pm 1$.
In a field (or a ring) of characteristic p we have $(x+y)^p = x^p + y^p$.
K^a: Algebraic closure of a field K; when $K = K^a$ is algebraically closed of characteristic 0, $\mu_n(K)$ is cyclic and isomorphic to $\mathbf{Z}/n\mathbf{Z}$.
$\mathbf{P}^1(K) = K \cup \{\infty\}$ denotes the projective line over the field K.

Topology, Metric Spaces

The closure of a subset $A \subset X$ (X being a topological space) is denoted by $\overline{A}$.
A *Hausdorff space* is a topological space X in which for every pair of distinct points, it is possible to find disjoint neighborhoods of these points: Equivalently, the diagonal Δ_X is closed in the product $X \times X$.
The diameter of a subset $A \subset X$ with respect to a metric d is

$$\operatorname{diam}(A) = \delta(X) = \sup_{x,y \in A} d(x, y) \leq \infty.$$

We say that A is *bounded* when $\operatorname{diam}(A) < \infty$.
The distance of a point $x \in X$ to a subset $A \subset X$ is $d(x, A) = \inf_{a \in A} d(x, a)$,

$$d(x, A) = 0 \Longleftrightarrow x \in \overline{A}.$$

The balls in a metric space (X, d) are denoted by

$$B_{\leq r}(a) = B_{\leq r}(a; X) = \{x \in X : d(x, a) \leq r\}\text{: closed (\textit{dressed}) ball,}$$
$$B_{<r}(a) = B_{<r}(a; X) = \{x \in X : d(x, a) < r\}\text{: open (\textit{stripped}) ball.}$$

For a ball with center a equal to the base point (the neutral element in a group, the 0 element in a ring), the notation will be just $B_{\leq r}$, $B_{<r}$.
The *sphere* of radius $r > 0$ and center a in the metric space (X, d) is

$$S_r(a) = \{x \in X : d(x, a) = r\} = B_{\leq r}(a) - B_{<r}(a).$$

A metric space is *separable* if it has a countable dense subset.
$C(X; K)$: Space of continuous functions $X \to K$, or simply $C(X)$ when K is understood; $C_b(X; K)$: Subspace consisting of the bounded continuous functions (when K is a valued field). The sup norm of a bounded function is

$$\|f\| = \|f\|_X = \sup_{x \in X} |f(x)| \quad (f \in C_b(X; K)).$$

Index

Q

R

S

T

U

V

W

Z

Graduate Texts in Mathematics

(continued from page ii)

72 STILLWELL. Classical Topology and Combinatorial Group Theory. 2nd ed.
73 HUNGERFORD. Algebra.
74 DAVENPORT. Multiplicative Number Theory. 2nd ed.
75 HOCHSCHILD. Basic Theory of Algebraic Groups and Lie Algebras.
76 IITAKA. Algebraic Geometry.
77 HECKE. Lectures on the Theory of Algebraic Numbers.
78 BURRIS/SANKAPPANAVAR. A Course in Universal Algebra.
79 WALTERS. An Introduction to Ergodic Theory.
80 ROBINSON. A Course in the Theory of Groups. 2nd ed.
81 FORSTER. Lectures on Riemann Surfaces.
82 BOTT/TU. Differential Forms in Algebraic Topology.
83 WASHINGTON. Introduction to Cyclotomic Fields. 2nd ed.
84 IRELAND/ROSEN. A Classical Introduction to Modern Number Theory. 2nd ed.
85 EDWARDS. Fourier Series. Vol. II. 2nd ed.
86 VAN LINT. Introduction to Coding Theory. 2nd ed.
87 BROWN. Cohomology of Groups.
88 PIERCE. Associative Algebras.
89 LANG. Introduction to Algebraic and Abelian Functions. 2nd ed.
90 BRØNDSTED. An Introduction to Convex Polytopes.
91 BEARDON. On the Geometry of Discrete Groups.
92 DIESTEL. Sequences and Series in Banach Spaces.
93 DUBROVIN/FOMENKO/NOVIKOV. Modern Geometry—Methods and Applications. Part I. 2nd ed.
94 WARNER. Foundations of Differentiable Manifolds and Lie Groups.
95 SHIRYAEV. Probability. 2nd ed.
96 CONWAY. A Course in Functional Analysis. 2nd ed.
97 KOBLITZ. Introduction to Elliptic Curves and Modular Forms. 2nd ed.
98 BRÖCKER/TOM DIECK. Representations of Compact Lie Groups.
99 GROVE/BENSON. Finite Reflection Groups. 2nd ed.
100 BERG/CHRISTENSEN/RESSEL. Harmonic Analysis on Semigroups: Theory of Positive Definite and Related Functions.
101 EDWARDS. Galois Theory.
102 VARADARAJAN. Lie Groups, Lie Algebras and Their Representations.
103 LANG. Complex Analysis. 3rd ed.
104 DUBROVIN/FOMENKO/NOVIKOV. Modern Geometry—Methods and Applications. Part II.
105 LANG. $SL_2(\mathbf{R})$.
106 SILVERMAN. The Arithmetic of Elliptic Curves.
107 OLVER. Applications of Lie Groups to Differential Equations. 2nd ed.
108 RANGE. Holomorphic Functions and Integral Representations in Several Complex Variables.
109 LEHTO. Univalent Functions and Teichmüller Spaces.
110 LANG. Algebraic Number Theory.
111 HUSEMÖLLER. Elliptic Curves.
112 LANG. Elliptic Functions.
113 KARATZAS/SHREVE. Brownian Motion and Stochastic Calculus. 2nd ed.
114 KOBLITZ. A Course in Number Theory and Cryptography. 2nd ed.
115 BERGER/GOSTIAUX. Differential Geometry: Manifolds, Curves, and Surfaces.
116 KELLEY/SRINIVASAN. Measure and Integral. Vol. I.
117 SERRE. Algebraic Groups and Class Fields.
118 PEDERSEN. Analysis Now.
119 ROTMAN. An Introduction to Algebraic Topology.
120 ZIEMER. Weakly Differentiable Functions: Sobolev Spaces and Functions of Bounded Variation.
121 LANG. Cyclotomic Fields I and II. Combined 2nd ed.
122 REMMERT. Theory of Complex Functions. *Readings in Mathematics*
123 EBBINGHAUS/HERMES et al. Numbers. *Readings in Mathematics*
124 DUBROVIN/FOMENKO/NOVIKOV. Modern Geometry—Methods and Applications. Part III.
125 BERENSTEIN/GAY. Complex Variables: An Introduction.
126 BOREL. Linear Algebraic Groups. 2nd ed.
127 MASSEY. A Basic Course in Algebraic Topology.
128 RAUCH. Partial Differential Equations.
129 FULTON/HARRIS. Representation Theory: A First Course. *Readings in Mathematics*
130 DODSON/POSTON. Tensor Geometry.
131 LAM. A First Course in Noncommutative Rings.
132 BEARDON. Iteration of Rational Functions.
133 HARRIS. Algebraic Geometry: A First Course.
134 ROMAN. Coding and Information Theory.
135 ROMAN. Advanced Linear Algebra.